Advances in vegetation science   4

Edited by
EDDY VAN DER MAAREL

Dr W. JUNK PUBLISHERS THE HAGUE – BOSTON – LONDON 1981

Vegetation dynamics in grasslands, heathlands and mediterranean ligneous formations

Symposium of the Working Groups for Succession research on permanent plots, and Data-processing in phytosociology of the International Society for Vegetation Science, held at Montpellier, France, September 1980

Edited by

P. POISSONET, F. ROMANE, M. A. AUSTIN, E. van der MAAREL & W. SCHMIDT

Reprinted from *Vegetatio*, Vols. 46/47, 1981

Dr W. JUNK PUBLISHERS THE HAGUE – BOSTON – LONDON 1981

Distributors:

for the United States and Canada

Kluwer Boston, Inc.

190 Old Derby Street

Hingham, MA 02043

USA

for all other countries

Kluwer Academic Publishers Group

Distribution Center

P.O. Box 322

3300 AH Dordrecht

The Netherlands

Library of Congress Cataloging in Publication Data
Main entry under title:

Vegetation dynamics in grasslands, heathlands, and
    Mediterranean ligneous formations.

    (Advances in vegetation science ; v. 4)
    "Reprinted from Vegetatio, vols. 46/47, 1981."
    1. Vegetation dynamics--Congresses. 2. Plant
succession--Congresses. I. Poissonet, P.
II. International Society for Vegetation Science.
III. Series.
QK910.V43   1982   581.5'264   81-20901
                                          AACR2

_____

ISBN-13: 978-94-009-7993-2   e-ISBN-13: 978-94-009-7991-8
DOI: 10.1007/978-94-009-7991-8

Reprinted from *Vegetatio,* Vols. 46/47, 1981

Foreword

This volume contains most of the contributions presented at the Symposium on Vegetation dynamics in grasslands, heathlands, and mediterranean ligneous formations, which took place at the Centre d'Etudes Phytosociologiques et Ecologiques 'Louis Emberger' (locally organized by the Department of General Ecology and the Directory Staff of this institute) at Montpellier. It was organized by the Working Group for Succession research on permanent plots, and the Working Group for Data-processing in phytosociology, both of the International Society for Vegetation Science.

The editors of this volume represent both working groups and the organizing institute. They acknowledge the considerable material assistance provided by the Centre National de la Recherche Scientifique, Direction des relations éxterieures de l'information, and Programme Interdisciplinaire de Recherche sur l'Environnement (PIREN); the Conseil général de l'Hérault; and the Ministère de l'Environnement et du Cadre de Vie.

The Symposium was opened by Prof. M. Godron. The many lectures and poster contributions were organized around five themes. A complete list is added to this volume (Appendix). Abstracts of these contributions were sent in prior to the Symposium and collected by the CEPE in a volume 'Actes du Symposium sur Dynamique de la Végétation dans les formations herbacées, les landes et les formations méditerranéennes ligneuses'. The 27 elaborated papers accepted for this volume are presented in roughly the same sequence.

The editors

Preface

M. Godron
Institut de Botanique, rue A. Broussonnet, 34000 Montpellier, France

Les communications présentées à ce Symposium mettent en évidence deux remarques simples, qui méritent pourtant d'être formulées avec précision pour que l'on puisse en tirer des conclusions correctes:
- les observations diachroniques sur des quadrats permanents sont une base irremplaçable pour l'étude de la *cinématique* des groupements végétaux;
- l'étude de la *dynamique* des groupements végétaux implique la recherche des 'forces' qui produisent les variations observées au cours du temps.

De la première remarque, nous devons tirer une conséquence: les études 'synchroniques' (c'est-à-dire fondées sur l'observation actuelle de groupements d'âges croissants situés en des lieux différents) ont l'avantage de conduire rapidement à des hypothèses très générales, mais elles n'ont pas une valeur absolument probante. Si l'on veut vérifier ces hypothèses, il est nécessaire de mettre en place des observations diachroniques. Il importe alors, en particulier, de tenir compte avec précision de la variation du 'grain' de la végétation (et de ses relations avec l'aire minimale et avec les diversités) tout au long des successions, en subdivisant la surface observée en fractions de taille égale, dont le nombre sera commodément une puissance de 2.

Le rôle de la diversité et de l'hétérogénéité dont le maintien est un des grands problèmes écologiques de la civilisation industrielle ne peut être correctement apprécié que grâce à de telles études, qui exigent autant de patience que de rigueur.

La deuxième remarque ouvre des perspectives de travail plus attirantes, mais aussi plus exigeantes encore. En effet, les forces qui produisent les variations cinématiques observées au cours du temps sont infiniment nombreuses, et nous nous trouvons dans la même situation que les physiciens qui étudient la thermodynamique des gaz et doivent intégrer l'ensemble des interactions entre molécules. Si nous voulons intégrer aussi l'ensemble des interactions entre les individus d'une communauté, nous devons, comme les thermodynamiciens, chercher les forces 'macroscopiques' qui contrôlent les variations.

La première de ces forces est la poussée ascensionnelle qui produit la stratification verticale de la végétation, et l'on peut prédire que l'étude de cette force très générale sera au cours des prochaines années l'un des sujets les plus importants de l'écologie. En effet, cette poussée est la source du stockage de phytomasse dont les forêts sont le témoin manifeste, et la crise pétrolière nous a fait prendre conscience que cette phytomasse peut encore jouer un rôle capital, en raison des risques produits par les centrales nucléaires.

La seconde tendance est l'augmentation générale de la métastabilité des systèmes écologiques spontanés, et il faut reconnaître que, dans ce domaine à peine exploré, les fausses pistes sont plus largement ouvertes que les cheminements rigoureux. Aussi est-il bon de préciser pourquoi les systèmes écologiques ne sont jamais vraiment stables, mais seulement métastables. Pour cela, nous prendrons une analogie mécanique simple, où le système sera comparé à une bille qui peut monter et descendre sur les creux et les bosses de 'montagnes russes' (Fig. 1). L'état le plus stable de ce système mécanique est celui où la bille est au point B; il correspond, en écologie, à la mort de tous les êtres vivants de la communauté étudiée. Quand la bille est en C ou D ou E, elle y reste tant que les perturbations qu'elle subit restent faibles, mais elle peut sortir de sa cuvette à l'occasion d'une forte perturbation. De même, une biocénose est toujours métastable en ce sens qu'elle oscille toujours saisonnièrement autour de son état d'équilibre, et qu'elle peut être projetée dans une autre cuvette, si une forte perturbation survient.

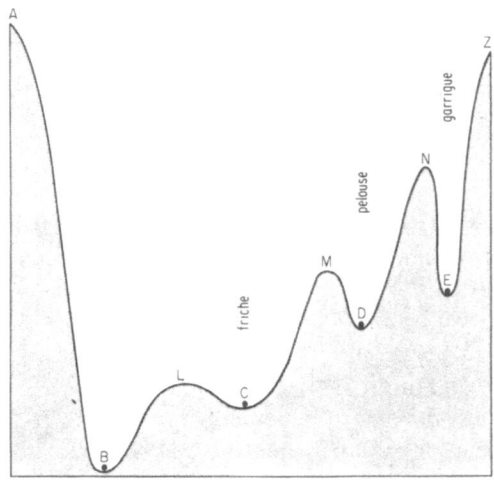

Fig. 1.

Il semble que la nature offre des positions d'équilibre métastable, telles que C, D et E, qui sont de plus en plus éloignées de l'état minéral (B), mais sont aussi de plus en plus profondément métastables. Ainsi, une garrigue de *Quercus ilex* est plus métastable qu'une pelouse de *Brachypodium phoenicoides* qui est elle-même plus métastable que la végétation d'une jeune friche à *Picris echioides*. Il est alors naturel que la succession des phénomènes conduise un jour ou l'autre la friche à se transformer en pelouse puis en garrigue, et l'on comprend pourquoi ces termes ambigus (tels que résilience) doivent être remplacés par des concepts précis dont la compréhension englobe l'ensemble des caractères liés la stabilité.

L'action consciente de l'Homme peut améliorer les rétro-actions qui augmentent la métastabilité des systèmes écologiques qui lui sont utiles. Ainsi, les grandes plaines consacrées à la culture des céréales sont dans certains cas plus métastables que les forêts, ou les steppes boisées que parcouraient les Ongulés au début du Quaternaire. Dans d'autres cas, nous diminuons plus ou moins consciemment la métastabilité de certaines biocénoses en risquant de les conduire au-delà de leur limite d'élasticité. C'est pourquoi il était très judicieux de consacrer cette réunion de septembre 1980 à la dynamique de la végétation.

L'impression finale qui ressort de l'ensemble des communications de ce Symposium est très encourageante, parce que les sujets étudiés sont certainement importants pour notre avenir, et parce que des progrès substantiels on été réalisés au cours des dernières années.

Il nous reste maintenant à travailler ensemble, au cours des années à venir aussi bien que pendant les journées du Symposium, pour apporter des réponses aux questions graves qui sont dans le prolongement de ce Symposium.

Contents

X

Permanent quadrats: An interface for theory and practice

M. P. Austin*

Division of Land Use Research, Institute of Earth Resources, CSIRO, P.O. Box 1666, Canberra City, A.C.T. 2601, Australia

Keywords: Nucleation, Permanent quadrats, Spatial heterogeneity, Succession, Vegetation dynamics

Abstract

Certain neglected concepts for studying vegetation dynamics are reviewed, particularly the incorporating of temporal and spatial heterogeneity; examples are given. Few studies have been of sufficient length to allow partition of the various types of temporal fluctuation influencing vegetation composition. Studies of changes in spatial pattern with time are similarly few. The formation of patches of vegetation and their change with time (nucleation) may determine the entire spatio-temporal structure of the ecosystem, e.g. vegetation groves in arid zones. Simultaneous study of temporal and spatial heterogeneity is required. Future studies will necessitate higher standards of evidence than previously accepted. The practice of equating a few spatially separated sites having different periods of time since a specific disturbance should not be accepted in the absence of evidence that their potential vegetation dynamics has been similar during those periods.

Permanent quadrats with an appropriate experimental design are needed to overcome these problems. The requirements to be met by such a design are discussed, including the role of location, contiguity, perturbation or synthetic experimentation and demographic measurement.

Introduction

Vegetation dynamics may be defined as the change with time of a vector of some suitable measurements of plant species performance. The study of vegetation dynamics is, and has been for some years in a state of flux. Many workers have studied succession, i.e. unidirectional change in vegetation, rather than vegetation dynamics. Numerous reviews of succession have been written in the last decade (Miles, 1979). Some have attempted to present a new conceptual framework: Odum's (1969) statement of trends in ecosystem properties has perhaps been the most influential. Others (McCormick, 1968; Dury & Nisbet, 1973; Colinvaux, 1973; Connell & Slatyer, 1977) have criticized the traditional Clementsian concept of succession towards a single climatic climax. MacIntosh (1980) has, in an extensive scholarly review of the literature, taken these authors to task for attacking an outmoded concept which few recent workers took seriously.

Students of succession often make the convenient assumption that temporal and spatial variability can be regarded as random noise and ignored. To study any form of vegetation dynamics, we need to be able to partition variability into its components whether they are successional trends, climatic fluctuations, cyclic changes, the persistent effect of episodic events or random noise. If not, how can we be sure that succession rather than a climatic fluctuation is responsible for the changes we observe?

I shall examine some of the research problems of vegetation dynamics and present a case that pro-

* I thank J. F. Angus, O. B. Williams and A. O. Nichols for comments on the manuscript and C. Helman for bibliographic assistance.

Vegetatio 46, 1–10 (1981).   0042-3106/81/0461 0001/$2.00.

gress is only likely if it is based on the use of permanent quadrats preferably with an imaginative experimental design.

There are three prerequisites for successful research: relevant concepts, suitable standards of evidence and appropriate research methods. There have been several reviews of methods (Slatyer, 1976, Austin, 1977; van der Maarel & Werger, 1978; MacIntosh, 1980; Noble & Slatyer, 1980), and I shall not review them there. It is necessary however to review the current concepts and standards of evidence used in the study of vegetation dynamics before discussing the appropriate use of permanent quadrats.

Relevant concepts

With the exception of Egler's (1954) ideas on 'initial floristic composition' and their extension by Connell & Slatyer (1977) (see also Noble & Slatyer, 1980; Noble in press), few positive suggestions for a new conceptual framework have been advanced (though see Odum, 1969).

Connell & Slatyer (1977) attempted to provide a more general framework for succession studies than the rigid Clementsian one. They recognized three possible types of vegetation sequences during succession.

(1) Facilitation or 'relay floristics' (classical Clementsian model)
(2) Tolerance
(3) Inhibition.

Both (2) and (3) are forms of the 'initial floristic composition' idea of Egler (1954); (2) where species may tolerate each other's presence and composition is determined by the earliest arrivals and (3) where species established earlier may inhibit the establishment of other species arriving later. The ideas have been developed further by Noble & Slatyer (1980) who suggest that certain vital attributes of individual species may determine succession after disturbance by fire. The use of these vital attributes for analysis of fire-induced succession in Australian conditions seems to have considerable potential, but if all aspects of vegetation dynamics are considered then the list of possible vital attributes becomes infinite for all practical purposes.

Two major concepts are often ignored but must be considered in any study of vegetation dynamics:

(1) the multitude of dynamic processes which may be operating, and
(2) the degree of spatial heterogeneity which these processes will produce at any given moment.

Temporal heterogeneity

The variability through time of the vegetation composition of specific sites has not been well-studied. Individual species populations have been examined in a few cases (Tamm, 1948, 1956, 1972; Sarukhan & Harper, 1973; Harper & White, 1974; Williams, 1970; Williams & Roe, 1975). Such population dynamics studies as reviewed by Harper (1977) are concerned with the age-class cohorts of the population rather than total numbers. Not all species are amenable to this type of study, and so far results are not sufficient to differentiate different types of temporal effects on the survival of particular age-classes. The detailed study required and limited applicability to certain growth forms means that the full complement of species in a community are unlikely to be examined in this way.

Studies using only total population estimates have shown that temporal 'fluctuations' can be relatively long-lasting in herbaceous communities. Albertson & Tomanek (1965) have shown the marked fluctuations which can occur in a thirty-year period in different mixed prairie communities in Kansas (Fig. 1a). Watt (1957, 1960, 1962, 1971) has presented some of the most detailed observations recording different types of temporal behaviour. Some examples are shown in Figure 1b, c. Van den Bergh (1979) has recently drawn attention to the periodic population eruptions and declines that can be observed in some of the few long-term studies of temperate grassland (e.g. Fig. 1b, c). Williams (1969, 1970) has examined similar changes in Australian semi-arid grasslands (see also Austin *et al.*, this symposium). However the problem of distinguishing long-term fluctuations from trends or long-term cyclic regeneration remains. The problem of having sufficient observations through time to partition the components is well seen in the figure (Fig. 1d) which shows variations in two species in one of the Park grass plots at Rothamsted for 125 yr. The period 1919 to 1948 contains many more observations than prior to 1919 and what may appear to be long-term trends in the earlier period

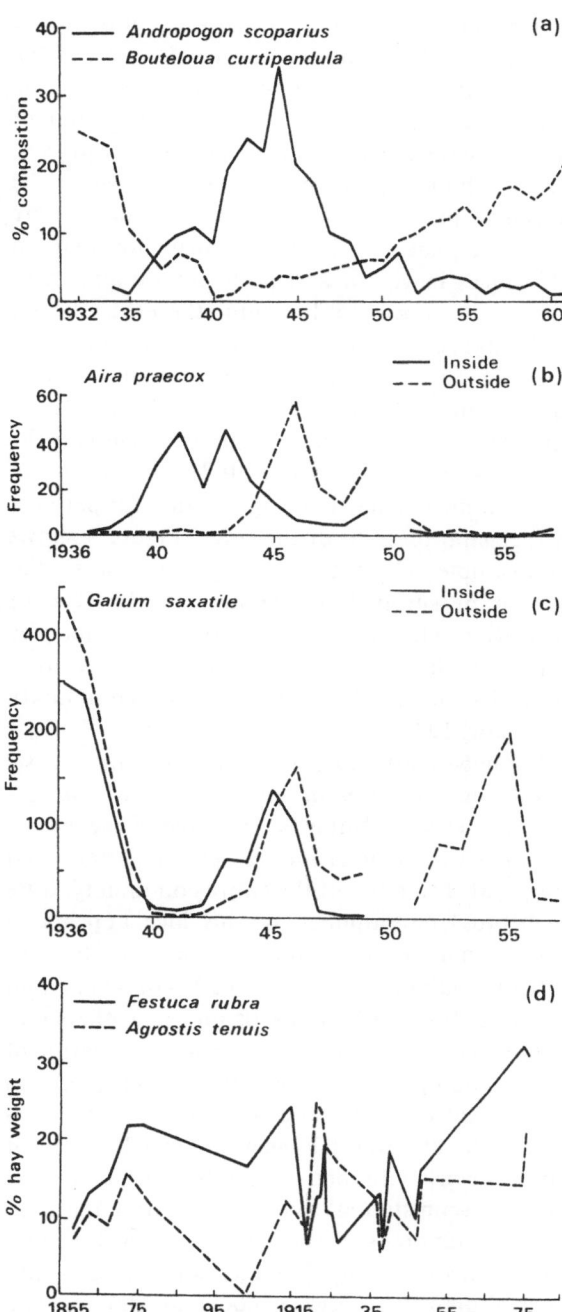

Fig. 1. Examples of species dynamics from long-term grassland studies. (a) Dominant grasses of the little blue-stem community (redrawn from Albertson & Tomarek, 1965). (b) Two populations of *Aira praecox* inside and outside enclosure showing increases at different times (redrawn from Watt 1960). (c) Changes in occurrence of Galium saxatile inside and outside enclosure showing in-phase dynamics (redrawn from Watt, 1960). (d) Changes in performance of two species over 125 years on unmanured, unlimed plots at Rothamsted (after Williams, 1978).

are seen to be simply year-to-year variability.

Episodic events which allow the establishment of an even-aged stand of grass (e.g. *Astrebla* (Mitchell grass) in northern Australia; pers. comm. O. B. Williams) once in approximately 30 years, or of *Callitris* (cypress pine) forest (Lacy, 1972) in various parts of semi-arid Australia once every fifty years may form part of the vegetation dynamics of communities in equilibrium with the environment. Failure to demonstrate the type of temporal change being examined in short-term studies of vegetation dynamics and the assumption that it is successional may limit the future use of many current studies.

Spatial heterogeneity

The spatial pattern that may arise during succession has been studied for many years (Watt, 1947a; Greig-Smith, 1964, 1979). Yarranton & Morrison (1974) have emphasized the spatial dynamics of succession drawing attention to a phenomenon they term 'nucleation', i.e. the initial establishment of a species (in their case *Juniperus virginiana*) which subsequently modifies the environment for establishment of other species. This is the 'reaction' concept of Clements as used in 'facilitation' succession but given a spatial aspect by Yarranton & Morrison. Cooper (1923, 1931, 1939; see also Lawrence, 1958) described the nucleation effect in

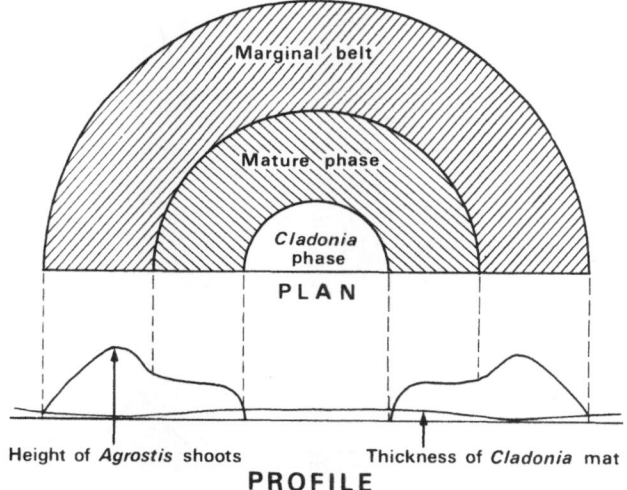

Fig. 2. Diagrammatic representation of the phases of an *Agrostis* ring (redrawn from Watt, 1947). The profile indicates differences in height growth between phases.

succession after glacial retreat with the establishment of patches of *Dryas drummondii.*

The development of a 'nucleus' from which individual species or later stages of a sere may spread need not be restricted to the 'facilitation' pathway of succession (Connell & Slatyer, 1977). Watt (1947a) in his classic paper 'Pattern and process in the plant community' describes many

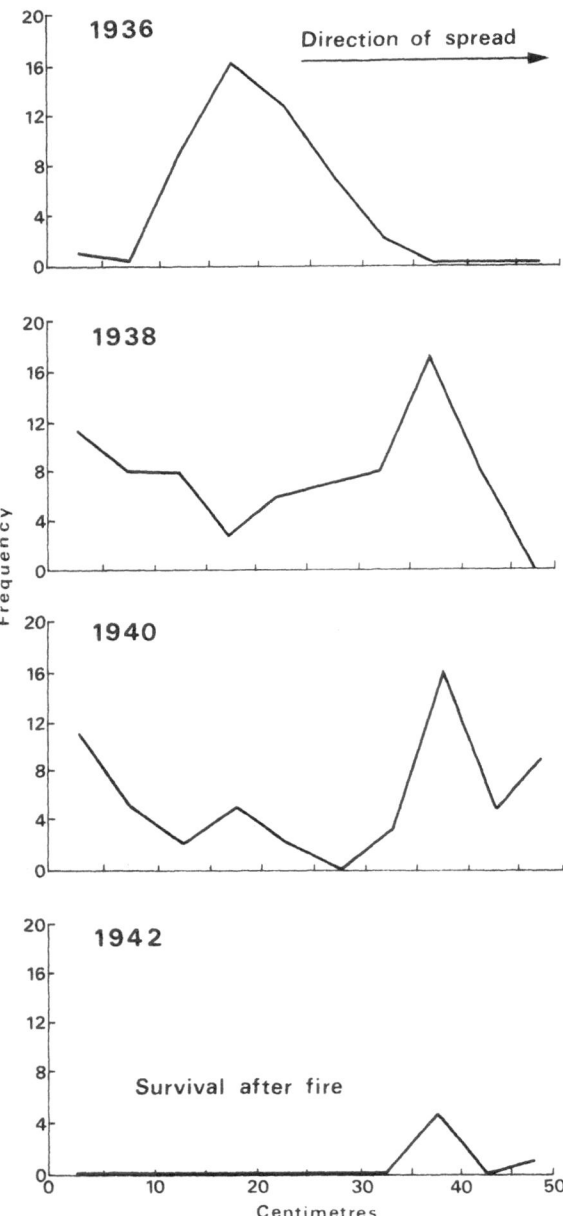

Fig. 3. Behaviour of a centrifugally spreading belt of an *Agrostis* spp. patch over time showing the impact of fire on survival (redrawn from Watt, 1960).

examples of 'nucleation' though without using the term where the dominant species determines the spatial structure and temporal behaviour of the community as a whole. In these cases, the patch of the dominant species shows four zones or phases, pioneer, building (these two are included in the marginal belt, Fig. 2), mature and degenerate (*Cladonia* phase Fig. 2). These zones show marked differences in vigour and performance and these differences appear to determine the presence and performance of other species. This phenomenon (e.g. Fig. 2) may be cyclic in communities in equilibrium with the environment (Watt, 1947a), with many of the subordinate species occupying the regeneration niche of Grubb (1977). The vigour of different phases may have significance for patterns of regeneration after disturbance. Figure 3 shows an example where fire may produce a spatial pattern of survival. Differences in the type of nucleation which occurs in a community may be important determinants of the type of vegetation dynamics which will be exhibited by the community (Kershaw, 1973).

Though Watt (1947b, 1955) with his classic studies on *Pteridium aquilinum* and *Calluna vulgaris* has clearly established the role of the 'patch' (or nucleus) in determining spatial pattern and temporal variability of the entire community, little conceptual development of this idea appears to have been done since. Some authors (e.g. Barclay-Estrup & Gimingham, 1969) have confirmed and extended Watt's observations on cyclic change in *Calluna vulgaris* communities. The combination of an expanding spatial front of building phase in *Pteridium aquilinum* and the existence of a persistent clonal patch has been indicated by recent Finnish work. Oinonen (1967a, b, c) has presented strong circumstantial evidence for occupancy of sites for hundreds of years by clones of bracken, *Pteridium aquilinum.* The largest identifiable clones have a diameter between 400 and 500 m. The evidence suggests that these clones became established after fires which were often associated with battles or warfare in the particular areas of Finland studied. The author presents similar evidence for *Lycopodium complanatum,* in some cases for the same sites as for *Pteridium.*

Other authors (Harberd, 1961; Harberd & Owen, 1969; Vasek, 1980) have presented evidence for other species, Harberd (1961) for large long-lived clones of *Festuca ovina* in upland Britain. He did not find such clones at a lowland site. Vasek (1980) describes small longlived clones of *Larrea* in arid regions in western U.S.A. Another example of the impact of nuclei or relict survivors of drought on dynamics can be provided by the lawn studies I have made (1980, and Austin & Belbin, this symposium). Persistent patches can inhibit possible dynamic pathways as well as facilitate a particular pathway. *Zerna (Bromus) erecta* dominated grasslands (Austin, 1968) may also show this effort; the persistent litter associated with the 'fairy-ring' of *Zerna* may limit the possibilities for invasion by shrubs. In many of these cases (*Pteridium* in Finland, *Zerna* in southern Britain), the species concerned can be shown to be at their limits of distribution relative to the environment concerned.

The most striking example of nucleation is the occurrence of vegetation strips or arcs in arid areas. These have been recorded from Sudan (Worrall, 1959, 1960), Somalia (Boaler & Hodge, 1962, 1964; Hemming, 1965), Jordan (White, 1969), Niger (White, 1970), and C. Australia (Slatyer, 1961, 1965). A general summary description is provided by White (1971, see also Greig-Smith, 1979). These arcs consist of a dense vegetation cover of trees, shrubs and/or grasses separated by bare (at least in a relative sense) areas often four times the width of the vegetation strip. These strips of vegetation run parallel to each other at right angles to the direction of slope, on very gently sloping homogeneous terrain which is subject to sheet wash during rain. There is evidence that these groves may migrate up the slope at about 0.3 m/yr though others appear stationary (Boaler & Hodge, 1964). The water flows across the bare areas and becomes ponded in front of the grove, slowly penetrating and infiltrating the soil in the grove; little if any water extends beyond the grove. Vegetation on the downslope part of the grove is depauperate and often contains dead trees and/or sparser grass cover. The grove soil has much greater infiltration capacity and the wetting depth is greater than for the intergroves (Boaler & Hodge, 1964).

These groves provide a most complex example of spatial pattern where the entire ecosystem has a spatial organization which results in higher biomass production, but for only part of the area. That this pattern may also migrate upslope with plants on the upslope successfully invading the bare area,

while those below cut off from supplies of moisture die, means that it may have a pronounced temporal pattern at any one spot. This interpretation of the hydrological processes associated with such groves has been giving striking support by an experimental study of *Acacia aneura* (mulga) groves in C. Australia by Perry (1970). By ripping up the inter-grove surface he improved infiltration in that area hence reducing the water-supply to the groves where the *Acacia aneura* trees then died.

Watt's description of the mosaic development of the acidophilous breckland grassland (1947a, 1960), determined by the senescent behaviour of *Festuca ovina* in which local cycles of erosion and deposition occur provides another complex example though on a finer scale, from a more temperate environment.

The study of vegetation dynamics must encompass both the temporal fluctuations which influence species populations and the spatial patterns which result from various forms of nucleation. To some extent every community contains fluctuations (runs of above or below average conditions of, say, five to ten yr.), secondary seres, regeneration cycles and persistent 'nuclear' patches. A diagram (Fig. 4) of the temporal behaviour types shows the complexity of the problem of separating the various simultaneous phenomena.

Recognition of the inter-related spatial and temporal components does allow the statement of certain questions regarding vegetation dynamics.

(1) Are certain environmental processes more important in determining vegetation dynamics in particular environments, e.g. high rates of evapotranspiration for development of arid-zone groves?

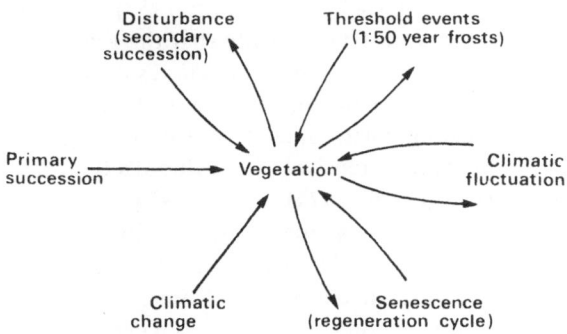

Fig. 4. Simultaneous vegetation dynamics. This shows some of the processes that may be influencing vegetation at any one time.

(2) Are the facilitation, tolerance or inhibition successional pathways of Connell & Slatyer (1977) characteristic of particular environments, e.g. facilitation by nitrogen-fixers in post-glacial primary succession (Cooper, 1939; Lawrence *et al.*, 1967)?

(3) Can different types of nucleation be associated with particular environments?

Answers to these questions would provide one input for an initial attempt at explaining differences between plant communities. Ultimately explanation must be in terms of mechanisms; at present we have few observations and even less firm evidence of the range of dynamic behaviour which may occur.

Standards of evidence

A major problem that faces students of vegetation dynamics is what constitutes sufficient evidence for acceptance of an interpretation of a sere, or a conclusion regarding changes in diversity properties with successional status. Examples of elaborate statistical analysis on only five early old-field successional sites can be found in the literature from which generalizations about succession are made. The static zonal approach where spatial zonations are equated to temporal sequences is now recognized as often misleading. The equating of a few spatially separated sites, having different numbers of years since a specific disturbance, requires the assumption that the sites are and have been equivalent in their potential vegetation dynamics. This may be equally misleading.

Some of the classic accounts of soil development during a post-glacial succession are based on nine observations (Crocker & Major, 1955). The original authors state very firmly that the graphs 'do not warrant the fitting of exact curves to the data, nor to more detailed considerations of any but the broader changes with time'. However, these graphs have appeared extensively in textbooks (e.g. Kershaw, 1973; Poole, 1974). Poole (1974) points out there have been few quantitative studies on succession, where individual quadrats have been followed over a sufficient period of time.

The study of long-term primary succession on sand dunes or lava using sites of different ages at different locations is particularly vulnerable to the

criticism that no account is taken of climatic change or fluctuation possibly confounding the interpretation. Current work on glacial moraine seres (Matthews, 1979a, b, c) is removing the objection of limited sample numbers by greatly increasing the number of observations on which analysis are undertaken. Ultimately, study of vegetation dynamics is dependent on the evidence of permanent quadrats (Watt, 1971) and, for longer periods, their imperfect but indispensable surrogate, the pollen profile, or stratification section (Walker, 1971).

The separation of different types of dynamic behaviour depends on having sufficient observations over a long enough period for a large contiguous area to have encompassed the major temporal behaviour patterns relevant to the vegetation under study. Analysis or model-building can then test whether the dynamics observed can be described by certain processes and whether the residual errors are random. If they are not, then other systematic effects remain to be discovered. No observational data currently known to me would measure up to these criteria; unfortunately permanent quadrats of the type used by Watt (1971) are no longer fashionable in ecology, at least in the Anglo-American countries (cf. Beeftink, 1979).

Vegetation management and permanent quadrats

The available studies which have collected any appreciable information about vegetation dynamics by use of permanent quadrats are often concerned with pasture management (Jones, 1932; Albertson & Tomanek, 1965), or forestry, though there are an increasing number involving vegetation management for conservation (van der Maarel, 1971).

The famous Park grass plots at Rothamsted (Williams, 1978) must constitute the longest running fertilizer experiment for hay meadows (125 yr), yet few results of ecological studies of the data have been published (though see Silvertown, 1980). Numerous grazing experiments with exclosures as controls have been done, as have fertilizer trials. Yet few have been run for long enough to demonstrate the independence of the treatment or control plot dynamics from successional trends or episodic events (though see Haecke this symposium). We (Austin *et al.*, this symposium) attempt to show some of the complexities of describing time-series data from such a grazing experiment where records over a twenty year period allow some partitioning of effects to be made.

In many cases, such experiments, though clearly demonstrating the sensitivity of the vegetation to change in some environmental factor, cannot be used to understand vegetation dynamics in terms of either rates or types of change whether spatial or temporal.

Experimental vegetation dynamics

Permanent quadrats must be an integral part of any study of vegetation dynamics, but how best to use them? If recorded for a long period of time, one can amass sufficient data for a very complete descriptive analysis, provided the data are in suitable detail. Watt's (1947a, 1960, 1971) work is without equal as an example. But can we assume all significant types of event for that community have been observed? How to analyse such data mathematically is another question, and despite the claims of ecosystem modellers, I do not believe we have yet seen an adequate analysis where an attempt has been made to separate the various dynamic components let alone a successful example of modelling with adequate validation and testing on independent data. The recent enthusiasm for Markov chains has serious problems (Usher, Austin, this symposium).

Sufficient time may not be available in one persons's lifetime to obtain such data. How can appropriate data be collected in a shorter period? Experiments are necessary (cf. Harper, 1977). The design of permanent quadrat experiments is a neglected topic (see also Connell, 1976). The usual response is a typical agronomic experimental design, where treatments and replicates are used in an analysis of variance. The purpose of such experiments is to determine whether there has been a significant effect of the treatment, i.e. are the means of treatments different? If there are complex interactions with unusual (episodic) weather events, then dramatic variations which last for years can occur (e.g. Van den Bergh, 1979; Albertson & Tomanek, 1965; van der Maarel, this symposium) and will bias the estimates of treatment effects depending on when the treatments were imposed

relative to the episodic event. If the spatial pattern (nucleation) phenomenon exists in the vegetation then progressive invasion of one plot by a species from another (Austin, 1980) or the effect of drought on the senescent phase (Watt, 1947a), may mean that the experiment yields little information.

The use of limited numbers of treatment levels may, in a successful experiment, indicate differences but one cannot necessarily define the shape of the response curve adequately (Boyd, 1973; Austin & Austin, 1980). Threshold effects in the multivariate dynamics of plant communities are quite possible, and may not be detectable with the usual three treatment levels. Replicates may give confidence but they also reduce the number of effects which are detectable.

In order to define the changes in vegetation dynamics which might occur in relation to position on an environmental gradient, numerous treatment levels will provide more information than numerous replicates.

There has been a tendency to collect only functional properties such as species diversity, or total biomass. Such data are only as good as the associated conceptual assumptions about succession. Watt (1947a, 1971) with very simple (presence in 2.5 × 2.5 cm grid) repetitive data collected in numerous adjacent quadrats has been able to demonstrate many phenomena which now need to be understood. The spatial context cannot be ignored, invasion, persistent clonal patches and local environmental heterogeneity need to be detected and allowed for.

One problem common even to the best permanent quadrat studies appears to be the confounding of analysis with location. If the study is made at a single location, chance local events may limit the success of any extrapolation to other localities (Swaine & Greig-Smith, 1980). Site replication is needed.

There are two possible types of experiment,
(a) perturbation where either the environment is changed by some treatment, e.g. fire, fertilizer, grazing or the biotic component is altered by removing/adding plant species or altering the growth characteristics with herbicides (Willis, 1972).
(b) Synthesis, where species are combined to form a specific assemblage whose behaviour is then studied. Ellenberg (1953, 1954) provides one of

the best examples of such an experimental analysis along a depth-to-watertable gradient, but the dynamic aspects are limited to only two years (Ellenberg, 1954). McCormick *et al.* (1974) provide another example with a well-defined ecosystem in depressions on granite outcrops with limited numbers of terrestrial plant species which can be studied.

Some possible components of a study of vegetation dynamics which should be considered as necessary for a suitable experiment would be as follows.

(a) Permanent quadrats. Some studies use only random quadrats at a permanent site. These can only determine mean changes, complex spatio-temporal mosaics will not be recognized without permanent quadrats.

(b) Spatial contiguity. Contiguous quadrats are necessary to recognize invasive processes or existing mosaics having a larger scale than a single quadrat.

(c) Temporal continuity of recording. Frequency of sampling will determine whether different types of temporal phenomena can be distinguished.

(d) Replicated at different locations. Without replication at distinct locations, local chance events can not be recognized.

(e) Efficient design involving perturbation and/or synthesis.

(f) Many treatment levels (or different positions on a gradient) rather than many different types of treatment.

(g) Recording of individual plants. The determination of plant demography by calculating the survival probabilities of individual plants presents the ultimate in detail for permanent quadrat studies. Only future research will determine the most effective trade-off between species performance estimates for all species in a community as against recording individuals of a few species.

The most practical as opposed to the most theoretically desirable experimental design has yet to be determined.

Any such design should allow recognition of relationships between dynamics and environment or treatment. Until such relationships can be established, questions regarding the relative importance of inhibition or tolerance pathways of succession are likely to be confounded either by environment, temporal fluctuations or spatial heterogeneity.

Conclusion

Egler (1977) provides two aphorisms which are appropriate to this symposium.

Vegetation science is the science of
1. nonhomogeneous spatial units
2. nonhomogeneous temporal units.

The mechanisms of community and ecosystem organization cannot be studied efficiently if they are not observed efficiently. The design and analysis methods for such a dynamic as opposed to a static study remain an important area in need of research. Many of the comments I have made are repetitions of points more elegantly expressed by Watt in 1947. He recognized the complexity of vegetation dynamics; permanent quadrats provide a vehicle for their study but we need to develop new methods for experimental vegetation dynamics in order to make the most of them.

References

Albertson, F. W. & Tomanek, G. W., 1965. Vegetation changes during a 30-year period in grassland communities near Hays, Kansas. Ecology 46: 714–720.

Austin, M. P., 1968. Pattern in a Zerna erecta dominated community. J. Ecol. 56: 197–218.

Austin, M. P., 1977. Use of ordination and other multivariate descriptive methods to study succession. Vegetatio 35: 165–175.

Austin, M. P., 1980. Exploratory analysis of grassland dynamics, an example of a lawn succession. Vegetatio 43: 87–94.

Austin, M. P. & Austin, B. O., 1980. Behaviour of experimental plant communities along a nutrient gradient. J. Ecol. 68: 891–918.

Austin, M. P. & Belbin, L., 1981. An analysis of succession along an environmental gradient using data from a lawn. Vegetatio 46: 19–30.

Austin, M. P., Williams, O. B. & Belbin, L., 1981. Grassland dynamics in an Australian Mediterranean type climate. Vegetatio 47: 201–211.

Barclay-Estrup, P. & Gimingham, C. H., 1969. The description and interpretation of cyclical processes in a heath community. I. Vegetational change in relation to the Calluna cycle. J. Ecol. 57: 737–758.

Beeftink, W. G., 1979. Vegetation dynamics in retrospect and prospect. Introduction to the Proceedings of the Second Symposium of the Working Groups on Succession Research on Permanent Plots. Vegetatio 40: 101–105.

Bergh, J. P. van den., 1979. Changes in the composition of mixed populations of grassland species. In: M. J. A. Werger (ed.) The study of vegetation, p. 57–80. Junk, The Hague.

Boaler, S. B. & Hodge, C. A. H., 1962. Vegetation stripes in Somaliland. J. Ecol. 50: 465–74.

Boaler, S. B. & Hodge, C. A. H., 1964. Observations on vegetation arcs in the Northern Region, Somali Republic. J. Ecol. 52: 511–44.

Boyd, D. A., 1973. Development in field experimentation with fertilizers. Phosphorus Agri. 61: 7–17.

Colinvaux, P. A., 1973. Introduction to Ecology. Wiley, New York, 621 pp.

Connell, J. H., 1976. Some mechanisms producing structure in natural communities. In: M. L. Cody & J. M. Diamond, (eds,) 'Ecology & evolution of communities' Belhamp Press, Cambridge, Mass. 545 pp.

Connell, J. H. & Slatyer, R. O., 1977. Mechanisms of succession in natural communities and their role in community stability and organisation. Am. Nat. 111: 1119–1144.

Cooper, W. S., 1923. The recent ecological history of Glacier Bay, Alaska. III. Permanent quadrats at Glacier Bay: an initial report upon a long period study. Ecology 4: 355–365.

Cooper, W. S., 1931. A third expedition to Glacier Bay. Alaska. J. Ecol. 12: 61–95.

Cooper, W. S., 1939. A fourth expedition to Glacier Bay, Alaska. Ecology 20: 130–155.

Crocker, R. L. & Major, J., 1955. Soil development in relation to vegetation and surface age at Glacier Bay, Alaska. J. Ecol. 43: 427–448.

Drury, W. H. & Nisbet, I. C. T., 1973. Succession. J. Arnold Arboretum 54: 331–368.

Egler, F. E., 1954. Vegetation science concepts. I. Initial floristic composition. A factor in old-field vegetation development. Vegetatio 4: 412–417.

Egler, F. E., 1977. The nature of vegetation, its management and mismanagement. An introduction to vegetation science. Aton Forest Norfolk, Connecticut. Privately published.

Ellenberg, H., 1953. Physiologisches und ökologisches Verhalten derselben Pflanzenarten. Ber. Dtsch. Bot. Ges. 65: 351–362.

Ellenberg, H., 1954. Über enige Fortschritte der kausalen Vegetationskunde. Vegetatio 5/6: 199–211.

Greig-Smith, P., 1964. Quantitative plant ecology. 2nd ed. Butterworths, London, 256 pp.

Greig-Smith, P., 1979. Presidential address 1979: pattern in vegetation. J. Ecol. 67: 755–779.

Grubb, P. J., 1977. The maintenance of species-richness in plant communities: the importance of the regeneration niche. Biol. Rev. 52: 107–145.

Harberd, D. J., 1961. Observations on population structure and longevity of Festuca rubra L. New Phytol. 60: 184–206.

Harberd, D. J. & Owen, M., 1969. Some experimental observations on the clone structure of a natural population of Festuca rubra L. New Phytol. 68: 93–104.

Harper, J. L., 1977. Population biology of plants. Academic Press, London. 892 pp.

Harper, J. L. & White, J., 1974. The demography of plants. Ann. Rev. Ecol. Syst. 5: 419–463.

Hemming, G. F., 1965. Vegetation arcs in Somaliland. J. Ecol 53: 57–68.

Jones, M. E., 1932. Grassland management and its influence on the sward. I. Factors influencing the growth of pasture plant. Empire J. Exp. Agr. 1: 43–57.

Kershaw, K. A., 1973. Quantitative and dynamic ecology. 2nd ed. Edward Arnold, London. 308 pp.

Lacy, C. J., 1972. Factors influencing occurrence of Cypress Pine vegetation in New South Wales. New South Wales For. Comm. Tech. Paper. 21.

Lawrence, D. B., 1958. Glaciers and vegetation in south-eastern Alaska. Am. Sci. 46: 89–122.

Lawrence, D. B., Shoenike, R. E., Quispel, A. & Bond, E., 1967. The role of Dryas drummondii in vegetation development following ice recession at Glacier Bay, Alaska with special reference to its nitrogen fixation by root nodules. J. Ecol. 55: 793–814.

Maarel, E. van der., 1971. Plant species diversity in relation to management. In: A. S. Watt & E. Duffey (eds.) The scientific management of animal and plant communities for conservation, pp. 45–63. Blackwell, Oxford.

Maarel, E. van der., 1981. Fluctuations in a coastal dune grassland due to fluctuations in rainfall: experimental evidence. Vegetatio 47: 259–265.

Maarel, E. van der. & Werger. M. J. A., 1978. On the treatment of succession data. Phytocoenosis 7: 257–258.

Matthews, J. A. 1979a. A study of the variability of some successional and climax plant assemblage-types using multiple discriminant analysis. J. Ecol. 67: 225–271.

Matthews, J. A., 1979b. The vegetation of Storbreen gletschervorfelt Jotunheimen, Norway. I. Introduction and approaches involving classification. J. Biogeogr. 6: 17–47.

Matthews, J. A., 1979c. The vegetation of Storbreen gletschervorfeld Jotunheimen, Norway. II. Approaches involving ordination and general conclusions. J. Biogeogr. 133–167.

McCormick, J. F., 1968. Succession. Via (Philadelphia) 1: 22–36.

McCormick, J. F., Lugo, A. E. & Charitz, R. R., 1974. Experimental analysis of ecosystems. In: B. R. Strain & W. D. Billings (eds.) Vegetation and environment, Handbook of vegetation science Part VI, Junk, The Hague. 195 pp.

MacIntosh, R. P., 1980. The relationship between succession and the recovery process in ecosystems. In: J. Cairns (ed.) The recovery process in damaged ecosystems. Ann Arbor Sci. Publ. Inc., Ann Arbor Mich.

Miles, J., 1979. Vegetation dynamics. Outline studies in ecology. Chapman and Hall, London, 80 pp.

Noble, I. R. in press. Predicting successional change. In: H. A. Mooney, J. M. Borricksen, N. L. Christensen, J. E. Lotan & W. A. Renness (eds.) Fire regimes and ecosystem properties. USDA For. Serv. Gen. Techn. Rep., Washington.

Noble, I. R. & Slatyer, R. O., 1980. The use of vital attributes to predict successional changes in plant communities subject to recurrent disturbances. Vegetatio 43: 5–21.

Odum, E. P., 1969. The strategy of ecosystem development. Science 164: 262–270.

Oinonen, E., 1967a. Sporal regeneration of bracken (Pteridium aquilinum (L.) Kuhn.) in Finland in the light of the dimensions and age of its clones. Acta For. Fenn. 83(1): 1–96.

Oinonen, E., 1967b. The correlation between the size of Finnish bracken (Pteridium aquilinum (L.) Kuhn.) clones and certain periods of site history. Acta For. Fenn. 83(2): 1–51.

Oinonen, E., 1967c. Sporal regeneration of ground pine (Lycopodium complanatum L.) in southern Finland in the light of the dimensions and ages of its clones. Acta For. Fenn. 83(3).

Perry, R. A., 1970. The effects on grass and browse production of various treatments on a mulga community in central Australia. In: M. J. T. Norman, (ed.) Proc. XIth Int. Grassland Congress.

Poole, R. W., 1974. An introduction to quantitative ecology. (McGraw Hill series in population biology) McGraw Hill, Tokyo. 532 pp.

Sarukhan, J. & Harper, J. L., 1973. Studies on plant demography: Ranunculus repens L., R. bulbosus L. and R. acris. I. Population flux and survivorship. J. Ecol. 61: 675–716.

Silvertown, J., 1980. The dynamics of a grassland ecosystem: botanical equilibrium in the Park Grass Experiment. J. Appl. Ecol. 17: 491–504.

Slatyer, R. O., 1961. Methodology of a water balance study conducted on a desert woodland (Acacia aneura F. Muell.) community in central Australia. Arid Zone Res. XVI Plant-water relationships in arid and semi-arid conditions. Proc. Madrid Symposium UNESCO.

Slatyer, R. O., 1965. Measurements of precipitation interception by an arid zone plant community (Acacia aneura F. Muell.). In: F. E. Eckart (ed.) 'Arid Zone Research XXV. Methodology of plant eco-physiology. Montpellier Symposium, UNESCO.

Slatyer, R. O. (ed.) 1976. Dynamic changes in terrestrial ecosystems: patterns of change, technique for study and application to management. UNESCO-MAB Techn. Notes 4.

Swaine M. & Greig-Smith, P., 1980. An application of principal components analysis to vegetation change in permanent plots. J. Ecol. 68: 33–41.

Tamm, C. O., 1948. Observations on reproduction and survival of some perennial herbs. Bot. Not. 3: 305–321.

Tamm, C. O., 1956. Further observations on the survival and flowering of some perennial herbs. Oikos 7: 274–292.

Tamm, C. O., 1972. Survival and flowering of some perennial herbs. III. The behaviour of Primula veris on permanent plots. Oikos 23: 159–166.

Usher, M. B., 1981. Modelling ecological succession, with particular reference to Markovian models. Vegetatio 46: 11–18.

Vasek, F. C., 1980. Creosote bush: long-lived clones in the Mojave Desert. Am. J. Bot. 67: 246–255.

Walker, D., 1971. Direction and rate in some British postglacial hydroseres. In: D. Walker & R. West (eds.) The vegetational history of the British Isles, p. 117–139. Cambridge Univ. Press.

Watt, A. S., 1947a. Pattern and process in the plant community. J. Ecol. 35: 1–22.

Watt, A. S., 1947b. Contributions to the ecology of bracken. IV. The structure of the community. New Phytol. 46: 97–121.

Watt, A. S., 1955. Bracken versus heather, a study in plant sociology. J. Ecol. 43: 490–506.

Watt, A. S., 1957. The effect of excluding rabbits from grassland B (Mesobrometum) in Breckland. J. Ecol. 45: 861–878.

Watt, A. S., 1960. Population changes in acidiphilous grassheath in Breckland, 1936–57. J. Ecol. 48: 605–629.

Watt, A. S., 1962. The effects of excluding rabbits from grassland A (Xerobrometum) in Breckland, 1937–60. J. Ecol. 50: 181–198.

Watt, A. S., 1971. Factors controlling the floristic composition of some plant communities in Breckland. In: E. Duffey & A. S. Watt (eds.) The scientific management of animal & plant communities for conservation, p. 137–152. Blackwell, Oxford.

White, L. P., 1969. Vegetation arcs in Jordan. J. Ecol. 57: 461–64.

White, L. P., 1970. Broussee tigree patterns in Southern Niger. J. Ecol. 58: 549–554.

White, L. P., 1971. Vegetation stripes on sheet wash surfaces. J. Ecol. 59: 615–22.

Williams, E. D., 1978. Botanical composition of the Park Grass plots at Rothamsted 1856–1976. Rothamsted Expt. Harpenden. 61 pp.

Williams, O. B., 1969. Studies in the ecology of the riverine plain. V. Plant density response of species in a Danthonia caespitosa grassland to sixteen years of grazing by Merino sheep. Aust. J. Bot. 17: 225–68.

Williams, O. B., 1970. Populations dynamics of two perennial grasses in Australian semi-arid grassland. J. Ecol. 58: 869–875.

Williams, O. B. & Roe, R., 1975. Management of arid zone grasslands for sheep. In: 'Managing terrestrial ecosystems'. Proc. Ecol. Soc. Aust. 9: 142–156.

Willis, A. J., 1972. Long-term ecological changes in sward composition following application of Maleic hydrozide and 2, 4-D. Proc. 11th Brit. Weed Control Conf. 360–367.

Worrall, G. A., 1959. The Butana grass patterns. J. Soil. Sci. 10: 34–53.

Worrall, G. A., 1960. Tree patterns in the Sudan. J. Soil Sci. 11: 63–7.

Yarranton, G. A. & Morrison, R. G., 1974. Spatial dynamics of a primary succession: nucleation. J. Ecol. 62: 417–428.

Accepted 30.6.1981.

Modelling ecological succession, with particular reference to Markovian models

M. B. Usher*
Department of Biology, University of York, York YO1 5DD, United Kingdom

Keywords: Community, Markov, Matrix, Models, Succession, Transition

Abstract

There is a brief review of models of succession: these are classified as verbal or descriptive, simulation, population dynamic, and Markovian. Many facets of the latter class of models are discussed, demonstrating that there are far more disadvantages to their use than apparent advantages. However, Markovian models do appear to have predictive ability, and it is also considered that the patterns of probabilities in Markovian matrices may have a role to play in interpreting opposing views on the mechanisms of succession. Data from the Breckland grasslands (Watt, 1960b) are used as an example.

Introduction to models

The earliest approach to modelling successional phenomena may be described as verbal or descriptive. The earlier writers interpreted the descriptions of successional sequences, and their verbal models have provided a basis both for scientific disagreement (which has proved to be a stimulus for scientific advance) and for mathematical modellers. The two contrasting views can be demonstrated in the works of Clements and Gleason. Clements (1916) stated:

'. . . the formation arises, grows, matures and dies, . . . each climax formation is able to reproduce itself, repeating with essential fidelity the stages of its development. The life history of a formation is a complex but definite process, comparable in its chief features with the life history of an individual plant'.

If Clements had written 50 years later, his verbal model would have been described as *deterministic*. Gleason (1926) clearly puts an opposing point of view when he states:

'. . . every species of plant is a law unto itself, the distribution of which in space depends upon its individual peculiarities of migration and environmental requirements . . . plant associations . . . depend solely on the coincidence of environmental selection and migration . . . a logical classification of associations into larger groups, or into successional series, has not yet been achieved'.

Views such as those of Gleason would form the basis of a *stochastic* model. How have these concepts been used in the half century, or more, since they were first published?

A concept that appeared to bring unity, in that it seemed to be basic to all aspects of theoretical ecology, was that of 'energy' or 'production ecology', and it was inevitable that attempts were made to reconcile the opposing points of view of succession. Odum (1969) described an orderly system of biomass accumulation, his diagrams showing a sigmoidal increase to an asymptote after 80–100

* This paper is an extension of part of a paper read to the Mathematical Ecology Group (jointly organised by the British Region of the Biometrics Society and the British Ecological Society) in March 1980 jointly with Miss K. E. Sparkes: her help and criticism is gratefully acknowledged.

Vegetatio 46, 11–18 (1981).   0042-3106/81/0461-0011/$1.60.

days in microcosms, and 80–100 years in forests. This increase in biomass was a function of the discrepancy between gross production and respiration, both of which became more or less equal towards the end of the successional sequence. Odum's model is pictorial rather than verbal, but it remains purely descriptive since the problem of why gross production reduces to the level of respiration is not addressed.

Mechanisms for what might cause the changes during the successional sequence are discussed by Drury & Nisbet (1973). They consider that environmental gradients are important, determining the differential growth rates and survival of the plant species. A rather different approach is taken by Grime (1977, 1979), who recognises three factors: disturbance, stress, and competition, the effects of which are modified by the general level of production. Competition is postulated to be most severe later in a successional sequence, and in Odum's (1969) model this may be the factor that reduces the gross production to the level of the respiration.

A rather more rigorous discussion of the theoretical basis of succession is given by Connell & Slatyer (1977). They consider that, following a disturbance which opens a relatively large space, releasing resources, succession follows one of three paths: facilitation, tolerance, or inhibition. These three paths, referred to as models, will be considered further below.

In all of these descriptive, either verbal or pictorial, models, the main challenge for the future is in formulating the ideas mathematically so that they can be rigorously tested both by experiment and by model.

Two other broad approaches to modelling successional phenomena should be mentioned. The advent of the digital computer with large amounts of backing store made simulation modelling of large ecosystems possible. The work of Leak (1970), Botkin et al. (1972) and Shugart et al. (1973) on woodlands demonstrates both the complexity of the computer program that is needed as well as the apparently good fit of the model to the situation being modelled. The models are clearly designed to have predictive value: Shugart et al., for example, predict the development of the various forest types in the western Great Lakes region of North America for the next 250 years, and Leak follows predicted

and actual changes over a 25 year period.

In simulation modelling, the main challenge for the future is probably in the art of validation: just how good does a model, and the predictions based on it, have to be?

A new development is in the application of *population dynamics* models. Van Hulst (1979a) considered two models, based on the sigmoidal increase in population growth,

$$dN_i/dt = r_i N_i (K_i - N_i)/K_i,$$

where N_i is the number of individuals (or biomass, etc.) of the ith species, r_i is its intrinsic rate of natural increase, K_i its carrying capacity, and t is time. With interspecific competition,

$$dN_i/dt = r_i N_i (K_i - \sum_j \alpha_{ij} N_j)/K_i,$$

where the α_{ij} terms are the (constant) competition coefficients. Independently, Sparkes (unpublished) was developing similar models, and asking such questions as 'How does one include the effects of allelopathy which persist after the death of an individual of species j?' or 'How does one allow for edaphic changes (e.g. accumulation of faecal material) caused by an individual of species j?'. Such questions are almost intractable mathematically.

With population dynamics models the main challenge for the future seems to be whether the model can be made simple enough to have any practical application.

The final group of models to be considered are the *Markovian* models. These can be viewed as essentially taking the concepts of Clements, and relying on the fact that if succession is an orderly process then probabilities for the transition from one state to another can be estimated (provided that there is no catastrophe). Alternatively, they can be viewed as building on the concepts of Gleason, since, if each individual is an entity in its own right, one should be able to estimate a series of probabilities which define all of the possible outcomes for the fate of that one individual. Thus, in many ways the Markovian models provide a unifying concept in the dichotomous approach to succession. Like all the other models except for those based on the dynamics of the individual species, they do not lead automatically to a consideration of the mechanisms of succession, although many authors do discuss the implications of the successional mechanisms (cf. Bledsoe & Van Dyne, 1971). Markovian models have been reviewed recently by van Hulst (1979b, 1980) and by Usher (1979), and some further aspects are discussed in the following sections.

Markovian models

Introduction

In order to use a Markovian model of succession, there are two stages in the collection of data suitable for the model. First, the successional sequence needs to be broken down into a series of states which can be written as S_1, S_2, ..., S_k, where S_1 is the initial state of a primary succession or the community resulting from a disturbance if the succession is secondary. Second, the probabilities, P_{ij}, for the transition from S_i to S_j need to be estimated. With repeated observation of permanent plots, estimation of P_{ij} is a simple matter of tallying the number of times the transition from S_i to S_j has occurred, but without repeated observation the estimation procedure is far more complex (see, for example, Cooke, in press). Estimation of P_{ij} for geological sequences is explained by Harbaugh & Bonham-Carter (1970), and for ecological sequences by Usher (1973).

The definition of the states, S_i, has generally not been seen as a problem, though careful consideration of these states is fundamental to the later development and interpretation of a model. The methods of defining states will be discussed in the next section, which will be followed by an example. The uses of a Markov model will then be enumerated and discussed.

States in a successional sequence

Working in a hardwood forest in Connecticut, Stephens & Waggoner (1970) used dominance by three genera of tree, maple (*Acer*, two species), oak (*Quercus*, five species) and birch (*Betula*, four species), and by two broader categories, 'other major hardwood species' and 'minor species', to define the five states that were included in their matrix of transition probabilities (Waggoner & Stephens, 1970). The approach of defining states in terms of the dominant species or genus, probably with a class or two for odds and ends, has been used by Horn (1975a, 1975b, 1976) for forest succession

in New Jersey, by Tucker (1979) for carr woodland in Yorkshire, and by Usher (1975) for termite succession in farm scrub in West Africa.

In any successional sequence there is a continuous replacement of species. At any point on this continuum one species may be dominant or a few species may be co-dominant. The problem with Markovian models is that this continuum has to be divided into a relatively small number of discrete states. Since these states are only arbitrary segments of the sequence, there is no *a priori* reason why each should be dominated by a single species. Usher (1979) suggested that the most appropriate method of defining states might be a successive approximation between multivariate techniques to cluster or ordinate the observations into states and the Markovian model to suggest redefinition of the states. A simple approach, using cluster analysis, is attempted in the example in the following section. A similar approach was adopted by Austin (1980) to simulate dynamics in a lawn vegetation.

Watt's data on the breckland grasslands

The data published by Watt (1960b) provide a sequence of yearly observations from 1936 until 1957 (1950 is missing) on two sets of 128 contiguous quadrats in a Breckland grassland. Recording in the field was by noting the presence or absence of higher plant species in permanent quadrats 1.25 cm², but groups of eight of these quadrats, perpendicular to the long axis of the transect, were summed in the published data. Thus, the following analyses are based on quadrats 10 × 1.25 cm, with plant frequences varying from 0 (absent from all eight sub-quadrats) to 8 (present in all eight sub-quadrats). Only six species or genera of higher plants occurred: these are *Agrostis* spp. (*A. canina* L. and *A. tenuis* Sibth.), *Aira praecox* L., *Festuca ovina* L., *Galium saxatile* L., *Luzula campestris* (L.) DC., and *Rumex tenuifolius* (Wallr.) Löve. There was abundant growth of lichens, but their frequencies were not published: quadrats with no higher plants can be assumed to be lichen dominated (Watt, pers. comm.).

The data were collected from two sites, and each set has been analysed separately. One, the control (CONT on the diagrams), was grazed by rabbits, and the other was enclosed (ENCL) by a rabbit-proof fence. Watt (1960a) has discussed the effects

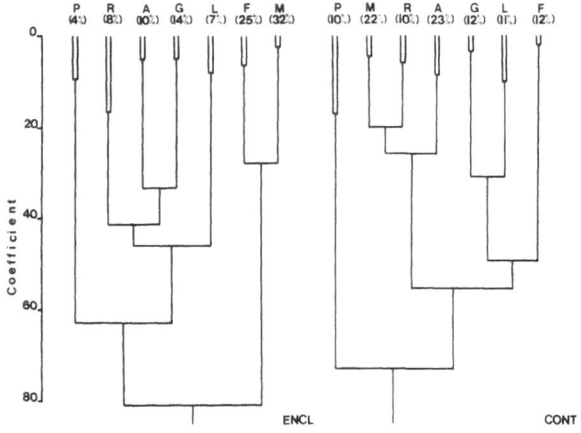

Fig. 1. Dendrograms generated by a cluster analysis of 200 quadrats of Watt's Breckland data, using Ward's method with the CLUSTAN package. The characteristic letters given to the clusters are shown in the top row, and the percentage of quadrats in that cluster in the second row. The points where the next divisions of the branches of the dendrogram would occur are shown by two parallel lines, drawn close together.

of rabbit grazing on the development of the plant communities.

The amount of data, two sets of 128 quadrats each assessed on 21 occasions, was too great for a simple analysis by the CLUSTAN package for numerical clustering, and hence a random set of 200 observations was selected. These were clustered, using Ward's method, investigating all possible numbers of clusters between 2 and 20 (arbitrarily selected). For both the CONT and ENCL data seven clusters were found to be satisfactory since none of these clusters were divisible until the sum of squares coefficient had been reduced to at least half of its former level (Fig. 1). These seven clusters could be characterised in terms of the frequencies of the six species, and an eighth group, consisting of zero frequencies for all higher plant species, was added. Most of the quadrats not included in the clustering process could be assigned to one of the seven clusters since their species scores were identical with one of the quadrats already in a cluster. The few remaining quadrats were easily assigned on the basis of the frequency of the species (see Table 1). The transition probabilities between clusters were estimated. The data fall into two parts: 13 yearly transitions for the 1936–49 period, and 6 yearly transitions for the 1951–57 period.

Table 1. The mean frequency of occurrence of plants in the eight clusters used for the Markovian model. As explained in the text, the maximum frequency is 8. The following symbols are used: –, no quadrat in the cluster included that species; +, the frequency is >0 by ≤0.05; CONT, the control plot; ENCL, the enclosed plot. The letters given to the clusters are shown in Fig. 1, and are discussed in the text.

Species	Clusters							
	F	G	P	L	R	A	M	N
CONT								
F. ovina	3.7	0.2	0.2	–	0.2	0.2	–	–
Agrostis spp.	1.0	0.4	2.0	0.2	1.2	2.9	0.5	–
L. campestris	0.3	+	0.3	2.0	0.4	0.2	0.2	–
G. saxatile	2.2	6.4	0.6	3.9	1.0	0.8	1.2	–
R. tenuifolius	0.1	0.1	1.3	0.1	1.8	0.4	–	–
A. praecox	–	–	1.2	–	–	–	–	–
ENCL								
F. ovina	5.9	0.5	0.3	0.3	0.6	0.4	1.5	–
Agrostis spp.	+	0.7	1.7	0.6	1.1	3.0	0.2	–
L. campestris	–	–	0.4	1.4	0.1	–	–	–
G. saxatile	–	4.8	5.2	0.8	0.8	0.4	0.3	–
R. tenuifolius	–	+	0.1	–	1.2	–	–	–
A. praecox	–	–	1.2	–	–	–	–	–

Watt's control plot

The seven clusters are shown in Figure 1 and in Table 1. In two of the clusters, F and G, there is a dominant species, *F. ovina* and *G. saxatile* respectively. In two other clusters, P and L, there is a characteristic species, *A. praecox* and *L. campestris* respectively. The remaining three clusters are essentially *Agrostis – Festuca* grassland in which the frequencies of *R. tenuifolius* and *Agrostis* spp. vary: *R. tenuifolius* is most frequent in cluster R, and *Agrostis* spp. is in cluster A. The matrix of transition probabilities, which is shown in Table 2, demonstrates three points. First, the largest entries are along the principal diagonal, indicating that it is generally commoner for a quadrat to remain in the same cluster than to change to a different cluster over the 1-year period. Second, none of the 64 probabilities is zero. Third, relatively few of the probabilities for transition from S_i to S_j ($i \neq j$) are large: the largest are shown diagrammatically in Figure 2, where there is some evidence of a cycle of change involving the three categories of *Agrostis – Festuca* grassland, the quadrats without higher plants, and the quadrats in which *A. praecox* is present.

Table 2. Matrices of the probabilities for transition between the eight states included in the Markovian models for Breckland grassland. The same symbols are used as in Table 1, except that + indicates a non-zero probability ≤0.005.

From cluster	To cluster							
	F	G	P	L	R	A	M	N
CONT								
F	.78	.02	.05	.03	.01	.04	.06	.01
G	.03	.37	.02	.09	.06	.13	.25	.05
P	.08	.03	.40	.02	.15	.19	.08	.05
L	.10	.05	.03	.46	.05	.06	.21	.04
R	.08	.08	.14	.05	.20	.16	.19	.10
A	.08	.03	.19	.01	.09	.43	.12	.05
M	.04	.12	.05	.06	.06	.18	.32	.17
N	+	.01	.04	+	.14	.09	.19	.53
ENCL								
F	.81	.02	–	.01	+	+	.11	.05
G	.07	.40	.07	.06	.03	.15	.16	.06
P	–	.27	.22	.07	.17	.15	.12	–
L	.03	.12	.01	.38	.10	.03	.21	.12
R	.05	.05	–	.16	.17	.21	.21	.15
A	.05	.05	.01	.09	.11	.39	.20	.10
M	.21	.07	+	.05	.06	.08	.39	.14
N	.06	.03	–	.03	.03	.02	.29	.54

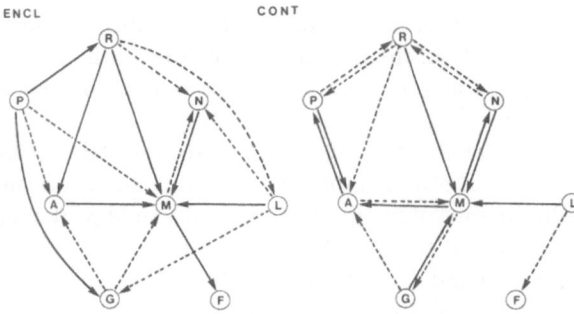

Fig. 2. Diagrams indicating the main transitions between states (based on the data in Table 2). The states are defined in Figure 1 and Table 1, ENCL and CONT refer to the enclosed and control plots respectively, continuous lines indicate the largest 15% of the probabilities and dashed lines the next 15% of the probabilities (the smallest 70% of probabilities are not shown).

The dominant eigenvector of the transition matrix in Table 2 is [27, 9, 15, 7, 11, 22, 21, 16] where the elements have been multiplied by a constant and rounded to integers so that their sum is 128 (the number of quadrats recorded each year). This is the prediction of what the grazed grassland would consist of after a long period of time. All

eight clusters remain represented in the grassland, and the whole community would continue to be relatively diverse. There is no evidence that any cluster would come to dominate the grazed grassland.

Watt's enclosed plot

The seven clusters shown in Figure 1 can be interpreted in a similar way to those of the control plot. Thus, in two of them, F and G, there is a dominant species, *F. ovina* and *G. saxatile* respectively, with all other species being absent or scarce. Two others, P and L, are characterised by the presence of *A. praecox* and by the relative abundance of *L. campestris,* respectively. The three other clusters, R, A and M, contain a mixture of *Agrostis* spp. and *F. ovina,* the former species being more abundant in A than in R and M, and *R. tenuifolius* occurring only in R. The mean frequencies of species in these clusters are shown in Table 1. Despite the similarities in the analyses of the ENCL and CONT data, this table demonstrates some differences between the clusters found by the two analyses. Thus, in both analyses, *A. praecox* only occurs in one cluster, but in the enclosed plot this species is more associated with *G. saxatile,* which has a mean frequency of 5.2, than it is in the control plot, where *G. saxatile* has a mean frequency of 0.6.

The transition matrix for the enclosed plot is shown in Table 2. This matrix is similar to that for the control plot: the elements along the principal diagonal are larger than those off the diagonal, and relatively few of the transition probabilities from S_i to S_j ($i \neq j$) are large. However, a few of these transition probabilities are zero: for example no transitions were observed in either direction between clusters F and P or between clusters N and P. The larger transition probabilities are shown diagrammatically in Figure 2, where it will be seen that there is no evidence of a cycle, but rather a general tendency towards cluster M and then towards cluster F.

The dominant eigenvector of the transition matrix in Table 2 is [45, 9, 1, 8, 6, 10, 28, 21] where again the elements have been adjusted to the nearest integer value such that they sum to 128. This vector differs from the control plot vector in that quadrats with *A. praecox* are predicted to become very rare, and those dominated by *G. saxatile* and those characterised by *L. campestris* and *R. tenuifolius* are all likely to become scarce, whilst quadrats dominated by *F. ovina* are predicted to become more abundant. The Markovian model thus predicts that the enclosed sward will become far less diverse, possibly with the loss of some of the higher plant species.

Uses of the Markov model

The first use of the Markovian model is to predict the future development of a successional sequence. The results of a prediction, based on the dominant eigenvectors of the transition matrices, are presented above and accord reasonably well with Watt's (1971) later observations where *F. ovina* developed, senesced and then left small hummocks of thick litter matted with lichen thalli.

A second question might relate to the randomness of the process being investigated. A test for randomness has been described by Harbaugh & Bonham-Carter (1970), in which

$$\chi^2 = -2 \sum_{i=1}^{m} \sum_{j=1}^{m} n_{ij} \ln(P_{ij}/P_j)$$

where m is the order of the transition matrix (8 in Table 2), P_{ij} is the element in the ith row and jth column of the probability transition matrix, n_{ij} is the corresponding element in the tally matrix (the matrix which enumerates the number of times that the transition from S_i to S_j has been observed), and P_j is the marginal probability of the jth column (i.e. n_j/n, where n_j is the column total and n is the grand total, both in the tally matrix). This is not strictly a test of the Markovian nature of the process, but rather a test of departure from randomness (see Usher, 1979, for the application of this test to the random process of selecting coloured sweets, with replacement, from a large box). The approximate χ^2 has $(m-1)^2$ degrees of freedom. Using the test on the two matrices in Table 2 gives $\chi^2 = 2\,015$ for the control plot and $\chi^2 = 2\,159$ for the enclosed plot, both with 49 degrees of freedom. These results indicate that the sequences observed on the Breckland Grassland cannot be considered as purely random processes.

Third, how quickly does the sequence converge on its final state, presumed to be the climax under the prevailing management conditions? The speed of convergence depends upon the ratio of the moduli of the dominant and the subdominant eigenvalues. Since all the rows of the transition probability matrices sum to unity, the dominant eigenvalue, λ_1, equals one. Using the control plot probabilities, the other seven eigenvalues are 0.732, $0.478 \pm 0.014i$, 0.337, 0.246, and $0.108 \pm 0.028i$. With the enclosed plot, the other eigenvalues are 0.725, 0.485, $0.336 \pm 0.052i$, 0.201, 0.132, and 0.073. The ratio of the first to second largest roots in absolute value, λ_1/λ_2, is 1.37 and 1.38 for the control and enclosed plots respectively: these results indicate that the speed of approach to the stable state will be more or less the same on both plots, and, since the ratio is not very large, it will be relatively slow.

Advantages and disadvantages of Markovian models

It would seem that the disadvantages rather outnumber the advantages. There are seven broad areas where the Markovian models are either difficult, impossible, or unrealistic from a biological point-of-view.

First, there is the difficulty of defining the states to include in the model. The obvious difficulty of dividing a continuum into discrete units has been outlined above, but clearly multivariate analyses or cluster analyses are likely to lead to logical divisions of a continuum rather than guesses based on the dominance of species or genera. It would probably be useful to iterate between the multivariate method and a Markovian model to find some kind of optimum division of the continuum.

Second, the data required are difficult to collect, and very time-consuming to collect. The example, using Watt's data, had to estimate 64 transition probabilities since eight states were recognised. In general, when m states are recognised, m^2 transition probabilities have to be estimated. There has been no attempt to fit errors to these estimates of P_{ij}, nor to investigate the effects that such errors might have on the dominant eigenvectors or on the ratio λ_1/λ_2.

Third, the model thus far discussed includes only single dependence. Thus, the probability of a transition from S_i to S_j depends only on the quadrat being in S_i and not on the history of the quadrat. This seems to be relatively unlikely biologically, especially since a change is more likely the longer that the quadrat has been in the same state. It is possible to build Markovian models with higher order dependence, but the number of transition probabilities to be estimated increases greatly. Thus, if double dependence were to investigated in Watt's data, 512 transition probabilities would have to be estimated (these estimates would probably be very imprecise since, although Watt's data are extensive, only 2 432 transitions were observed on each plot). In general, if there are m states, for a model with double dependence m^3 transition probabilities have to be estimated.

Fourth, the transition probabilities may not remain constant in time. Thus, during the course of a succession the probability of the transition from S_i to S_j may either increase or decrease. Such changes have been documented by Usher (1979), who suggested that non-stationarity may be the norm rather than the exception in ecological processes. This view is confirmed by the work of Austin (1980) on lawn succession.

Fifth, these Markovian models average out all the spatial effects, and hence the patterns which are known to occur, especially in plant communities, are ignored. This is particularly serious in the example using Watt's data, since the Markovian model cannot include any terms relating to the pattern of spread of *Agrostis* spp. and *F. ovina*, as discussed by Watt (1971).

Sixth, the data that are collected must be fully representative of all the species, all the suites of species, as well as of all of the possible transitions. It is impossible to include in the model the arrival of a new species, or suite of species, and hence the model will always have the element of uncertainty: will something else arrive later? Extinctions can, however, be predicted.

Finally, there is the philosophical problem of using this type of model. If you have enough data to build a really good Markovian model, getting round the problems of non-stationarity and double dependence, and knowing that no unexpected species is going to arrive, do you need a model at all? If one has enough information, one knows more-or-less all that there is to know about the course of the succession, one knows what will happen and what the history of many parts are, and hence the creation of a model is a waste of time. The use of a model is really only to explore the

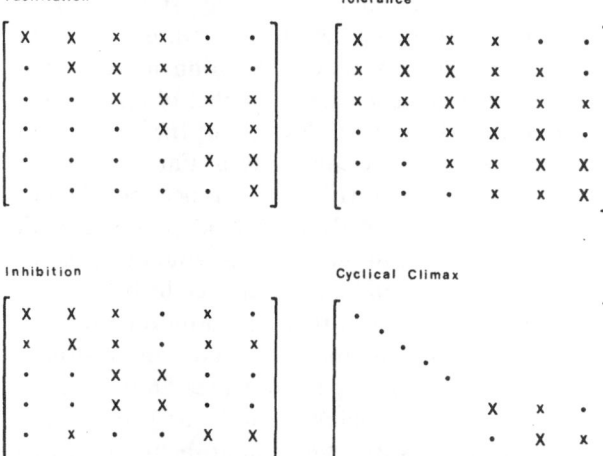

Fig. 3. Pattern matrices for theoretical concepts of the nature of succession. In the 'facilitation' model a forward movement occurs, whereas in the 'tolerance' model some backward movement may also occur. Larger transitions probabilities are denoted by X, smaller ones by x, and zero or near-zero probabilities by · (a dot). In the inhibition model the system can enter the third or fourth states from either the first or second, but once in one of these states there is no escape: the third and fourth states thus form an absorbing set. In the cyclical climax only the last three states are shown: if the last state is the nth, then the system can only move around the cycle (n-2)th, (n-1)th, nth, (n-2)th, and so on, in that order.

behaviour of a system which is basically unknown, and doubt must remain as to whether Markovian models are sufficiently robust for this sort of use.

However, to be more positive, there are two areas where these models would seem to have considerable use. The first concerns prediction. In the example using Watt's data, predictions were made about the development of the vegetation sward in the future. Usher (1975) used some observations of termite species on baitwood blocks to predict that the population of wood feeding species would increase: such an increase was desired since the site was to be used for testing building material against attack by subterranean termites, and the experiment was designed so that a forecast could be made. The models seem, therefore, to have predictive use, though more case histories are required before this can be anything more than a hypothesis.

The second area of application could be in the testing of concepts expressed as descriptive or verbal models. In Figure 3 the three models of Connell & Slatyer (1977) are expressed in matrix format, where larger and smaller probabilities are expressed by X and x respectively. In the facilitation model there is generally a forward movement of probabilities, so that the transition matrix would be expected to have a large number of zeros below the principal diagonal. Neither of the matrices in Table 2 conform to this pattern. Far fewer zeros would be expected with the tolerance model (similar to Drury & Nisbet's, 1973, model), though with this model many of the elements would be close to zero. This situation is approximated by both of the matrices in Table 2. If there is inhibition, then this would be shown by the presence of an absorbing state, or an absorbing set of states, in the matrix model. A cyclical climax would be shown by a group of states with transition probabilities which form a cycle.

It seems essential for the development of the theory of ecological succession that the concepts in the literature are formalised in mathematical terms so that they can be tested and compared experimentally. Markovian models are well suited for this kind of testing and comparison. This view is not wholly shared by van Hulst (1980), who advocates the need to explore in much greater depth other models of succession.

References

Austin, M. P., 1980. An exploratory analysis of grassland dynamics: an example of a lawn succession. Vegetatio 43: 87–94.

Bledsoe, L. J. & Van Dyne, G. M., 1971. A compartment model simulation of secondary succession. In: B. C. Patten (ed.) Systems analysis and simulation in ecology, Vol. 1: 497–511. Academic Press, New York.

Botkin, D. B., Janak, J. F. & Willis, J. R., 1972. Some ecological consequences of a computer model of forest growth. J. Ecol. 60: 849–872.

Clements, F. E., 1916. Plant succession: an analysis of the development of vegetation. Carnegie Inst. Washington Publ. 242, 512 pp.

Connell, J. H. & Slatyer, R. O., 1977. Mechanisms of succession in natural communities and their role in community stability and organisation. Am. Nat. 111: 1119–1144.

Cooke, D., in press. A Markov chain model of plant succession. To be published in the proceedings of the July 1980 meeting of the Institute of Mathematics and its Applications.

Drury, W. H. & Nisbet, I. C. T., 1973. Succession. J. Arnold Arboretum 54: 331–368.

Gleason, H. A., 1926. The individualistic concept of the plant association. Torrey Bot. Club Bull. 53: 7–26.

Grime, J. P., 1977. Evidence for the existence of three primary strategies in plants and its relevance to ecological and evolutionary theory. Am. Nat. 111: 1169–1194.

Grime, J. P., 1979. Plant strategies and vegetation processes. Wiley, Chichester, 222 pp.

Harbaugh, J. W. & Bonham-Cater, G., 1970. Computer simulation in geology. Wiley-Interscience, New York, 575 pp.

Horn, H. S., 1975a. Markovian properties of forest succession. In: M. L. Cody & J. M. Diamond (eds.). Ecology and evolution of communities, p. 196–211. Belknap Press Harvard University Press, Cambridge.

Horn, H. S., 1975b. Forest succession. Scient. Amer. 232: 90–98.

Horn, H. S., 1976. Succession. In: R. M. May (ed.). Theoretical ecology: principles and applications, p. 187–204. Blackwell, Oxford.

Hulst, R. van, 1979a. On the dynamics of vegetation: succession in model communities. Vegetatio 39: 85–96.

Hulst, R. van, 1979b. On the dynamics of vegetation: Markov chains as models of succession. Vegetatio 40: 3–14.

Hulst, R. van, 1980. Vegetation dynamics or ecosystem dynamics. Dynamic sufficiency in succession theory. Vegetatio 43: 147–151.

Leak, W. B., 1970. Successional change in northern hardwoods predicted by birth and death simulation. Ecology 51: 794–801.

Odum, E. P., 1969. The strategy of ecosystem development. Science 164: 262–270.

Shugart, H. H., Crow, T. R. & Hett, J. M., 1973. Forest succession models: a rationale and methodology for modelling forest succession over large regions. Forest Science 19: 203–212.

Stephens, G. R. & Waggoner, P. E., 1970. The forest anticipated from 40 years of natural transitions in mixed hardwoods. Bull. of the Connecticut Agric. Exp. Station, 707, 58 pp.

Tucker, J. J., 1979. Age structure of the tree population of Far Wood, Askham Bog, in relation to succession. Unpubl. B.Sc. Thesis, University of York, 160 pp.

Usher, M. B., 1973. Biological management and conservation. Chapman & Hall, London, 394 pp.

Usher, M. B., 1975. Studies on a wood-feeding termite community in Ghana, West Africa. Biotropica 7: 217–233.

Usher, M. B., 1979. Markovian approaches to ecological succession. J. Anim. Ecol. 48: 413–426.

Waggoner, P. E. & Stephens, G. R., 1970. Transition probabilities for a forest. Nature, London 225: 1160–1161.

Watt, A. S., 1960a. The effect of excluding rabbits from acidiphilous grassland in Breckland. J. Ecol. 48: 601–604.

Watt, A. S., 1960b. Population changes in acidiphilous grass-heath in Breckland, 1936–57. J. Ecol. 48: 605–629.

Watt, A. S., 1971. Factors controlling the floristic composition of some plant communities in Breckland. In: E. Duffey & A. S. Watt (eds.). The scientific management of animal and plant communities for conservation, p. 137–152. Blackwell, Oxford.

Accepted 6.7.1981.

An analysis of succession along an environmental gradient using data from a lawn

M. P. Austin & L. Belbin*

Division of Land Use Research, Institute of Earth Resources, CSIRO, P.O. Box 1666, Canberra City, A.C.T. 2601, Australia

Keywords: Environmental gradient, Generalized linear model, Lawn, Numerical classification, Succession, Transition matrix

Abstract

Differences in vegetation dynamics over a period of two years along an environmental gradient of shading on a lawn are examined. 'Communities' (groups) recognized by numerical classification are correlated with degree of shading, season and differences between years by means of generalized linear model analysis. More of the variance is explained if environmental position is used instead of degree of shading as spatial distribution of a strongly competitive species (*Trifolium repens*) is confounded with shading. Transition probabilities are related to environmental position and season. Simulation with Markov matrices for each season and position demonstrate markedly different successions for different positions. These simulations have no predictive value however as accurate estimate of transition probabilities requires knowledge of the state of adjacent quadrats, i.e. individual observations of transition probabilities are not independent.

Transition matrices are unlikely to be useful for predictive analysis of succession when spatial pattern is of significance in the community.

Introduction

The problem of how to analyse succession in vegetation with numerous species populations, environmental gradients and complex spatial patterns has not been solved. Recent attempts (Horn, 1976; van Hulst, 1979b; Shugart *et al.*, 1973; Enright & Ogden, 1979) have emphasized the possible use of transition matrices (Markov chains) or linear differential equations, though others have expressed reservations regarding the assumptions necessary (Noble & Slatyer, in press; Noble, in press; Austin, 1980; Usher, this symposium). The approach of using simplified physiological process models (Botkin *et al.*, 1972; Shugart & Noble, in press) has had some apparent success. The problem

of obtaining reasonable estimates for the physiological parameters of the models for large numbers of species may be expected to provide difficulties. Shugart *et al.* (in press) appear to have tackled this problem with some success for a rainforest site; though critical tests of predictions have yet to be made. These studies have all concerned forests where individuals can be easily recognized. Grassland succession has involved extensive simplification (Bledsoe & van Dyne, 1971) because of the difficulty of defining individuals and the numbers of species of diverse growth forms involved.

Analysing grassland succession by simplification using multivariate descriptive procedures (Austin, 1977) and combining these with transition matrix simulation (Austin, 1980) have been attempted. This later study used a divisive information statistic procedure to classify small contiguous quadrats on a lawn recorded over a period of six years into a series of 'communities' (here refered to as

* We thank S. Kendall, S. Witts, A. Howard and C. Helman for their assistance with the data preparation and analysis, and R. Cunningham for advice on statistical analysis.

Vegetatio 46, 19–30 (1981). 0042-3106/81/0461-0019/$2.40.
© Dr W. Junk Publishers, The Hague.

groups). Yearly transitions between groups were then used to develop a simulation of group changes through time. It was demonstrated that the matrices were *not* stationary during the period of study due to changes in the life history characteristics of certain dominant species and it was suggested that the failure of simulation after allowing for this change was probably due to environmental and spatial heterogeneity effects. There were several limitations:

(1) The environment of the lawn was assumed to be uniform and succession was assumed to be similar over the whole lawn, though it had been shown (Austin, 1977) that major differences existed.

(2) Quadrats had been aggregated for the numerical classification because of computer limitations, leading to very low numbers of transitions for estimating the transition probabilities.

(3) The behaviour of the groups with respect to season was ignored though evidence was presented to show that season had an important impact on one of the principal species, *Sagina procumbens.*

This study examines the implications of relaxing these limitations on the use of transition matrices when applied to grassland.

Data and methods

The floristic composition of a lawn in a London (U.K.) suburb had been observed over the period 1957–1967 with varying degrees of detail (see Austin, 1977, 1980). Presence or absence of species was recorded for 6 × 6 inch (15.25 × 15.25 cm) contiguous quadrats for a lawn of 35.5 m². The lawn formed a rectangle with the long axis north-south. Five blocks (positions) each of 288 quadrats reflecting different shading conditions were recognized along this axis for purposes of analysis (Austin, 1977). During the period 1966–67, the lawn vegetation was recorded four times/yr, late March/early April, June, late July/early August, late September/early October. The data (29 species × 11 232 quadrats) for this period were subjected to divisive information analysis classification (DIVINF, Lance & Williams, 1968). The changes with season and year in the groups defined by the classification for each quadrat were

determined and transition matrices and linkage diagrams calculated. A radiation index (measure of degree of shading, Fleming, 1971) was determined for fourteen points on the lawn, and mean values obtained for each block of quadrats by interpolation. The trampled western edge was excluded from subsequent analysis leaving 216 quadrats for each block. The frequency of different groups in different seasons, years and levels of radiation index was examined using a Generalized Linear Model (GLM) analysis (Nelder & Wedderburn, 1972). This procedure allows for a more flexible form of regression analysis than the usual approach which assumes a normal error function (e.g. Austin & Cunningham, in press).

Results

Numerical classification to define communities

The results of the DIVINF classification are shown in Figure 1. Twelve groups have been recognized from the dendrogram, though interpretable groups could be recognized down to the 30-group level. The first division is on *Festuca ovina* and reflects the environmental effect of shading. The lawn has a house immediately to the south which reduces direct solar radiation to a minimum (maximum of one hour about midday in June) on the southern part of the lawn. *F. ovina* does not occur in the most densely shaded areas, nor in the trampled region on the western edge of the lawn (Austin, 1977). The second division on the positive side of the dendrogram is on *Trifolium repens* which in this habitat, is a strong competitor suppressing many of the composite rosette weeds but growing much more slowly in the low radiation index (high percentage shade) areas of the lawn (Austin, 1980). The division species on the negative side is *Sagina procumbens,* a species whose life-history characteristics have changed dramatically during the full period of observation (Austin, 1980), and which shows a marked seasonal variation in its presence in the shaded areas of the lawn where it was previously dominant throughout the year. The other divisions similarly reflect either, shading and/or competition, e.g. *Leontodon autumnalis* which generally occurs in areas with high radiation index values and no *T. repens.* There is also the

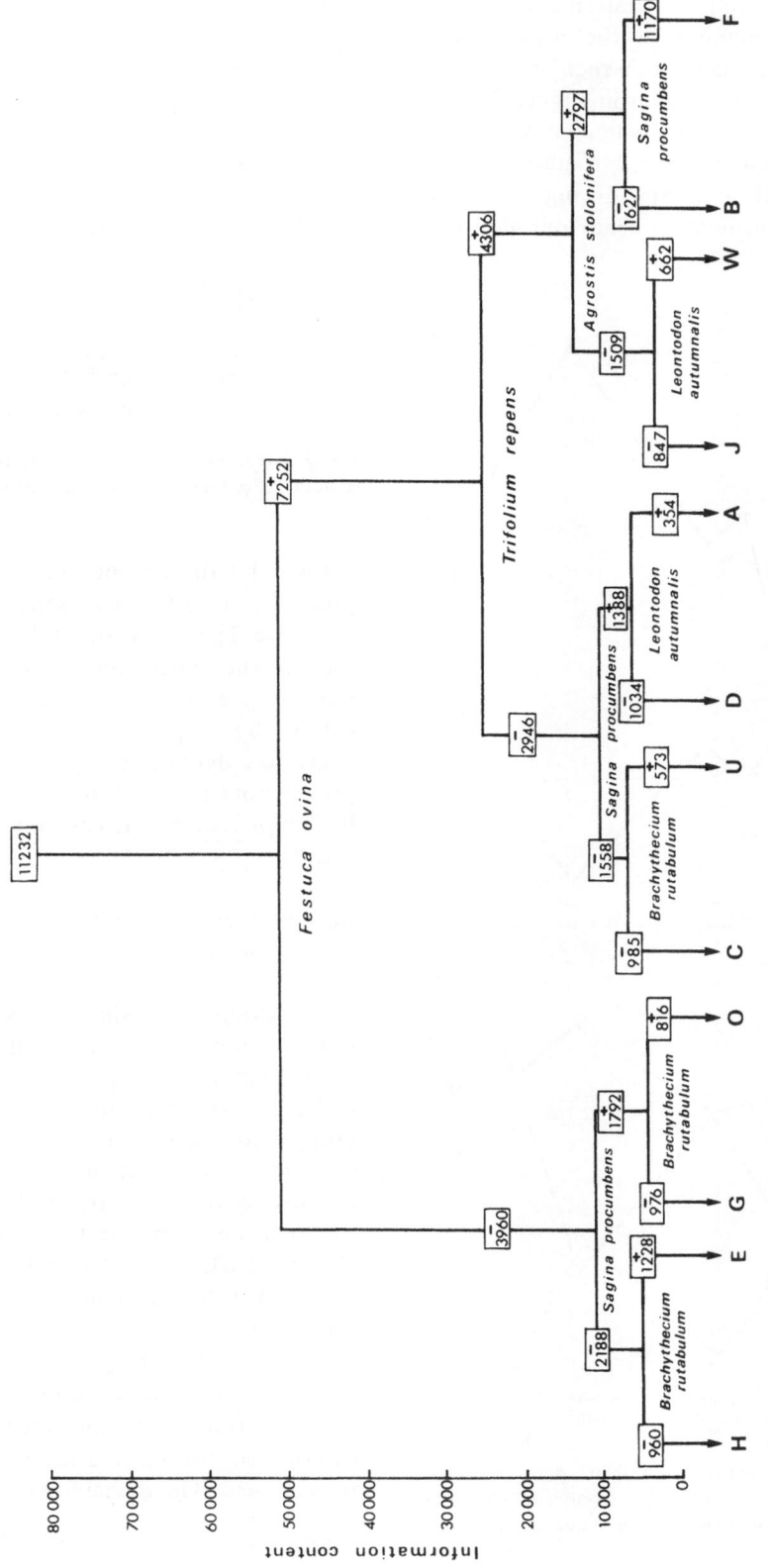

Fig. 1. Numerical classification (divisive information analysis) of lawn quadrats over eight seasons April 1966 – October 1967.

effect of trampling along the western edge.

The seasonal dynamics of the major group subdivisions for the shaded (absence of *F. ovina*) regions of the lawn are shown in Figure 2a. The population buildup in early summer of *S. procumbens* and its decline during late summer with a pronounced population collapse during winter (e.g. group 0) with complementary behaviour of groups

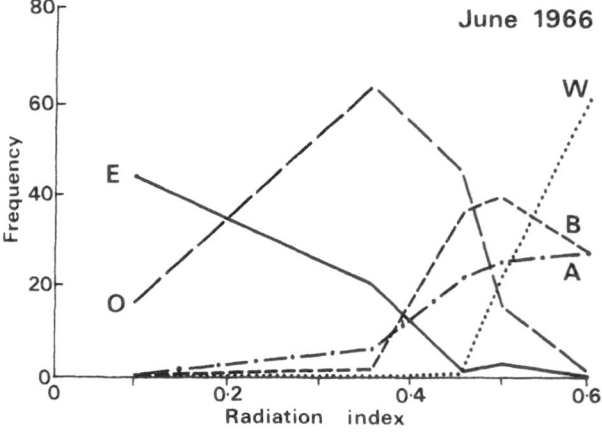

Fig. 3. Distribution of groups in relation to radiation index (shading). See Figure 1 for community definitions of letters.

defined by the absence of *S. procumbens* (e.g. group E), produces a seasonal variation in group structure. The behaviour of the groups in the sunny areas can be seen in Figure 2b where selected groups indicate the progressive replacement of groups defined by *L. autumnalis* (e.g. group A) and the negatively defined group C characterized by the presence of *Ceratodon purpureus* (cf. Austin, 1980), by *T. repens* dominated communities (e.g. groups B and F).

Influence of environment and season on group distribution

The gradient in shading as measured by the radiation index (a level unshaded site has a value of 1.0) runs ca. N-S along the long axis of the lawn and is clearly correlated with group distribution. Many groups are also influenced by season and are changing with time. In order to examine the relative importance of these different factors in relation to the distribution of the twelve groups recognized (Figure 1), a GLM analysis was carried out for each group. The factors considered were season, expressed as four classes, year as two classes, and environment as average radiation index. The frequency of the twelve groups for each of five positions (blocks) as indicated in Figure 3 were determined for each season in each year and environment. The model used was:

$\log p$ = linear function of environmental variables
assuming a Poisson error function.

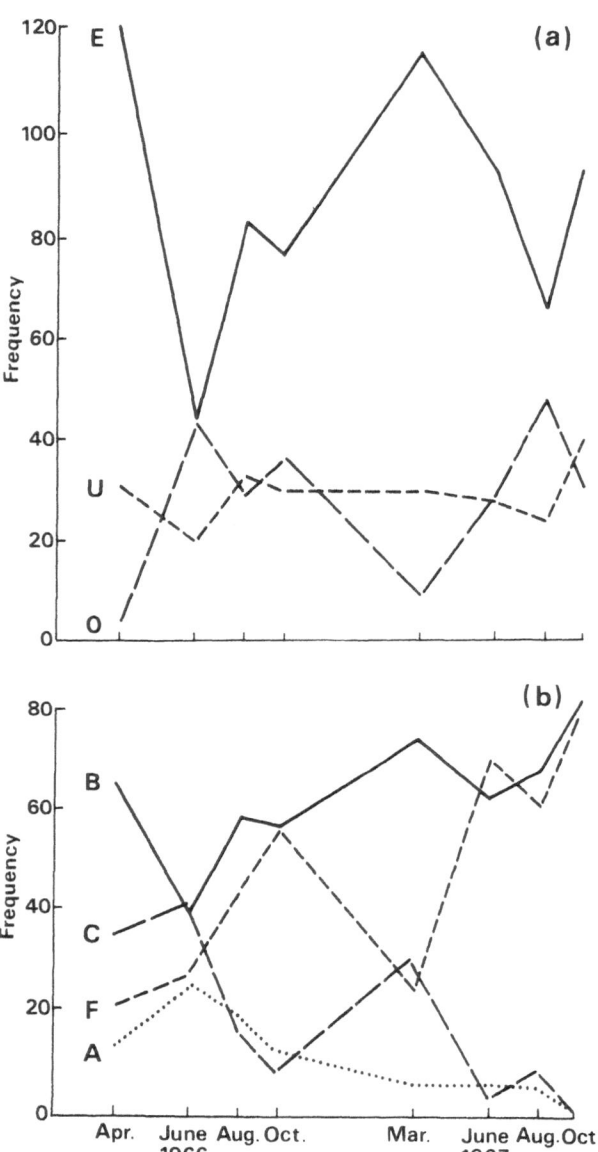

Fig. 2. Dynamics of selected groups in shade, position one (a) and sun, position four (b). See Figure 1 for group definitions of letters. Frequency is the observed number of quadrats occupied by a community.

where *p* is number of quadrats containing community 'x'.

The factors and their interactions were analysed; the results for one group, A, are shown in Table 1. The linear relationship with radiation is significant, and there are suggestive though not significant differences between seasons. There are also significant changes from 1966 to 1967 ('year' factor) indicating that succession is continuing or the years are different in weather conditions (cf. Fig. 2a & b). If the model is an appropriate fit to the data, the expected mean residual deviance for a Poisson error function is 1.0. The observed value of 2.46 indicates that there are sources of heterogeneity in the data which have not been accounted for.

For many of the groups, the relationship with radiation index is not a simple linear trend over all five levels. Some groups, e.g. W, show a linear trend plus a threshold effect at about the 0.45 level for the radiation index (Fig. 3), while others, B and D, have complex functions with respect to radiation index. The radiation index is strongly correlated with position along the long axis of the lawn (see Austin, 1977), while the 'patch' group phenomenon associated with the spread of *T. repens* (group B) (Fig. 3) occurs particularly in the central and sunny regions of the lawn. Group 'competition' and radiation index tend to be confounded. If the GLM is fitted to the data with five blocks (classes) representing five environmental positions, then the spatial distribution of the groups can be accounted for in terms of a single environmental variable, though radiation index effects and group competition can not be distinguished. The GLM analysis using environmental position as a series of classes allows the fitting of linear and threshold effects of the type described for group W more readily than for a continuous variable such as radiation index. Table 2 shows the analysis for group A; additional information is now available regarding the significance of the season effect and interaction between environmental position and year in determining group distribution; the mean residual deviance (0.961) now indicates that the residual error is random. There is a significance difference between the radiation index and position models (Table 3).

Table 1. Relationship between distribution of group A (+ *Festuca ovina* – *Trifolium repens* + *Sagina procumbens* + *Leontodon autumnalis*) and radiation index, season and year.

Factor	Deviance	DF	Change of deviance	'F' ratio
Initial Model	305.0	39		
Radiation index	152.09	1	152.89	62.2***
Year	102.00	1	50.10	20.4***
Season	83.63	3	18.37	2.49 10% < P < 5%
	Mean residual deviance = 2.460			

Table 2. Relationship between distribution of group A and environmental position, season and year.

Factor	Deviance	DF	Change of deviance	'F' ratio
Initial Model	305.0	39		
Environmental position	107.10	4	197.88	51.5***
Year	57.01	1	50.10	52.2***
Season	38.64	3	18.37	6.4**
Position/year interaction	25.94	4	12.80	3.3*
	Mean residual deviance = 0.961			

24

Table 3. Test of radiation index model and environmental position model for community A.

	Deviance	DF	Difference in deviance
Radiation index model	83.63	34	57.69*** (F with 7,27 DF)
Environmental position model	25.94	27	

Marked differences exist between the groups in the factors which influence their distribution in time and space (Table 4). Position and season are significant factors in all the positional model analyses. There are numerous significant interaction terms for many groups indicating the dynamic changes which are taking place in certain parts of the lawn. Group H is the completely negatively defined group characteristic of shaded areas which are subject to trampling. Exclusion of the trampled western section from the analysis has reduced the frequency of the group and left only a positional (radiation index) effect as significant. The residual mean square values are now all within acceptable range of the expected value of 1.0. Values below one indicate the limitations of the method where numerous zero values exist in the data, e.g. group W in Figure 3.

Use of transition matrices

Transition probabilities can not be directly correlated with environment; if a group is absent, then the transition probability remains undefined. The expected and observed occurrence of groups can be compared using the GLM analysis presented above. Using the estimated values from the GLM analysis as marginal totals for those blocks where the groups are absent, the expected transition probabilities can be calculated on a random expectation basis and compared with the observed. There is clear evidence indicating that the transition probabilities are not a simple random variable dependent on the number of groups present at the times studied. There are 'preferred' transitions; the succession is not a series of random events. Comparison of transition probabilities for different environmental positions do not require detailed tests to indicate they are different. This can be clearly shown by linkage diagrams (Fig. 4) which show the major transitions (links) and their direction for three of the five environmental positions studied.

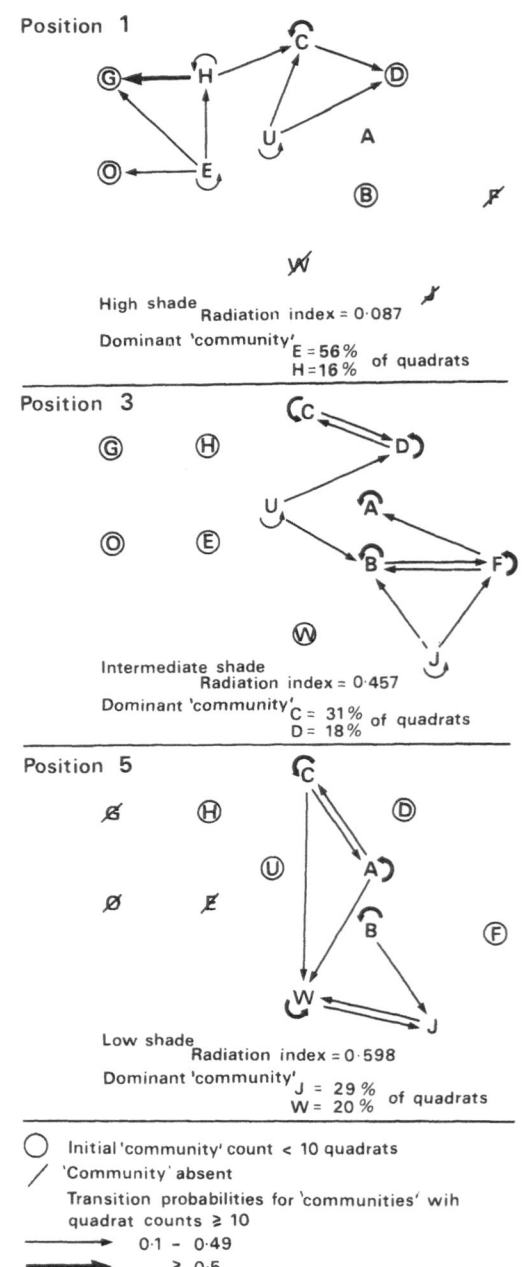

Fig. 4. Linkage diagrams for different environmental positions for the period April to June 1966.

Table 4. Models of relationship between groups and environmental variables (details of regression coefficients are not shown).

Community	Original mean deviance	Radiation index model	Residual mean deviation	Environmental position model	Residual mean deviation
A	7.82	Radiation*** + year*** (+ season)	2.46	Position*** + year*** + season** + position · year*	0.961
B	17.26	Radiation*** + year*	10.06	Position*** + year*** + season** + position · year*	1.195
C	12.43	Season*** (+ year)	8.60	Position*** + year*** + season*** + position · year** + year · season*	1.401
D	21.19	No significant terms	21.19	Position*** + season**	1.194
E	44.72	Radiation***	9.27	Position*** + year** + season*** + year · season***	0.783
F	27.57	Radiation*** (+ year + season)	15.76	Position*** + year*** + season*** + position · year* + year · season***	0.829
G	11.67	Radiation*** + year* + season*	6.18	Position*** + year*** + season*** + year · season*	0.799
H	10.41	Radiation***	4.61	Position***	1.068
J	30.62	Radiation*** + season** (+ radiation · year)	4.40	Position*** + season** + position · year*** + year*** + year (N.S.)	0.987
O	14.06	Radiation*** + season**	3.20	Position*** + year* + season*** + position · season*** + year · season*** + position · year**	0.274
U	9.66	Radiation***	3.80	Position*** + season** + position · season* + position · year* + (year · season)	0.514
W	36.05	Radiation*** + year** + season**	2.95	Position*** + year** + season***	0.633

See Fig. 1 for defining species for each group.

( ) Factors in brackets significant at 5% < P < 10% level.

Under high shade (position one), in the spring the dominant groups E (*F. ovina – S. procumbens + Brachythecium rutabulum*) and H (*F. ovina – S. procumbens – B. rutabulum*) become invaded by *S. procumbens* and are then classified as either G. (*F. ovina + S. procumbens – B. rutabulum*) or O (*F. ovina + S. procumbens + B. rutabulum*) as indicated in Figure 2a. A minor series of groups in this environment are C, D and U which show their own separate pattern of major transitions during this season (Figure 4). Three groups are absent from this environment in this season. With the reduction of shade at position two, the same groups and their transitions remain important.

The further decreases in shade at position three results in a sharp shift of the structure of the transition matrix. The G, O, H, E groups are now present in very low numbers, while all the recognized groups occur to some extent in this environment. The major spring transitions concern C and D (*F. ovina* groups without *T. repens*) and separately B, F and J, (*T. repens* defined groups). The lower shade at position four indicates a similar pattern of transitions. This may have more to do with the position of the *T. repens* patches than with the environmental gradient. The lowest shade area position five, shows a slightly different structure of changes with the three groups J. W., and A being of greater importance here than at higher shade levels. Group distribution and dynamics are a function of environment.

The seasonal changes in transition probabilities are confounded with the successional or yearly differences changes (Fig. 5). Some groups e.g., C, progressively decline, so that seasonal transitions in 1967 are not comparable with those for 1966. The seasonal behaviour of W (+ *F. ovina + T. repens – Agrostis stolonifera + Leontodon autumnalis*) is well-defined reflecting the seasonal behaviour of the component species particularly *L. autumnalis,* as a summer growing species and *A. stolonifera* which in this habitat tends to suffer from drought in the summer months. The winter period however shows a sharp reversal in the direction of change indicated by the transition probabilities. In 1967, there is an increase in changes from W to J reflecting the elimination of *L. autumnalis* by *T. repens.*

The possible behaviour of the groups in different environments can be examined by simulation. Figure 6 shows the behaviour of some groups over a period of years in three environmental positions for the four seasons (April-June; June-August; August-October; October 1966-March 1967). It is assumed that the recording period late March to early April

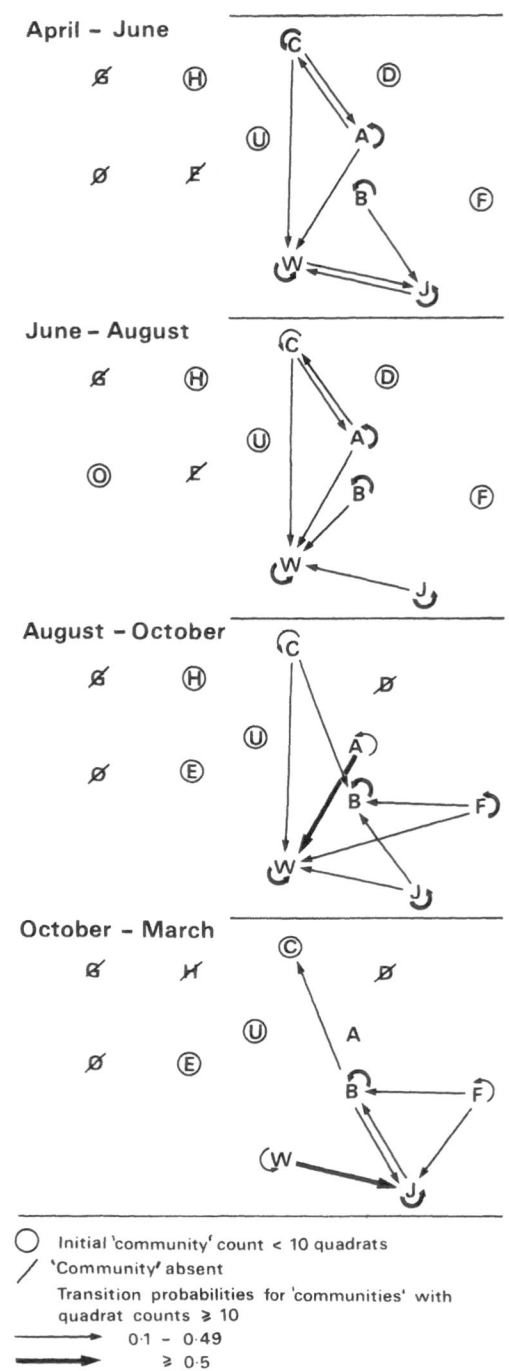

Fig. 5. Linkage diagrams for different seasons for environmental position five. Symbols as in Figure 4.

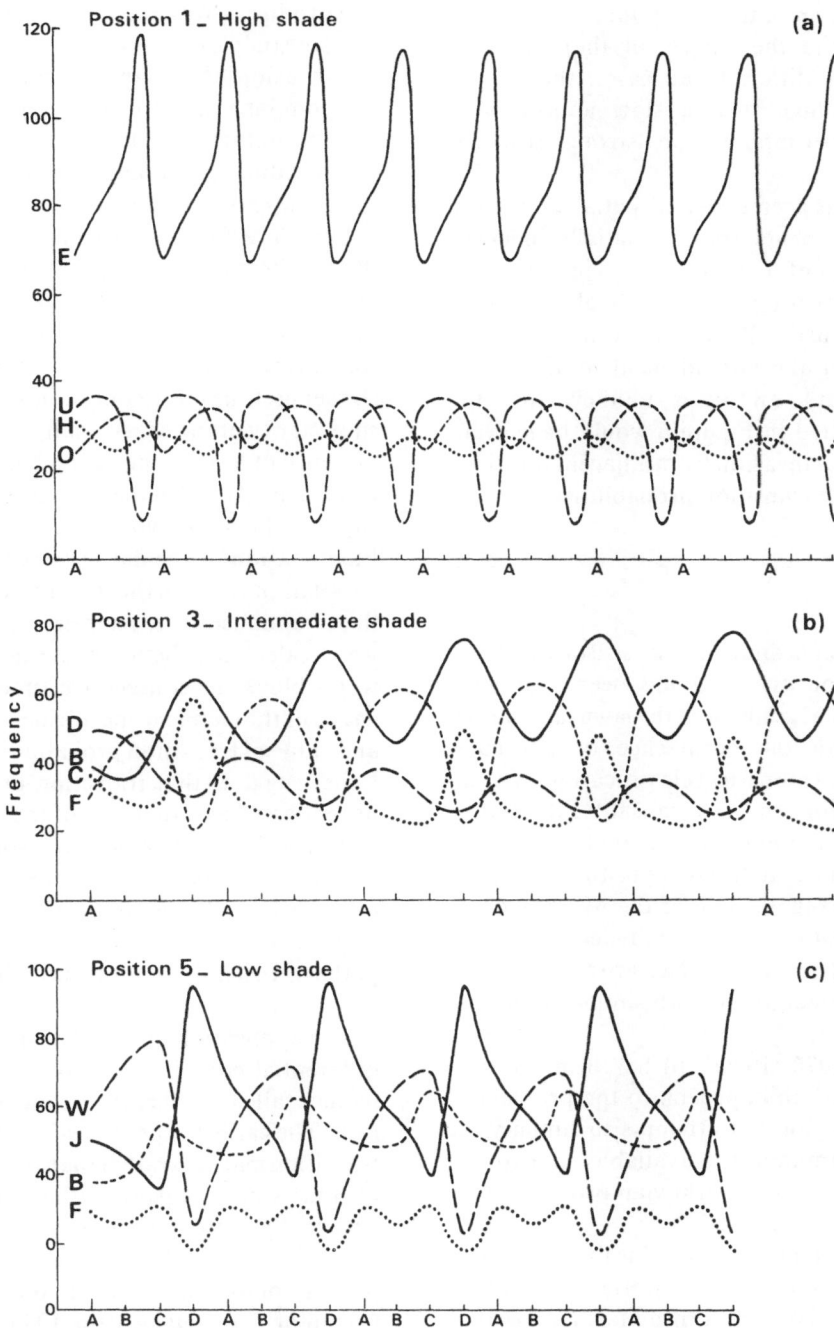

Fig. 6. Simulation of dynamics of the principal groups in three different environmental positions as in Figure 4. Note difference in time taken to reach equilibrium for environmental position one. Letters as in Figure 1. Each unit of time corresponds to a multiplication by one of the seasonal transition matrices, i.e. A, B, C, D.

(1967) can be equated with that of late April (1966) and improved estimates of the transition probabilities can be obtained by averaging the equivalent values from the two years. Each environmental position exhibits different group dynamics as would be expected from Table 4 and Figures 4 and 5.

Different groups are numerically more frequent in each position and the shapes of the curves of temporal change differ. In all cases, there is an initial rapid change then a slow approach to equilibrium superimposed on strong seasonal changes.

No account has been taken of spatial effects, the different environments are not spatially independent, the patches of *T. repens* surviving from the 1959 drought have progressively spread across four of the blocks (Austin, 1980). The transition probability for a quadrat is not independent of the state of its neighbours, and it is unlikely that the dynamics predicted in Figure 6 would have eventuated, because of invasion from adjacent positions would change the transition probabilities.

Discussion

The role of environment and season in determining vegetation dynamics has been examined. The environmental gradient of the lawn has marked correlations with the occurrence of particular groups defined by numerical classification. Assuming the correlations indicate causality, the transition matrix simulation predicts markedly different vegetation dynamics at different positions along the shade gradient (Fig. 6). Use of the Markov chain method of simulation has been sufficient to indicate the distinct patterns which may arise with differences in environment even with simple vegetation such as a lawn.

Van Hulst (1978, 1979a, b) has in a series of papers developed an approach to the problem of modelling succession but attempts to simplify the problem by examining the available data to test whether succession is a Markovian process. There are two important assumptions:

(1) The current state will determine the future state of any quadrat. Historical effects are of no importance. The senescence of *T. repens* after occupying a site for five years (Austin, 1980) and numerous other similar observations (Watt, 1947, 1960) contradict this without the need for statistical tests.

(2) The transition probabilities are calculated from independent observations. The spatial autocorrelation of the quadrats in this study is high due to the progressive wave of group change

produced by the spread of patches of *T. repens*. No study of local succession can presume that invasion and change occur without a spatial component. Even large regional events have this nature, e.g. the many examples of species invading a continent where they had not previously occurred (Elton, 1958).

Usher (1979) examined a number of data sets for the Markovian property that successive steps in a sequence are not independent, and demonstrated their dependency. Temporal events are not random transitions between all possible communities. (Usher uses the term 'collection'). He then tested for stationarity of the Markovian process with data on presence of termite species on baitwood blocks in a disturbed site in Ghana (Usher, 1979). The results suggest the null hypothesis should be rejected. Examination of the data showed that there was a correlation between the transition probabilities and the abundance of a recipient species (transitions were calculated between species occupying the wood blocks at consecutive times). As the abundance of the recipient species increased, the number and value of transition probabilities to it increased. Usher suggests that transition probabilities could be expressed as a function of the abundance of the recipient class. The same effect would arise if spatial heterogeneity existed with one species expanding over the area though initially present in low numbers. The blocks were distributed in a spatial pattern; Usher does not mention this as a possibility.

The application of the Markov chain (and linear differential equation) approach to vegetation dynamics fails to reflect the behaviour of the vegetation. The earlier paper (Austin, 1980) using annual transition matrices rather than the seasonal matrices of this study suggested the simulated changes occurred prior to their actual observation due to the transition probabilities being assumed to be independent of the state of neighbouring quadrats. The spread of *T. repens* (Austin, 1980) appears to effect the rosette species (*Hypochaeris radicata, Taraxacum officinale* and *Leontodon autumnalis*). Once these species have been eliminated, even if the *T. repens* dies, the transition probabilities *for this lawn* can not be those estimated from the original data for changes from *T. repens* defined communities to those in which it is absent. Because of competition, the transition probability matrix can not be stationary. Many would argue (McIntosh, 1980; Noble & Slatyer, in press; Austin, 1980) that the failure of the Markov chain approach is not at all surprising given the method's assumptions. Others have been more willing to see whether the approach is sufficient for practical purposes (Godron & Lepart, 1973; Usher, 1979; Stephens & Waggoner, 1970) and some have argued that the method is sufficiently robust to allow conclusions regarding succession to be drawn (Horn, 1976). Walker (1970) used the transition matrix as a descriptive tool to demonstrate the limitations of many successional assumptions about autogenic hydroseres. Van Noordwijk-Puijk *et al.* (1979) have also used the approach descriptively for studying saltmarsh dynamics.

The lawn study shows that marked differences in transition probabilities may exist along an environmental gradient. The testing of these differences when classes are absent from certain matrices has been rejected as inappropriate statistically and unnecessary on any rational basis. The effect of seasonal variation is more difficult to assess but due to the progressive changes occurring over the two years studied (Tables 1 and 2) and the associated disappearance of certain classes, e.g. group C in 1967, the validity of testing transition matrices with such undefined zero values seems questionable.

Nucleation (Yarranton & Morrison, 1974), the formation of patches in certain locations which modify the biotic and/or the physical environment is the key process in determining the vegetation dynamics of an area. The nucleation process in *T. repens* consisted of a slow expansion (five years, 1960–1965), then during the spring of 1966, the original centre died due to senescence or an interaction of senescence with unfavourable conditions (drought? disease?). This creation of gaps in the community cover allowed some limited re-establishment of the rosette species and expansion of some survivors. The dependency of the competition effect on spatial adjacency is very clear.

Analysis has demonstrated the role of nucleation and spatial heterogeneity even when differences in environment are taken into account. Theoreticians (Horn, 1976; van Hulst, 1978, 1979a, b; Usher, 1979) have tended to ignore spatial effects, assuming that they can be averaged out if a sufficiently large area is taken into account. Such large environmentally and climatically *homogeneous*

areas are unlikely to exist in reality and uncertainty over relationships between theory and practice will remain until methods which explicitly take account of spatial heterogeneity are incorporated into both the theory and analysis of succession. Gimingham (this symposium) has reported successful simulation with a Markovian chain approach for a particular *Calluna* heath. In this case, the initial community was relatively homogeneous and transitions at the quadrat size used reflected phases in the small scale patch cycle described by Watt (1955) for individual *Calluna* plants. Success will depend on the particular circumstances of each study.

The present study used 216 quadrats for 12 groups for each of 5 environmental positions for each of four seasons. The differential distribution of groups along the shade gradient meant that numbers of observations on which to base estimates of transition probabilities are very low. The influence of misclassifications in the numerical analysis on these probabilities may be considerable. Consistency between environmental positions and seasons suggested however that the difficulties of using transition matrix simulation were not due to poor estimates resulting from low numbers of observations.

Seasonal effects on the occurrence of different communities were established by the GLM analysis. The incorporation of these effects into the Markov chain simulation (Fig. 6) does not appear likely to overcome the previously recognized limitations of transition matrices. Although no direct comparison with the previous simulation (Austin, 1980) was possible due to the different time period and types of data used, all the circumstantial evidence indicates that spatial heterogeneity is likely to be responsible for failures to predict future vegetation dynamics even if different environmental conditions are recognized as playing a role.

References

Austin, M. P., 1977. Use of ordination and other multivariate descriptive methods to study succession. Vegetatio 35: 165–175.

Austin, M. P., 1980. An exploratory analysis of grassland dynamics: an example of a lawn succession. Vegetatio 43: 87–94.

Austin, M. P. & Cunningham, R. B., in press. Observational analysis of environmental gradients. Proc. Ecol. Soc. Aust.

Bledsoe, L. J. & van Dyne, G. M., 1971. A compartment model simulation of secondary succession. In: B. C. Patten (ed.), Systems analysis and simulation in ecology Vol. 1. Academic Press, New York.

Botkin, D. B., Janak, J. F. & Wallis, J. R., 1972. Some ecological consequences of a computer model of forest growth. J. Ecol. 60: 849–872.

Elton, C. S., 1958. The ecology of invasions by animals and plants. Methuen, London.

Enright, N. & Ogden, J., 1979. Application of transition matrix models in forest dynamics: Araucaria in Papua New Guinea and Nothofagus in New Zealand. Aust. J. Ecol. 4: 3–23.

Fleming, P. M., 1971. The calculation of clear-day solar radiation on any surface. Paper presented at AIRAH Federal Conference, Perth.

Godron, M. & Lepart, J., 1973. Sur la représentation dynamique de la végétation au moyen de matrices de succession. In: W. Schmidt (ed.), Sukzessionsforschung, Ber. Symp. Int. Ver. Vegetationskunde, p. 269–287. Cramer, Vaduz.

Horn, H. S., 1976. Markovian properties of forest succession. In: M. Cody & J. Diamond (eds.), Ecology and evolution of communities, pp. 196–211. Harvard University Press, Cambridge.

Hulst, R. van, 1978. On the dynamics of vegetation: patterns of environmental and vegetation change. Vegetatio 38: 65–75.

Hulst, R. van, 1979a. On the dynamics of vegetation: succession in model communities. Vegetatio 39: 85–96.

Hulst, R. van, 1979b. On the dynamics of vegetation: markov chains as models of succession. Vegetatio 40: 3–14.

Lance, G. N. & Williams, W. T., 1968. Note on a new information-statistic classificatory program. Comput. J. 11: 195.

McIntosh, R. P., 1980. The relationship between succession and the recovery process in ecosystems. In: J. Cairns (ed.), The recovery process in damaged ecosystems. Ann Arbor, Michigan.

Nelder, J. A. & Wedderburn, R. W. M., 1972. Generalized linear models. J.R. Statist. Soc. A 135: 370–84.

Noble, I. R., in press. The role of past and present fire frequency and intensity on the evolution of plant attributes: species interactions, competition, succession, predation. SCOPE/MAB EAST/WEST Centre, Honolulu Meeting 1978.

Noble, I. R. & Slatyer, R. D., in press. Concepts and models of succession in vascular plant communities subject to recurrent fire. In: A. M. Gill, R. H., Groves & I. R. Noble (eds.), Fire and the Australian biota, ANU Press, Canberra.

Noordwijk-Puijk, K. van, Beeftink, W. G. & Hogeweg, P., 1979. Vegetation development on salt-marsh flats after disappearance of the tidal factor. Vegetatio 39: 1–13.

Shugart, H. H., Crow, Jr. T. R. & Hett, J. M., 1973. Forest succession models: a rationale and methodology for modelling forest succession over large regions. For. Sci. 19: 203–212.

Shugart, H. H. & Noble, I. R., in press. A computer model of succession and fire response of the high altitude *Eucalyptus* forest of the Brindabella Range, Australian Capital Territory. Aust. J. Ecol.

Shugart, H. H., Hopkins, M. S., Burgess, I. P. & Morlock, A. T., in press. The development of a succession model for subtropical rain forest and its application to assess the effects of timber harvest at Wiangaree State Forest, New South Wales. J. Environ. Management.

Stephens, G. R. & Waggoner, P. E., 1970. The forests anticipated from forty years of natural transitions in mixed hardwoods. Bull. Connecticut Agricul. Exp. Stat., New Haven, 707.

Usher, M. B., 1979. Markovian approaches to ecological succession. J. Animal Ecol. 48: 413–426.

Walker, D., 1970. Direction and rate in some British post-glacial hydroseres. In: D. Walker & R. G. West (eds.), Studies in the vegetational history of the British Isles, Essays in honour of Harry Godwin, pp. 117–189. Cambridge University Press.

Watt, A. S., 1947. Pattern and process in the plant community. J. Ecol. 35: 1–22.

Watt, A. S., 1955. Bracken versus heather, a study in plant sociology. J. Ecol. 43: 490–506.

Watt, A. S., 1960. Population changes in acidiphilous grass-heath in Breckland, 1936–57. J. Ecol. 48: 605–629.

Yarranton, G. A. & Morrison, R. G., 1974. Spatial dynamics of a primary succession: nucleation. J. Ecol. 62: 417–428.

Accepted 27.7.1981.

Probing time series vegetation data for evidence of succession*

L. Orlóci

Department of Plant Sciences, University of Western Ontario, London, Ontario, Canada N6A 5 B7

Keywords: Canonical variates, Chi square, Succession, Time series, Vegetation

Abstract

A subset of Stephens & Waggoner's (1970) data, spanning 40 years of recording at the same site, is analysed by AOC for trends in density fluctuations. The results suggest that the recording period is sufficiently long for trends to be detected. A dominant trend depicts density changes as simple monotone functions of time. Other lesser trends signify cyclic changes of different lengths. The fact that a dominant monotone trend exists is interpreted as evidencè of succession.

Introduction

To render the idea of succession operational, succession should be defined with regard to some measurable property of the vegetation. If the classical theories of succession are the reference (see review by van Hulst, 1978), an operational definition should specify compositional changes which form a monotone progression in time from some designated original state. Should an examination turn up no evidence for monotone progression, succession would have to be ruled out within the time span involved and a steady state be declared. Even in such a state some compositional changes could not be discounted, but they would be cyclic or completely random.

That there is a time constraint on the interpretation of successional evidence is obvious, for changes which are actually cyclic in the long run may appear monotone in the short run. It then follows that statements about succession have to be conditional, limited in their validity to the time span within which changes are observed. A consequence of this is that the evidence has to be reinterpreted should

the time span be broadened.

Inferences about succession would have to be speculative in the absence of hard data. Indirect evidence may of course help to prevent drawing ficticious conclusions. For example, vegetation zonation in a pond may sometimes serve as an analogue pattern of hydrarch succession. The use of analogue patterns is common practice in vegetation studies, attempting to gain insight into the nature of possible long term compositional changes in a community.

For an accurate description of succession, however, there must be access to time series data. The data set of Stephens & Waggoner (1970), spanning 40 years of density changes, is an example. Such data may be expected to contain complex monotone and cyclic trends which cannot be unraveled by unaided inspection. In principle, any trend seeking algorithm should help if it can decompose complex trends. Analysis of concentration (AOC) (Feoli & Orlóci, 1978) is particularly well-suited for the task in categorial data such as the densities reported by Stephens & Waggoner (1970). AOC incorporates a device for testing of global trends and another for decomposing complex trends into independent linear components. The ith compo-

* Nomenclature follows Little (1953)

Vegetatio 46, 31–35 (1981). 0042-3106/81/0461-0031/$1.00.
© Dr W. Junk Publishers, The Hague.

Table 1. Estimated number of stems of major species (per acre) in the Muck site of Stephens & Waggoner (1970). Stand age is 20–30 years in 1927. This published data set is the sole source of information analysed in the present paper.

Species	Stem counts/acre			
	1927	1937	1957	1967
Acer saccharum	0	0	0	6.13
A. rubrum	218.40	219.17	281.63	268.93
Quercus rubra	0	0	6.13	6.13
Betula alleghaniensis	6.24	6.34	24.97	18.83
Betula lenta	0	0	6.13	12.70
Fraxinus americana	31.20	37.44	74.90	68.77
Fraxinus nigra	6.24	6.34	12.70	25.00
Ulmus americana	37.44	18.72	31.10	31.10
Nyssa sylvatica	12.48	0	0	0
Totals estimated	312	288	438	438

nent trend is defined by a pair of synthetic variables (X_i, Y_i) called canonical variates. The X variable is specific to the recording periods (relevés) and the Y variable to the species.

The decision to analyse Stephens & Waggoner's data by AOC arose after preliminary examination of the raw data (Table 1) revealed that trends are present suggestive of monotone and cyclic changes. Results are presented only for Stephens & Waggoner's Muck site. 'Muck' is the usual technical term for well decomposed organic matter with high sand and silt content.

Testing for global trends

Two properties are of interest: existence and sharpness. Existence is tested based on chi square,

$$\chi^2 = \sum_{h=1}^{q} \sum_{j=1}^{t} \frac{(F_{hj} - F_{hj}^0)^2}{F_{hj}^0} \qquad (1)$$

where

q = number of observational periods (4 in Table 1)

t = number of species (9 in Table 1)

F_{hj} = density of species h at time period j

F_{hj}^0 = a random expectation of density.

Existence is declared should the value of (1) be at least as large as $\chi^2_{\alpha; \nu}$, the α probability point of the chi square distribution with $\nu = (q\text{-}1)(t\text{-}1)$ degrees of freedom. The observed value ($\chi^2 = 127.126$) is clearly very large in such terms ($\chi^2_{0.05; 24} = 36.415$), suggesting that the time series in the Muck site

incorporates one or more trends. This justifies undertaking the next step in which density changes are decomposed into independent components.

Probing for component trends

AOC assumes that density changes are $m \leqslant \min (q, t)$ dimensional. These decompose the total chi square into an equal number of additive components,

$$\chi^2 = F_{..} R_l^2 + \ldots + F_{..} R_m^2 \qquad (2)$$

$F_{..}$ represents the grand total of the data (1 475, Table 1). The R_i^2 are canonical correlation coefficients (defined in the Appendix); the numerical values of $F_{..} R_i^2$ are given in Table 2. These reveal the existence of a dominant trend, accounting for 69% of the total chi square, and two lesser trends. The latter account respectively for 22% and 9% of the total.

X, Y variables

Each component in (2) has a pair of variables

Table 2. Canonical components of the total χ^2 in Table 1.

Component i	$F_{..} R_i^2$	%
1	88.016	69
2	27.738	22
3	11.371	9

Table 3. Canonical scores for recording periods.

Canonical variate	Scores				Correlation with time
	1927	1937	1957	1967	
X_1	1.7	0.3	−0.5	−1.0	monotone
X_2	−0.8	1.5	0.6	−1.0	cyclic
X_3	0.2	−1.3	1.3	−0.6	cyclic

Table 4. Canonical scores for species.

Canonical variate	Scores								
	1	2	3	4	5	6	7	8	9
Y_1	−0.4	0.2	−3.0	−1.2	−3.3	−0.7	−1.4	0.9	7.1
Y_2	−7.4	0.4	−1.4	0.1	−3.5	0.2	−1.9	−0.9	−6.0
Y_3	−7.0	−0.4	4.1	2.9	0.2	0.8	−1.2	0.6	2.8

associated with it. The variables in the ith pair,

$$X_i = (X_{1i} \ X_{2i} \ldots X_{qi})$$

$$Y_i = (Y_{1i} \ Y_{2i} \ldots Y_{ti})$$

order the q recording periods and the t species on the ith dimension of (2). The elements (X_{ji}, Y_{hi}) are the canonical scores. The X scores, given in Table 3, clearly establish the presence of a definite monotone trend (the major component X_1) and cyclic changes (the lesser components X_2, X_3). The cycle lengths are 40 and 20 years. The Y scores (Table 4) reveal groups of species with similar or contrasting contributions to the monotone trend (Y_1) as specified by X_1 or to the cyclic changes (Y_2, Y_3) as specified by X_2 and X_3.

Lattices of deviations

That the individual species populations undergo density changes is clear from the raw data on first inspection, and that those changes have monotone and cyclic components has been established. The contributions of the individual species to particular components can be measured. For this purpose, the data elements are partitioned according to the components in (1).

The actual partitioning involves deviations from expectation. As shown in the Appendix, the ith component of any deviation $F_{hj} - F^0_{hj}$ is given by

$$\Delta_{hj(i)} = F_{hj(i)} - F^0_{hj(i)} = X_{ji} Y_{hi} R_i (F_{h.} F_{.j})^{1/2}$$

Symbols $F_{h.}$ and $F_{.j}$ represent respectively the total density of the hth species over all recording periods and the total density at the jth recording period over all species. The numerical values of the $\Delta_{hj(i)}$ are given in Table 5. The monotone and cyclic components isolated, their significance is largely self-evident. *Betula alleghaniensis* is a typical case:

Year	1927	1937	1957	1967
M	-6.3	-1.1	2.4	5.0
C1	-0.1	0.2	0.1	-0.2
C2	0.7	-3.8	5.7	-2.6
Total	-5.7	-4.7	8.2	2.1

It can be seen based on the totals that the global

Table 5. The monotone (M) and cyclic (C) components of density changes within species. Entries represent deviations from random expectation.

Species	Component	Components of deviations 1927	1937	1957	1967
Acer saccharum	M	-2.2	-0.4	0.8	1.7
	C1	1.1	-1.8	-1.2	1.9
	C2	-0.2	1.0	-1.5	0.7
	Total	-1.3	-1.2	-1.8	4.3
Acer rubrum	M	19.6	3.3	-7.4	-15.5
	C1	-8.5	14.1	9.1	-14.8
	C2	-1.7	8.9	-13.2	6.1
	Total	9.4	26.2	-11.5	-24.2
Quercus rubra	M	-3.2	-0.5	1.2	2.6
	C1	0.4	-0.7	-0.4	0.7
	C2	0.2	-1.2	1.7	-0.8
	Total	-2.6	-2.4	2.5	2.5
Betula alleghaniensis	M	-6.3	-1.1	2.4	5.0
	C1	-0.1	0.2	0.1	-0.2
	C2	0.7	-3.8	5.7	-2.6
	Total	-5.7	-4.7	8.2	2.1
Betula lenta	M	-5.6	-0.9	2.1	4.4
	C1	1.6	-2.6	-1.7	2.8
	C2	0.0	-0.1	0.1	-0.1
	Total	-4.0	-3.7	0.5	7.1
Fraxinus americana	M	-13.2	-2.2	5.0	10.5
	C1	-1.2	2.0	1.3	-2.1
	C2	0.7	-3.8	5.6	-2.6
	Total	-13.7	-4.0	11.9	5.8
Fraxinus nigra	M	-6.4	-1.1	2.5	5.1
	C1	2.3	-3.8	-2.5	4.0
	C2	-0.3	1.4	-2.2	1.0
	Total	-4.4	-3.5	-2.2	10.1
Ulmus americana	M	9.5	1.6	-3.6	-7.5
	C1	2.6	-4.4	-2.8	4.6
	C2	0.3	-1.6	2.4	-1.1
	Total	12.4	-4.4	-4.0	-4.0
Nyssa sylvatica	M	7.9	1.3	-3.0	-6.3
	C1	1.8	-3.0	-1.9	3.1
	C2	0.2	-0.8	1.2	-0.6
	Total	9.8	-2.4	-3.7	-3.7

trend is non-monotone. It is a convolute trend in fact, reflecting the combined influence of three components: an accentuated monotone trend (M), a long cycle (C1), and a short cycle (C2). The population begins in 1927 with a density much below random expectation (M). The long-cycle changes (C1) tend to widen the difference, but this is more than compensated for by short-cycle changes. Progressing in time, the deviation from random expectation narrows, after density peaking far above expectation at the top of the short cycle in 1957. Following this, the monotone component keeps density above random expectation, notwithstanding reductions owing to cyclic changes.

Discussion

The results make it doubtful that visual inspection of the raw data could do justice to the objectives of a succession study. The time series is likely to be complex where the data embody complex trends, not detectable without the use of analytical techniques.

In the Stephens & Waggoner's data, trends are found associated with monotone and cyclic changes of density. The monotone trend is most intensive (69%) and affects all species. In some species, density undergoes a steady increase from below to above random expectation. The species involved are those invading the site (*A. saccharum, Q. rubra, B. lenta, F. americana, F. nigra*) or just past a peak in proportional increase (*B. alleghaniensis*). In another case (*A. rubrum*), density actually increases but not rapidly enough to prevent the proportional representation of the species from undergoing an actual monotone decline as density drops below random expectation. In still other cases (*U. americana, N. sylvatica*), a definite monotone reduction sets in with time, leading to decline or even obliteration of the species.

Density changes in the long cycle are less intensive (22%) and affect species in two ways. In convex long cycles (*A. rubrum, B. alleghaniensis, F. americana*), density rises above expectation at midrange in the time series. In the concave long cycles, such as in *F. nigra* and *U. americana*, density falls below expectation at midrange. Invading species (*A. saccharum, Q. rubra, B. lenta*) and *N. sylvatica* behave similarly. Short-cycle changes are the least intensive (9%). The species exhibit one of two patterns (+-+- or -+-+).

To pinpoint the exact causes for the observed trends would require facts not available from the published record. The conclusions drawn thus have to be confined to the statement that the presence of a dominant monotone trend suggests a strong successional affect. If cyclic changes dominated, the conclusion would be: the vegetation reached a steady state at the scale of the cycle.

Trend seeking multivariate methods offer definite advantages in time series analysis over the more conventional methods in succession studies. When the data are categorical, AOC is an appropriate method. It can be used to test for global trends and to decompose complex trends into simple components, suitable for interpretation. The results presented in the foregoing sections clearly support this conclusion.

References

Feoli, E. & Orlóci, L., 1979. Analysis of concentration and detection of underlying factors in structured tables. Vegetatio 40: 49–54.

Hulst, van R., 1978. On the dynamics of vegetation: pattern of environmental and vegetational changes. Vegetatio 38: 65–75.

Little, E. L. Jr., 1953. Check list of native and naturalized trees of the United States (including Alaska). U.S.D.A. Agr. Handbook 41. 472 pp.

Stephens, G. R. & Waggoner, P. E., 1970. The forests anticipated from 40 years of natural transitions in mixed hardwoods. Bull. 707. Connecticut Agricult. Stat., New Haven.

Accepted 12.5.1981.

Appendix

1. Extraction of canonical correlations and scores

The $R_1^2, R_2^2, \ldots, R_m^2$ are eigenvalues of the cross product matrix

$$S = U'U$$

where

$$U_{hj} = \frac{F_{hj}}{(F_{h.} F_{.j})^{1/2}} - \frac{(F_{h.} F_{.j})^{1/2}}{F_{..}}$$

Since $\alpha_1, \alpha_2, \ldots, \alpha_m$ are the associated eigenvectors, the quantity

$$X_{ji} = \frac{\alpha_{ji} - \bar{\alpha}_i}{\sum\limits_{e=1}^{q} (\alpha_{ei} - \bar{\alpha}_i)^{1/2}} \left[\frac{F_{..}}{F_{.j}}\right]^{1/2}$$

is a canonical score for recording period j on the ith canonical variate and

$$Y_{hi} = \sum_{j=1}^{q} F_{hj} X_{ji} / R_i F_{h.}$$

is a canonical score for species h.

2. Decomposition of U_{hj}

The m components of U_{hj},

$$U_{hj} = U_{hj(1)} + U_{hj(2)} + \ldots + U_{hj(m)}$$

are generated by the product

$$U_{hj(i)} = Y_{hi} \, X_{ji} \, R_i.$$

Since

$$U_{hj} = \frac{F_{hj}}{(F_{h.} \, F_{.j})^{1/2}} - \frac{(F_{h.} \, F_{.j})^{1/2}}{F_{..}}$$

we have

$$F_{hj} = \left[U_{hj} + \frac{(F_{h.} \, F_{.j})^{1/2}}{F_{..}} \right] \left[F_{h.} \, F_{.j} \right]^{1/2}$$

$$= \left[U_{hj(1)} + U_{hj(2)} + \ldots + U_{hj(m)} \right] \left[F_{h.} \, F_{.j} \right]^{1/2} +$$

$$+ \frac{F_{h.} \, F_{.j}}{F_{..}}$$

and further,

$$F_{hj} - F_{hj}^{o} = U_{hj(1)} \left[F_{h.} \, F_{.j} \right]^{1/2} + U_{hj(2)} \left[F_{h.} \, F_{.j} \right]^{1/2} + \ldots +$$

$$+ U_{hj(m)} \left[F_{h.} \, F_{.j} \right]^{1/2}.$$

The (X_i, Y_i) lattice of component deviations have $q \times t$ cells and elements.

$$F_{hj(i)} - F_{hj(i)}^{o} = U_{hj(i)} \left[F_{h.} \, F_{.j} \right]^{1/2}, \; h = 1, 2, \ldots, q; j = 1, 2, \ldots, t$$

Towards the analysis of vegetation succession

R. Gittins*

Department of Plant Sciences, University of Western Ontario, London, Ontario, Canada N6A 5B7

Keywords: Association in *r* x *c* tables, Canonical analysis, Canonical variates analysis, Dummy variable, Hydrosere, Tropical rain forest, Vegetation dynamics

Abstract

Attention is drawn to canonical analysis as a plausible model for analyzing vegetation succession. An assessment of the opportunities afforded by canonical analysis for this purpose is then made by reference to two applications of the method. The applications deal with characteristics of hydroseral processes and with the dynamic status of an area of lowland tropical rain forest. On the basis of these and other studies the conclusion is drawn that canonical analysis could contribute usefully in efforts to place the study of dynamic ecosystem processes on a more analytic footing.

Introduction

Succession refers to cumulative change in vegetation over time. In view of the long-standing interest of ecologists in studies of this kind it is surprising to find that it is less widely recognized than it deserves to be that:

a) the definitive study of succession calls for sequential observations over time, say at times, t_0, t_1, ..., t_m; and that

b) the field observations in successional studies are *vector-valued,* $\mathbf{x}_j = [x_{1j}, ..., x_{pj}]$, where x_{ij} represents the abundance of species $i(i = 1, ..., p)$ in the jth stand.

In connection with the first point we observe that, where the age-span of the communities or vegetation of interest exceeds say 2–3 years, sequences of temporally related observations are often difficult to obtain. Moreover, the field observations themselves almost always involve vector rather than scalar quantities. Put another way, the observed data are *multivariate* in character. Indeed, the data in successional studies are not merely multivariate but in a sense are actually doubly multivariate, generating a 3-dimensional array $\mathbf{Z}$ (N x p x m), say, where N refers to stands, p to species and m to time-points (see Fig. 1). Thus, the need for some means of efficiently summarizing the data will largely be self-evident. In particular, since the observations themselves are multivariate, specifically some form of multivariate analysis would appear to be called for. Paradoxically, applications of multivariate analysis in successional studies are comparatively rare. Nevertheless, it is clear from the literature that a need for improved analytical methods exists.

These considerations indicate that the study of succession is likely to prove more than ordinarily demanding with respect to both data acquisition

*This work was supported by a Natural Sciences and Engineering Research Council award to L. Orlóci. It is a pleasure to acknowledge my indebtedness both to Dr. Orlóci for his support and encouragement and to the Department of Plant Sciences at the University of Western Ontario for hospitality and help. Appreciation is expressed also to Dr. John Ogden, Research School of Pacific Studies, The Australian National University, Canberra for kindly placing his rain forest data at my disposal and for his invaluable help in interpreting the results of analyses of these data.

Vegetatio 46, 37–59 (1981). 0042–3106/81/0461–0037/$4.60.

and analysis. The present paper addresses the second of these two issues and has the principal objective of drawing attention to one of a number of plausible methods for the analysis of data pertaining to vegetation succession.

A model for analysis of vegetation succession

Model identification

From Figure 1 it is apparent that successional studies are open to formulation in terms of the analysis of relationships between realizations of p-variate quantities over m points in time ($m>2$). Expressed in this way, the problem of investigating succession is seen to be formally equivalent to the familiar task of exploring relationships generally between any two or more sets of variables. We will make use of this observation to identify an algebraic model by means of which the study of succession may be made analytic.

For simplicity, consider first the case when the number of timepoints, or, in ecological terms, seral stages, is just two ($m = 2$). It will be clear that in this case ecological interest centers on the overall compositional similarity of the vegetation at times t_0 and t_1. Close similarity would be equivalent to declaring the vegetation to be dynamically stable, while weak similarity would mark the occurrence of temporal change. Information on vegetational similarity will be conveyed by the matrix of correlation coefficients between the p observed variables (species or plant communities) at times t_0 and t_1, or by some related scalar-products matrix. Denoting the partitioned vector of observations for the jth sample ($j=1, ..., N$) standardized to zero mean and unit variance at t_0 and t_1 by $\mathbf{z}_j = [z_{01} ... z_{0p} | z_{11} ... z_{1p}]$ and the correlation matrix of the $\mathbf{z}_j$ by $\mathbf{R}$, we may write

$$\mathbf{z}_j = \begin{bmatrix} \mathbf{z}_{0j} \\ --- \\ \mathbf{z}_{1j} \end{bmatrix} \quad \mathbf{R} = \frac{1}{N-1}\Sigma \mathbf{z}_j \mathbf{z}_j = \begin{bmatrix} \mathbf{R}_{00} & | & \mathbf{R}_{01} \\ ----- & | & ----- \\ \mathbf{R}_{10} & | & \mathbf{R}_{11} \end{bmatrix}$$

In these expressions $\mathbf{z}_{0j}$ and $\mathbf{z}_{1j}$ are p-component subvectors of observations on the same p species or

communities at times t_0 and t_1, respectively, while

$\mathbf{R}_{00}$: p x p matrix of correlations between the observed variables at t_0

$\mathbf{R}_{11}$: p x p matrix of correlations between the observed variables at t_1

$\mathbf{R}_{01} = \mathbf{R}_{10}$: p x p matrix of intercorrelations between the variables at times t_0 and t_1.

The internal structure of a partitioned correlation matrix of this kind is rarely evident on inspection. We are therefore led to search for some means of making the internal structure of $\mathbf{R}$ manifest. Now it is well known (e.g. see Anderson, 1958; Timm, 1975; Morrison, 1976) that canonical analysis provides a means of reducing the linear correlation structure between any two sets of variables to its simplest possible form. Thus, it would appear that canonical analysis could in principle, at least, provide a suitable algebraic or statistical model for exploratory studies of succession where $m = 2$.

If we now relax the condition that $m = 2$, we need only observe that *generalized* canonical analysis can be used to simplify the linear correlation structure between any m ($m > 2$) sets of variables (Horst, 1961; Kettenring, 1971) to see that canonical analysis might prove to be a useful general procedure for the exploratory investigation of temporal change in vegetation ($m \geqslant 2$).

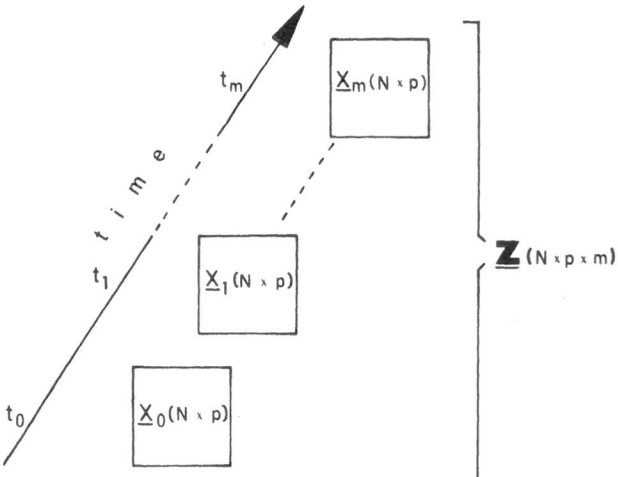

Fig. 1. Definitive studies of succession call for sequential observations over time, say at time-points $t_0, t_1, ..., t_m$. Collectively, the resulting field observations give rise to a 3-dimensional array $\mathbf{Z}$ (N $\times$ p $\times$ m) of N stands $\times$ p species $\times$ m time 'slices'.

Having succeeded in identifying what may turn out to be a suitable algebraic model for achieving our ecological goal we have in effect translated the problem from its initial expression in purely ecological terms to an equivalent operational expression. In other words, given two or more sets of sequentially related observations, it should be possible to clarify the dynamic relationships among them by the routine application of a standard algebraic or statistical procedure. The remainder of this contribution is directed chiefly towards assessing the extent to which canonical analysis fulfills this expectation in practice. Before turning to consider some practical applications, however, we first summarize something of the nature and properties of canonical analysis itself.

Canonical analysis

Let $X_0(N \times p)$ and $X_1(N \times q)$ denote observations on two sets of variables with a joint $(p + q)$-variate distribution $(p \geqslant q)$. As a special case we include the situation in which X_0 and X_1 consist of observations on the same p jointly distributed variables at times t_0 and t_1 $(p = q)$. For convenience, suppose each variable is measured about its respective mean and denote the sample mean vector and covariance matrix of the jth observation $x_j = [x_{0j} \ x_{1j}]$ $(j = 1, ..., N)$ by

$$m = \begin{bmatrix} O_0 \\ O_1 \end{bmatrix} \qquad V = \begin{bmatrix} V_{00} & V_{01} \\ V_{10} & V_{11} \end{bmatrix}$$

The purpose of canonical analysis is to transform the observed variables to new variables, $u_k = a'_k x_0$ and $v_k = b'_k x_1$ $(k = 1, ..., q)$, respectively, of unit variance, for which the covariance or correlation matrix has the simplest possible form. The analysis requires determination of those matrices of coefficients

$$A = [a_1 | ... | a_p]$$
$$B = [b_1 | ... | b_q]$$

which carry V into the tridiagonal form

$$\begin{bmatrix} A & O \\ O & B \end{bmatrix} \begin{bmatrix} V_{00} & V_{01} \\ V_{10} & V_{11} \end{bmatrix} \begin{bmatrix} A & O \\ O & B \end{bmatrix} = \begin{bmatrix} I_p & \Gamma \\ \Gamma' & I_q \end{bmatrix}$$

where I_p and I_q are identity matrices of order p and q, respectively

$$\Gamma = \begin{bmatrix} \mathrm{diag}\ (r_k) \\ \hline O \end{bmatrix},$$

$r_1 > ... > r_q$ and 0 is the null matrix of order $q \times (p - q)$.

The r_k are the canonical correlation coefficients, the a_k and b_k are vectors of canonical weights while the transformed variables $u_k = a'_k x_0$ and $v_k = b'_k x_1$ are the canonical variates of x_0 and x_1, resp. To obtain the $\{r_k; a_k, b_k; u_k, v_k\}$ it is well-known (e.g. see Anderson, 1958; Timm, 1975; Morrison, 1976) that it is necessary to solve the generalized eigenvalue problem

$$\begin{bmatrix} O & V_{01} \\ V_{10} & O \end{bmatrix} \begin{bmatrix} a \\ b \end{bmatrix} = \lambda \begin{bmatrix} V_{00} & O \\ O & V_{11} \end{bmatrix} \begin{bmatrix} a \\ b \end{bmatrix}. \qquad (1)$$

The square-roots of the eigenvalues λ of (1) yield the canonical correlation coefficients r_k while the eigenvectors $[a' \ b']$ normalized to satisfy the side-conditions

$$a' V_{00} a = b' V_{11} b = 1$$

provide the vectors of canonical weights.

Canonical analysis reduces the problem of interpreting the $p \times q$ correlation coefficients between two sets of p and q observed variables (or, alternatively, of the p^2 correlations between the same set of p observed variables at times t_0 and t_1) into the interpretation of at most q canonical correlation coefficients between q pairs of linear transformations of the observed variables – the canonical variates. The canonical correlation coefficients themselves provide a basis for assessing the independence of the two sets of variables, such as a test of the hypothesis

$$H_o : \Sigma_{01} = O ,$$

for example, where Σ_{01} is the population counterpart of V_{01}. Furthermore, where an hypothesis of this kind proves untenable then the number of large or noteworthy canonical correlations enables the intrinsic dimensionality of the dependence structure to be assessed. Commonly, the effective dimen-

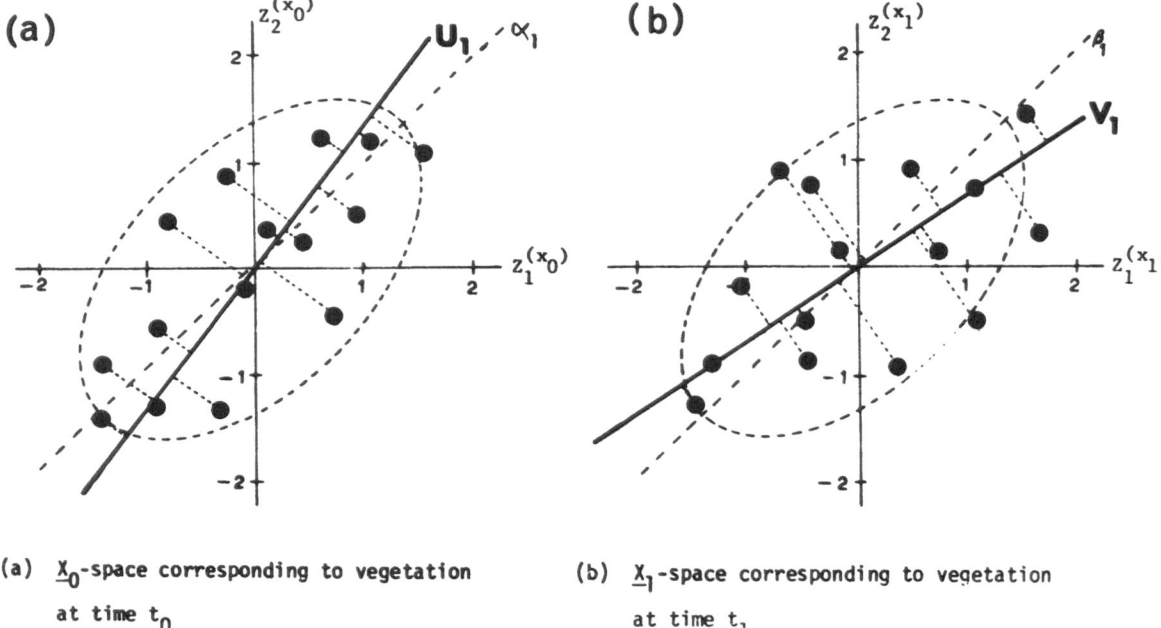

(a) $\underline{X}_0$-space corresponding to vegetation
     at time t_0

(b) $\underline{X}_1$-space corresponding to vegetation
     at time t_1

Fig. 2. Diagrammatic illustration of the principle of canonical analysis. A sample (N = 15) and concentration ellipse are shown (a) in $\mathbf{X}_0$-space corresponding to vegetation at time t_0 $(p = 2)$; and (b) in $\mathbf{X}_1$-space corresponding to vegetation at time t_1 $(q = 2)$. The canonical variates $\mathbf{u}_1$ and $\mathbf{v}_1$ are by definition those linear combinations of the variables of each space for which the simple correlation coefficient, r_1, between the projected points is a maximum over all possible choices for $\mathbf{u}$ and $\mathbf{v}$; r_1 is the first or largest canonical correlation. For comparison, the first principal components, α_1 and β_1, of each measurement domain are also indicated. In general, there will be $s = \min(p,q)$ pairs of such canonical variates, $\mathbf{u}_k$ and $\mathbf{v}_k$, and canonical correlation coefficients, r_k.

sionality proves in practice to be as small as three or four, and, in such cases, generally represents an appreciable condensation of the original data. Interest then centers on interpreting the new variables, that is the canonical variates, $\mathbf{u}_k$ and $\mathbf{v}_k$ corresponding to the trustworthy r_k.

In geometric terms, canonical analysis seeks those dimensions of $\mathbf{X}_0$-space and $\mathbf{X}_1$-space for which the successive correlation coefficients between the projected sample-points are maximized (Fig. 2). Like principal components analysis, canonical analysis may be conceived of as a rotation of coordinate axes to new positions, which, in the case of canonical analysis, emphasize the correlation structure of the data. Usually most of the useful information is found to be concentrated in the first few pairs of rotated axes, so that as in components analysis, the dimensionality of the system may be reduced, often appreciably. Moreover, successive canonical variates $\mathbf{u}_k$, $\mathbf{v}_\ell$ $(\ell, k = 1, ..., q)$ within and between sets, with the important exception of

those, one from each set, corresponding to a particular root $(k = \ell)$, are mutually uncorrelated. The transformation to canonical variates therefore has the effect of disentangling the within-set correlations between the original variables while channeling all correlation between sets through $s \leqslant q$ canonical correlation coefficients between the new variables. Where the intrinsic dimensionality does not exceed three, the canonical variates themselves provide a convenient reference frame within which to display and examine the covariance structure of interest. In such cases, scattergrams and stereograms of the sample mapped into one or more of the subspaces defined by the canonical variates provide useful descriptions of the data. The resulting displays are frequently found to aid comprehension and interpretation. On the other hand, where the intrinsic dimensionality is greater than three, some other form of graphical representation is called for. Andrews (1972) has proposed a general method for plotting high-dimensional data by mapping each

, $\mathbf{v}_\ell$ $(\ell, k = 1, ..., q)$ within and
the important exception of

Andrews (1972) has proposed a general method for
plotting high-dimensional data by mapping each

data point, say $\mathbf{u}'_{kj} = (u_{1j}, ..., u_{sj})$, where $\mathbf{u}_{kj}$ may for example represent the score of the jth sample on the kth canonical variate of $\mathbf{X}_0$-space, into the arbitrary sine-cosine function

$$f_{\mathbf{u}j}(t) = u_{1j}/\sqrt{2} + u_{2j} \sin t + u_{3j} \cos t + u_{4j} \sin 2t + u_{5j} \cos 2t + ... ,$$

and plotting the function over the range $-\pi < t < \pi$. In this way each stand in $\mathbf{u}_k$- or $\mathbf{v}_k$-space comes to be represented by a curve in 2-space (see Fig. 4). The advantage of so doing is that no constraint on the dimensionality of the data which may be treated arises. The appeal of the procedure in connection with applications of canonical analysis in which the intrinsic dimensionality of the analysis exceeds three will be obvious. The resulting plots have several other useful properties. The most important of these in the present context is that similarities in location and shape between curves are directly related to the proximity of the corresponding sample-points in the full $\mathbf{u}_k$- or $\mathbf{v}_k$- spaces of the canonical analysis. Put another way, the closer the sample-points in the space of the canonical variates the greater the proximity and overall similarity of the corresponding curves on the function-plot.

The above sketch of canonical analysis focuses on relationships between the same set of p variables at two points in time, or more generally on relationships between any two sets of variables. The application of canonical analysis, however, is not restricted to applications of this kind. For an account of generalized canonical analysis which may be useful in connection with observations over m time periods, $t_0, t_1, ..., t_m$ reference should be made to Kettenring (1971).

Applications

Having identified and described what in principle appears to be a suitable model for the exploratory analysis of succession in vegetation, we are now in a position to consider how well the model performs in practice. With this objective in mind, two applications of canonical analysis are described and discussed below. The ecological content of the analyses is deliberately varied with the aim of providing a reasonably substantial basis for assessment while at the same time illustrating something of the

flexibility of canonical analysis itself. The examples concern: a) the direction of succession in a number of British Post-glacial hydroseres; and b) the dynamic status of an area of lowland tropical rain forest as indicated by separate assessments of the contribution of mature trees and seedlings, respectively, to overall forest composition.

The orders of the three-dimensional arrays $\mathbf{Z}$ on which the analyses were based were ($66 \times 10 \times 2$) and ($25 \times 170 \times 2$), resp. As we shall see, the ecological objective of each study can be formulated in terms of relationships between two sets of variables. For this reason the objectives may be pursued by means of canonical analysis. Before turning to the applications, it is necessary to draw attention to certain features of the available data which, in the context of canonical analysis at least, undermine their quality. The most fundamental deficiency is that the data do not consist of sequential observations at times $t_0, t_1, ..., t_m$, and is common to both studies. In each case, however, the variables used can be considered to be surrogates for time and therefore may be acceptable for present purposes. A second deficiency of the data is that they relate to the equivalent of only two time-points, t_0 and t_1 ($m = 2$). It is therefore clear that, at best, the applications will exemplify only comparatively degenerate cases of the definitive scheme shown in Figure 1. Yet a further shortcoming is that in each case the data derive from small samples ($N \leqslant 66$). These various constraints reflect difficulties in the acquisition of a sound data base which is all too common in field studies of vegetation succession generally. For this reason the principal value of the examples may reside more in the indication they provide of ways in which canonical analysis could contribute towards making the study of vegetation succession analytic, than in any features of strictly ecological interest which emerge.

Direction of some British post-glacial hydroseres

Introduction

Hydroseres are successions associated with lakes and slow rivers which characteristically begin with open water and are generally thought to progress through a number of stages to culminate in bog or mesic forest. Convincing evidence for such postulated hydroseres however is scanty. In the absence of direct evidence it has commonly been supposed that the spatial zonation of vegetation around lakes recapitulates hydrosere history. As hydroseres leave behind them in the form of partially preserved and often identifiable remains of the plant communities which contributed to them, Walker (1970) has suggested that the stratigraphic record might help to resolve some of the issues concerning the hydroseral process. By identifying plant and pollen remains from a number of lake or basin deposits in Britain, Walker was able to recognize 12 vegetation stages or ecological units in terms of which the process of vegetation change could be described. These stages, in what seem to be their proper order, were as follows.

1. Biologically unproductive water
2. Micro-organisms in open water
3. Submerged macrophytes
4. Floating-leaved macrophytes
5. Reedswamp
6. Sedge tussock
7. Fen dominated by grasses
8. Swamp carr formed by trees
9. Fen carr dominated by trees
10. Aquatic *Sphagna*
11. Bog
12. Marsh

Walker (1970) also compiled records for a total of $N = 66$ transitions between stages, and went on to classify the transitions in terms of the pair of vegetation stages related. These data are reproduced here as Table 1. Among other things, Walker was interested in the extent to which the commonly observed lake-side zonation consisting of submerged macrophytes, floating-leaved macrophytes, reedswamp, fen, fen carr, and its variants, parallels an autogenic succession. We address this question here by means of a canonical analysis of the data of Table 1.

One way of viewing the entries of Table 1 is to regard them as a sample of $N = 66$ observations simultaneously classified with respect to a pair of variables – namely, antecedent and succeeding vegetation stages. The two variables are divided into r and c ordered, discrete states $\{A_i\}$ and $\{S_j\}$, resp. $(i = 1, ..., r; j = 1, ..., c)$. Regarded in this way, the ijth entry to the Table specifies the frequency with which the ith antecedent stage or ecological unit was replaced by the jth succeeding unit.

In the classical theory of the hydrosere (e.g. see Tansley, 1939), an orderly process is described in which each 'antecedent' community or stage is replaced by the next, following 'succeeding' unit. In terms of the $r \times c$ table, this process is equivalent to requiring the ith antecedent state $A_i (i = 1, ..., r\text{-}1)$ to be replaced by state S_{i+1}. The end-point of the sequence, A_r, may be regarded as being either replaced by itself or by the ultimate succeeding state, S_c. Accordingly, under conditions of strict regularity, succession would give rise to a two-way table in which the non-zero entries were strictly confined to the superdiagonal, except for those of

Table 1. Hydrarch succession in the British Isles. Frequencies of transitions between twelve ecological units recognized from plant remains in lake deposits. Units are identified on this page. Reproduced from Walker (1970).

					Succeeding unit								
	1	2	3	4	5	6	7	8	9	10	11	12	Total
1	.	.	.	.	.	.	.	.	.	.	.	.	0
2	.	.	.	1	1	.	.	.	.	.	.	.	2
3	.	.	.	2	5	.	.	.	.	.	.	.	7
4	.	.	1	.	3	.	1	.	2	1	.	.	8
5	.	.	.	.	.	1	4	3	5	2	4	.	19
6	.	.	.	.	.	.	1	.	.	.	.	.	1
7	.	.	.	.	.	.	.	.	4	1	3	.	8
8	.	.	.	.	.	.	1	.	1	1	.	.	3
9	.	.	.	.	1	.	.	1	.	1	5	.	8
10	.	.	.	.	.	.	.	.	.	.	4	.	4
11	.	.	.	.	.	.	.	.	2	1	.	.	3
12	.	.	.	.	2	.	1	.	.	.	.	.	3
Total	0	0	1	3	12	1	8	4	14	7	16	0	66

the 'climax' stage, A_r, which would be located on the principal diagonal. On the other hand, in the absence of all regularity, the process would generate a table whose non-zero entries were scattered about the table as a whole in a way which would defy description in simple terms. From Table 1 it is clear that the observations, which for the most part are concentrated in the upper right-hand triangle, fall somewhere between the extreme situations described. This in itself could lead us to suppose that succession of a somewhat less than perfectly regular kind is indeed indicated by the data. In order to gain further insight as to how the different $\{A_i\}$ and $\{S_j\}$ categories are related, some form of analysis would obviously be desirable.

By viewing Table 1 as a two-way table of frequencies, our ecological objective is recognisably equivalent to the standard problem of testing or otherwise assessing the independence of two categorical variables. Moreover, it is well-known that canonical analysis provides a means of testing for independence in $r \times c$ tables (e.g. see Bartlett, 1965; Kshirsagar, 1972; Kendall & Stuart, 1973). Before embarking on such an analysis, however, it will advantageous to re-express the data so as to facilitate analysis of this kind. First, notice that a number of the ecological units appearing in Table 1 were not in fact realized in the sample. Nothing will be gained by burdening the analysis with hypothetical communities of this kind. We may therefore delete the first row of the Table together with columns, 1, 2 and 12 without jeopardizing the analysis. This step leads to a revised table of $r = 11$ rows and $c = 9$ columns. Secondly, it will be helpful to re-express the revised table as a data matrix rather than a frequency table, i.e. as a two-dimensional array in which the rows correspond to sample-observations and the columns to variables, or vice-versa. For this purpose we will require two sets of binary-valued dummy variables which will enable us to operationalize the row and column categories underlying the Table (see Lancaster, 1958, p. 719). Accordingly, corresponding to the row categories we define the variables:

$$_{ki}X_0 = \begin{cases} 1 \text{ if the } k\text{th observation belonged to the} \\ j\text{th succeeding category} \\ 0 \text{ otherwise,} \end{cases}$$

where $k = 1, ..., N, i = 1, ..., p$ and $p = r - 1$.

Similarly, for the column classification we define:

$$_{kj}X_1 = \begin{cases} 1 \text{ if the } k\text{th observation belonged to the} \\ j\text{th succeeding category} \\ 0 \text{ otherwise,} \end{cases}$$

where, $k = 1, ..., N, j = 1, ..., q$ and $q = c - 1$. The dummy variables created in this way may be collected together to form two matrices $\mathbf{X}_0(N \times p)$ and $\mathbf{X}_1(N \times q)$. By appropriately 'stacking' the matrices, the revised table may be expressed as the three-dimensional array $\mathbf{Z}$ of order $(N \times p \times m)$ referred to earlier. From the point of view of canonical analysis, however, an alternative view of the data may be preferred. By suitably combining the matrices of dummy variables, the revised table of frequencies can be represented by the partitioned matrix $\mathbf{Z} = [\mathbf{X}_0 \, \mathbf{X}_1]$ of order $N \times (p + q)$, where the rows correspond to samples and the columns to ecological units. The relationship between the $\{A_i\}$ and $\{S_j\}$ is now that between the variables comprising $\mathbf{X}_0$ and $\mathbf{X}_1$. It will perhaps be appreciated from what has been said above that canonical analysis can be used to investigate the relationship between $\mathbf{X}_0$ and $\mathbf{X}_1$. Interest would center in particular on the extent and nature of any departure from independence between $\mathbf{X}_0$ and $\mathbf{X}_1$. If a null hypothesis of independence was found on statistical or other grounds to be untenable, then in ecological terms, it would seem that the antecedent and succeeding units were indeed likely to be temporally dependent in some way. Interest would then focus on the nature and ecological interpretation of the association.

Proceeding to the canonical analysis, therefore, we have a biased sample of $N = 66$ transitions between stratigraphic vegetational stages or units, $p = 10$ indicator variables representing eleven antecedent ecological units and $q = 8$ indicator variables corresponding to nine succeeding ecological units. The ecological objective of the analysis is to assess the evidence for association between the $\{A_i\}$ and $\{S_j\}$ and to characterize the association, if present.

The results of the canonical analysis are summarized in Tables 2–5.

Results

Independence

The canonical correlation coefficients are reported in Table 2. The first correlation, $r_1 = 0.87$, departs appreciably from zero. It appears from the largest-root test of Table 2(a) that r_1 is significant beyond the 1 percent level ($r_1^2 = 0.75 > \Theta_\alpha(8, \frac{1}{2}, 23) = 0.56$ for $\alpha = 0.01$). A similar result is provided by Bartlett's approximate likelihood ratio test of Table 2(b), where we find $\chi_s^2(80) = 147.65 > \chi_\alpha^2(80) = 124.84$ for $\alpha = 0.001$. Evidently, the data do not support the null hypothesis of independence. Accordingly, we may be led to conclude that the antecedent and succeeding ecological units are indeed associated, or, in ecological terms, that is, are temporally related.

In view of the lack of independence among the samples, a strictly informal assessment of the null hypothesis, however, may in fact be more appropriate than the tests just considered. An assessment of this kind can be made based principally on the magnitude of r_1. In this case it is important to

Table 2. Hydrarch succession in the British Isles. Canonical analysis of relationships between antecedent and succeeding ecological stages.

a. The canonical correlation coefficients r_k and nominal tests of their significance. Approximate critical values of Roy's largest-root criterion $\Theta_\alpha(s,m,n)$ are shown for $\alpha = 0.05$ and $\alpha = 0.01$; ($m = 1/2$; $n = 23$).

k	s	r_k	r_k^2	$\Theta.05^{(s,m,n)}$	$\Theta.01^{(s,m,n)}$	p
1	8	0.869	0.755	0.507	0.556	≪0.01
2	7	0.623	0.388	0.473	0.524	>0.05
3	6	0.560	0.314	0.435	0.489	>0.05
4	5	0.420	0.176	0.393	0.449	>0.05
5	4	0.322	0.104	0.345	0.403	>0.05
6	3	0.204	0.042	0.290	0.350	>0.05
7	2	0.178	0.032	0.226	0.387	>0.05
8	1	0.075	0.006	–		>0.10*

* From the F-distribution (cf. Morrison (1976, p. 178, eqn. (41)): $F_s(3,48) = 0.097 < F_\alpha(3,48) = 2.210$ for $\alpha = 0.1$.

b. Bartlett's approximate test of the joint nullity of the smallest s-k canonical correlations.

k	Roots	X^2	df	p
0	1,2,3,4,5,6,7,8	147.65	80	<0.001
1	2,3,4,5,6,7,8	69.46	63	>0.10
2	3,4,5,6,7,8	42.19	48	>0.75
3	4,5,6,7,8	21.28	35	>0.90
4	5,6,7,8	10.53	24	>0.99
5	6,7,8	4.47	15	>0.99
6	7,8	2.10	8	>0.97
7	8	0.31	3	>0.95

attempt to make allowance for the effect of the number of variables and samples entering the analysis on the magnitude of r_1. In view of the comparatively large variable/sample ratio ($v/N = 18/66 = 0.27$) it seems likely that r_1 will be biased upward to an unknown extent by a factor attributable to this source. Nevertheless, the magnitude of r_1 is such that we may feel reasonably confident in declaring the antecedent and succeeding domains to be associated on this basis alone.

Having concluded that the two domains are indeed likely to be temporally connected we proceed to enquire into the dimensionality of the relationship. For this purpose the sequential tests of the $r_k(k > 1)$ of Table 2a and b will be helpful.

Dimensionality

Sequential tests based on the individual roots r_k^2 ($k = 1, ..., q$) using Roy's criterion (Table 2a) suggest that r_1 alone departs significantly from zero. Similarly, Barlett's approximate tests of the joint nullity of the smallest s-k roots ($k = 1, ..., q-1$) indicate that $r_2 = \cdots = r_8 = 0$ (Table 2b). These tests therefore suggest that the dimensionality of the association is one.

Informal assessments of dimensionality based on the goodness-of-fit criteria of Table 3 can also be made. It is apparent from Table 3a and from Table 3b and c, resp., that the percentage trace and percentage explained variance associated with the root r_1^2 is consistently roughly twice that associated with r_2^2 and r_3^2, which themselves are of approximately equal explanatory power. These results, are not inconsistent with the unique position of r_1^2 arrived at earlier. In addition, however, they provide some grounds for supposing that the explanatory power attributable to the roots r_2^2 and r_3^2 may not necessarily be entirely negligible.

In order to obtain insight into the nature of the relationship between the antecedent and succeeding domains we shall require the canonical variates $\mathbf{u}_k$ and $\mathbf{v}_k$ corresponding to the kth root, r_k^2. We have seen that while the canonical variates corresponding to the first root r_1^2 are likely to prove most informative, it may nevertheless be advantageous to examine also the canonical variates associated with r_2^2 and r_3^2.

The canonical variates

The scores of the $N = 66$ samples on the

Table 3. Hydrarch succession in the British Isles. Canonical analysis of relationships between antecedent and succeeding ecological stages. Indices of goodness of fit.

a. Percentage $\mathrm{tr}(R_{22}^{-1} R_{21} R_{11}^{-1} R_{12})$ absorbed by the canonical roots, $r_k^2 (k = 1, ..., q)$, individually (%tr) and cummultively (C%).

k	r_k^2	%tr	C%
1	0.755	41.55	41.55
2	0.388	21.35	62.90
3	0.314	17.28	80.18
4	0.176	9.69	89,87
5	0.104	5.72	95.59
6	0.042	2.31	97.90
7	0.032	1.76	99.66
8	0.006	0.33	99.99
Total	1.817	99.99	–

b. Redundancy in the succeeding vegetation domain, $V_{x|v_k}^2$ ($k = 1, ..., 3$), attributable to the canonical variates v_k of the antecendent demain.

k	$V_{x|v_k}^2$	% Explained variance	Cumulative %
1	0.098	51.58	51.85
2	0.052	27.51	79.36
3	0.039	20.63	99.99
Total	0.189	99.99	–

c. Redundancy in the antecedent vegetation domain, $U_{y|u_k}^2$ ($k = 1, ..., 3$), attributable to the canonical variates u_k of the succeeding domain.

k	$U_{y|u_k}^2$	% Explained variance	Cumulative %
1	0.078	52.70	52.70
2	0.039	26.35	79.05
3	0.031	20.95	10.00
Total	0.148	100.00	–

canonical variates are not themselves reported. We have chosen instead to direct attention to the correlations between the canonical variates and the original variables as well as to a number of interpretive indices derived from these correlations. These quantities, which are reported in Table 4, often prove to be more informative in practice than the canonical variates themselves. In the present study it is the canonical variates v_k of the antecedent domain and the interpretive indices associated with them which are of special interest. We consider first the adequacy of the v_k in character-

izing the domain on which they are defined.

From Table 4 (lower-left) it can be seen from the variances $V_k^2 (k = 1, ..., 3)$ of the first three canonical variates v_k that the v_k each account for some 10% of the total variance of the antecedent domain. Collectively, therefore these three canonical variates account for 30.2% of the total variance of $\mathbf{X}_0$. It therefore appears that the descriptive power of the v_k in relation to the domain on which they are defined is weak. It follows immediately that the canonical correlation coefficients themselves will not be trustworthy indices of the strength of the dependency between domains. While the canonical correlations (Table 2a) show corresponding members of a pair of canonical variates to be at least moderately strongly correlated across domains, it is plain from Table 4 that the v_k themselves (and the u_k also) can scarcely be considered to adequately characterize their respective domains. It is also apparent from the V_k^2 that there is little to distinguish between the v_k ($k = 1, ..., 3$) as descriptive or summarizing constructs; all three are of roughly equal worth in this respect.

In order to characterize the v_k in ecological terms we consider next the correlations between these canonical variates and the variables on which they are defined. From the sign of the intraset correlation coefficients of the antecedent ecological units with v_1 (Table 4, lower-left) we note first of all that the units differ with respect to the direction of their covariation with v_1. The units contributing most to this canonical variate are A_2, open water (–0.42), A_3, submerged macrophytes (–0.74) and A_{12}, marsh (–0.24), towards the negative pole, and A_5, reedswamp (0.40) and A_7, fen (0.24), towards the positive pole. The second canonical variate, v_2, is similarly characterized by A_9, fen carr (0.51) and A_{10}, aquatic *Sphagna* (0.63) with positive correlations and by A_6, sedge tussock (–0.39) and A_5, reedswamp (–0.37) with negative correlations; similarly, v_3 is characterized principally by A_6, sedge tussock (–0.68) and A_4, floating-leaved macrophytes (0.42).

Having broadly characterized the v_k in terms of the ecological units on which they are defined we turn to consider the variances or explanatory power of the v_k ($k = 1, ..., 3$) in the succeeding vegetation domain (Table 4, upper-right). From the redundancies, $V_{x_1|v_k}^2$, generated by the v_k in the X_1-domain it is apparent that the v_k are even weaker

Table 4. Hydrarch succession in the British Isles. Canonical analysis of relationships between antecedent and succeeding vegetation stages. Correlations between the original variables and the canonical variates.

Canonical variate	U_1	U_2	U_3	h_w^2	V_1	V_2	V_3	h_b^2
Succeeding Vegetation								
S_3, subm. macrophytes	-0.074	-0.130	0.249	0.084	-0.064	-0.081	0.139	0.030
S_4, flt-lvd. macrophytes	-0.556	0.082	-0.215	0.362	-0.483	0.051	-0.120	0.250
S_5, reedswamp	-0.745	0.009	0.031	0.556	-0.647	0.006	0.017	0.419
S_6, sedge tussock	0.089	-0.116	-0.103	0.032	0.077	-0.072	-0.058	0.014
S_7, fen	0.091	-0.205	0.022	0.051	0.079	-0.128	0.012	0.023
S_8, swamp carr	0.205	-0.499	-0.779	0.898	0.178	-0.311	-0.436	0.319
S_9, fen carr	0.271	-0.331	0.488	0.421	0.235	-0.206	0.273	0.172
S_{10}, aquatic *Sphagna*	0.158	-0.150	0.307	0.142	0.137	-0.093	0.172	0.057
S_{11}, bog	0.381	0.880	-0.234	0.974	0.331	0.548	-0.131	0.427
Variance extracted	0.130	0.137	0.124	0.391	0.098	0.053	0.039	0.190
Redundancy	0.098	0.052	0.039	0.189	0.098	0.053	0.039	0.190

Canonical variate	V_1	V_2	V_3	h_w^2	U_1	U_2	U_3	h_b^2
Antecedent Vegetation								
A_2, open water	-0.420	0.056	-0.145	0.201	-0.365	0.035	-0.081	0.141
A_3, subm. macrophytes	-0.736	0.067	-0.144	0.567	-0.640	0.042	-0.081	0.418
A_4, flt-lvd. macrohytes	-0.192	-0.243	0.417	0.270	-0.167	-0.151	0.234	0.105
A_5, reedswamp	0.398	-0.369	-0.297	0.383	0.346	-0.230	-0.166	0.200
A_6, sedge tussock	0.115	-0.391	-0.679	0.627	0.100	-0.244	-0.380	0.214
A_7, fen	0.244	0.125	0.283	0.155	0.212	0.078	0.158	0.076
A_8, swamp carr	0.103	-0.190	0.245	0.107	0.090	-0.118	0.137	0.041
A_9, fen carr	0.133	0.507	-0.087	0.282	0.116	0.316	-0.049	0.116
A_{10}, aquatic *Sphagna*	0.197	0.634	-0.187	0.476	0.171	0.394	-0.105	0.196
A_{11}, bog	0.126	-0.200	0.360	0.185	0.109	-0.125	0.202	0.168
A_{12}, marsh	-0.244	-0.060	0.025	0.064	-0.212	-0.037	0.014	0.047
Variance extracted	0.103	0.101	0.098	0.302	0.078	0.039	0.031	0.148
Redundancy	0.078	0.039	0.031	0.148	0.078	0.039	0.031	0.148

explanatory constructs across domains than within the antecedent domain ($V_{x_1|v_1}^2 = 0.098$; $V_{x_1|v_2}^2 = 0.053$; $V_{x_1|v_3}^2 = 0.039$). It can also be seen that the explanatory power of the canonical variates falls systematically across the v_k, v_1 being the strongest and v_3 the weakest. Thus, it emerges that individually the v_k account resp. for only some 10, 5 and 4% of the total variance of the succeeding vegetation domain, while collectively they account for 19% of the total variance of this domain. Although it is beginning to emerge that the fit of the model is poor, it may nevertheless prove worthwhile to pursue the ecological nature of the relationships defined by the canonical variates corresponding to r_1^2 and r_2^2, at least.

From the correlations between v_1 and the ecological units of the succeeding domain (Table 4, upper-right) it is apparent that v_1 contributes above all to the variance of S_5, reedswamp (-0.65).

Bearing in mind the sign and magnitude of the $\{A_i\}$ units with v_1, it seems clear that A_2 (-0.42), A_3 (-0.74), A_4 (-0.19) and A_{12} (-0.24) all contribute in a direct sense towards the explained variance of reedswamp, and, to a lesser extent, towards that of S_4, floating-leaved macrophytes (-0.48) also. In addition, v_1 makes subsidiary contributions to the variance of several other $\{S_j\}$. This follows from the positive sign of the correlations common to several $\{A_i\}$ and $\{S_j\}$ units. Among these, we see that A_5 (0.40) and A_7 (0.24) contribute most towards the explained variance of S_9, fen carr (0.23) and S_{11}, bog (0.33). We note also that a number of inverse relationships exist between those $\{A_i\}$ and $\{S_j\}$ whose correlations with v_1 differ in sign. Thus A_2 (-0.42), A_3 (-0.74) and A_{12} (-0.24) vary inversely across domains with S_9 (0.23) and S_{11} (0.33), while A_5 (0.40) and A_7 (0.24) are similarly related to S_4 (-0.48) and S_5 (-0.65).

The interset correlations between v_2 and the $\{S_j\}$ show the correlations to be dominated by that of S_{11}, bog (0.55). Recalling the sign and magnitude of the intraset correlations of the $\{A_i\}$ with v_2, it is clear that A_9 (0.51) and A_{10} (0.63) contribute most towards the variance of bog. Moreover, the intraset correlations of A_5 (−0.37) and A_6 (−0.39) indicate that v_2 makes subsidiary contributions also towards the variances of S_8, carr (−0.31) and S_9, fen carr (−0.21). Having regard for the sign of the intraset and interset correlations, we see that A_5 (−0.37) and A_6 (−0.39) are inversely related to bog S_{11} (0.55), while A_9 (0.51) and A_{10} (0.63) vary inversely with S_8 (−0.31) and S_9 (−0.21).

In view of the negligible explanatory power of the third canonical variate, v_3, ($V^2_{\mathbf{x}_1|v_3} = 0.039$) in the $\{S_j\}$ domain it appears that there would be little to gain from pursuing any possible ecological significance which it may process.

Before proceeding, we pause to draw together the salient points to have emerged so far. One means of accomplishing this is to specify the temporal sequences of vegetation types indicated by the directions of the structure correlations between the $\{A_i\}$ and $\{S_j\}$ with v_1 and v_2 (Table 5). By heeding the strengths of the correlations as well, major and subsidiary pathways may also be distinguished and such pathways are also summarized in Table 5. Attention is drawn particularly to the following features of the Table:
a) the diversity of successional pathways;
b) the comparatively short nature of the implied successions; and
c) the special significance of reedswamp and bog as focal-points in the process.

We are now in a position to consider the fit of the rank $r = 2$ model. The total redundancy in the $\{S_j\}$ domain attributable to the v_k ($k = 1, 2$) is $V^2_{\mathbf{x}_1|v_1, v_2} = 0.15$ (Table 3(b)). In other words, the first two

canonical variates of the antecedent domain together account for 15% of the total variance of the vegetation stages of the succeeding domain. Thus, the fit of the model is manifestly poor. To see how the explanatory power of the model is distributed among the $\{S_j\}$ we require the interset communalities h_b^2 (Table 4, upper-right). Reedswamp (0.42) and bog (0.43) are the $\{S_j\}$ stages whose variance is best accounted for. On the other hand, the variance of a majority of the $\{S_j\}$ is scarcely accounted for at all, notably those of submerged macrophytes (0.03), sedge tussock (0.01) and fen (0.02).

Ecological assessment of the results

The results clearly point towards the operation of a successional process of some kind. Yet the process can hardly be described as either systematic or convergent. Rather there appear to be a number of short, alternative pathways which lead towards multiple stable states among which reedswamp and bog are prominent. The lack of fit of the model itself is a clear reflection of the indeterminate nature of the process. Thus, the view of the hydrosere which emerges is at variance with that of systematic convergence towards a stable end-stage as postulated in the classical theory. Moreover, the hydroseral process appears to be too varied and too fragmented for the familiar lakeside zonation of vegetation types to be regarded as a reliable analogue of it.

It is instructive to compare the results described with those of Walker's (1970) original study. As the two sets of results are expressed in different terms detailed comparison is hardly possible. Nevertheless, a number of general remarks are worth making. On the basis of his investigation, Walker was led to conclude, among other things, that: a) the data

Table 5. Hydrarch succession in the British Isles. Canonical analysis. Temporal sequences of vegetation types indicated by the structure correlations between the $\{A_i\}$, $\{S_j\}$ and v_k ($k = 1, 2$).

Canonical variate	Sequence
(a) *Major pathways*	
v_1	Open-water (2) ⟶ subm. macrophytes (3) ⟶ f-l. macrophytes (4) ⟶ reedswamp (5)
v_2	Fen carr (9) ⟶ aquatic *Sphagna* (10) ⟶ bog (11)
(b) *Subsidiary pathways*	
v_1	Reedswamp (5) ⟶ fen (7) ⟶ fen carr (9) ⟶ bog (11)
v_2	F-l. macrophytes (4) ⟶ reedswamp (5) ⟶ sedge tussock (6) ⟶ swamp carr (8) ⟶ bog (11)

clearly confirm the operation of hydroseral processes during the Post-glacial; and that b) the zonation of vegetation observed in lakes and mires at any one time is in a sense an analogue of the sequence of vegetation types which may occur at any place.

These statements were, however, qualified by noting first that, while certain sequences of transitions are 'preferred', variety is the keynote of the hydroseral succession; and secondly that the range of extant vegetation types at any one locality does not necessarily reflect the sequence which has led to the current pattern at that site. Thus, the conclusions reached by Walker tend to be at variance with the classical view of the hydrosere. Furthermore, Walker was also led to recognize the crucial roles of reedswamp and bog in the process at the sites examined.

The overall picture of the hydroseral process which emerges from the present study is therefore similar in its essentials to that described by Walker.

Summary and conclusions

The salient points of ecological interest to have emerged may be summarized as follows:

a) antecedent ecological units have tended to be replaced by units subsequent to them in the hypothesized sequence, so that hydroseral processes of a sort do appear to have operated;

b) nevertheless, connections between antecedent and succeeding stages are comparatively weak so that temporal change is characterized more by diversity than by uniformity, with short sequences of community types culminating in local endstages predominating over orderly convergence towards a single, global end-point;

c) the most pronounced trends are those consisting of (i) open water, submerged macrophytes and marsh, which culminates in reedswamp; and of (ii) fen carr and aquatic *Sphagna,* which culminates in bog;

d) lakeside zonation of vegetation may at best be regarded only as an imperfect analogue of temporal change at any given locality.

The similarity of these results to those reported by Walker goes some way towards validating the method proposed here. Nevertheless, the results of the canonical analysis differ importantly in several respects from those of Walker. In particular we mention that canonical analysis in addition to drawing attention to the existence of temporal trends, i.e. to the direction of relationships, also provided some indication of the strength of these as well as of the fit or explanatory power achieved by the model as a whole. Finally, in wider context we remark that the general view of succession which emerged as a less determinate process than it was once thought to be is supported by the results of a number of recent studies. The notion that temporal change in vegetation often proceeds by means of alternative successional pathways has been expressed by Olson (1958), Drury & Nisbet (1973) and Mathews (1979). Moreover, Sutherland (1974) has referred to the possibility of the existence of multiple stable states.

In short, therefore, it seems justifiable to conclude from the results of the present analysis that canonical analysis may provide a useful systematic, exploratory approach to successional studies of the kind exemplified by the application considered.

Dynamic status of a lowland tropical rain forest

Introduction

Direct observation of successional change in forest vegetation is scarcely possible owing to the length of the time-scale over which forest succession generally operates. One means of circumventing this problem is to regard size-class measurements, s_0, s_1, ..., s_m as if they constituted observations at times $t_0, t_1, ..., t_m$. Differences in size correspond to differences in age which, cautiously interpreted, may be substituted for observations at particular points in time or of particular seral stages. In this way it may be possible to gain insight into the dynamic status of forest vegetation over the short-term, at least. The account which follows describes such an analysis.

Field observations collected by Dr. John Ogden of the Australian National University, Canberra and kindly placed at my disposal consisted of estimates of the composition of $N = 25$ 100×100 m stands of lowland rain forest in Guyana, South America. In all $p = 170$ species of woody plant were encountered for which density estimates of $m = 2$ size-classes has been obtained ('trees' > 5 cm dbh; 'seedlings' < 30 cm in height). Thus in terms of the scheme illustrated in Figure 1, the data constituted a three-dimensional array **Z** of order $(25 \times 170 \times 2)$. The ecological objective of the study was to arrive at a preliminary assessment of the dynamic status of the forest. More specifically, the objective was to discover whether the field observations were consistent with the view that the forest was likely to be replaced by vegetation of substantially the same composition, as the trees composing the forest aged and died and were progressively replaced by the present generation of seedlings. Thus attention focused on tree/seedling relationships; if composition was changing with time we might expect these relationships to be weaker than would be the case if the vegetation were dynamically stable. In operational terms, therefore, we are led to an interest in the correlation structure of the tree/seedling density estimates. Canonical analysis is readily identified as an appropriate algebraic or statistical model for exploring the correlation structure between m sets of variables $(m \geq 2)$.

Before turning to the canonical analysis we pause to outline steps which were taken in order to render the original field data better-suited to analysis of this kind. Denoting the $N \times p$ matrix of tree density estimates by $\mathbf{A}_0^*$ (25×170) and the corresponding matrix of seedling densities by $\mathbf{A}_1^*$ (25×170), the most pressing need in view of the small size of the sample and large number of variables was to reduce the number of variables in some way consistent with the overall objective of the study. It was also desirable to arrive at some more comprehensive description of site vegetation than that provided by a simple statement of species' representation, given the stated aim of the investigation. A reduction in species number was achieved by ranking species in terms of their independent contribution to the total sum of squares, using a modified version of the computer program RANK developed by Orlóci (1978, p. 26) for this purpose. After examining the ordering of species obtained in the light of field knowledge of the species (Ogden, pers. comm.) and in relation to a stress index calculated from the data (Orlóci, 1978, p. 34), all but the top-ranking thirty-three species were eliminated. The reduced matrices arrived at in this way were denoted $\mathbf{A}_0$ (25×33) and $\mathbf{A}_1$ (25×33), resp. To obtain numerical indices or descriptors of site vegetation, separate non-centered principal components analyses of $\mathbf{A}_0$ and $\mathbf{A}_1$ were performed, followed by varimax rotation and

postnormalization of the components. In each case, six components were found to adequately characterize the tree and seedling counterparts of the vegetation, accounting for 93% and 88% of the total sum of squares of the cross-products matrices, $\mathbf{A}_0\mathbf{A}'_0$ and $\mathbf{A}_1\mathbf{A}'_1$, resp. Each rotated component could be equated with and was named after a more or less clearly-defined forest community. The names of the communities together with an indication of the importance of each, as expressed by the percentage $\text{tr}(\mathbf{A}_i\mathbf{A}'_i)$ associated with its defining component, appear in Table 6. The rotated components were collected to yield the matrices $\mathbf{A}_0$ (25×6) and $\mathbf{X}_1$ (25×6) which themselves were associated to give a partitioned matrix $\mathbf{Z} = [\mathbf{X}_0]\mathbf{X}_1]$ of order $N \times (p + q)$ suitable for canonical analysis. The formal objective of analysis was to examine the relationship between the two sets of components, that is, in ecological terms, between tree and seedling counterparts of the forest communities.

Thus, the canonical analysis came to be based on a systematic sample of $N = 25$ 100×100 m stands of rain forest vegetation, each characterized by $p = 6$ and $q = 6$ generalized variables expressing the contribution of tree and seedling communities, resp. to the total vegetation. The ecological objective was to arrive at a preliminary assessment of the dynamic status of the vegetation.

Table 6. Lowland tropical rain forest, Guyana. Communities identified following non-centered components analyses of tree and seedling matrices $\mathbf{A}_0$ (25×33) and $\mathbf{A}_1$ (25×33), resp. The relative size or importance of communities is indicated by the % trace $(\mathbf{A}_i\mathbf{A}'_i)$ associated with the defining component.

Community*	%tri$(\mathbf{A}_0\mathbf{A}'1_0)$	%tr$(\mathbf{A}_1\mathbf{A}'_1)$
Greenheart	23	23
Wallaba	20	17
Pentaclethra	13	9
Morabukea	13	14
Mora		6
Eschweilera	17	20
Jessenia	7	–
Total	93	89

* More specifically, the species after which the communities were named are: Greenheart, *Ocotea rodiaei* (Schomb.) Mex.; Wallaba, *Eperua falcata* Aubl.; *Pentaclethra, Pentaclethra macroloba* (Willd.) Kze.; Morabukea, *Mora gonggrijpii* (Kleinh.) Sandw.; *Mora, Mora excelsa* Bth.; Eschweilera, *Eschweilera sagotiana* Miers.; *Jessenia, Jessenia bataua* (Mart.) Burret.

Results

Independence

The canonical correlation coefficients are reported in Table 7. In the absence of independent samples and hence of significance tests, we assess their worth informally. From the magnitude of $r_1 = 0.99$, in particular, it appears at first sight as though the independence of the two sets of variables is unsupported by the data. However, in informally assessing the correlations in this way, it is necessary to bear in mind any possible effect of the variable/sample ratio on their magnitude. The ratio of variables to samples is unusually high ($12/25 = 0.48$) and undoubtedly will have resulted in r_1, and, indeed in the r_k ($k = 1, ..., 6$) generally, being positively biased, though to an unknown extent. The departure of r_1 from zero is nevertheless such as to be rather convincing in rendering the hypothesis of independence untenable. Accordingly, we declare the data to be consistent with the view that the tree and seedling counterparts of the vegetation are likely to be associated. Interest therefore turns to establishing the dimensionality of the association.

Dimensionality

It is clear from Table 7 that the first four canonical correlations, which vary between $r_1 = 0.99$ and $r_4 = 0.87$, are large. The fifth canonical correlation, $r_5 = 0.65$, also departs appreciably from zero and therefore is hardly negligible. From the sizes of the r_k ($k = 1, ..., 5$) the existence of one or more linear relationships between tree and seedling counterparts of the vegetation can scarcely be doubted. Only $r_6 = 0.06$ is sufficiently close to zero

Table 7. Lowland tropical rain forest, Guyana. Canonical analysis of tree-seedling relationships. Canonical correlation coefficients, $r_k(k=1,...,6)$, canonical roots, r_k^2, and % $tr(R_{22}^{-1}R_{21}R_{11}^{-1}R_{12})$ absorbed by the r_k^2 individually (%tr) and cumulatively (C%).

k	r_k	r_k^2	%tr	C%
1	0.988	0.977	24.60	24.60
2	0.956	0.914	23.01	47.61
3	0.946	0.894	22.51	70.12
4	0.871	0.760	19.13	89.25
5	0.651	0.424	10.67	99.92
6	0.056	0.003	0.08	100.00
Total	–	3.972	100.00	–

to suggest that it may be neglected with impunity. The goodness-of-fit criteria of Tables 7 and 8a, b are also helpful in clarifying the dimensionality. From Table 7 it can be seen that the first three roots each absorb some 23–24% of the $tr(R_{22}^{-1}R_{21}R_{11}^{-1}R_{12})$, while r_4^2 is only slightly less efficient, accounting for 19%. The fifth root, r_5^2, is rather weaker, absorbing 11% of the total dispersion. A broadly similar picture is provided by the percentage of the total variation of the seedling and tree domains, $tr(R_{11})$ and $tr(R_{22})$, resp., predictable from the 'other' domain (Table 8a and b).

Collectively, these various indices point towards a dimensionality of four or five. After considering the ecological interpretability of the corresponding canonical variates, I have opted for five as the effective dimensionality of the linear association between the tree and seedling communities.

We turn now to consider the interpretation of the relationships associated with the roots, r_k^2, $k = 1, ..., 5$. For this purpose the canonical variates u_k and v_k ($k = 1, ..., 5$) will be required.

The canonical variates

In interpretation, it is the ecological meaning or significance of the canonical variates on which interest centers. Interpretation may be accomplished by: a) first plotting the sample in various subspaces defined by the canonical variates and then using field knowledge and familiarity with the vegetation to search for meaningful patterns and relationships; b) examining correlation coefficients between the canonical variates and the original variables.

For simplicity we shall confine our attention here principally to the first of these, although for completeness the structure correlations are also reported (Table 9). Fig. 3 shows the sample ($N = 25$) plotted in the 2-dimensional subspaces associated with the first two canonical correlation coefficients, r_1 and r_2.

The most striking feature of Figure 3a and b is the tendency of sample-points to fall in each case close to a diagonal line through the origin. The configurations very clearly express the respective canonical correlations ($r_1 = 0.99$; $r_2 = 0.96$). It is easy to appreciate from the Figure that the analysis does in fact emphasize the correlation structure of the data.

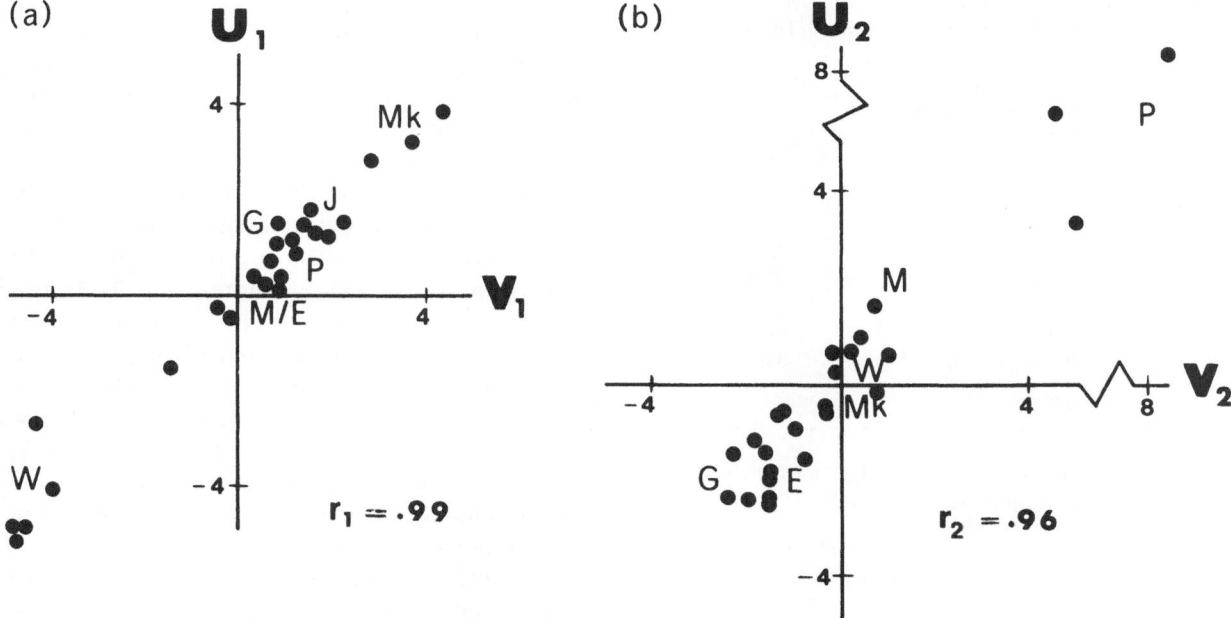

Fig. 3. Lowland tropical rainforest, Guyana. Canonical analysis of tree-seedling relationships. N = 25 stands mapped into subspaces defined by canonical variates (a) u_1 and v_1; (b) u_2 and v_2. E: *Eschweilera* forest; G: Greenheart, M: *Mora;* M/E: *Mora/Eschweilera;* Mk: Morabukea, P: *Pentaclethra;* W: Wallaba.

Figure 3a shows the sample projected onto the first pair of canonical variates, u_1 and v_1. It is plain that Morabukea and Wallaba forests contribute most to the relationship shown. From field knowledge of the vegetation, and also by reference back to the original data, it is known that: a) stands with high positive values on u_1 and v_1 are characterized principally by Morabukea tree vegetation and that it is precisely these stands whose seedling vegetation is also characterized principally by Morabukea; b) stands with sizeable negative scores on u_1 and v_1 consist predominantly of Wallaba forest and that such stands are precisely those whose seedling vegetation is also composed chiefly of Wallaba.

The tree and seedling counterparts resp. of Morabukea and Wallaba forest are evidently each closely related in a direct sense. Furthermore, a well-defined inverse compositional relationship exists between the two communities in terms of both trees and seedlings. In ecological terms, therefore, the canonical variates v_1 and u_1 appear to correspond to the existence of strong direct compositional relationships between tree and seedling consitutents of Morabukea and Walleba forests, resp. as well as to a sharp compositional distinction

between the two communities in terms of both tree and seedling species assemblages.

Figure 3b shows the sample projected onto the (u_2, v_2)-plane and presents an entirely different aspect of the sample. It is readily seen that *Pentaclethra* and Greenheart forests contribute most to the linear relationship corresponding to r_2^2. By reference to the original data, it can be shown that stands with high positive scores on u_2 and v_2 consist chiefly of *Pentaclethra* tree vegetation and that it is precisely these stands whose seedling vegetation is also made up principally of *Pentaclethra;* stands with sizeable negative scores are similarly characterized in terms of Greenheart. Therefore, u_2 and v_2 may be interpreted as reflecting the compositional similarity of tree and seedling counterparts of *Pentaclethra* and Greenheart forests, resp. as well as the existence of a sharp contrast in the floristic composition of the two communities.

Continuing in the same way, the sample could be mapped into a variety of other subspaces associated with the non-zero r_k^2. However, the picture of tree/seedling relationships which would emerge would necessarily be fragmented. For this reason, a single, comprehensive representation of the sample

Table 8. Lowland tropical rain forest, Guyana. Canonical analysis of tree-seedling relationships. Indices of goodness of fit.

a. Redundancy in the seedling domain $V^2_{x|v_k}$, attributable to the canonical variates v_k of the tree domain.

k	$V^2_{x|v_k}$	% Explained variance	Cumulative %
1	0.248	32.21	32.21
2	0.193	25.06	57.27
3	0.180	23.38	80.65
4	0.136	17.66	98.31
5	0.013	1.69	100.00
6	0.000	0.00	100.00
Total	0.770	100.00	

b. Redundancy in the tree domain, $U^2_{y|u_k}$, attributable to the canonical variates u_k of the seedling domain.

k	$U^2_{y|u_k}$	% Explained variance	Cumulative %
1	0.356	44.00	44.00
2	0.159	19.65	63.65
3	0.141	17.43	81.08
4	0.111	13.72	94.80
5	0.042	5.19	99.99
6	0.000	0.00	99.99
Total	0.809	99.99	–

which simultaneously provided insight into the disposition of the sample in the full 5-dimensional tree and seedling spaces of the analysis would obviously be advantageous. It is to such a representation that we shall turn shortly. Before doing so, however, it will be useful to summarize the ecological implications of the results to have emerged so far.

From the overwhelming similarities between the tree and seedling species assemblages of Morabukea, Wallaba, *Pentaclethra* and Greenheart forests it seems clear that each of these communities has the potential at least to be succeeded in the short-term by forest of substantially the same composition, as the current seedling generation matures. Moreover, from the strength of the tree/seedling relationships in each case, it is abundantly clear that the communities in question are most unlikely to be in any way dynamically related among themselves. The tree/seedling relationships defined by the smaller roots, r^2_k ($k > 2$), although not described here and involving other forest communities, are open to interpretation in sub-

stantially the same terms. Table 9 provides a comprehensive statement of the main and subsidiary relationships between tree and seedling counterparts of the vegetation, as defined by the fitted rank five model.

A canonical representation

A unified canonical representation of the sample in the full 5-dimensional tree and seedling spaces of the analysis is readily constructed using the procedure introduced by Andrews (1972). In this way, as remarked earlier, each stand in u_k – or v_k-space comes to be represented by a curve in 2-space (see Figure 4). To avoid overburdening the sine-cosine plots we have elected to work below with the centroids of particular communities, rather than with the individual samples composing the communities themselves. This step reduces the total number of curves for consideration from 50 to 12. Figure 4 shows the sine-cosine curves corresponding to the centroids of the tree communities, and Figure 5 the curves of the tree and seedling community centroids plotted simultaneously.

The most striking feature of Figure 4 is the wide separation and distinctiveness of each of the curves. Compare, for example, the curves corresponding to Morabukea (Mk), Wallaba (W), *Pentaclethra* (P) and Greenheart (G) forests. Their distinctiveness indicates clearly that in v_k-space ($k = 1, ..., 5$), the mean separation between communities is appreciable. It seems clear, therefore, that in ecological terms the forest as a whole tends to be made up of a number of floristically distinct communities.

Turning to the joint plot of the sample in both tree and seedling spaces of the canonical analysis (Fig. 5), attention is drawn to three points:

a) the well-separated curves corresponding to tree and seedling counterparts of different communities;

b) the proximity and overall similarity of curves corresponding to tree and seedling counterparts of the same community;

c) the singular nature of the curve for *Jessenia* (J) which alone is comparatively unrelated to the curve of any other community.

Thus, the separation of the curves for different communities noted in connection with the tree component of the vegetation is equally well-defined in terms of the seedling component. On the other hand, the remarkable fidelity of the relationship

Table 9. Lowland tropical rain forest, Guyana. Canonical analysis of tree-seedling relationships. Correlations between the original variables and the canonical variates.

Canonical variate	U_1	U_2	U_3	U_4	U_5	h_w^2	V_1	V_2	V_3	V_4	V_5	h_b^2
Seedling community												
Greenheart	0.038	-0.542	0.763	0.286	0.118	0.971	0.038	-0.519	0.721	0.249	0.072	0.858
Wallaba	-0.952	0.047	-0.256	0.153	0.002	0.997	-0.941	0.045	-0.242	0.133	0.001	0.964
Pentaclethra	0.198	0.847	0.326	-0.205	0.300	0.995	0.196	0.810	0.308	-0.179	0.195	0.859
Marabukea	0.639	-0.179	-0.624	0.405	0.054	0.997	0.631	-0.171	-0.590	0.353	0.035	0.901
Mora	0.026	0.040	-0.010	-0.662	0.277	0.518	0.026	0.038	-0.009	-0.577	0.180	0.367
Eschweilera	0.411	-0.469	0.250	-0.567	0.058	0.776	0.406	-0.449	0.236	-0.494	0.038	0.667
Variance extracted	0.254	0.211	0.201	0.179	0.031	0.876	0.248	0.193	0.180	0.136	0.013	0.770
Redundancy	0.248	0.193	0.180	0.136	0.013	0.770	0.248	0.193	0.180	0.136	0.013	0.770

Canonical variate	V_1	V_2	V_3	V_4	V_5	h_w^2	U_1	U_2	U_3	U_4	U_5	h_b^2
Tree community												
Greenheart	0.292	-0.560	0.730	0.211	-0.115	0.989	0.289	-0.536	0.690	0.184	-0.075	0.887
Wallaba	-0.936	0.000	-0.264	0.206	0.088	0.996	-0.925	0.000	-0.250	0.180	0.057	0.954
Pentaclethra	0.438	0.742	0.285	-0.300	0.277	0.991	0.433	0.710	0.269	-0.261	0.180	0.864
Morabukea	0.754	-0.173	-0.504	0.372	-0.067	0.995	0.745	-0.165	-0.477	0.324	-0.044	0.917
Mora / Eschweilera	0.548	-0.297	0.078	-0.661	-0.100	0.842	0.542	-0.284	0.074	-0.576	-0.065	0.716
Jessenia	0.410	-0.241	0.000	-0.355	0.699	0.841	0.405	-0.230	0.000	-0.309	0.455	0.519
Variance extracted	0.365	0.173	0.157	0.146	0.100	0.942	0.357	0.159	0.141	0.111	0.042	0.810
Redundancy	0.356	0.159	0.141	0.111	0.042	0.809	0.357	0.159	0.141	0.111	0.042	0.810

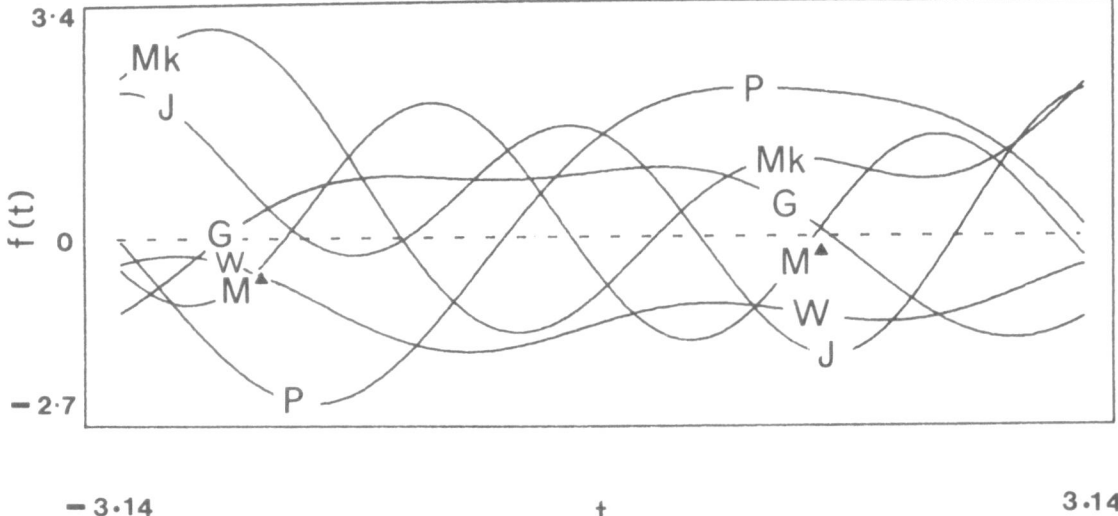

Fig. 4. Lowland tropical rain forest, Guyana. Canonical analysis of tree-seedling relationships. High-dimensional plot of the canonical variates. Centroids of tree communities of v_k-space ($k = 1, ..., 5$) mapped into an arbitrary trigonometric function. G: Greenheart forest; J: *Jessenia;* M*: *Mora/Eschweilera;* Mk: Morabukea; P: *Pentaclethra;* W: Wallaba.

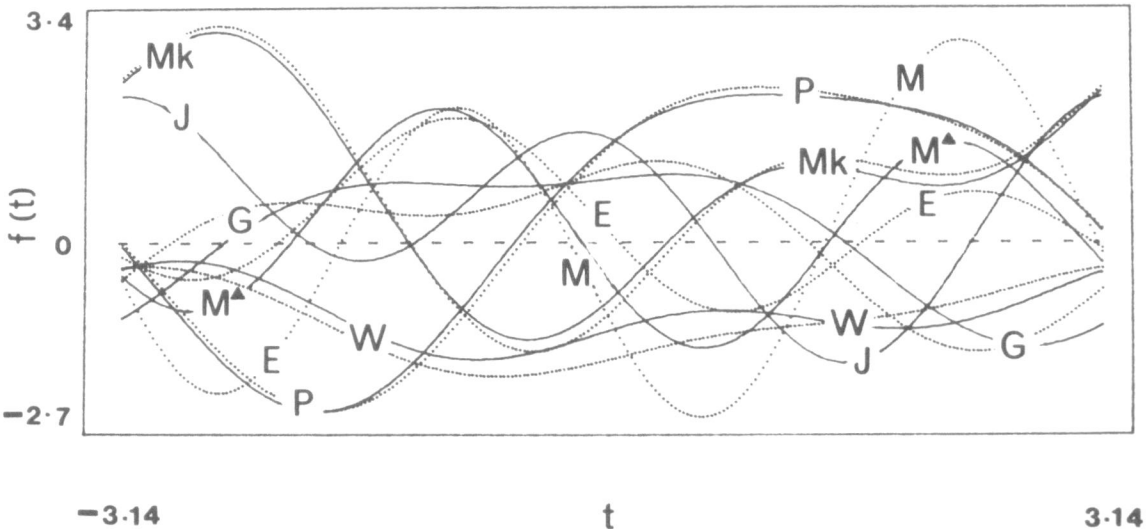

Fig. 5. Lowland tropical rain forest, Guyana. Canonical analysis of tree-seedling relationships. High-dimensional plot of the canonical variates. Simultaneous plot of tree and seedling community centroids in v_k- and u_k-spaces ($k = 1, ..., 5$), resp. mapped into an arbitrary trigonometric function. Solid curves: tree communities; dotted curves: seedling communities. E: *Eschweilera* forest; G: Greenheart; J: *Jessenia,* M: *Mora;* M*: *Mora/Eschweilera;* Mb: Morabukea; P: *Pentachlethra;* W: Wallaba.

between curves for tree and seedling counterparts of particular communities as *t* varies over its entire range, stands in marked contrast to the highly distinctive curves of different communities. The affinity between tree and seedling curves is es-

pecially strong in the case of Morabukea and *Pentaclethra* forests, rather less so in the case of Wallaba and Greenheart and somewhat weaker still in the case of *Mora/Eschweilera* (M*) forest and its seedling affiliates, *Mora* and *Eschweilera.* Finally, in the case of *Jessenia,* which is represented by forest vegetation only, the relationship vanishes entirely. In connection with *Mora/Eschweilera* forest, it is of some interest to observe that the curve for this community maintains a position mid-way between and symmetrically related to the comparatively well-separated *Mora* and *Eschweilera* seedling curves.

Ecological significance of the results

The salient point to have emerged is that the vegetation examined appears to consist of communities which tend to be compositionally distinct but between tree and seedling counterparts of which there exists for the most part a considerable degree of floristic and vegetational similarity. Broadly speaking, therefore, each community appears potentially able at least to be replaced in due course by a community of substantially the same composition. In a word, therefore, the dynamic status of the vegetation seems to be essentially stable. The curves for *Mora/Eschweilera* forest and its seedling affiliates and that of *Jessenia* forest, however, are open to interpretation in rather different terms. From Figure 5 it is conceivable that more or less distinctive *Mora* and *Eschweilera* seedling communities may be in the process of segregating from *Mora/Eschweilera* forest. It emerges also that *Jessenia* forest has little affinity with any seedling community. Thus, it may be that *Jessenia* represents a forest community which is not maintaining itself and which therefore may shortly come to be replaced by a community of a different kind. Thus, of the communities recognised, *Jessenia* and *Mora/Eschweilera* are those for which the likelihood of dynamic changes in composition seem most plausible.

The structure correlations and other interpretive indices of the canonical analysis (Table 9) provide means by which the nature of the dynamic relationships described can be pursued further. For our present purpose, however, sufficient has been said to illustrate the kind and extent of the information which canonical analysis can provide. On the basis of the results described, it seems reasonable to conclude: a) that in the short-term at least, the dynamic status of the vegetation appears to be essentially stable; and b) that *Jessenia* and *Mora/Eschweilera* are the communities which

above all may be those most susceptible to departure from stable equilibrium.

Several of the most striking points to have emerged from the present analysis have been commented on by previous workers. In particular, Ogden (1966, p. 33) in an analysis of the same data by quite different methods, arrived at essentially the same view of the close compositional similarity between tree and seedling counterparts of the various communities as expressed here. In addition, several workers have commented on the remarkable compositional similarity of the tree and seedling species' assemblages of Morabukea and *Pentaclethra* forests, which are notable features of the present analysis (Davis & Richards, 1933; Ogden, 1966, 1977). Ogden (1966, p. 33) has further remarked that the seedling population of Wallaba forest, which is itself a highly distinctive forest community, contains elements from adjacent forest communities which, although germinating, fail to survive to maturity. This feature may well account for the small but nevertheless discernable separation between the Wallaba tree and seedling curves apparent on Figure 5. Such independent substantiation of certain aspects of the results goes some way towards validating the form of analysis introduced here.

Summary and conclusions

Canonical analysis was used to summarize estimates of forest composition expressed in terms of two size components, trees and seedlings, while simultaneously emphasizing interrelationships between these components. The results obtained provided insight into the dynamic status of the vegetation studied. Briefly, the results enabled a distinction to be made between communities which appeared to be in stable equilibrium from communities which seemed less likely to be so, and, in the case of the latter, provided some indication of the nature of the changes which might be in progress. It seems justifiable to conclude on the strength of these results that canonical analysis might prove helpful in exploratory investigations of vegetation dynamics of a similar nature.

Two aspects of the analysis call for further comment. These concern the role of size-class data in successional studies and certain consequences of the preliminary components analyses. The use of size-classes as a means of investigating vegetation dynamics has several pitfalls (Austin, 1977). In particular, the interpretation of results ought, strictly speaking, to be guided by knowledge of the reproductive biology of the species involved. In the present study, unfortunately, as in cases which deal with the vegetation of remote areas generally, the required biological information may simply be lacking. It is also worth noticing that the time-span to which size-class data relate is very short in comparison with the life-span of a forest community as a whole. It must therefore be accepted that at best only short-term changes in dynamic status will be detectable. The use of size-class data here is not intended as an endorsement of the practice. On the contrary, it is recognized that whenever possible the acquisition of temporally related observations is almost always likely to prove advantageous. In the case of forest vegetation, however, the problem arises that, because of the length of the time-scale over which dynamic changes progress, adequate data are often likely to be quite simply beyond reach. Awareness of this difficulty suggests that simulation could provide a worthwhile alternative approach in studies of forest succession especially (e.g. see Botkin, Janak & Wallis, 1972; Shugart & West, 1975; Emanuel, Shugart & West, 1978). A digital computer can readily be programmed to generate sequentially related variables under various assumptions.

Preliminary components analyses were used to obtain descriptive indices of site vegetation. This step is not an essential part of the procedure described. Nevertheless, when it is applied it can have farreaching effects on the outcome of analysis simply because it provides the input variables for the subsequent canonical analysis. Thus some attempt to assess the stability of the results of the components analyses in a replicate sample would generally be worthwhile. This remark is equally applicable to the results of the canonical analysis itself. In other words, efforts directed towards validation of the outcome of both analyses would increase confidence in the results themselves and in the conclusions drawn from them. For this reason, the value of securing large samples can hardly be overemphasized.

Finally, attention is drawn to certain extentions and generalizations of the procedure described. Though the analysis employed size-class data, its application to temporally dependent observations would be straightforward. Moreover, although the analysis involved size-classes of just two kinds ($m = 2$), canonical analysis itself generalizes readily to the analysis of $m > 2$ $N \times p$ arrays. One further generalization is worth mentioning. This is that, while interest in the above investigation focused on community rather than species dynamics, the method is equally applicable to species, provided that the number of species involved is not excessive in relation to sample size.

Discussion

Two applications of canonical analysis were described, dealing with the nature of succession in some British hydroseres and with the dynamic status of a lowland tropical rain forest. The results in each case proved to be open to ecological interpretation. Neither study, however, was based on strictly sequential observations over time. Moreover, although canonical analysis itself generalizes readily to cases in which there are $m > 2$ $N \times p$ arrays $X_0, X_1, ..., X_m$ corresponding for example to the time-points, seral stages or size-classes $t_0, t_1, ..., t_m$, the applications described involved just two sets of variables ($m = 2$). The applications therefore can hardly be considered to provide anything other than a very preliminary basis for assessment. Obviously, there is scope for investigating the properties and the extent of the applicability of canonical analysis in successional work further. As with any analytical procedure, responsible use of the method would require proper attention to the design and implementation of the study as well as to interpretation of the results obtained (Gittins, 1979).

The varied nature of the ecological problems addressed, illustrates something of the versatility of canonical analysis. The applicability of the procedure is nevertheless considerably wider than the examples described might suggest. For example, in situations where some estimate of the relative ages of a number of serally related communities is known a priori, then canonical variates analysis may be used to clarify the relationships among the stages. Thus, in studies of vegetation succession on glacial moraines where the age of either the moraines themselves or of their supporting vegetation can be established with some accuracy, canonical variates analysis has proved informative even though the field observations themselves strictly correspond to just one point in time ($m = 1$). Mathews (1979) and Birks (1980) have each employed canonical variates analysis in these circumstances. More generally, even where only the beginning and final stages of a sere can be recognized with confidence, the use of canonical variates analysis may still be rewarding, as Grigal & Ohman (1975) have shown. Other quite different uses of canonical analysis in successional contexts have been made. Reyment (1976, 1978, 1979), for example, has described a longitudinal study based on samples collected from different horizons of a borehole core. The canonical variate scores of the sequentially related samples were then used to obtain an ecolog for the core – that is a graphical statement of the stability of the organisms or communities studied over time in relation to changing physical characteristics of the host environment. Madsen (1977) has also made use of canonical analysis in a comparable longitudinal study. Furthermore, canonical analysis has been used to reconstruct paleoclimatic patterns from tree-ring width data (Blasing, 1978). In this way, canonical analysis may sometimes enable dynamic changes in vegetation composition to be related to climatic shifts which may have been at least in part responsible for initiating and maintaining the changes. Altogether, therefore, it appears from these studies and from the examples described above that the potential applicability of canonical analysis in successional studies may be considerable.

A difficulty common to many successional studies is that temporal and spatial trends are likely to be confounded in the data. There is no general answer as to whether it is better to separate the two components before examining their effects or to deal with them simultaneously. Swaine & Greig-Smith (1980), for example, have advocated the elimination of spatial trends before going on to examine temporal change, and show how this can be achieved simply by centering the data appropriately. Such a procedure however leads to a major source of variation, as well as to the interaction between spatial and temporal components, being discarded without first having assessed their ecological effects. Conceivably, the impact of the spatial x temporal interaction could exceed that of either spatial or temporal variation alone. Indeed,

Mathews (1978, 1979) has remarked with reference to a particular study that 'succession in general, and divergence in particular, are greatly influenced by environment'. Wu & Botkin (1978) and Gutierrez & Fey (1980, p. 34) have expressed similar views. Thus, it would appear that if dynamic processes are to be understood and interpreted, it may be advisable to study the system as a whole rather than its dismembered dynamic aspect alone. It is therefore worth remarking that the adoption of canonical analysis would not in itself commit one to either strategy. On the one hand generalized canonical analysis would enable temporal and spatial variation to be analysed jointly, while on the other, spatial trends could if desired first be identified by trend-surface analysis (Gittins, 1968; Mathews, 1978) and their effects then removed by the use of partial canonical analysis (Timm & Carlson, 1976).

In order to clarify the nature of the contribution which it appears canonical analysis could perhaps make to the investigation of succession, it will be instructive finally to briefly consider relationships between canonical analysis and models of a different kind, namely dynamic models. We have seen that canonical analysis has the useful properties of parsimony, wide applicability and ready incorporation of probabalistic elements. On the other hand, it constitutes an essentially static model which represents the structure of the system examined, rather than the process going on in it. What is likely to be required ultimately, however, is a mathematical formulation of the dynamics of successional processes. Further, we have remarked that obtaining an adequate data base for successional studies by field observation or by experimentation commonly proves to be difficult or impossible. It is for precisely these reasons that models in which structure and function are both represented and which exploit the opportunities afforded by computer simulation appear to have so much to offer. Dynamic models of this kind have several desirable properties not shared by static models. Although they are ecosystem-specific and lack the parsimony of static models, dynamic simulation models have considerable freedom from constraints and assumptions and allow for the introduction of the non-linearity and feed-back which seem to be characteristic of ecological systems generally (Jeffers, 1978). In addition, dynamic models have the further advantage of taking account of rates of change

58

and hence of being able to predict the course of succession, at least in relation to a chosen ecosystem. It is in the context of this class of model that the role and contribution of static models such as canonical analysis is best appreciated.

Dynamic models require for their construction a set of sufficient parameters which are considered to interact to generate the successional process of interest. In aiding identification of the presence of the smallest possible number of ecologically significant effects and relationships, and in specifying the direction of these, canonical analysis may enable the essential components of the process to be recognized, at least in a provisional way. The ecological significance of these elements, once provisionally identified, may then be confirmed and refined through controlled experiment before being incorporated as parameters in a dynamic model of the ecosystem in question. From results so far obtained, the promise of computer simulation models for investigating and predicting succession appears to be considerable (Shugart & West, 1977; Emanuel, Shugart & West, 1978; Gutierrez & Fey, 1980; Solmon, Delcourt, West & Blasing, 1980).

In short, it seems that it is to dynamic models that we shall increasingly turn to further our understanding of successional processes. Canonical analysis nevertheless may well prove useful in the exploratory stage of such endeavors by facilitating recognition of a minimal subset of elements necessary for the development of models of this kind.

References

Anderson, T. W., 1958. An introduction to multivariate statistical analysis. Wiley, New York, 374 pp.

Andrews, D. F., 1972. Plots of high-dimensional data. Biometrics 28: 125–136.

Austin, M. P., 1977. Use of ordination and other multivariate descriptive methods to study succession. Vegetatio 35: 165–175.

Bartlett, M. S., 1965. Multivariate statistics. In theoretical and mathematical biology, T. H. Waterman & H. J. Morowitz (eds). Blaisdell, New York, pp. 201 224.

Birks, H. J. B., 1980. Modern pollen assemblages and vegetational history of the moraines of the Klutlan Glacier and its surroundings, Yukon Territory, Canada. Quaternary Res. 14: 101–129.

Blasing, T., 1978. Time series and multivariate analysis in paleoclimatology. In H. H. Shugart Jr. (ed.) Time series and ecological processes. Proceedings of the Society of industrial

and applied mathematics conference, Utah, SIAM, Philadelphia, 211–226.

Davis, T. A. W. & Richards, P. W., 1933. The vegetation of Moraballi Creek, British Guiana: an ecological study of a limited area of tropical rain forest. Part I. J. Ecol. 21: 350–384

Drury, W. H. & Nisbet, I. C. T., 1973. Succession. J. Arnold Arboretum, Harvard University 54: 331–368.

Emanuel, W. R., Shugart, H. H. Jr. & West, D. C., 1978. Spectral analysis and forest dynamics: the effects of perturbations on long-term dynamics. In: H. H. Shugart Jr. (ed.) Time series and ecological processes. Proc. Soc. Industr. Appl. Math. Conf. Utah. SIAM, Philadelphia, p. 193–208.

Gittins, R., 1968. Trend-surface analysis of ecological data. J. Ecol. 56: 845–869.

Gittins, R., 1979. Ecological applications of canonical analysis. In L. Orlóci, C. R. Rao & W. M. Stiteler (eds.) Multivariate methods in ecological work. International Co-operative Publishing House, Fairland, Maryland, p. 309–535.

Grigal, D. F. & Ohmann, L. F., 1975. Classification, description and dynamics of upland plant communities within a Minnesota wilderness area. Ecol. Monogr. 45: 389–407.

Gutierrez, L. T. & Fey, W. R., 1975b. Simulation of successional dynamics in ecosystems. In: G. S. Innis (ed.) New directions in the analysis of ecosystems. Simulation Councils Proc. Ser. 5: 73–82.

Gutierrez, L. T. & Fey, W. R., 1975a. Simulation of secondary autogenic succession in the short grass prairie ecosystem. Simulation 24: 113–125.

Gutierrez, L. T. & Fey, W. R., 1980. Ecosystem succession: a general hypothesis and a test model of a grassland. The MIT Press, Cambridge, Mass., 231 pp.

Horst, P., 1961. Relations among m sets of measures. Psychometrika 26: 129–150.

Jeffers, J. N. R., 1978. An introduction to systems analysis: with ecological applications. Arnold, London, 198 pp.

Kendall, M. G. & Stuart, A., 1973. The advanced theory of statistics, Vol. 2 Inference and Relationship, 3rd ed. Griffin, London, 723 pp.

Kettenring, J. R., 1971. Canonical analysis of several sets of variables. Biometrika 58: 433–451.

Kshirsagar, A. M., 1972. Multivariate analysis. Marcel Dekker, New York, 534 pp.

Lancaster, H. O., 1958. The structure of bivariate distributions. Ann. Math. Stat. 29: 719–736.

Madsen, K. S., 1977. A growth curve model for studies in morphometrics. Biometrics 33: 659–669.

Mathews, J. A., 1978. Plant colonisation patterns on a gletschervorfeld, southern Norway: a meso-scale geographical approach to vegetation change and phytometric dating. Boreas 7: 155–178.

Mathews, J. A., 1979. A study of the variability of some successional and climax plant assemblage-types using multiple discriminant analysis. J. Ecol. 67: 255–271.

Morrison, D. F., 1976. Multivariate statistical methods. 2nd ed. McGraw-Hill, New York, 415 pp.

Ogden, J., 1966. Ordination studies on a small area of tropical rain forest. M.Sc. thesis. University of Wales, 129 pp.

Ogden, J., 1977. Tropical rain forest in the Guyana lowlands: description and dynamics. Unpubl. rept. Australian National University, Canberra, 11 pp.

Olson, J. S., 1958. Rates of succession and soil changes on southern Lake Michigan sand dunes. Bot. Gaz. 119: 125–170.

Orlóci, L., 1978. Multivariate analysis in vegetation research. 2nd ed. Junk, The Hague, 451 pp.

Reyment, R. A., 1976. Chemical components of the environment and Late Campanian microfossil sequences. Geol. Foren. Stockholm Forh. 98: 322–328.

Reyment, R. A., 1978. Graphical display of growth-free variation in the Cretaceous benthonic foraminifer Afrobolovina afra. Palaeogeography, Palaeoclimatology, Palaeoecology 25: 267–276.

Reyment, R. A., 1979. Multivariate analysis in statistical paleoecology. In: L. Orlóci, C. R. Rao & W. M. Stiteler (eds.) Multivariate methods in ecological work. Intern. Co-o Publ. House, Fairland, Maryland, pp. 211–235.

Shugart, H. H., Jr. & West, D. C., 1977. Development of an Appalachian deciduous forest succession model and its application to assessment of the impact of the chestnut blight. J. Environm. Manag. 5: 161–179.

Solomon, A. M., Delcourt, H. R., West, D. C. & Blasing, T. J., 1980. Testing a simulation model for reconstruction of prehistoric forest-stand dynamics. Quaternary Res. 14: 275–293.

Sutherland, J. P., 1974. Multiple stable points in natural communities. Amer. Nat. 108: 859–873.

Swaine, M. D. & Greig-Smith, P., 1980. An application of principal components analysis to vegetation change in permanent plots. J. Ecol. 68: 33–41.

Tansley, A. G., 1939. The British Islands and their vegetation. Cambridge University Press, 930 pp.

Timm, N. H., 1975. Multivariate analysis with applications in education and psychology. Wadsworth California, 689 pp.

Timm, N. H. & Carlson, J. E., 1976. Part and bipartial canonical correlation analysis. Psychometrika 41: 159–176.

Walker, D., 1970. Direction and rate in some British Postglacial hydroseres. In: D. Walker and R. G. West (eds.) Studies in the vegetational history of the British Isles. Cambridge University Press, 117–139.

Wu, L. S.-Y. & Botkin, D. B., 1978. On population processes of longlived species. In: H. H. Shugart, Jr., (ed.) Time series and ecological processes, Proc. Soc. Industr. Appl. Math. Conf. Utah. SIAM, Philadelphia, p. 245–254.

Accepted 23.5.1981.

Problems in heathland and grassland dynamics*

J. Miles

Institute of Terrestrial Ecology, Hill of Brathens, Banchory, Kincardineshire AB3 4BY, Scotland

Keywords: Allelopathy, Competition, Disease, Dispersal, Fire, Grassland, Grazing, Heathland, Soil, Stability

Abstract

A review of factors governing vegetation change in heathlands and grasslands is presented, with emphasis on soil factors. Climate, microclimate, fire, fauna, grazing by vertebrates, grazing by invertebrates, disease, dispersal, establishment, competition, allelopathy, stabilization, soil formation and podzolization are discussed with emphasis on inconsistencies and lacks in our present knowledge. Examples are mainly from NW European *Calluna* heath and related woodlands.

Introduction

In recent years there has been a quiet revolution in attitudes towards succession, with pragmatism replacing dogmatism, a movement helped in particular by the seminal papers of Egler (1954), McCormick (1968) and Drury & Nisbet (1971, 1973). We are now aware that succession is not an orderly process, and that changes stemming from many sources occur concurrently. We are slowly unravelling the complexities of vegetation change by identifying and studying the causes.

However, it is inevitable that as more knowledge is gained, new areas of ignorance are indicated. 'An effect is but the cause for another effect' (Egler, 1970).

In this paper I briefly discuss a selection of the factors governing vegetation change, with greater than usual emphasis on the soil. Instead of trying just to review what is known however, I try to highlight areas where our knowledge stops.

Vegetation and environment

A simplified scheme of the interactions between vegetation and the determinants of its composition discussed here is given in Figure 1. This ignores the direct effects of man, such as fertilizing and mowing, but recognises that many of the biological processes that are functions of the vegetation are themselves important determinants of vegetation composition. It is worth stressing that the extrinsic factors listed do not simply influence the vegetation, but also mostly show interactions with it and

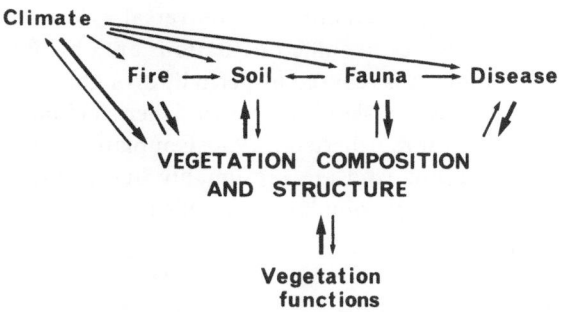

Fig. 1. Simplified scheme of interactions between vegetation and certain factors influencing its composition. Influences shown by bold arrows are those discussed in the text.

---

* Nomenclature follows Clapham, Tutin & Warburg (1962) for vascular plants; species not included in this work have authorities cited in the text.

Vegetatio 46, 61–74 (1981).   0042-3106/81/0462-0061/$2.80.
© Dr W. Junk Publishers, The Hague.

between themselves. The influences discussed in this paper are shown by bold arrows.

Climate and microclimate

Climatic variables influence all the other important environmental factors. Fire is influenced particularly by the evapo-transpiration balance (determining fuel water content and thus combustibility), by wind speed (determining rate of spread of fires, and influencing fire temperature and thus its biological effects), and by lightning (a source of ignition). Climatic and microclimatic conditions also influence inter alia rates of mineral weathering and organic matter decomposition in the soil, herbivore distribution and population sizes, and the rates of spread of many plant diseases.

The direct effects of climate have long been accepted as the primary control of plant distribution and thus of vegetation. However, precisely how such control operates is known for very few species. Climatic extremes probably control the extent of many other species, but if the effects are less plainly evident, they are unlikely to be recognized by casual observations. A good example is Dahl's (1951) conclusion that warm summers controlled the distribution of many arctic-alpine plants at lower altitudes in Fennoscandia.

Year to year fluctuations in the composition of grasslands were recognized over 60 yr ago (Rabotnov, 1974). Now, with the insights from long term studies (Brenchley & Warrington, 1958; Watt, 1960; Albertson & Tomanek, 1965; Rabotnov, 1966; Williams, 1978), as well as a priori evidence, we realize that fluctuations are universal. However, while the causes of some fluctuations are obvious, for example those caused by periodic drought in the Great Plains grasslands of North America (Malin, 1956; Weaver & Albertson, 1956; Coupland, 1958), others are not, and are presumably due to much more subtle and complex variations in climate.

Fire

Most heathlands and many grasslands are periodically burnt (e.g. Grant, Hunter & Cross, 1963; Daubenmire, 1968; Lloyd, 1968; Miles, 1971; Vogl, 1974; Gimingham, Chapman & Webb, 1979;

Specht, 1979). Such fires cause significant short term shifts in the competitive balance between the component species. While many case history studies have now provided a general picture of the kinds of change occurring in heathland after fire (see citations in Hansen, 1964, 1976; Gimingham, 1972), very little is known about the differential effects of burning at varying frequencies.

A short time interval between fires may have profound effects on floristic composition. For example, it is believed that the great abundance of *Molinia caerulea* on the moorlands of northwest Scotland is a result of frequent burning in the past (McVean & Ratcliffe, 1962), though the critical periodicity of burning is not known. Miller & Miles' (1971) finding that vegetative regeneration of *Calluna vulgaris* dominated heathland is greatest from 6–10 yr old stands in NE Scotland, and declines progressively in older stands, suggests that burning every five years or less might cause such change. Miles (unpubl.) found that when a patch of *Calluna-Erica tetralix-Molinia* dominated heathland in southern England was burnt in May 1964, 98% of *Calluna* stools and 100% of *Erica cinerea, E. tetralix* and *Ulex minor* stools regenerated vegetatively. Yet on part of this patch burnt again by accident in the following February (fuelled by one season's production of *Molinia* litter), only 39%, 57%, 73% and 73% respectively of stools regenerated again, and all seedlings were killed.

A very long time interval between burning *Calluna* heathland permits the development of an uneven-aged stand, with all growth phases (pioneer, building, mature, degenerate) present (Watt, 1955; Barclay-Estrup & Gimingham, 1969). Watt (1955) studied a site in the Breckland of eastern England and stated that in the final degenerate phase the *Calluna* branches collapsed outwards and gradually died back, forming a progressively enlarging central gap in which the surface mor slowly decayed, eventually exposing mineral soil in which *Calluna* seedlings established, so beginning a new cycle. This statement has been widely accepted and quoted as a generality, yet only after I had quoted it (Miles, 1979) did I realize that I had never seen *Calluna* seedlings in the middle of degenerate bushes! Searches of all old *Calluna* stands I encountered during 1978–80 in England and Scotland failed to reveal a single seedling in such gaps. It was commonplace to find pioneer phase *Calluna* growing in the centre of degenerate bushes. However, dissection showed that in every instance such pioneer *Calluna* had originated from a trailing branch from an adjacent bush which had rooted adventitiously and so formed a new plant. Also, no gaps were seen in which the mor had decayed to expose mineral soil, or which were otherwise suitable for establishment of *Calluna* seedlings. All gaps had a complete cover of plants, mostly of pleurocarpous mosses or grasses. Given these findings, it is possible that conditions in Watt's Breckland heath may be fairly unique. The establishment of *Calluna* seedlings in the gaps formed by degenerate bushes may be the exception rather than the rule.

Another effect of fire is to give a mulch of nutrien-rich ash to the soil (Haines, 1926; Allen, 1964; Daubenmire, 1968; Viro, 1974). Although the effect of this ash fertilization is short-lived, it may account in large part for the rapid and luxurient development of mosses commonly found on heathland after a fire. Similar flushes of moss growth have been obtained by experimentally fertilizing heathland (Miles, 1968, 1973b). Whether such moss swards have any significant effects on the development of heathland vegetation after fire is unknown. They may however inhibit seedling establishment (Miles, 1973b), and so slow down succession. Equally, they may stabilize unstable surfaces and so speed up succession (Leach, 1931).

Fauna

Animals influence vegetation in many ways. Through the effects of grazing and browsing, trampling, seed eating and seed dispersal, animals affect plant establishment and development, and thus vegetation composition. Dunging and urination by larger animals increases soil fertility locally, and acts through the soil microflora to speed up organic matter decomposition and nutrient cycling. The soil fauna promote organic matter decomposition through comminution of litter, and can bare soil, thus creating niches for establishment, for example the formation of earth mounds by some ants and by moles (*Talpa europaea* L.). Soil mixing by the soil fauna affects inter alia aeration, infiltration rates, and the distribution of viable buried seeds. For example, the heathland ant *Tetramorium cespitum* Latreille collects and stores seeds in underground nest galleries as food, mainly of *Calluna vulgaris* and *Erica cinerea*, though seeds of *Agrostis setacea*, *Molinia caerulea* and *Ulex minor* have also been found (Brian, Hibble & Stradling, 1965; Brian, Elmes & Kelley, 1967). The discussion here is however confined to the effects of grazing.

Grazing by vertebrates

Grazing by domestic live-stock and wild herbivores is often the most important factor regulating the species composition of grasslands, and is also a significant influence on many heathlands. Gross changes in the relative abundance of species can occur rapidly with drastic changes in grazing pressures. Thus when sheep were excluded from an upland grassland in Wales that had developed under free range grazing by sheep, the proportion of *Calluna vulgaris* in the sward increased in two years to 20% from an initial value of less than 1% (Fig. 2). After 14 years the proportion of *Calluna* had increased to 85%. Allowing sheep free access again showed the changes to be reversible however; after 12 yr *Calluna* had declined to only 5% (Jones, 1967).

The known successional relationships with varying grazing pressures of the main upland vegetation types in England and Wales have been summarized by Miles, Welch & Chapman (1978). They also

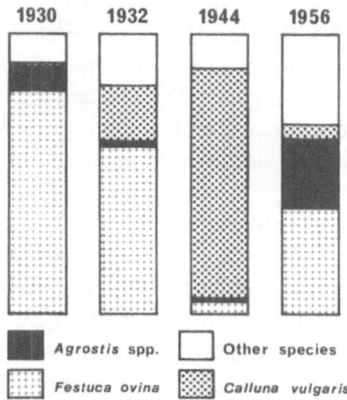

Fig. 2. Changes in composition (by % weight) of an upland grassland on mineral soil in Wales: in 1930 after sward development under open grazing by sheep, cattle and horses; in 1932 after 2 years without grazing; in 1944 after 14 years without grazing; in 1956 after 12 years of subsequent open grazing. Data from Jones (1967).

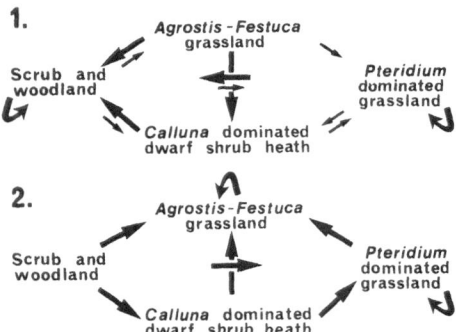

Fig. 3. Generalized sequences of successional relationships between four widespread types of vegetation on well-drained upland soils in Britain, given 1) low or negligible, and 2) high grazing pressures. Large arrows represent usual sequences, small arrows less common sequences. The grasslands fall within the order *Nardetalia,* the *Calluna* heaths within *Calluno-Ulicetalia,* *Pteridium* stands within both these orders, and the scrub and woodland mostly within *Quercetalia robori-petraeae* and *Vaccinio-Piceetalia.* After Miles (1979), courtesy of Chapman & Hall.

attempted to give time scales for the duration of each change and the sheep densities which can cause a particular change in a particular time. Because of the poverty of information, the relationships given are partly hypotheses. That they seem to be the most comprehensive synthesis yet published just reflects our rudimentary understanding of grazing influences. For example, Figure 3 summarizes the successional changes between four widespread types of vegetation on well-drained acid soils. Most of the thresholds of change are unknown. In NE Scotland however, Welch (1974) has estimated that *Calluna* dominated heathland seems to change to *Agrostis-Festuca* grassland with a grazing intensity of 2.5 sheep/ha/yr, but just how this critical limit may vary elsewhere is not known.

Grazing by invertebrates

Invertebrate herbivores do not seem able to control the physiognomy and life form of vegetation in the way that many vertebrates can. Perhaps it is because their effects are generally less evident, as well as because of the taxonomic difficulties encountered, that they have been so little studied. However, most plants are variously consumed by a large array of invertebrates, mainly insects. Thus Phipps (1930) listed 292 species of insects that feed on *Vaccinium* spp. in the State of Maine, while

Franklin (1948, 1950, 1952) listed 51 species (many of them different from the Maine list) feeding on *Vaccinium* (mainly *V. macrocarpon* Ait.) in Massachusetts, including eight species recorded as killing large patches of *Vaccinium.* In comparison, rather few insect feeders have been noted on most heathland plants in Europe; Table 1 gives examples from the British Isles. This probably reflects a lack of systematic effort, perhaps fostered by disbelief in the significance of insect grazing.

Most detailed studies have been on those insects that cause damage on a catastrophic and/or economic scale. The only insect known to cause similarly catastrophic damage in European heathlands or grasslands is *Lochmaea suturalis* Thoms. (heather beetle). The larvae and adults of this species feed on the young shoots of *Calluna vulgaris,* especially of mature and old plants, and sometimes on *Erica cinerea* and *E. tetralix* (Cameron, McHardy & Bennett, 1944; Morison, 1963). In severely grazed plants the remaining shoots also die back, and populations of *L. suturalis* often locally reach sizes such that extensive patches of *Calluna* are killed (cf. de Smidt, 1977). This phenomenon is apparently favoured by warm dry summers (Morison, 1963), which would account for the greater frequency of catastrophic damage in southern England, Holland and north Germany for example (Betrem, 1929; Cameron, McHardy & Bennett, 1944), than in Scotland.

Apart from the rather few case histories like those already noted, several pieces of circumstantial evidence suggest that insects can have very considerable effects, albeit generally inconspicuous, on the population size of plants. First, there are the records of typically non-gregarious species being introduced to other countries where, without their normal complement of insect herbivores, they then increased to cause major weed problems. The introduction of *Euphorbia cyparsissias* to Ontario and *Linaria vulgaris* to Saskatchewan are examples (Harris, 1973). Second, there are records of introductions like these which were subsequently reduced from weed status to scattered plants and clumps by introduction of their natural insect pests. Classic examples are the control of *Hypericum perforatum* in California by the beetle *Chrysolina quadrigemina* (Suffr.) (Huffaker & Kennett, 1959), and of *Opuntia* spp. in Australia by the moth *Cactoblastis cactorum* (Berg) (Wilson, 1960). Third, there are interference studies like that of Cantlon (1969). He estimated the effect of insect herbivores on a population of *Melampyrum lineare* Desr. by repeatedly treating plots with the organochlorine insecticides aldrin and DDT, and the organophosphorus insecticide malathion. After 2 yr the density in spring of young *M. lineare* plants was three times that in the untreated control plots. Large numbers of seedlings in the field regularly disappear from unknown causes (Miles, 1972, 1973a), and Cantlon's study suggests insects cause much of this mortality. Molluscs also graze and kill seedlings (Miles & Kinnaird, 1979). Taken together therefore, the evidence suggests that the population size of many plants may be controlled by insects and other invertebrate herbivores.

The effect of insect grazing on plants is not always a negative one however. Defoliation may under certain circumstances stimulate growth and reproduction (Harris, 1973). Dempster (1975) for example noted that grazing by caterpillars of *Tyria jacobaeae* L. (cinnabar moth) on basal leaves of *Senecio jacobaea* produced up to an eight-fold multiplication of the plant by stimulating root buds to develop into new rosettes. Positive growth responses of this magnitude must be of competitive significance. Their frequency of occurrence is quite unknown, but if they are not infrequent there exists the intriguing possibility that their overall significance may exceed the apparently rather small effects on average of allelopathy.

Disease

Excluding the man-made and managed vegetation of forestry, agriculture and horticulture, even less seems to be known about the role of pathogens and parasites in controlling vegetation composition than of the roles of invertebrate herbivores and allelopathy. While a number of fungal infections have been recorded for particular species (see Table 1 for example), the importance of most has not been evaluated.

In heathland however, die-back of *Calluna vulgaris* plants caused by infection of older branches by the usually saprophytic agaric *Marasmius androsaceus* (L. ex Fr.) Fr. is not uncommon (MacDonald, 1949). *M. androsaceus* also kills *Erica cinerea* and *E. tetralix*, though less frequently, and has been recorded on *Juncus squarrosus* (Welch, 1966).

Fusarium nivale (Fr.) Ces., which is an important winter-killing disease of forage grasses and cereals in northern regions, can cause extensive die-back in *Agrostis-Festuca* grasslands (Gray & Nicholson, 1957). The incidence of die-back is greatest when there is prolonged snow cover in spring. In the Scottish Highlands, patches of *Calluna vulgaris* covered by long lying snow patches often exhibit die-back when exposed, and show extensive mycelial infection (Miles, unpublished). This suggests attack by *F. nivale* or an allied species, though the matter does not seem to have been critically investigated.

Table 1. Numbers of species of insect feeders and fungal pathogens and parasites of heathland vascular plants described to date in the 'Biological Flora of the British Isles'.

Species	Insect feeders	Fungal pathogens	References
Calluna vulgaris	41	2	Gimingham 1960
Erica cinerea	33	7	Bannister 1965
E. tetralix	33	7	Bannister 1966
Vaccinium myrtillus	30	17	Ritchie 1956
V. vitis-idaea	24	15	Ritchie 1955
Juncus squarrosus	8	5	Welch 1966
Empetrum nigrum	7	7	Bell & Tallis 1973
Deschampsia flexuosa	4	1	Scurfield 1954
Nardus stricta	1	3	Chadwick 1960
Agrostis setacea	0	0	Ivimey-Cook 1959
Erica mackaiana	0	0	Webb 1955

However, though fungal or other pathogens may rarely cause widespread death of mature plants in natural vegetation, soil-borne fungi may be an important cause of seedling mortality, especially when seedlings are shaded (Vaartaja, 1962; Taher & Cooke, 1975). There are many genera of normally saprophytic fungi with species able to act as facultative parasites, and to infect and kill susceptible young seedlings. Certainly damping-off by fungi can occur in seedlings in heathland (Miles, 1972), but the extent and importance of damping-off in controlling vegetation composition seems never to have been systematically investigated.

Vegetation factors

The plant dependent processes touched upon here are those first recognized by Clements (1905): dispersal (or migration), establishment (or ecesis), competition and the newer ideas of allelopathy, and stabilization. Site modification (or reaction) is discussed later in the section on soil.

Dispersal

The species arriving at a site during the early years of succession are usually those that determine the short term outcome. In most species the bulk of seeds and fruits fall fairly near to the parent plant (Miles, 1979). The effective dispersal ranges are known for relatively few species however. Yet such knowledge is essential if we wish to attempt to predict the course and rate of change during succession at particular sites. It is equally necessary in creative nature conservation. For example, starting from forest, how long did it take for many of our ancient anthropogenic plant communities in Europe, such as species-rich hay meadows, to acquire their present assemblage of species? If we ever wish to recreate examples of these fast vanishing types through natural processes rather than large scale gardening, then knowledge of dispersal ranges is highly desirable.

An even more difficult problem is the significance of the small proportion of seed that is transported over long distances (Ridley, 1930, van der Pijl, 1972). The dilution effect means that there is no immediate significance to the landscape receiving such propagules. But events such as the deposition of a wind-borne *Pinus* seed, or the accidental dropping of a *Quercus* fruit by an overflying jay (*Garrulus glandarius* L.), are of enormous potential importance to the surrounding few square metres.

Establishment

Milthorpe (1961) remarked that 'It is reasonably certain that the establishment of plants from seed in vegetation occurs only in 'bare areas' arising from the death of previous occupants or from incomplete coverage.' This points to the field conditions where we may expect to find seedlings, but highlights the enduring problem of just how to define the establishment niche. There are many difficulties. The optimum conditions for growth and development during establishment may change for many species as plants mature (Grubb, 1977). Further, because vegetation is in a state of constant flux, especially evident in herbaceous swards, the microsite in which a particular seed germinated is itself short-lived. It may even have significantly changed by the time the seedling is evident, thereby posing the dilemma that while a microsite for germination in the field can be recognized only after the event, it can be described only before the event.

The dynamic nature of vegetation also poses problems for the young plants. For example, an experiment in *Calluna* heathland showed that the conditions most favourable for germination and early seedling survival were disastrous for most species (Miles, 1974b). This was because very small gaps closed over during the growing season through vegetative spread of the surrounding plants.

Clearly, definition of the establishment niche for any species requires a combination of field description and experimentation together with definition of the physical and chemical conditions for germination and seedling growth (optimum and amplitude) in controlled environments.

Competition

Probably no other process is so widely invoked yet so poorly understood under field conditions as competition. This is especially true of competition between roots, where the accessibility problem makes experimentation so much more difficult than with studies of competition between shoots.

As a result, the significance of root competition in controlling the performance of most heathland and grassland species is unknown. Yet since Fricke's (1904) pioneer root trenching experiment, workers have repeatedly shown the extent to which root competition from trees inhibits the growth of subordinate woody and herbaceous species. It thus seems very likely that root competition will be important in other habitats. The tendency to dominance in species such as *Brachypodium pinnatum, Calluna vulgaris, Deschampsia cespitosa* and *Pteridium aquilinum* may be partly due to this, as well as to the more commonly invoked processes of shoot competition, smothering by litter, and perhaps also allelopathy.

Allelopathy

Allelopathic interactions have been claimed or suggested for a wide range of natural and man-made plant communities, varying from xeric to hydric, and from temperate zones to the tropics (Rice, 1974). However, it seems that no-one has yet been able firstly to demonstrate that a plant in the field releases a toxic chemical or chemicals into the environment of another plant in sufficient quantities to influence its growth, and then to differentiate this clearly from competitive or other effects.

Claims for the reality of allelopathy have thus necessarily relied primarily on in vitro bioassays, backed up by correlations with plant performance in the field. However, in many studies such correlations have been poor or absent (Stowe, 1979). Even with the well known claim for allelopathic effects by shrubs of the Californian chaparral, the chemical effects are confounded with the possibly highly significant effects of grazing by small mammals and of drought (Bartholomew, 1970, 1971). Further, even cursory examination of the literature leads to the conclusion that probably every plant species has at least one chemical, or break-down product resulting from the extraction method, in at least one of its organs, that can be shown by in vitro tests to have a potential allelopathic effect on at least one other species!

The component species of temperate grasslands show complex patterns of associations between species (de Vries, 1954; Agnew, 1961; Turkington, Cavers & Aarsen, 1977), many of which are stable through time (Aarsen, Turkington & Cavers, 1979).

While this phenomenon presumably in part reflects differences and similarities of use of the minutely varying soil environment, allelopathic interactions could also be involved. For example, Newman & Rovira (1975) found from in vitro studies that the root exudates of *Anthoxanthum odoratum, Cynosurus cristatus* and *Holcus lanatus,* all species that tend to be strongly gregarious, stimulated their own growth. In contrast, those of *Hypochoeris radicata* and *Plantago lanceolata,* species tending to grow naturally as isolated individuals, were autotoxic. Are such effects of competitive significance under field conditions?

Lichens are particularly rich in secondary chemicals (Rundel, 1978), and there have been many reports of different species, especially of *Cladonia,* inhibiting germination (Pyatt, 1967; Ramaut & Thonar, 1972; Ramaut & Corvisier, 1975; Lemee & Arluison, 1976). It has also been suggested that *Cladonia* spp. can slow down the growth of young conifers by inhibiting their mycorrhizal fungi (Brown & Mikola, 1974) and thus reducing phosphate uptake (Fisher, 1979). Lichens are often important components of heathlands but their effects if any on the floristic composition are not known.

Perhaps of greater importance for heathlands is that many of the dominants have also been claimed to have allelopathic potential: *Calluna vulgaris* (Mantilla, Arines & Vieitez, 1975), *Deschampsia flexuosa* (Jarvis, 1964), *Erica australis* L. (Carballeira & Cuervo, 1977), *E. cinerea* (Ballester & Vieitez, 1971), *E. scoparia* L. (Ballester, Albo & Vieitez, 1977), *E. umbellata* L. (Salas, Ballester & Vieitez, 1973), *E. vagans* (Arines, Vieitez & Mantilla, 1974) and *Pteridium aquilinum* (Gliessman & Muller, 1971; Glass, 1976). It is also known that *Calluna* roots produce a toxin inhibitory to the ectomycorrhizal fungi of certain conifers, especially *Picea* spp. (Handley, 1963; Robinson, 1971, 1972). This is presumed to be the reason for the well known inhibition of growth of these conifers on *Calluna* heathland (Weatherell, 1953; Handley, 1963; Malcolm, 1975), termed 'checking' by foresters. It is not known whether allelopathy plays any significant role in suppressing other species during the *Calluna* growth cycle, or whether the potential allelopathic effects of the other heathland species exist in the field.

Newman (1978) found no pattern in the published reports of allelopathy. He noted that on average (1) the production of toxic chemicals is not beneficial to the producer plant, (2) competition does not result in increased toxin production, and (3) long co-existence does not result in species becoming more tolerant of each others toxic exudates. He concluded that there is often not specific selection for allelopathic ability or tolerance of allelopathy, but that these plant properties are better viewed as the fortuitous by-products of the evolution of other characteristics. The occurrence of allelopathy, if indeed it does exist in the field, may thus be quite unpredictable.

Stabilization

Floristic stability can have many different causes. Climatic or edaphic extremes may so limit the variety of species that can exist, in desert, tundra or coastal mud flats for example, that on disturbance only the original species can recolonize. Stability can be imposed by management which arrests succession, for example the grazing, mowing and burning of grasslands and heathlands. It may also reflect past land use and management, such as on the extensive *Calluna* heathlands of the Scottish Highlands where near total deforestation gives a measure of stability simply due to lack of seed of potential invaders (Miles, 1973b).

However, stability also arises through the ability of some species to persist in time. Some vegetation can effectively resist invasion by trees for long periods. Examples include patches of suckering shrubs such as *Symphoricarpos rivularis*, the fern *Pteridium aquilinum*, and the tropical grass *Imperata cylindrica* (L.) Beauv. (Niering & Goodwin, 1974; Miles, 1979). This is a fascinating phenomenon, yet no detailed study of the mechanisms by which these species inhibit tree establishment, thus arresting succession, seems to have been published.

Clonal populations of many species persist through surprisingly long periods of time. Harberd (1961) identified a clonal patch of *Festuca rubra* that he estimated to be between 400 and >1 000 yr old. Oinonen (1967a, b & c, 1968, 1969) has estimated clonal longevity in several species, and found patches of *Pteridium aquilinum* reckoned to be >500 yr old.

Sagar & Harper (1961) noted that '. . . correlations between the distribution of mature plants and existing environmental factors may be spurious, since the crucial factors determining the present position of individuals were those operating in the seedling stages'. This is even more likely to be the case when hundreds or even thousands of years have passed rather than just months or years. Thus Summerfield (1972) recorded a small patch of *Narthecium ossifragum* that had persisted in a *Sphagnum recurvum* P. Beauv. lawn for at least 50 yr, growing in acid, very nutrient poor and oxygen deficient peat quite unsuitable for present day establishment of seedlings. These are examples where the clonal habit has enabled species to persist in time despite changing environmental conditions during succession. Summerfield (1972) termed this 'biological inertia'.

Soil

The effects of soil variability as a determinant of vegetation composition are well known, though they are not as well understood (Ernst, 1978). However, one aspect of soil-plant relationships that has been comparatively ignored is the question of the extent to which such variability in soil is the result of the plants growing there rather than the cause. This neglect is puzzling, because a century ago Müller (1879, 1884) demonstrated the profound changes from mull to mor soils that occurred when *Quercus* woodland was replaced by *Calluna vulgaris* and *Fagus sylvatica*. He also suggested that these changes in vegetation were causing podzolization. Tüxen (1957, 1975) repeatedly pointed out the influence of vegetation on the formation of the soil profile under *Calluna* heathland in NW Germany.

There are two main reasons why *a priori* we should now expect the soil conditions under a given plant to change if it is replaced by a plant of a different species. First, it has long been known that the litter from different species can vary markedly in chemical composition. It varies not just in those elements important to plant growth, but also in secondary chemicals. Both *inter alia* help determine rates of decomposition and nutrient cycling (Heal, 1979), while some of the latter also directly acidify the soil and accelerate eluviation (Duchaufour, 1977). Second, different species vary in the extent to which they modify the chemical composition of rain dripping off their leaves and canalized as stemflow (Ernst, 1978). Such nutrient enriched throughfall would be expected to modify the soil under the plant's canopy, and particularly around the stem base where canalization as stemflow is considerable. Such patterns of soil properties have in fact been found under individual trees and shrubs (Zinke, 1962; Grubb, Green & Merrifield, 1969; Gersper & Holowaychuck, 1970a & b; Grubb & Suter, 1971; Lodhi, 1977). Only a few woody species have yet been examined in this way however, and apparently no herbaceous species, although tussock forming species in particular might repay investigation. In most studies of the effects of plants on the soil, soil sampling has been carried out so that only the average soil conditions under a stand were determined.

However, studies made this century have suggested that marked changes in soil pH and chemical status can occur under particular species, though again most investigations have been on trees (see Handley, 1954; Birkeland, 1974; Rastorova, 1974; Grieve, 1975; Miles, 1978, for partial reviews). Of particular relevance here are those studies that bear out Müller's hypothesis that the typically acid, podzolized soils of *Calluna* heathlands are the result of the *Calluna* monoculture (Gimingham, 1960, 1972; Dimbleby, 1962; Grubb, Green & Merrifield, 1969; McVean & Lockie, 1969; Grubb & Suter, 1971). Most heathland soils are of course intrinsically poor in nutrients, being developed in highly siliceous parent materials. But the evidence suggests that most (cf. de Smidt, 1979) *Calluna* heathlands in NW Europe developed under man's influence in place of broad-leaved woodlands on brown forest soils or brown podzolic soils (the 'sols bruns ocreux' and 'sols ocre podzoliques' of Duchaufour (1977)). Apart from the effects of acidification and greater podzolization, typical heathland mor humus has about twice the lipid content of the mull humus of broad-leaved woodlands, and these lipids are generally inhibitory to soil micro-organisms and to germination and plant growth (Stevenson, 1966; Fustec-Mathon, Righi & Jambu, 1975).

These results suggest that many of the plant species present under former woodland conditions (Dimbleby, 1962) may have disappeared not just because of the loss of the woodland canopy and the effects of repeated burning, but because of the *Calluna*-induced changes in the soil. Certainly,

when sown experimentally many species fail to establish on heathlands although they may naturally have grown there before deforestation occurred (Miles, 1974a, 1975).

In so far as most heathlands are not of natural origin, have reduced floristic and thus presumably faunistic diversity, and more strongly podzolized and hence generally less productive soils (Låg, 1962; Pyatt, 1970; Page, 1971; Miles, 1974a) than the former natural woodlands, they are degraded ecosystems. It is thus particularly interesting that several workers have claimed that the growth of *Betula pendula* and *B. pubescens* is associated with increased soil pH and base status, mull formation in place of mor, and even depodzolization (see the review by Gardiner (1968)). The published reports contain very little substantive evidence however, yet are clearly of great potential significance if true. For example, although heathlands in NW Europe now cover only a fraction of their former area, there are probably still at least 1 000 000 ha of *Calluna* dominant heath in Scotland and perhaps at least an equal area with abundant *Calluna* (Miles & Young, 1980). For this reason I have been studying the effects of *Betula* spp. on heathland soils since 1974 (Miles, 1978; Miles & Young, 1980).

The main findings of this work are that there is indeed a trend of fairly profound soil change under *Betula*, in particular:

(1) increases in pH and exchangeable calcium;
(2) increases in rates of cellulose decomposition and nitrogen mineralization;
(3) increasing growth responses of test plants grown in the glasshouse;
(4) production of mull or mull-like moder humus in place of mor;
(5) an increase in earthworm (lumbricid) numbers, with a change in the species composition towards that characteristic of mull soils.

There is also a suggestion that depodzolization may occur. If it does it must come about through soil mixing, with downward movement of surface organic matter and upward movement of iron-rich B horizon material obliterating the bleached A_2 (or A_e, E) horizon. It seems that the mixing agents will be roots and the soil mesofauna, perhaps earthworms in particular.

However, it must be stressed that the Scottish sites where depodzolization may occur have podzols with thin A_2 and soft B_h / B_s horizons, and also relatively base-rich soil parent materials. It is doubtful if either the thicker A_2 horizons commonly found further south, for example those of the Bagshot district of S England or the Lorraine Plateau or Vosges hills of NE France, or the hard 'iron pan' ('Ortstein') characteristic of the heaths on highly siliceous sands in Jutland and N Germany, could be so easily destroyed.

There was also evidence that the soil changes found under developing *Betula* stands tend to be reversed as the stands open up with tree senescence and death. This suggests a cyclic trend, in the opposite direction to that found under certain conifers (Page, 1968, 1974). In each case the maximum effects, of mull formation and de-acidification under *Betula*, and of mor formation and acidification under conifers such as *Picea* spp. and *Pinus* spp., occurs at the stage of greatest litter input. The effects increase as the stands develop and decline as they thin out. Such cyclic changes in soil conditions from point to point should a priori be expected, reflecting the growth cycles of dominant species (Heal, 1979).

Similarly, it would seem *a priori* that mull-forming vegetation other than *Betula* stands should be associated with the same soil trends, and mor-forming vegetation other than *Calluna* stands with the same tendency towards acidification and accelerated podzolization. The evidence suggests this is so. Thus Bornebusch (quoted in Handley (1954)) described apparent depodzolization under *Quercus borealis* in Denmark. Conversion of *Calluna* heathland in the Southern Uplands of Scotland to *Agrostis-Festuca* grassland, and its subsequent maintenance as such, by sheep grazing, also appears to bring about depodzolization (I. A. Nicholson & I. S. Paterson, unpubl.). Conversely, there is evidence that mor-forming conifers bring about podzolization in susceptible soils (Fisher, 1928; Marzhan, 1959; Page, 1968; Troedsson, 1972; Grieve, 1978).

However, while these may be two opposing trends of soil change on poor soils in NW Europe, the extent to which they will occur will be governed *inter alia* by:

(1) the species concerned, with its own intrinsic chemical composition, palatability to soil fauna, and quality as an energy and nutrient source for microbial activity (or rather the range of these properties within the population); also the degree to which these parameters vary with differences in soil properties;

(2) variations in site and soil properties, especially aspect and slope, drainage, ground water chemistry and movements, degree of podzolization and of iron pan development, particle size composition, and mineralogy (the last determining the rates of nutrient supply from mineral weathering);

(3) the previous site history, which significantly influences labile soil characteristics such as pH, exchangeable nutrients and organic matter content.

The influence that plants can have on the soil has a potential significance at least as great as anything mentioned earlier in this paper, and arguably greater.

There is clearly a need for more and more detailed studies of the effects of different species on the soil. Brief investigations such as that of Ten Khak Mun & Fedorova (1975) on *Polygonum* spp. and Faille's (1977) on *Calamagrostis epigejos* cry out for data on other species. Above all perhaps, there is a need for experimental studies. Almost all investigations to date have just compared the contemporary soil conditions under different adjacent vegetation types, and it was usually just presumed that these soils were identical at some earlier point in time. The hazards of interpretation inherent in this approach have been pointed out by Stone (1975). However, while efforts can be made to demonstrate antecedent similarity (Miles & Young, 1980), it can never be shown conclusively. Comparative measurements made at the same point in time provide hypotheses which must be tested by making sequential measurements in time.

References

Aarsen, L. W., Turkington, R. & Cavers, P. B., 1979. Neighbour relationships in grass-legume communities. II. Temporal stability and community evolution. Can. J. Bot. 57: 2695–2703.

Agnew, A. D. Q., 1961. The ecology of Juncus effusus L. in North Wales. J. Ecol. 49: 83–102.

Albertson, F. W. & Tomanek, G. W., 1965. Vegetation changes during a 30-year period in grassland communities near Hays, Kansas. Ecology 46: 714–720.

Allen, S. E., 1964. Chemical aspects of heather burning. J. Appl. Ecol. 1: 347–367.

Arines, J., Vieitez, E. & Mantilla, J. L. F., 1974. Estudio de la actividad biologica de Erica vagans L. Anal. Edafol. Agrobiol. 33: 689–695.

Ballester, A., Albo, J. M. & Vieitez, E., 1977. The allelopathic potential of Erica scoparia L. Oecologia 30: 55–61.

Ballester, A. & Vieitez, E., 1971. Estudio de sustancias de crecimiento aisladas de Erica cinerea L. Acta Cient. Compost. 8: 79–84.

Bannister, P., 1965. Biological flora of the British Isles. Erica cinerea L. J. Ecol. 53: 527–542.

Bannister, P., 1966. Biological flora of the British Isles. Erica tetralix L. J. Ecol. 54: 795–813.

Barclay-Estrup, P. & Gimingham, C. H., 1969. The description and interpretation of cyclical processes in a heath community. I. Vegetation change in relation to the Calluna cycle. J. Ecol. 57: 737–758.

Bartholomew, B., 1970. Bare zone between California shrub and grassland communities: the role of animals. Science 170: 1210–1212.

Bartholomew, B., 1971. Role of animals in suppression of herbs by shrubs. Science 173: 462–463.

Bell, J. N. B. & Tallis, J. H., 1973. Biological flora of the British Isles. Empetrum nigrum L. J. Ecol. 61: 289–305.

Betrem, J. C., 1929. De heidekever en zijn biologie. Tijdschr. Plziekt. 35: 150–180.

Birkeland, P. W., 1974. Pedology, weathering, and geomorphological research. Oxford University Press, New York.

Brenchley, W. E. & Warington, K., 1958. The Park Grass Plots at Rothamsted 1856–1949. Rothamsted Experimental Station, Harpenden.

Brian, M. V., Elmes, G. & Kelley, A. F., 1967. Populations of the ant Tetramorium caespitum Latreille. J. Anim. Ecol. 36: 337–342.

Brian, M. V., Hibble, J. & Stradling, D. J., 1965. Ant pattern and density in a southern English heath. J. Anim. Ecol. 34: 545–555.

Brown, R. T. & Mikola, P., 1974. The influence of fruticose soil lichens upon the mycorrhizae and seedling growth of forest trees. Acta for. fenn. 141.

Cameron, A. E., McHardy, J. W. & Bennett, A. H., 1944. The heather beetle (Lochmaea suturalis). An enquiry into its biology and control made on behalf of the British Field Sports Society. British Field Sports Society, Petworth.

Cantlon, J. E., 1969. The stability of natural populations and their sensitivity to technology. In: G. M. Woodwell & H. H. Smith (eds.), Diversity and stability in ecological systems. Brookhaven Symp. Biol. 22: 197–205.

Carballeira, A. & Cuervo, A., 1980. Seasonal variation in allelopathic potential of soils from Erica australis L. heathland. Acta Oecol., Oecol. Plant. 1: 345–353.

Chadwick, M. J., 1960. Biological flora of the British Isles. Nardus stricta L. J. Ecol. 48: 255–267.

Clapham, A. R., Tutin, T. G. & Warburg, E. F., 1962. Flora of the British Isles. 2nd ed. Cambridge University Press, Cambridge.

Clements, F. E., 1905. Research methods in ecology. University Publ. Comp., Lincoln, Nebraska.

Coupland, R. T., 1958. The effects of fluctuations in weather upon the grasslands of the Great Plains. Bot. Rev. 24: 273–317.

Dahl, E., 1951. On the relation between summer temperature and the distribution of alpine vascular plants in the lowlands of Fennoscandia. Oikos 3: 22–52.

Daubenmire, R., 1968. Ecology of fire in grasslands. Adv. Ecol. Res. 5: 209–266.

Dempster, J. P., 1975. Animal population ecology. Academic Press, London.

Dimbleby, G. W., 1962. The development of British heathlands and their soils. Oxf. For. Mem. 23.

Drury, W. H. & Nisbet, I. C. T., 1971. Inter-relations between developmental models in geomorphology, plant ecology, and animal ecology. General Systems 16: 57–68.

Drury, W. H. & Nisbet, I. C. T., 1973. Succession. J. Arnold Arbor. 54: 331–368.

Duchaufour, P., 1977. Pédologie. I. Pédogenèse et Classification. Masson, Paris.

Egler, F. E., 1954. Vegetation science concepts. I. Initial floristic composition – a factor in old-field vegetation development. Vegetatio 4: 412–417.

Egler, F. E., 1970. The way of science. A philosophy of science for the layman. Hafner, New York.

Ernst, W., 1978. Chemical soil factors determining plant growth. In: A. H. J. Freysen & J. W. Woldendorp (eds.), The structure and functioning of plant populations, p. 155–187. North-Holland Publ. Comp., Amsterdam.

Faille, A., 1977. Action des peuplements de Calamagrostis epigeios (L.) Roth dans la dynamique des écosystèmes de la forêt de Fontainebleau. II. Influence sur quelques caractères des humus et leurs activités microbiennes. Rev. Ecol. Biol. Sol 14: 289–306.

Fisher, R. F., 1979. Possible allelopathic effects of reindeer-moss (Cladonia) on jack pine and white spruce. Forest Sci. 25: 256–260.

Fisher, R. T., 1928. Soil changes and silviculture on the Harvard Forest. Ecology 9: 6–11.

Franklin, H. J., 1948. Cranberry insects in Massachusetts. Bull. Mass. agric. Exp. Stn. 445.

Franklink, H. J., 1950. Cranberry insects in Massachusetts, Parts II–VII. Bull. Mass. agric. Exp. Stn. 445 (Continued).

Franklin, H. J., 1952. Cranberry insects in Massachusetts, Supplement. Bull. Mass. agric. Exp. Stn. 445 (Supplement).

Fricke, K., 1904. 'Licht und Schattenholzarten', ein wissenschaftlichen nicht begründetes Dogma. Zentl. ges. Forstw. 30: 315–325.

Fustec-Mathon, E., Righi, D. & Jambu, P., 1975. Influence des bitumes extraits de podzols humiques hydromorphes des Landes du Médoc sur la microflore tellurique. Rev. Ecol. Biol. Sol 12: 393–404.

Gardiner, A. S., 1968. The reputation of birch for soil improvement. A literature review. Res. Dev. Pap. For. Commn, Lond. 67.

Gersper, P. L. & Holowaychuk, N., 1970. Effects of stemflow water on a Miami soil under a beech tree: I. Morphological and physical properties. Proc. Soil Sci. Soc. Amer. 34: 779–786.

Gersper, P. L. & Holowaychuk, N., 1970. Effects of stemflow water on a Miami soil under a beech tree: II. Chemical properties. Proc. Soil Sci. Soc. Amer. 34: 786–794.

Gimingham, C. H., 1960. Biological flora of the British Isles. Calluna vulgaris (L.) Hull. J. Ecol. 48: 455–483.

Gimingham, C. H., 1972. Ecology of heathlands. Chapman & Hall, London.

Gimingham, C. H., Chapman, S. B. & Webb, N. R., 1979. European heathlands. In: R. L. Specht (ed.), Heathlands and related shrublands, A. Descriptive studies. Elsevier Sc. Publ. Comp., Amsterdam, p. 365–413.

Glass, A. D. M., 1976. The allelopathic potential of phenolic acids associated with the rhizosphere of Pteridium aquilinum. Can. J. Bot. 54: 2440–2444.

Gliessman, S. R. & Muller, C. H., 1971. The phytotoxic potential of bracken Pteridium aquilinum (L.) Kuhn. Madroño 21: 299–304.

Grant, S. A., Hunter, R. F. & Cross, C., 1963. The effects of muirburning Molinia-dominant communities. J. Br. Grassld Soc. 18: 249–257.

Gray, E. G. & Nicholson, I. A., 1957. Snow mould on upland pasture in North Scotland. Trans. Bot. Soc. Edinb. 37: 123–128.

Grieve, I. C., 1975. A study of the effects of some tree species on selected physical and morphological properties of the soil. Ph.D. thesis, University of Bristol.

Grieve, I. C., 1978. Some effects of the plantation of conifers on a freely drained lowland soil, Forest of Dean, U.K. Forestry 51: 21–28.

Grubb, P. J., 1977. The maintenance of species-richness in plant communities: the importance of the regeneration niche. Biol. Rev. 52: 107–145.

Grubb, P. J., Green, H. E. & Merrifield, R. C. J., 1969. The ecology of chalk heath: its relevance to the calcicole-calcifuge and soil acidification problems. J. Ecol. 57: 175–212.

Grubb, P. J. & Suter, M. B., 1971. The mechanism of acidification by Calluna and Ulex and the significance for conservation. In: E. Duffey & A. S. Watt (eds.), The scientific management of animal and plant communities for conservation, p. 115–133. Blackwell Sc. Publ., Oxford.

Haines, F. M., 1926. A soil survey of Hindhead Common. J. Ecol. 14: 33–71.

Handley, W. R. C., 1954. Mull and mor formation in relation to forest soils. Bull. For. Commn, Lond. 23.

Handley, W. R. C., 1963. Mycorrhizal associations and Calluna heathland afforestation. Bull. For. Commn, Lond. 36.

Hansen, K., 1976. Studies on the regeneration of heath vegetation after burning-off. Bot. Tidsskr. 60: 1–41.

Hansen, K., 1964. Studies on the regeneration of heath vegetation Dansk Bot. Arkiv 31.2, 118 pp.

Harberd, D. J., 1961. Observations on the population structure and longevity of Festuca rubra L. New Phytol. 60: 184–206.

Harris, P., 1973. Insects in the population dynamics of plants. In: H. F. van Emden (ed.), Insect plant relationships. Symp. R. Entomol. Soc. Lond. 6: 201–209.

Heal, O. W., 1979. Decomposition and nutrient release in even-aged plantations. In: E. D. Ford, D. C. Malcolm & J. Atterson (eds.), The ecology of even-aged plantations. Institute of Terrestrial Ecology, Cambridge, p. 257–291.

Huffaker, C. B. & Kennett, C. E., 1959. A ten-year study of vegetational changes associated with biological control of Klamath weed. J. Range Mgmt 12: 69–82.

Ivimey-Cook, R. B., 1959. Biological flora of the British Isles. Agrostis setacea Curt. J. Ecol. 47: 697–706.

Jarvis, P. G., 1964. Interference by Deschampsia flexuosa (L.) Trin. Oikos 15: 56–78.

Jones, L. I., 1967. Studies on hill land in Wales. Tech. Bull. Welsh Pl. Breed. Stn. 2.

Leach, W., 1931. On the importance of some mosses as pioneers on unstable soils. J. Ecol. 19: 98–102.

Lemee, G. & Arluison, M., 1976. Action inhibitrice du Cladonia rangiformis Hoffm. sur la germination de phanerogames des pelouses xérophiles. In: Etudes de Biologie Végétale. Hommage au Professeur P. Chouard, pp. 87–100. Gauthier Villars, Paris.

Lloyd, P. S., 1968. The ecological significance of fire in limestone grassland communities of the Derbyshire Dales. J. Ecol. 56: 811–826.

Lodhi, M. A. K., 1977. The influence and comparison of individual forest trees on soil properties and possible inhibition of nitrification due to intact vegetation. Amer. J. Bot. 64: 260–264.

Låg, J., 1962. Studies on the influence of some edaphic growth factors on the distribution of various forest vegetation in Norway. Adv. Front. Plant Sci. 1: 87–96.

MacDonald, J. A., 1949. Heather rhizomorph fungus in Scotland. Proc. R. Soc. Edinb. B. 63: 230–241.

McCormick, J., 1968. Succession. Via (Student publication, Graduate School of Fine Arts, Univ. of Pennsylvania), 1: 22–35, 131–132.

McVean, D. N. & Lockie, J. D., 1969. Ecology and land use in Upland Scotland. Edinburgh Univ. Press, Edinburgh.

McVean, D. N. & Ratcliffe, D. A., 1962. Plant communities of the Scottish Highlands. A study of Scottish mountain, moorland and forest vegetation. H.M.S.O., London.

Malcolm, D. C., 1975. The influence of heather on silvicultural practice – an appraisal. Scott. For. 29: 14–24.

Malin, J. C., 1956. The grassland of North America. Prolegomena to its history, with addenda. J. C. Malin, Lawrence, Kansas.

Mantilla, J. L. G., Arines, J. & Vieitez, E., 1975. Actividad biologica sobre el crecimiento y germinacion de extractos de Calluna vulgaris (L.) Hull. Ann. Edafol. Agrobiol. 34: 789–795.

Marzhan, B., 1959. Degradatsiya lesnykh pochv v chekhoslovakii. Vestnik. Sel'skokhoz. Nauki 1959: 87–98.

Miles, J., 1968. Invasion by Bryum bornholmense Winkelm. & Ruthe of heathland treated with calcium carbonate. Trans. Br. Bryol. Soc. 5: 587.

Miles, J., 1971. Burning Molinia-dominant vegetation for grazing by red deer. J. Br. Grassld Soc. 26: 247–250.

Miles, J., 1972. Experimental establishment of seedlings on a southern English heath. J. Ecol. 60: 225–234.

Miles, J., 1973a. Early mortality and survival of self-sown seedlings in Glenfeshie, Inverness-shire. J. Ecol. 61: 93–98.

Miles, J., 1973b. Natural recolonization of experimentally bared soil in Callunetum in north-east Scotland. J. Ecol. 61: 399–412.

Miles, J., 1974a. Experimental establishment of new species from seed in Callunetum in north-east Scotland. J. Ecol. 62: 527–551.

Miles, J., 1974b. Effects of experimental interference with stand structure on establishment of seedlings in Callunetum. J. Ecol. 62: 675–687.

Miles, J., 1975. Performance after six growing seasons of new species established from seed in Callunetum in north-east Scotland. J. Ecol. 63: 891–901.

Miles, J., 1978. The influence of trees on soil properties. In: Institute of Terrestrial Ecology Annual Report 1977. Institute of Terrestrial Ecology, Cambridge, p. 7–11.

Miles, J., 1979. Vegetation Dynamics. Chapman & Hall, London.

Miles, J. & Kinnaird, J. W., 1979. Grazing: with particular reference to birch, juniper and Scots pine in the Scottish Highlands. Scott. For. 33: 280–289.

Miles, J., Welch, D. & Chapman, S. B., 1978. Vegetation and management in the uplands. In: Upland land use in England and Wales. Countryside Comm., Cheltenham, Publ. CCPIII, p. 77–95.

Miles, J. & Young, W. F., 1980. The effects on heathland and moorland soils in Scotland and northern England following colonization by birch (Betula spp.). Bull. Ecol. 11: 233–242.

Miller, G. R. & Miles, J., 1970. Regeneration of heather (Calluna vulgaris (L.) Hull) at different ages and seasons in north-east Scotland. J. Appl. Ecol. 7: 51–60.

Milthorpe, F. L., 1961. The nature and analysis of competition between plants of different species. In: F. L. Milthorpe (ed.), Mechanisms in biological competition. Symp. Soc. Exp. Biol. 15: 330–355.

Morison, G. D., 1963. The heather beetle (Lochmaea suturalis Thomson). The North of Scotland College of Agriculture, Aberdeen.

Müller, P. E., 1879. Studier over Skovjord, som bidrag til skovdyrkningens theori. I. Om bøgemuld og bøgemor paa sand og ler. Tiddskr. Skovbrug 3: 1–124.

Müller, P. E., 1884. Studier over Skovjord, som bidrag til skovdyrkningens theori. II. Om muld og mor i egeskove og paa heder. Tidsskr. Skovbrug 7: 1–232.

Newman, E. I., 1978. Allelopathy: adaptation or accident? In: J. B. Harborne (ed.), Biochemical aspects of plant and animal coevolution pp. 327–342. Academic Press, London.

Newman, E. I. & Rovira, A. D., 1975. Allelopathy among some British grassland species. J. Ecol. 63: 727–737.

Niering, W. A. & Goodwin, R. H., 1974. Creation of relatively stable shrublands with herbicides arresting 'succession' on rights-of-way and pastureland. Ecology 55: 784–795.

Oinonen, E., 1967a. Sporal regeneration of bracken (Pteridium aquilinum (L.) Kuhn) in Finland in the light of the dimensions and the age of its clones. Acta For. Fenn. 83.1.

Oinonen, E., 1967b. The correlation between the size of Finnish bracken (Pteridium aquilinum (L.) Kuhn) clones and certain periods of site history. Acta For. Fenn. 83.2.

Oinonen, E., 1967c. Sporal regeneration of ground pine (Lycopodium complanatum L.) in southern Finland in the light of the dimensions and the ages of its clones. Acta For. Fenn. 83.3.

Oinonen, E., 1968. The size of Lycopodium clavatum L. and L. annotinum L. stands as compared to that of L. complanatum and Pteridium aquilinum (L.) Kuhn stands, the age of the tree stand, and the dates of fire on the site. Acta For. Fenn. 87.

Oinonen, E., 1969. The time table of vegetative spreading of the lily-of-the-valley (Convallaria majalis L.) and the wood small-reed (Calamagrostis epigeios (L.) Roth) in southern Finland. Acta For. Fenn. 97.

Page, G., 1968. Some effects of conifer crops on soil properties. Commonw. For. Rev. 47: 52–62.

Page, G., 1971. Properties of some common Newfoundland forest soils and their relation to forest growth. Can. J. For. Res. 1: 174–192.

Page, G., 1974. Effects of forest cover on the properties of some Newfoundland forest soils. Publs Can. For. Serv. 1332.

Phipps, C. R., 1930. Blueberry and huckleberry insects. Bull. Maine Agric. Exp. Stn. 356.

Pijl, L. van der, 1972. Principles of dispersal in higher plants. 2nd ed. Springer, Berlin.

Pyatt, F. B., 1967. The inhibitory influence of Peltigera canina on the germination of graminaceous seeds and the subsequent growth of the seedlings. Bryologist, 70: 326–329.

Pyatt, D. G., 1970. Soil groups of upland forests. Forest Rec., Lond. 71.

Rabotnov, T. A., 1966. Pecularities of the structure of polydominant meadow communities. Vegetatio 13: 109–116.

Rabotnov, T. A., 1974. Differences between fluctuations and successions. Examples in grassland phytocoenoses of the U.S.S.R. In: R. Knapp (ed.), Vegetation dynamics, pp. 19–24. Dr. W. Junk, The Hague.

Ramaut, J. L. & Corvisier, M., 1975. Effets inhibiteurs des extraits de Cladonia impexa Harm., C. gracilis (L.) Willd. et Cornicularia muricata (Ach.) Ach. sur la germination des graines de Pinus sylvestris L. Oecol. Plant. 10: 295–299.

Ramaut, J. L. & Thonar, J., 1972. Inhibition de la germination de differentes graines par Evernia prunastri (L.) Ach. I. An. Quim. 68: 575–595.

Rastorova, O. G., 1974. O vzaimodeistvii drevesnykh porod s pochvami. Tr. Petergof. Biol. Inst. Leningr. Gos. Univ. 23: 53–74.

Rice, E. L., 1974. Allelopathy. Acad. Press, New York.

Ridley, H. N., 1930. The dispersal of plants throughout the world. Reeves, Ashford.

Ritchie, J. C., 1955. Biological flora of the British Isles. Vaccinium vitis-idaea L. J. Ecol. 43: 701–708.

Ritchie, J. C., 1966. Biological flora of the British Isles. Vaccinium myrtillus L. J. Ecol. 44: 291–299.

Robinson, R. K., 1971. Importance of soil toxicity in relation to the stability of plant communities. In: E. Duffey & A. S. Watt (eds.), The scientific management of animal and plant communities for conservation, p. 105–113. Blackwell Sc. Publ., Oxford.

Robinson, R. K., 1972. The production by roots of Calluna vulgaris of a factor inhibitory to growth of some mycorrhizal fungi. J. Ecol. 60: 219–224.

Rundel, P. W., 1978. The ecological role of secondary lichen substances. Biochem. Syst. Ecol. 6: 157–170.

Sagar, G. R. & Harper, J. L., 1961. Controlled interference with natural populations of Plantago lanceolata, P. major and P. media. Weed Res. 1: 163–176.

Salas, C., Ballester, A. & Vieitez, E., 1973. Estudio biologico y quimico de Erica umbellata L. Anal. Edafol. Agrobiol. 32: 807–814.

Scurfield, G., 1954. Biological flora of the British Isles. Deschampsia flexuosa (L.) Trin. J. Ecol. 42: 225–233.

Smidt, J. T. de, 1977. Interaction of Calluna vulgaris and the heather beetle (Lochmaea suturalis). In: R. Tüxen (ed.) Vegetation und Fauna, p. 179–186, Cramer, Vaduz.

Smidt, J. T. de, 1979. Origin and destruction of northwest European heath vegetation. In: O. Wilmanns & R. Tüxen (eds.), Werden und Vergehen von Pflanzengesellschaften, p. 411–435. Cramer, Vaduz.

Specht, R. L., 1979. Heathlands and related shrublands of the world. In: R. L. Specht (ed.), Heathlands and related shrublands, A. Descriptive studies p. 1–18. Elsevier Sc. Publ. Comp., Amsterdam.

Stevenson, F. J., 1966. Lipids in soil. J. Am. Oil Chem. Soc. 43: 203–210.

Stone, E. L., 1975. Effects of species on nutrient cycles and soil change. Phil. Trans. Roy. Soc. B271: 149–162.

Stowe, L. G., 1979. Allelopathy and its influence on the distribution of plants in an Illinois old-field. J. Ecol. 67: 1065–1085.

Summerfield, R. J., 1972. Biological inertia – an example. J. Ecol. 60: 793–798.

Taher, M. M. & Cooke, R. C., 1975. Shade-induced damping-off in conifer seedlings. I. Effects of reduced light intensity on infection by necrotrophic fungi. New Phytol. 75: 567–572.

Ten Khak Mun & Fedorova, L. V., 1975. Role of tall herbaceous plants in soil formation. Soviet Soil Sci. 7: 41–45.

Troedsson, T., 1972. Betydelsen av markens egenskaper i modern samhällsplanering. K. Skogs- o. LandbrAkad. Tidskr. 111: 250–262.

Turkington, R. A., Cavers, P. B. & Aarssen, L. W., 1977. Neighbour relationships in grass-legume communities. 1. Interspecific contacts in four grassland communities near London, Ontario. Can. J. Bot. 55: 2701–2711.

Tüxen, R., 1957. Die Schrift des Bodens. Angew. Pfl. Soz. Stolzenau 14.

Tüxen, R., 1975. La Lüneburger Heide (Lande de Lunebourg), origin et fin d'un paysage endémique. Coll. Phytosoc. 2, La végétation des landes d'Europe occidentale. p. 379–396 Lille, Cramer, Vaduz.

Vaartaja, O., 1962. On the relationship of fungi to survival of shaded tree seedlings. Ecology 43: 547–549.

Viro, P. J., 1974. Effects of forest fire on soil. In: T. T. Kozlowski & C. E. Ahlgren (eds.), Fire and ecosystems, p. 1–45. Acad. Press, New York.

Vogl, R. J., 1974. Effects of fire on grasslands. In: T. T. Kozlowski & C. E. Ahlgren (eds.), Fire and ecosystems, p. 139–194. Acad. Press, New York.

Vries, D. M. de, 1954. Constellation of frequent herbage plants, based on their correlation in occurrence. Vegetatio 5–6: 105–111.

Watt, A. S., 1955. Bracken versus heather. A study in plant sociology. J. Ecol. 43: 490–506.

Watt, A. S., 1960. Population changes in acidophilous grass-heath in Breckland, 1936–57. J. Ecol. 48: 605–629.

Weatherell, J., 1953. The checking of forest trees by heather. Forestry 26: 37–41.

Weaver, J. E. & Albertson, F. W., 1956. Grasslands of the Great Plains, Johnsen Publ. Comp., Lincoln, Nebraska.

Webb, D. A., 1955. Biological flora of the British Isles. Erica mackaiana Bab. J. Ecol. 43: 319–330.

Welch, D., 1966. Biological flora of the British Isles. Juncus squarrosus L. J. Ecol. 54: 535–548.

Welch, D., 1974. The floristic composition of British upland vegetation in relation to grazing. Land 1: 59–68.

Williams, E. D., 1978. Botanical composition of the Park Grass plots at Rothamsted 1856–1976. Rothamsted Exp. Stat. Harpenden.

Wilson, F., 1960. A review of the biological control of insects and weeds in Australia and Australian New Guinea. Tech. Commun. Commonw. Inst. Biol. Control 1.

Zinke, P. J., 1962. The pattern of influence of individual forest trees on soil properties. Ecology 43: 130–133.

Accepted 10.7.1981.

Zur Sukzession der mediterranen Vegetation auf der Insel Lokrum bei Dubrovnik*

Lj. Ilijanić[1] & S. Hećimović[2]
[1] *Botanisches Institut der Naturwissenschaftlichen Fakultät der Universität, Marulićev trg 20/II, 41000 Zagreb, Jugoslawien*
[2] *Biologisches Institut, 50000 Dubrovnik, Jugoslawien*

Keywords: Dauerflächen, Mediterrane Vegetation, Naturschutz, *Quercetum ilicis*, Sukzession, Vegetationskartierung

Abstract

Through vegetation mapping in 1959 and 1979 of the eumediterranean Adriatic island of Lokrum (under nature conservation since 1948) a progressive succession could be indicated. Progression is most pronounced on former grasslands and garrigue, the latter still being assigned to the *Orno-Quercetum ilicis myrtetosum*. In the S part of the island an extension of the *Orno-Quercetum ilicis typicum* could be indicated on areas where the first map showed *Orno-Quercetum ilicis myrtetosum*.

For further detailed quantitative investigations of dynamics and structure, permanent quadrats were established. Degradation phases, with typical own plant species, will be maintained through adaequate human interference, such as small-scale chopping and strictly controlled burning.

Einleitung

Die Insel Lokrum (0,72 km²) in der unmittelbaren Nähe von Dubrovnik (Kroatien) liegt pflanzengeographisch im südlichen Gebiet der eumediterranen immergrünen *Quercion ilicis*-Zone des ostadriatischen Küstenlandes im Sinne von Horvatić (1957, 1963b, vgl. auch Horvat, Glavač & Ellenberg, 1974). In klimatischer Hinsicht zeichnet sich das Gebiet mit einer durchschnittlichen Jahrestemperatur von über 16 °C (Dubrovnik 16,4 °C), und mit verhältnismäßig sehr hohen Niederschlägen von über 1 200 mm aus (Dubrovnik 1 256 mm jährlich).

Die klimazonale Vegetation des Gebietes, der *Orno-Quercetum ilicis*-Wald, ist auf großen Flächen des Küstenlandes vernichtet bzw. in regressiver Sukzession in verschiedene Degradationsstadien (Garrigue-, Trockenrasen- und Steintriften-Gesellschaften) umgewandelt. Die Degrada-

tion fand auch auf Lokrum statt, war aber nie so katastrophal und verursachte nicht solche Folgeerscheinungen wie in vielen anderen Gebieten (vgl. auch den Beitrag von Visiani (1863), der die Vegetation der Insel schon vor 120 Jahren untersuchte), so daß die Insel mit ihrer dunkelgrünen Farbe, gesehen gegen die hellen Stadtmauern Dubrovniks, die alte Stadt wie eine Perle verziert.

Seit 1948 steht die Insel unter Naturschutz, und auf dem grössten Teil der Insel sind keine unmittelbaren menschlichen Einflüsse mehrgestattet. Auf solchen Flächen entwickelte sich die Vegetation in einer progressiven Sukzession in Richtung auf die potenzielle natürliche Schlußgesellschaft, den Steineichenwald (*Orno-Quercetum ilicis*). Vor zwei Jahrzehnten (1959) wurden auf der Insel eingehende Vegetationsuntersuchungen und -kartierungen durchgeführt (Horvatić, 1962). Die Resultate sind nur teilweise veröffentlicht, die Vegetationskarte liegt im Manuskript vor.

Im Jahr 1979 wurden die Untersuchungen fortgesetzt und die Kartierungen wiederholt, um die Veränderungen festzustellen, die in der Pflanzen-

---

* Nomenklatur der Pflanzensippen nach Ehrendorfer (1973), Nomenklatur der Pflanzengesellschaften nach Horvatić (1963a).

Vegetatio 46, 75–81 (1981).   0042-3106/81/0462-0075/$1.40.

decke im Zeitraum von 20 Jahren entstanden sind. Hier möchten wir die wichtigsten Resultate kurz darstellen.

Methodik

Die Vegetationsuntersuchungen wurden nach den Prinzipien und Methoden der pflanzensoziologischen Schule Zürich-Montpellier durchgeführt (Braun-Blanquet, 1964) und die festgestellten Pflanzengesellschaften auf der Karte im Maßstab 1:2 000 aufgenommen.

Kurzer Überblick über die Vegetationseinheiten

Die gesamte Vegetation der Insel wird eingehend mit der pflanzensoziologischen Tabellen an anderer Stelle dargestellt werden. Deshalb verzichten wir hier auf eine ausführlichere pflanzensoziologische Beschreibung der Gesellschaften, umso mehr, als die hier angegebenen Vegetationseinheiten, außer dem *Limonietum anfracti,* schon bekannt sind und früher beschrieben wurden (vgl. Horvatić, 1957, 1958, 1962, 1963).

Wir bringen nur einen kurzen Überblick über die auf der Insel festgestellten Pflanzengesellschaften nach der systematischen Zugehörigkeit, wie sie von Horvatić (1963a) gefaßt wurden:

Klasse: *Quercetea ilicis*
  Ordnung: *Quercetalia ilicis*
    Verband: *Quercion ilicis*
      Ass.: *Orno-Quercetum ilicis*
        Subass.: *typicum*
        Subass.: *myrtetosum*
            Fazies *Pinus halepensis*
  Ordnung: *Cisto-Ericetalia*
    Verband: *Cisto-Ericion*
      Ass.: *Erico-Calicotometum infestae*

Klasse: *Thero-Brachypodietea*
  Ordnung: *Thero-Brachypodietalia*
    Verband: *Cymbopogo-Brachypodion ramosi*
      Ass.: *Oryzopsetum miliaceae*
    Verband: *Vulpio-Lotion*
      Ass.: *Ornithopodo-Vulpietum*
      Ass.: *Gastridio-Brachypodietum ramosi*
Klasse: *Crithmo-Staticetea*
  Ordnung: *Crithmo-Staticetalia*
    Verband: *Crithmo-Staticion*
      Ass.: *Limonietum anfracti* Ass.nov. (n.n.)
          (= *Plantagini-Staticetum cancellatae* H-ić p.p.)

Klasse: *Asplenietea rupestria*
  Ordnung: *Asplenietalia glandulosi*
    Verband: *Centaureo-Campanulion*
      Ass.: *Seslerio-Putorietum calabricae*
      Ass.: *Asplenio-Cotyledonetum horizontalis*
Klasse: *Chenopodietea*
  Ordnung: *Chenopodietalia*
    Verband: *Diplotaxidion*
      Ass.: *Tribulo-Amaranthetum*
      Ass.: *Fumario-Cyperetum rotundi*
    Verband: *Chenopodion muralis* (fragm.)

Die Veränderungen der Vegetation 1959–1979

Die in den letzten zwei Jahrzehnten entstandenen Veränderungen in der Pflanzendecke der Insel sind aus den beigegebenen vergleichenden Vegetationskarten zu ersehen (Abb. 1 und 2).

Auf den Vegetationskarten sind Parkanlagen, Olivengärten und die auf solchen Flächen entwickelten Trockenrasen-, Unkraut- und Ruderal-Vegetation sowie die Weg- und Gebäudeflächen weggelassen worden (weisse Flächen). Es werden hier nur die Flächen betrachtet, auf denen sich die Vegetation in der genannten Zeit mehr oder weniger ohne deutliche menschliche Eingriffe entwickelte. Besonders gilt das für die nördliche Hälfte der Insel, wo viele Touristen, die besonders zahlreich im Sommer die Insel besuchen, die Spazierwege nicht verlassen.

Wie zu erwarten war, traten die größten Veränderungen dort ein, wo früher eine offene und niedrige Rasen- und Garrigue-Vegetation entwickelt war. Alle kleinen Rasenbestände und die größten Flächen der damaligen Übergangsbestände zwischen *Erico-Calicotometum* und *Orno-Quercetum ilicis myrtetosum,* die 1959 noch existierten (Abb. 1), wandelten sich in *Orno-Quercetum ilicis myrtetosum*-Macchie um (Abb. 2).

Auf der größten Garrigue-Fläche, die auf der Südwestseite der N Inselhälfte im Jahr 1959 vorhanden war, schritt die Vegetation – außer kleiner Flächen, auf denen noch heute die *Erico-Calicotometum*-Garrigue existiert – in diesem Zeitraum bis zum Übergangsstadium zwischen dem *Erico-Calicotometum* und dem *Orno-Quercetum ilicis myrtetosum* fort.

Ähnlich sieht man die fortschreitende Entwicklung auf der S und SE spitze der Insel, wo Macchienpflanzen in die frühere Garrigue-Vegetation, die hier größtenteils in der Fazies mit *Pinus halepensis* vertreten ist, vorgedrungen sind.

Die progressive Vegetationsentwicklung verläuft hier aber langsamer als auf der nördlichen Hälfte der Insel. Die Ursache dafür ist nach unserer Meinung die stärker erodierte Kalkunterlage sowie der sehr starke Windeinfluß und die damit verbundene Sprühversalzung, was auch aus der Breite der Halophyten-Zone geschlossen werden kann (vgl. Abb. 1 und 2). Die Sukzession schreitet jedoch auch hier weiter fort, und die Garrigue-Flächen werden im Vergleich mit den früheren kleiner. Die Rasenvegetation, die vor zwanzig Jahren bereits nur sehr kleine Flächen bedeckte, ist ganz verschwunden.

Im S Teil der Insel konnte auch eine Vergrösserung des *Orno-Quercetum ilicis typicum*-Bestandes festgestellt werden.

Die Festlegung der Dauerflächen

Auf den dargestellten Vegetationskarten sind zuerst die größeren bzw. mehr qualitativen Veränderungen festzustellen, d.h. die Vegetationsveränderungen, die pflanzensoziologisch-systematisch ausgedrückt werden können. Deswegen sieht man auf der Vegetationskarte nicht die Veränderungen, die in den früheren Macchienbeständen – außer der genannten kleineren Vergrößerung des *Orno-Quercetum typicum* – entstanden sind, da sie pflanzensoziologisch noch immer derselben Vegetationseinheit, d.h. dem *Orno-Quercetum ilicis myrtetosum* im Sinne von Horvatić (1963a) angegliedert werden können. Veränderungen der Dynamik, der Struktur und der Mengenverhältnisse können dabei nicht genau bzw. nur teilweise verfolgt werden. Um diese zu erfassen, sind langjährige Untersuchungen auf kleineren Dauerflächen notwendig. Den besten Einblick in die Vegetationssukzession auf größeren Flächen gibt jedoch die Kombination beider Methoden, wie Londo (1978) mit Recht betont.

Schon vor 15 Jahren haben wir die Festlegung von Dauerflächen für solche Untersuchungen vorgeschlagen (Ilijanić, 1965, vgl. auch Ilijanić & Meštrov, 1975). Dabei würde auch die Insel Lokrum genannt. Die Insel ist für solche Untersuchungen besonders geeignet, da sie unter Naturschutz steht, leicht zugänglich ist und auf der Insel

im Botanischen Garten des Biologischen Instituts eine Wohn- und Arbeitsmöglichkeit besteht.

Es wurden bisher vier Dauerflächen von einer Grösse von ca. 1 ha ausgewählt und die Vegetation hier phytozoenologisch aufgenommen. Um alle wichtigen Vegetationseinheiten und Expositionen der Insel zu erfassen, haben wir noch drei weitere Dauerflächen vorgesehen. Die Zusammensetzung der Vegetation auf den Dauerflächen ist in den Tabellen 1 und 2 dargestellt.

Auf der ersten Dauerfläche sind die *Orno-Quercetum ilicis myrtetosum* Macchie, die *Erico-Calicotometum*-Garrigue (sehr kleine Fläche) und auf

Tabelle 1. Die Zusammensetzung der Vegetation auf der Dauerfläche Nr. 1, in 1979 Synthetische Tabelle (10 Aufn.). Die Arten mit Stetigkeit <60% sind weggelassen (vollständige Tabellen sind bei den Autoren zu befragen).
A = *Erico-Calicotometum infestae* (2 Aufn.).
B = Übergang *Erico-Calicotometum → Orno-Quercetum ilicis myrtetosum* (5 Aufn.).
C = *Orno-Quercetum ilicis myrtetosum* (3 Aufn.).

Obere Strauchschicht bis 4 m	A	B	C
Phillyrea latifolia	2^{1-2}	5^{2-3}	3^{2-3}
Coronilla emerus subsp. emeroides	2^{+-1}	5^{+-1}	3^{+-1}
Juniperus phoenicea	2^{2-3}	5^{+-1}	1^+
Arbutus unedo	1^+	4^{+-3}	3^{2-3}
Cistus incanus	2^{1-3}	4^{+-1}	1^+
Lonicera implexa Ait	1^+	3^+	3^+
Erica manipuliflora	2^{2-3}	3^{+-2}	1^1
Calicotome villosa (= infesta)	1^+	5^{+-1}	1^+
Juniperus oxycedrus	2^+	3^{+-1}	1^+
Pistacia lentiscus	–	4^{+-2}	3^1
Myrtus communis	–	3^{1-2}	3^1
Smilax aspera	1^+	2^+	2^+
Rhamnus alaternus	1^+	1^+	2^+
Viburnum tinus	1^+	–	3^+

Untere Strauchschicht und Krautschicht bis 1 m			
Brachypodium retusum	2^4	5^{1-4}	3^1
Teucrium flavum	2^{+-1}	5^{+-1}	2^+
Asparagus acutifolius	1^+	4^+	2^+
Gladiolus illyricus	1^+	5^+	1^+
Ruscus aculeatus	–	4^{1-2}	3^{+-3}
Teucrium polium	2^+	3^+	–
Rubia peregrina	–	3^+	2^+
Bromus erectus	1^+	2^{+-3}	1^+
Allium dalmaticum	2^+	2^+	–
Coronilla emerus subsp. emeroides	–	2^+	2^+
Viburnum tinus	–	1^+	3^+
Salvia officinalis	–	1^+	3^+
Phillyrea latifolia	–	1^+	2^+
Myrtus communis	–	1^+	2^+
Centaurea glaberrima	2^+	–	–

78

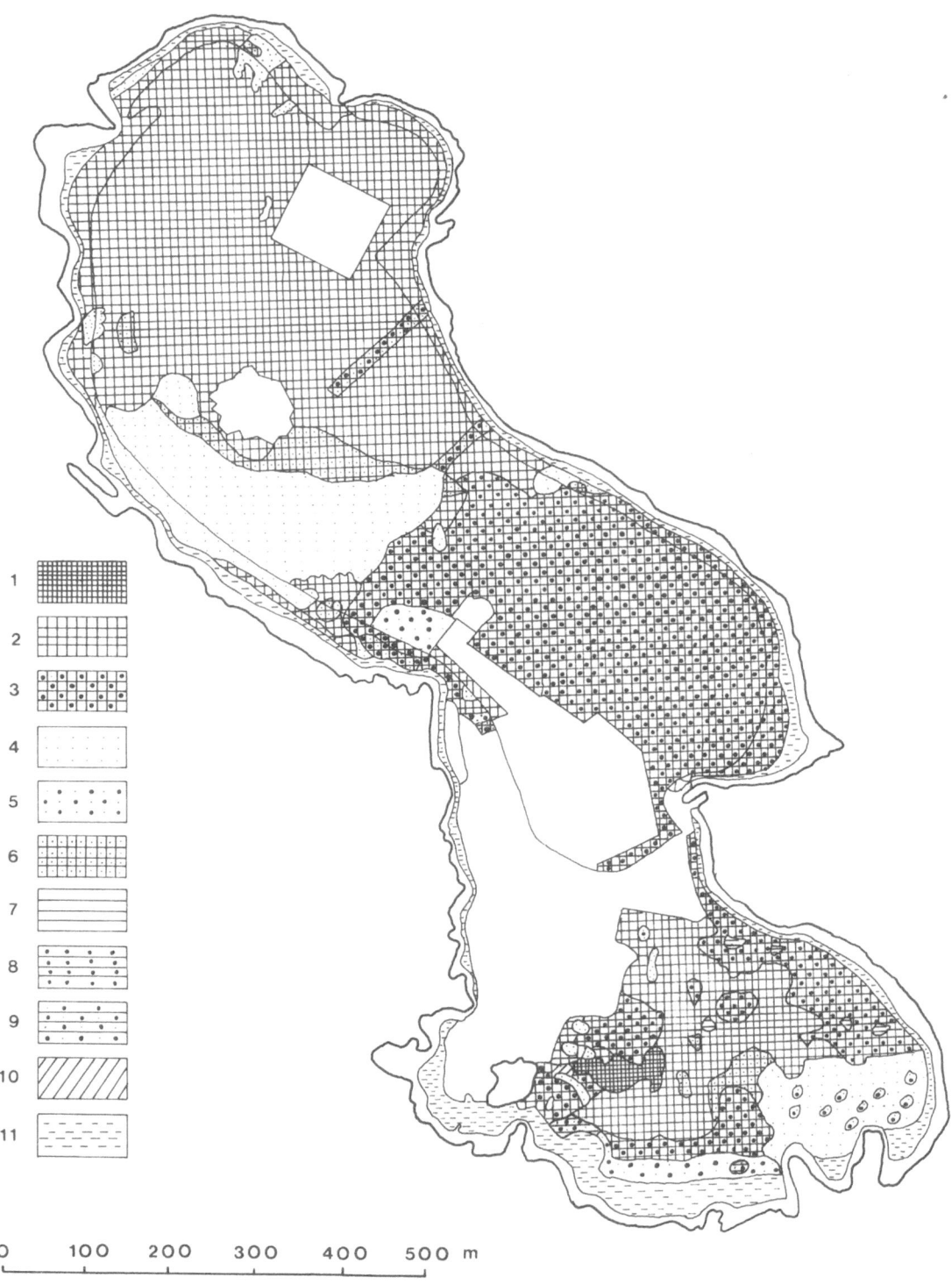

Abb. 1. Vegetationskarte der Insel Lokrum im Jahr 1959. Legende: 1. *Orno-Quercetum ilicis typicum,* 2. *Orno-Quercetum ilicis myrtetosum,* 3. *Orno-Quercetum ilicis myrtetosum* Fazies *Pinus halepensis,* 4. *Erico-Calicotometum infestae,* 5. *Erico-Calicotometum infestae* Fazies *Pinus halepensis,* 6. Übergang *Erico-Calicotometum → Orno-Quercetum ilicis myrtetosum,* 7. *Gastridio-Brachypodietum ramosi,* 8. *Gastridio-Brachypodietum ramosi* Fazies *Pinus halepensis,* 9. Übergang *Gastridio-Brachypodietum → Erico-Calicotometum* Fazies *Pinus halepensis.* 10. *Oryzopsetum miliaceae,* 11. *Limonietum anfracti.*

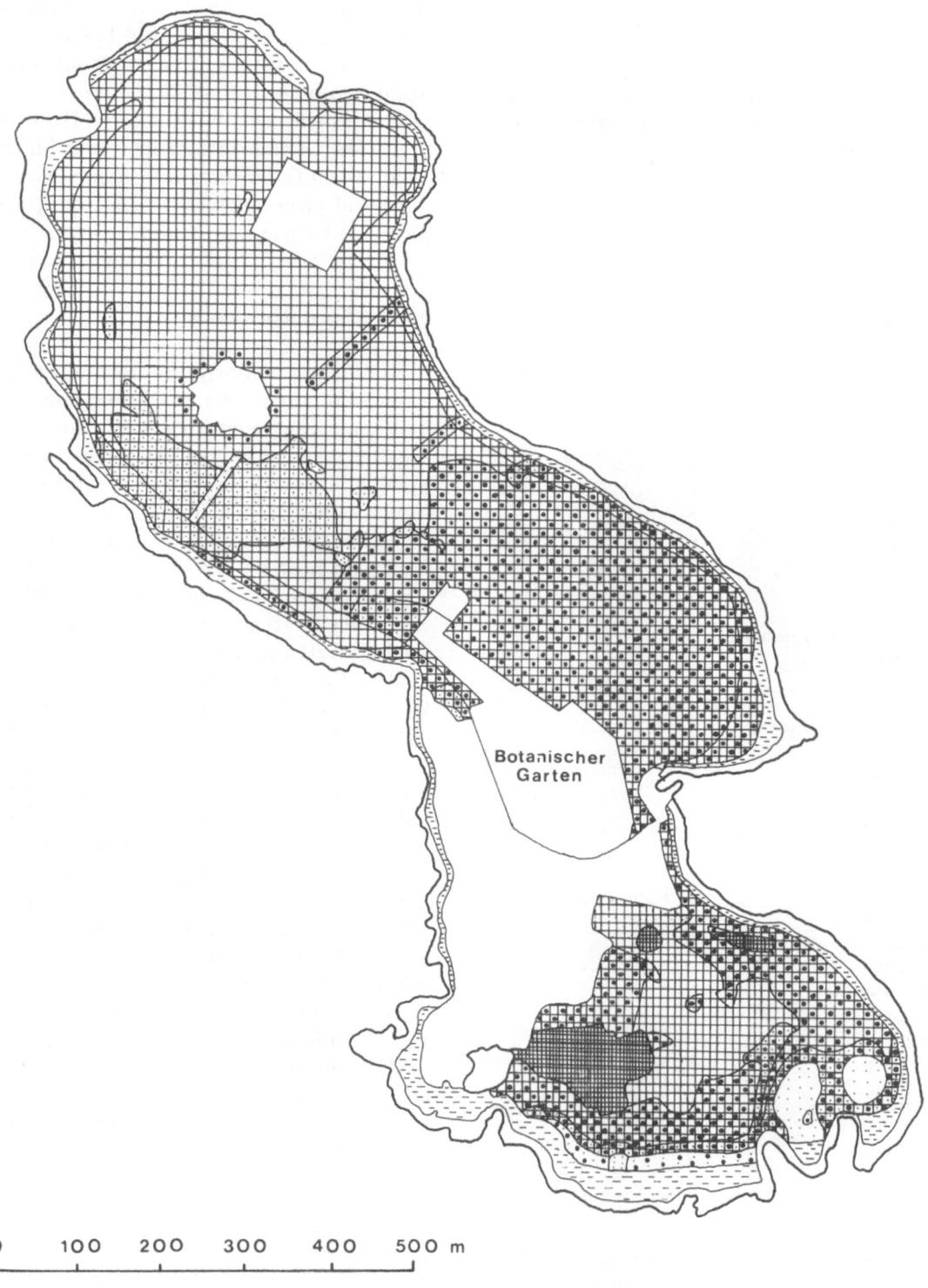

Botanischer
Garten

0   100   200   300   400   500 m

Abb. 2. Vegetationskarte der Insel Lokrum im Jahr 1979. Legende wie auf der Abbildung 1.

Tabelle 2. Die Zusammensetzung der Vegetation auf den Dauerflächen 2–4.

A. Dauerfläche Nr. 2, *Orno-Quercetum ilicis myrtetosum* (9 Aufn.).

B. Dauerfläche Nr. 3, Ibid. Fazies mit *Pinus halepensis* (9 Aufn.).

C. Dauerfläche Nr. 4, *Orno-Quercetum ilicis typicum* (5 Aufn.).

D. Ibid. *Orno-Quercetum ilicis myrtetosum* Fazies mit *Pinus halepensis* (2 Aufn.).

Baumschicht	A –6 m	B –15 m	C –10 m	D –10 m
Phillyrea latifolia	9$^{1\ 4}$	6$^{1\ 2}$	3$^{+\ 4}$	2^{1}
Arbutus unedo	8$^{+\ 3}$	4^{1}	3$^{+\ 1}$	1^{1}
Quercus ilex	6$^{+\ 1}$	3$^{+\ 3}$	5$^{2\ 4}$	2$^{+\ 1}$
Juniperus oxycedrus	6$^{+\ 1}$	–	–	–
Erica arborea	2^{+}	3$^{+\ 3}$	3$^{+\ 1}$	–
Pinus halepensis	–	7$^{1\ 4}$	1^{+}	2$^{2\ 3}$
Laurus nobilis	–	–	5$^{+\ 3}$	1^{3}

Obere Strauchschicht (–4 m)				
Myrtus communis	9$^{1\ 4}$	9$^{1\ 4}$	2^{+}	2$^{+\ 1}$
Smilax aspera	9$^{1\ 2}$	9$^{+\ 2}$	4$^{+\ 1}$	2$^{+\ 1}$
Phillyrea latifolia	9$^{1\ 2}$	9$^{+\ 3}$	4$^{+\ 3}$	2$^{+\ 3}$
Viburnum tinus	9$^{+\ 2}$	9$^{+\ 2}$	4$^{+\ 1}$	2$^{+\ 2}$
Pistacia lentiscus	9$^{+\ 1}$	9$^{+\ 3}$	2^{+}	2$^{1\ 2}$
Arbutus unedo	9$^{+\ 1}$	7$^{-\ 1}$	2^{+}	1^{+}
Lonicera implexa	9^{+}	6^{+}	2^{+}	2^{+}
Coronilla emerus ssp. emeroides	9$^{+\ 1}$	–	2^{+}	1^{+}
Erica arborea	4$^{+\ 1}$	8$^{+\ 2}$	2^{+}	1^{+}
Quercus ilex	3^{+}	–	5$^{+\ 1}$	1^{+}
Laurus nobilis	2^{+}	9$^{1\ 4}$	5$^{+\ 2}$	2$^{+\ 1}$
Frangula rupestris	–	8$^{+\ 1}$	–	–
Fraxinus ornus	–	6$^{+\ 1}$	–	–
Clematis flammula	–	–	–	2^{+}
Hedera helix	–	–	–	2^{+}

Untere Strauch- und Krautschicht (–1 m)				
Ruscus aculeatus	9$^{+\ 4}$	9$^{1\ 3}$	5$^{3\ 4}$	2$^{2\ 3}$
Viburnum tinus	9$^{+\ 1}$	9$^{+\ 1}$	2^{+}	2$^{+\ 1}$
Rubia peregrina	9$^{+\ 1}$	9^{+}	3^{+}	2$^{+\ 1}$
Myrtus communis	9^{+}	9^{+}	–	2^{+}
Phillyrea latifolia	9^{-}	–	4^{+}	–
Asparagus acutifolius	6^{+}	–	3^{+}	2$^{+\ 1}$
Coronilla emerus subsp. emeroides	6^{-}	4^{+}	–	–
Quercus ilex	6^{+}	5^{+}	5^{+}	2^{+}
Smilax aspera	5^{-}	–	2^{+}	–
Hedera helix	1^{-}	4$^{+\ 1}$	3$^{+\ 1}$	1^{+}
Laurus nobilis	–	9$^{+\ 1}$	5$^{+\ 1}$	2^{1}
Fraxinus ornus	–	6^{+}	–	–
Frangula rupestris	–	6^{+}	–	–
Famus communis	–	–	5$^{+\ 2}$	2$^{+\ 1}$

dem größten Teil der Fläche ein Übergang zwischen Garrigue und Macchie (*Erico-Calicotometum* – *Orno-Quercetum ilicis myrtetosum*) vertreten.

Die zweite und die dritte Dauerfläche sind pflanzensoziologisch einheitlicher. Die zweite ist mit dem *Orno-Quercetum ilicis myrtetosum* bewachsen, das schon als Niederwald betrachtet werden darf und die dritte mit einem *Pinus halepensis*-Wald (*Orno-Quercetum ilicis myrtetosum*, Fazies von *Pinus halepensis* im Sinne von Horvatić.

Auf der vierten Dauerfläche sind Steineichenwald, *Orno-Quercetum ilicis typicum* vertreten, dabei auf einem kleineren Teil *Orno-Quercetum ilicis* in der Fazies mit *Pinus halepensis*.

Naturschutzmaßnahmen

Durch die progressive Sukzession werden viele Arten verdrängt, die an Degradationsstadien wie Trockenrasen- Steintriften- und Garrigue-Gesellschaften gebunden sind. Das gilt auch für die prachtvollen Bestände mit *Pinus halepensis*. Der Nachwuchs dieser heliophilen Art kann sich im tiefen Schatten der Macchie nicht entwickeln und findet nur noch an der Meeresküste Zuflucht, wo *Pinus halepensis* als salzertragende Pflanze am Rand der Macchie genug Licht bekommt und konkurrenzfähiger ist.

Deswegen schlagen wir vor, daß auf bestimmten, kleineren Flächen durch entsprechende Eingriffe (regelmäßigen Holzschlag und eventuell auch streng kontrollierter Brand sowie Mahd) die gewünschten Degradationsstandien als Dauergesellschaften aufrechterhalten bleiben.

In diesem Falle könnte diese wunderschöne Insel als ausgezeichnetes wissenschaftliches und Unterrichtsobjekt noch besser als heute dienen, und mit seiner bereits bestehenden sehr wertvollen Botanischen Sammlung lebender Pflanzen aus den wärmeren Gebieten der Erde noch mehr Besucher das ganze Jahr hindurch anziehen.

Anderenfalls wird die Vegetationsentwicklung zur Schlußgesellschaft, d.h. zum *Orno-Quercetum ilicis*-Wald führen und die Insel wäre nur ein Beispiel für die Artenarmut des immergrünen eumediterranen Steineichenwaldes.

Literatur

Braun-Blanquet, J., 1964. Pflanzensoziologie. 3. Aufl. Springer, Wien-New York.

Ehrendorfer, F. (ed.), 1973. Liste der Gefäßpflanzen Mitteleuropas. 2. Aufl. Fischer, Stuttgart.

Horvat, I., Glavač, V. & Ellenberg, H., 1974. Vegetation Südosteuropas. Geobot. sel. IV, pp. 768, Fischer, Stuttgart.

Horvatić, S., 1957. Pflanzengeographische Gliederung des Karstes Kroatiens und der angrenzenden Gebiete Jugoslawiens. Acta Bot. Croat. 16: 33–61.

Horvatić, S., 1962. Beiträge zur Kenntnis der Vegetation des südkroatischen Küstenlandes (Vorl. Bericht über die Untersuchungen im Jahre 1959, kroat.). Ljetopis Jugoslav. Akad. Znan. Umjetn. 66: 302–308.

Horvatić, S., 1963a. Carte des groupements végétaux de l'ile de Pag avec un aperçu général des unité végétales du Littoral Croate (en croate avec un résumé en français). Prirodoslov. istraživ. Jugoslav. Akad. Znan. Umjetn. 33, p. 1 184, Zagreb.

Horvatić, S., 1963b. Pflanzengeographische Stellung und Gliederung des ostadriatischen Küstenlandes im Lichte neuesten phytozoenologischen Untersuchungen (Kroatisch mit deutscher Zusammenfass.). Acta Bot. Croat. 22: 27–81.

Ilijanić, Lj., 1965. Die Notwendigkeit zur Festlegung von Dauerflächen und Ihre Bedeutung für Untersuchungen der Pflanzendecke Jugoslawiens (Kroatisch mit deutscher Zusammenfass.). Acta Bot. Croat. 24: 83–90.

Ilijanić, Lj., Meštrov, M., 1975. Dauerflächen für langfristige Ökosystemforschungen. Ekologija (Beograd) 10: 107–113.

Londo, G., 1978. Möglichkeiten zur Anwendung von Vegetationskundlichen Untersuchungen auf Dauerflächen. Vegetatio 38: 185–190.

Visiani, R., 1863. Sulla vegetazione e sul clima dell'isola di Lacroma in Dalmazia. Trieste.

Accepted 15.4.1981.

Processus dynamiques de reconstitution dans la série du *Quercus ilex* en Corse

C. Allier[1] & A. Lacoste[2]*

[1] *Laboratoire de Botanique, FST Nice et Laboratoire de Taxonomie Végétale Expérimentale et Numérique associé au CNRS, Université de Paris-Sud, bât. 362, 91405 Orsay Cedex, France*
[2] *Laboratoire de Taxonomie Végétale Expérimentale et Numérique associé au CNRS, Université de Paris-Sud, bât. 362, 91405 Orsay Cedex, France*

Keywords: *Quercus ilex* series, Permanent plots, Transects, Diachronic study

Abstract

The use of permanent plots in representative sites of various phytosociologically defined stages of the *Quercus ilex* series in NW Corsica allows us to analyse the natural dynamical process occurring in the development sequence. The regular floristic analysis of stands and transects (mainly using a linear cover measure) during 5 yr not only shows variations of composition and structure but also results in a diachronic understanding of the temporal relations between the stages. The study of forest plots (structure, number and development of seedlings) enables us to generalize the principle of ponctual and cyclic self-transformation of climatic and subclimatic stages.

Introduction

L'objet de la présente étude concerne l'évolution des formations ligneuses de l'étage mésoméditerranéen (Ozenda, 1975) sur silice s'insérant dans la série du *Quercus ilex*. Initiées dès 1973 dans le cadre des actions interdisciplinaires concertées de la Délégation Générale à la Recherche Scientifique et technique (D.G.R.S.T., Comités 'Equilibres et Lutte Biologiques' puis 'Gestion des Ressources Naturelles Renouvelables'), ces recherches portaient plus spécialement sur le thème 'maquis' en Corse nord-occidentale (Bassin du Fango). Cette formation arbustive dense est l'un des éléments majeurs des paysages de l'île dont l'importance économique, réelle ou potentielle, restait à analyser. Toutefois le maquis ne pouvait être logi-

quement appréhendé indépendamment des stades antérieurs ou ultérieurs de la série évolutive et sans une définition préalable de leur statut phytosociologique (Allier & Lacoste, 1980).

Conditions générales de l'étude

Localisation et contexte phytosociologique

Le Bassin du Fango, vaste unité naturelle de plus de 300 km², s'ouvre au NE dans le Golfe de Galéria, entre Calvi et Porto. Il est développé essentiellement dans les rhyolites ignimbritiques du complexe du Cinto, lequel ferme le fond de la vallée à l'est.

La majeure partie du bassin jusqu'à 600 m d'altitude est le domaine du maquis ou de la forêt de *Quercus ilex* (Forêt de Tetti, Forêt du Fango, Forêt du Filosorma). Le laboratoire de Pirio est situé au coeur de de domaine.

Le contexte retenu s'avère exemplaire à plusieurs titres:

Le nombre d'unités phytosociologiques (syntaxons) constituant la série dynamique est élevé, six

* Avec la collaboration de Mme D. Chiaverini pour l'illustration. Il nous est agréable de remercier ici l'A.P.E.E.M. (Association pour l'Etude Ecologique du Maquis) pour la possibilité qui nous a été donnée d'utiliser les installation du laboratoire de Pirio, ce qui a grandement facilité les campagnes successives de terrain.

Vegetatio 46, 83–91 (1981). 0042-3106/81/0462-0083/$1.80.
© Dr W. Junk Publishers, The Hague.

bien caractérisées (Tableau 1) sans prendre en compte dans les stades initiaux les groupements cryptogamiques sur sols pelliculaires. Ces unités s'insèrent dans une dynamique orthogénétique sans ramifications ni boucles.

Malgré la variation altitudinale (de 0 à 600-700 m), chaque formation reste constante, au niveau de l'étage mésoméditerranéen, sur l'ensemble du bassin. Par ailleurs, les résultats semblent valables pour une grande partie de la Corse septentrionale, compte tenu de la relative uniformité des conditions climatiques et édaphiques (Simi, 1964). A chaque type de formation correspond une seule unité syntaxonomique. Il n'y a donc pas la diversité phytosociologique des groupements calcifuges décrits en Provence cristalline (Loisel, 1971, 1976).

La description de ces unités ainsi que la discussion de leur statut phytosociologique ne sont pas à nouveau envisagées ici. Le tableau 1 précise les éléments nécessaires ultérieurement et renvoie aux numéros des relevés correspondants (Allier & Lacoste, 1980). La multiplication des stades ligneux ou sous-ligneux et la réduction des stades herbacés, ceci étant lié au caractère méditerranéen, sont à noter.

Les placettes permanentes

Elles ont été mises en place dès 1975 au sein de stations représentatives des divers syntaxons constitutifs de la série. Parallèlement à l'étude de la végétation, y étaient conduites d'autres observations (pédologie et pédofaune, zoocénoses, etc.) destinées à permettre une compréhension globale de l'écosystème. De forme carrée, leur surface, fonction du type de groupement, était choisie supérieure à l'aire minimale. Des repères permettaient de concrétiser les limites de chaque placette. Si une mise en défens intégrale n'a pu être envisagée, aucune action anthropozoogène d'importance n'a pu y être décelée pendant la période d'observation,

La placette 1 correspondant à la pelouse à thérophytes (*Helianthemo-Plantaginetum bellardii* Aubert & Loisel, 1971) a été choisie en bordure de cistaie. En fait, ces pelouses n'occupent jamais de grandes surfaces, quelques m² au maximum en mosaïque dans les autres stades. Le fait de jouxter une cistaie n'est donc pas un accident pouvant entraîner des erreurs dans l'interprétation mais une régle générale.

Pour la cistaie (*Helichryso-Cistetum villosi* Allier & Lacoste, 1980) deux placettes ont été retenues, une de cistaie pure (2a), l'autre où commençaient à apparaître des espèces du maquis. De même pour la forêt (*Quercetum ilicis ornetosum* Allier & Lacoste, 1980), la variabilité étant plus grande à cause de la diversité des traitements suivis, ont été retenues trois placettes, mais deux seulement (6a et 6c) ont pu être conservées.

Les méthodes

Etant donné l'extrême diversité floristique et structurale, de la pelouse à thérophytes à la forêt climacique, une méthode unique n'a pu être retenue pour le suivi des placettes.

Pour la pelouse, la dominance d'espèces annuelles ne permet pas l'utilisation des méthodes classiques d'analyse des groupements herbacés (Daget, 1969; Poissonet J. & P., 1969). En effet la répartition des germinations l'année suivante est plus ou moins aléatoire, dépendant du type de semences, de leur période de germination, etc. Seule a été notée l'apparition des sous-ligneux dans la pelouse.

Pour les stades arbustifs ont été mis en place des transects linéaires constants de 20 m (Parker & Savage, 1944) le long desquels était relevée la projection du recouvrement linéaire global (L) de chaque espèce indépendamment de la distinction des strates (Anderson, 1942; Brown, 1954). Les superpositions d'espèces amènent à obtenir une somme totale des recouvrements supérieure à la longueur du transect et qui rendent compte de la complexification du groupement. Conjointement était relevée la hauteur moyenne (H) de la couronne du végétal (Canfield, 1942).

Pour les placettes forestières ont été notés les diamètres des arbres et arbustes à 1,5 m tandis que les germinations étaient repérées et comptées dans deux transects de 2 m de large sur 20 m de long, parallèlement à la pente et perpendiculairement.

En fait pour les deux stades les plus fermés du maquis, maquis haut et maquis arboré (*Erico-Arbutetum phillyreetosum* et *Erico-Arbutetum quercetosum* Allier & Lacoste, 1980), ces méthodes ont trouvé leur limite. La projection horizontale d'arbustes à couronne haute et intriquée est très peu fiable, de plus très sensible au déplacement latéral (vent, effet de pente). Le relevé détaillé des pousses (sur plusieurs années) était très difficile sinon impossible dans l'entrelacs de ces formations. Si une approche statique est possible, le suivi dynamique pose de gros problèmes.

Enfin une approche des biomasses des cistaies et maquis a été envisagée selon deux techniques (Feuillas, 1979): destructive pour la cistaie, semi-destructive et par construction de tarifs pour le maquis (Pardé, 1961).

Les résultats

La période d'observation de 5 ans s'avère suffisante pour les stades initiaux, herbacés ou arbustifs, mais ne permet pas d'apprécier de transformations significatives pour les stades proches du climax. Il est en effet classique d'observer que la vitesse d'évolution décroît avec le temps c'est-à-dire avec la maturation sériale.

L'analyse des transects n'a pu être poursuivie que dans les stades sous-ligneux bas. Seul le recouvrement linéaire a pu être conservé, le produit H × L devenant sans signification pour les espèces dominées: bien qu'en voie d'élimination, leur encombrement aérien semblait s'accroître (H augmentant par étiolement).

La pelouse de l'Helianthemo-Plantaginetum bellardii

Elle se transforme rapidement par envahissement latéral et étalement des couronnes des suffrutes-

Tableau 1. Caractère des placettes expérimentales.

| Formation | N° parcelle | N° relevé phyto. | Cadre syntaxonomique | | Hauteur moy. (m) | Recouv. moyen (%) |
			Association	Unités sup.		
Forêt	6c	66	*Quercetum ilicis ornetosum*	QUERCETEA ILICIS	>15	95
↑	6a	62				
Maquis arboré	5	51	*Erico-Arbutetum quercetosum*	Myrtion communi Pistacio-Rhamnetalia QUERCETEA ILICIS	6–8	100
↑						
Maquis haut	4	37	*Erico-Arbutetum phillyreetosum*		4–5	90
↑						
Maquis bas	3	29	*Erico-Arbutetum cistetosum*		2	80
↑	2b	17				
Cistaie			*Helichryso-Cistetum villosi*	CISTO-LAVANDULETEA	1	77
↑	2a	19				
Pelouse à Thérophytes	1	8	*Helianthemo-Plantaginetum bellardi*	HELIANTHEMETEA	0,10	70

centes proches, *Cistus monspeliensis* la plupart du temps, et aussi par apparition de germinations de ce ciste. Pour les diverses stations observées en sus de la placette 1, dans le délai de 5 ans la plupart des pelouses étaient envahies. Apparemment vite colonisées et éliminées, elles se réinstallent à la faveur de la mort ou de l'arrachage d'un ciste, d'incendies ou autres accidents. Il y a ainsi en permanence une mosaïque de petites pelouses à proximité ou à l'intérieur des cistaies.

Les espèces constitutives de la pelouse sont progressivement éliminées par ombrage ou compétition mais certaines des annuelles se maintiennent plus facilement sous la cistaie alors que les vivaces à bulbe (*Allium, Romulea, Leucoium, Poa* . . .) disparaissent. Ces annuelles, aptes à se resemer sous les cistes, figurent d'ailleurs sur les cartes d'analyse factorielle en position intermédiaire entre *Helianthemo-Plantaginetum bellardii* et *Helichryso-Cistetum villosi* (Allier & Lacoste, 1980). Elles trouvent sous les cistes des trouées suffisantes pour accomplir leur cycle.

Les seules stations de pelouse à thérophytes où la vitesse d'évolution est beaucoup plus faible sont celles sur sol très réduit, presque pelliculaire, sur roche inclinée. Le rajeunissement permanent du profil par l'érosion bloque partiellement l'évolution en éliminant les sous-ligneux.

Cistaies et maquis bas

Pour ces deux stades, les résultats ont été regroupés dans le Tableau 2. L'examen de ces résultats appelle plusieurs remarques.

L'évolution se fait par élimination des herbacées de petite taille (thérophytes, *Carlina*) et accroissement du recouvrement global des ligneux de grande taille (*Erica, Arbutus*) pour chaque placette. Toutefois il n'est pas possible de comparer l'importance d'une espèce donnée entre les diverses placettes, celles-ci ayant une origine différente. Ainsi le maquis bas choisi (parcelle 3) était assez ouvert, avec 1/4 du transect occupé par des transgressives de la pelouse à thérophytes. Ceci explique que le rôle du *Cistus monspeliensis* dans le recouvrement de celle-ci ne soit pas négligeable et s'accroisse pendant le temps d'observation.

La complexité structurale augmente entre les stades mais aussi dans chaque stade au cours du temps. Cela se traduit dans le recouvrement cumulé

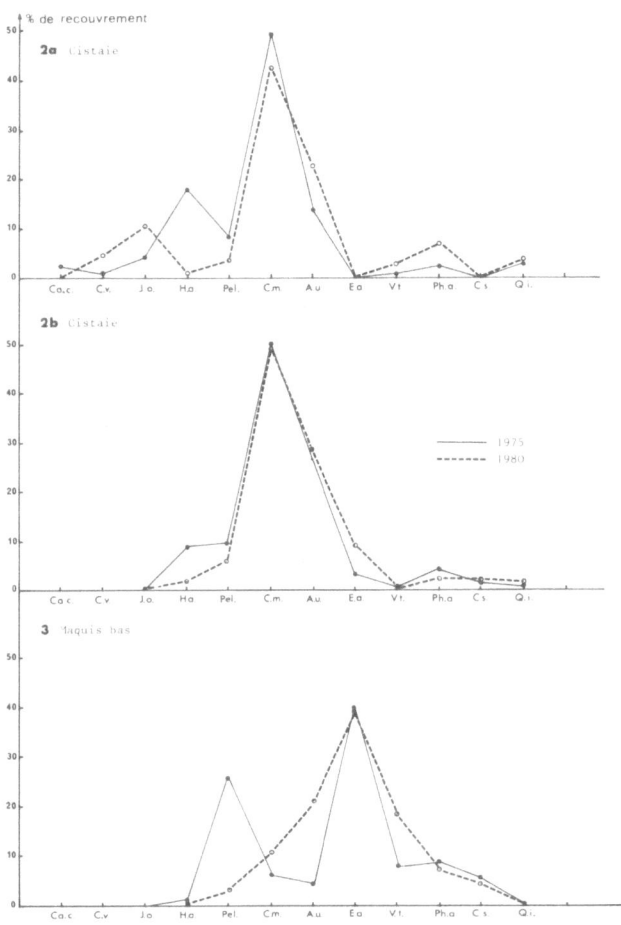

Fig. 1. Evolution du recouvrement des parcelles.

R (somme des projections sur le transect) par une augmentation comprise entre 18% (3) et 47% (2a). Il est logique de constater que c'est le groupement le plus ouvert et le moins ligneux (*Helichryso-Cistetum*) qui en 5 ans subit le plus fort accroissement. Le maquis bas par contre (*Erico-Arbutetum cistetosum*), déjà pluristrate, ne manifeste qu'un accroissement modeste du recouvrement cumulé.

Du fait de cette augmentation du recouvrement, les parts relatives (non pondérées) et absolues des participations des espèces peuvent sembler diverger. C'est le cas de *Cistus monspeliensis* qui dans la parcelle 2a augmente son recouvrement linéaire (1300 cm contre 1060) mais dont la part relative tombe de 49 à 42%, ou d'*Erica arborea* qui dans le maquis (parcelle 3) joue un rôle sub-constant: 40%

Tableau 2. Evolution du recouvrement des espèces (r) et du recouvrement cumulé (R) sur les placettes de cistaie et de maquis bas.

PLACETTE	Abréviations	2a				2b				3			
Date		1975		1980		1975		1980		1975		1980	
Segment de recouvrement en cm		cm	r %	cm	r %	cm	r %	cm	r %	cm	r %	cm	r %
Carlina corymbosa	Ca.c.	30	2	0	0	–	–	–	–	–	–	–	–
Cistus villosus	C.v.	20	1	120	4	–	–	–	–	–	–	–	–
Juniperus oxyced.	J.o.	100	4,5	330	10,1	–	–	–	–	–	–	–	–.
Helichrysum ang.	H.a.	335	18	20	0,7	190	8,8	40	1,6	30	1	–	0
Pelouse à thér.	Pel.	160	7,5	100	3,3	200	9,2	140	5,5	700	26	100	3,2
Cistus monspel.	C.m.	1060	49	1300	42	1080	50	1240	49	160	6	320	10,4
Arbutus unedo	A.u.	290	13,5	700	22,5	570	26,5	700	28	120	4,4	650	21
Erica arborea	E.a.					70	3,2	230	9	1080	40	1210	39
Viburnum tinus	V.t.	10	0,5	70	2,4	10	0,5	10	0,4	210	7,5	570	18,5
Phillyrea angust.	Ph.a.	40	2	180	5,8	100	4,6	60	2,4	240	8,8	250	7
Cistus salviaefol.	C.s.					25	1,2	50	2	150	5,5	170	4
Quercus ilex	Q.i.	40	2	100	3,3	10	0,5	40	1,6	5	0,2	10	0,3

Evolution du recouvrement linéaire des espèces

$$r \% = \frac{\Sigma \text{ des recouvrements de l'espèce}}{\Sigma \text{ du recouvrement de toutes les espèces } (= \text{ recouvrement global})}$$

PLACETTE	2a				2b				3			
Date	1975		1980		1975		1980		1975		1980	
	cm	R %	cm	R %	cm	R %	cm	R %	cm	R %	cm	R %
Rec. linéaire cumulé R (transects de 2000 cm)	2155	108	3100	155	2150	107	2520	126	2720	136	3080	154
Augmentation de R	47%				19%				18%			

$$R = \frac{\Sigma \text{ du recouvrement de toutes les espèces}}{\text{longueur du transect (200 cm)}}$$

Evolution du recouvrement cumulé

et 39% alors que son recouvrement a augmenté, mais beaucoup moins que celui d'*Arbutus unedo*.

La Figure 1 résume le rôle des espèces principales dans cette succession. Le rôle édificateur dans la cistaie est dévolu à *Cistus monspeliensis* et à un degré moindre à l'*Helichrysum angustifolium*. Ce dernier, de taille maximale réduite, est plus rapidement éliminé (parcelle 3). Dans les stades ultérieurs, c'est *Erica arborea* et *Arbutus unedo* et plus discrètement *Viburnum tinus* qui ont un rôle prépondérant. *Cistus salviaefolius* n'apparaît que tardivement et a un rôle réduit. Il reste une espèce dominée, de caractère plus mésophile que les autres

cistes et poussant donc plus facilement à l'abri. *Quercus ilex* peut apparaître très tôt (2a) mais son développement est très lent initialement. Son rôle n'apparaît vraiment que dans le maquis haut. Enfin *Phillyrea angustifolia* et *Juniperus oxycedrus* sont présentes de façon plus aléatoire et leur rôle édificateur est faible.

La biomasse aérienne des ligneux est un reflet bien plus précis de l'état du groupement et de la contribution des diverses espèces (Fig. 2). On constate que *Cistus monspeliensis* passe par un maximum dans la cistaie, fournissant dans certaines stations (non analysées dans ce travail), la

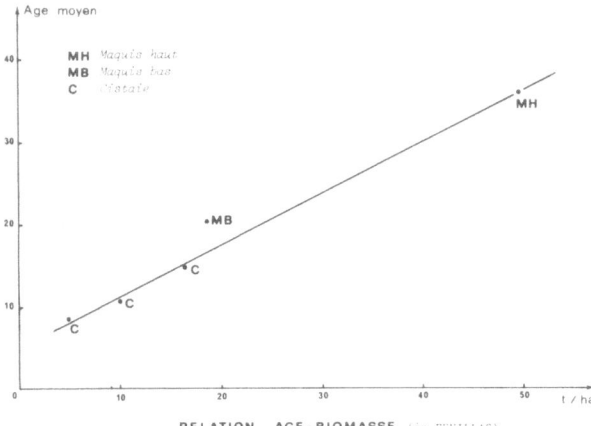

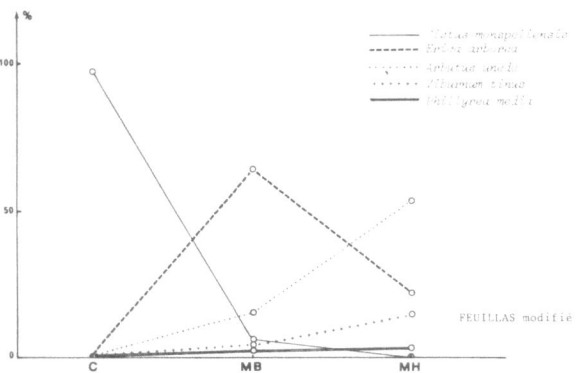

Fig. 2. Contribution relative des espèces à la biomasse.

des arbustes d'une formation s'il est corrélé à l'ancienneté n'intègre pas la totalité de la durée. Dans les placettes retenues, le passage cistaie-maquis bas demande entre 10 et 20 ans, celui entre maquis bas et maquis haut une trentaine d'années. Mais ces chiffres ne sont pas généralisables à la totalité des situations et représentent plus un cas favorable qu'une moyenne, l'échantillonnage ayant porté sur des parcelles bien développées et représentatives du stade.

La forêt

Son analyse, comme il a été signalé plus haut, a porté plus sur les problèmes de régénération et de structure que sur le passage maquis arboré – forêt. Ce stade climacique (*Quercetum ilicis ornetosum*) pose des problèmes tant pour l'étude des recouvrements que celle des biomasses. De plus la période de 5 ans est considérablement brève pour que des modifications autres que catastrophiques (chablis, parasitose . . .) puissent être mises en évidence.

La Figure 3 regroupe la structure ligneuse des deux parcelles forestières 6a et 6c. Il est net que:

6a est plus jeune que 6c, les classes de circonférence étant décalées. 6c représente une placette vieillissante avec la seule persistance de *Quercus ilex* âgés, certains même en voie de dépérissement. Dans les stades les plus jeunes, les arbustes persistent mais confinés dans des classes de faible circonférence.

Ces placettes 'climaciques' sont loin d'être vierges de toute action humaine. L'irrégularité du diagramme de 6c montre que des classes ont été préférentiellement exploitées autrefois (classes 160–200 et 100–120). Cependant toute intervention a cessé depuis 35 ans au moins (documents oraux). D'autre part, ont aussi été utilisés jadis pour le charbon de bois, des arbustes comme *Erica, Arbutus, Phillyrea*. Cet usage peut être tenu pour partiellement responsable de la régression de ces espèces (6c) ou de leur état physiologique: diamètre faible, grand étirement (6a). Mais l'étiolement que provoque le couvert de la futaie de chênes donne des formes comparables et intervient certainement dans l'élimination des arbres à croissance plus faible.

Il y a un vide important dans les classes de diamètre correspondant aux jeunes *Quercus ilex,* très marqué en 6c (Omita). Cette absence pose le

quasi totalité de la biomasse. Son rôle devient très faible dès le maquis bas et nul dans le maquis haut. L'optimum de participation d'*Erica arborea* est dans le maquis bas qu'elle domine largement (Feuillas, 1979). Par contre les contributions d'*Arbutus unedo, Viburnum tinus* et à un degré moindre de *Phillyrea media* augmentent régulièrement. Dans le maquis haut, *Arbutus unedo* représente 52% de la biomasse, c'est donc lui qui joue le rôle essentiel dans l'évolution de la formation.

Enfin la vitesse des transformations entre stades varie selon le substrat. Assez rapide sur les anciens sols culturaux (2a, 3), elle est faible sur les substrats compacts à sol réduit (pentes sur rhyolites de la vallée du Fango). Sur la période des cinq années d'observation, la mesure de cette vitesse ne peut être tentée. On peut tourner la difficulté en évaluant l'âge (Fig. 2) des espèces ligneuses par comptage des cernes annuels (Feuillas, 1979). Mais l'âge modal

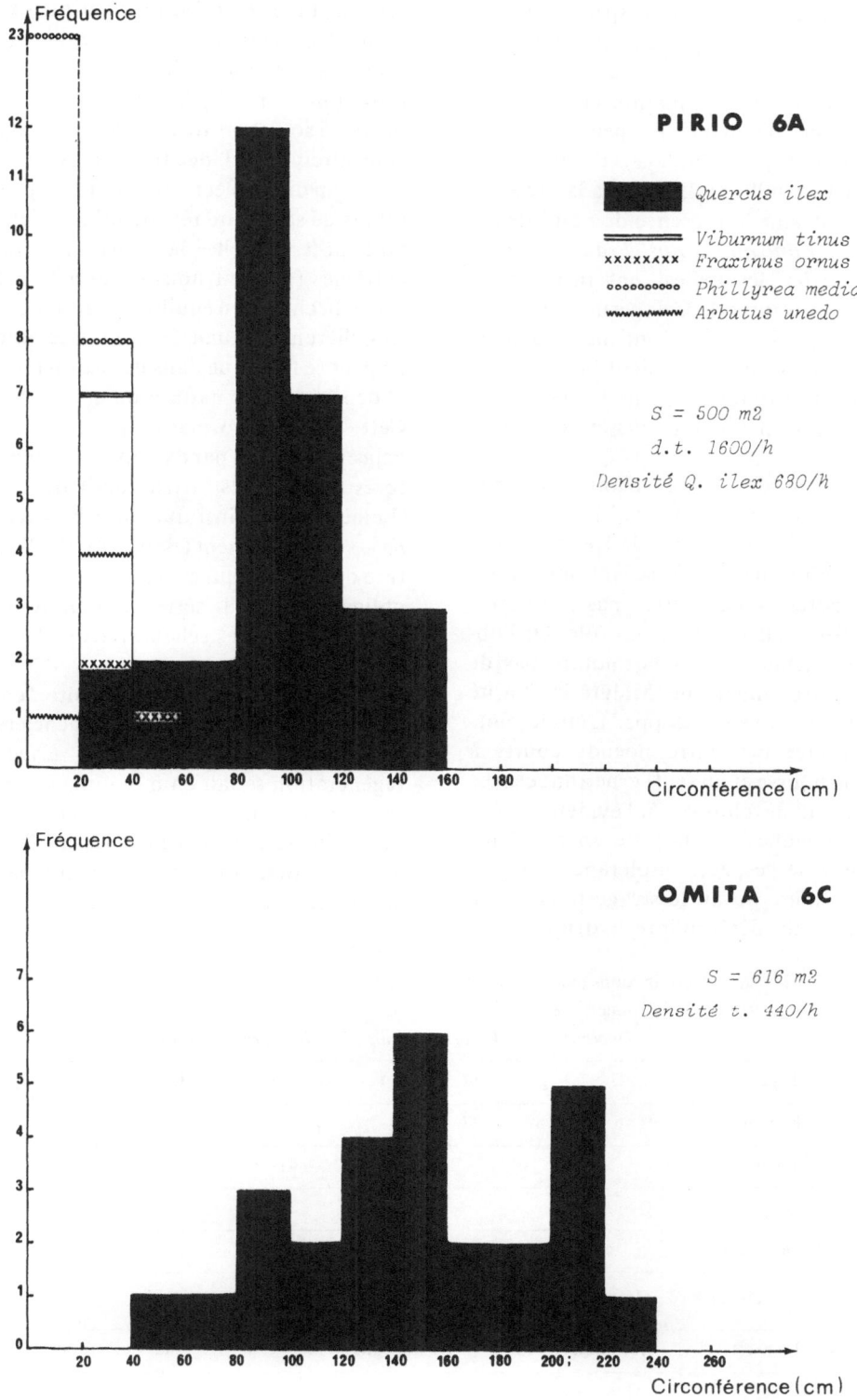

Fig. 3. Structure ligneuse des parcelles forestières.

problème de la régénération de l'espèce sous son propre couvert. Le décompte des germinations (Tableau 3) permet de formuler des hypothèses.

Les jeunes plants et germinations sont abondants. Nous avons distingué, un peu arbitrairement, les individus de petite taille (g) et ceux de plus d'un an, avec 4 à 6 feuilles (G). Dans 6a la présence de semenciers entraîne la présence de nombreuses germinations de *Viburnum tinus, Fraxinus ornus* ou *Phillyrea media,* lesquelles sont rares en 6c (absentes dans le transect). Ces germinations ne dépassent pas l'âge de 3 ans. Sont inclus dans le décompte, car on ne peut les identifier sans les détruire, les rejets sur racines. Sur les bases des troncs les rejets sont peu abondants et peu vigoureux.

Le dépérissement des jeunes plants n'est pas imputable à l'action des animaux. En effet, en particulier pour la forêt d'Omita, la fréquentation des chèvres est faible, les vaches ne stationnent pas en forêt et les porcs ne parcourent pas ce secteur. Seuls les sangliers peuvent avoir un rôle. Or l'observation de ces jeunes plants ne montre pas de broutage ou de sectionnement. Malgré la densité assez forte, aucun ne se développe. L'étude morphologique montre des entre noeuds courts à l'extrémité de la pousse et un seul cerne annuel avec bois normalement développé. A l'évidence, l'élimination de ces jeunes plants pose un problème d'écophysiologie de l'espèce. Intolérance des germinations à la litière ou au lessivage foliaire, ou bien sensibilité à un déséquilibre hydrique à un certain stade? Une étude expérimentale ultérieure permettra de donner la réponse. Mais au-delà de la cause de ces éliminations se pose le problème de l'équilibre climacique. Ce *Quercetum ilicis ornetosum,* à sous-bois très pauvre en espèces ligneuses, contrairement à l'idée trop classique du climax en étage méditerranéen, ne contient pas les constituants de sa propre régénération. Même en admettant que les parcelles 6a et surtout 6c constituent des extrêmes (ce dont nous doutons), la notion classique du climax en équilibre, autorégénéré, doit être singulièrement nuancée. En fait le renouvellement ne peut se faire que dans des clairières ouvertes par le dépérissement naturel ou accidentel des arbres. Cette autotransformation ponctuelle et cyclique a déjà été signalée, par exemple à propos des réserves forestières (séries 'artistiques') de Fontainebleau (Lemée, 1966) ainsi que pour la forêt de *Quercus ilex,* sur le continent (Baudière, 1972). Cette ouverture de la forêt, qui est la seule régression naturelle obligatoire dans la série, ne conduit pas au maquis arboré comme les schémas, trop classiques, l'indiquent.

Les clairières voient réapparaître, en fonction de leur taille et la proximité des semenciers, les espèces arbustives héliophiles (*Cistus, Erica* . . .). Cette régénération se fait en mosaïque à l'intérieur de la forêt et ne peut donc faire l'objet d'une estimation de la vitesse comme cela pourrait être fait sur les autres stades, son caractère étant essentiellement accidentel.

Tableau 3. Germinations dans les placettes forestières.

6 A. PIRIO   d. totale   8,1 / m²

Type	*Quercus ilex*		*Viburnum tinus*		*Fraxinus ornus*		*Phillyrea media*	
	G	g	G	g	G	g	G	g
Nombre	5	18	37	59	14	27	–	2
Total	23		96		41		2	
%	14		59		27		1	
Densité / m²	0.53		2.4		1		0.05	

6 C. OMITA   d. totale   1,35 / m²

Transect	Horizontal 50 m²		Descendant 50 m²		Total	
Type	G	g	G	g	G	g
Nombre	36	28	31	40	67	68
%					50	50

Conclusions

L'étude de la dynamique dans la série calcifuge du *Quercus ilex* en Corse nord-occidentale, réalisée sur une période de cinq années au sein de placettes expérimentales, fournit des résultats quantitatifs intéressants pour les premiers stades (pelouse, cistaie, maquis bas). Cette période d'observation s'avère par contre insuffisante pour les stades proches de la maturité (forêt subclimacique ou climacique).

Une méthodologie unique ne peut être appliquée à des stades aussi différents. Si la mesure du recouvrement linéaire fournit des informations suffisantes dans les fruticées basses, cette méthode est inutilisable pour les formations pluristrates hautes ou complexes. L'étude des biomasses est plus instructive dans ces stades.

Le rôle dynamique essentiel est joué par un nombre réduit d'espèces: *Cistus monspeliensis* pour la cistaie, *Erica arborea* pour le maquis bas et *Arbutus unedo* pour le maquis haut constituent la partie la plus importante du recouvrement ou de la biomasse.

Le stade climacique (*Quercetum ilicis ornetosum*) n'est pas véritablement auto-entretenu par ses constituants (équilibre interne) mais fait intervenir des phénomènes 'catastrophiques' qui n'existaient pas dans les stades antérieurs où les phénomènes de compétition et de coopération semblent prédominants. Le maintien du climax se réalise donc par des régressions locales, naturelles, qui permettent la réinstallation d'espèces sous-ligneuses caractéristiques de stades sériaux jeunes.

Références

Allier, C. & Lacoste, A., 1980. Maquis et groupements végétaux de la série du chêne vert dans le bassin du Fango (Corse). Ecol. Medit. 5: 59–82.

Anderson, K. L., 1942. A comparison of line transects and permanent quadrats in evaluating composition and density of pasture vegetation on the tall prairie grasse type. J. Amer. Soc. Agron. 34: 805–922.

Baudiere, A., 1972. La forêt de chênes verts dans les Gorges d'Héric et sa signification biogéographique sur les limites de son aire. Bull. Soc. Bot. Fr. 119: 19–64.

Brown, D., 1954. Methods of surveying and measuring vegetation. Comm. Bur. Pastures and Fields Crops 42, Hurley, 233 pp.

Canfield, R. H., 1942. Sampling ranges by the line interception method. Plantcover composition-density-degree of forage use. Res. Rep. 4, U.S.D.A., For. Serv., South-western For. Range Exp. Sta., 28 pp.

Daget, Ph., 1969. Méthodes d'inventaire phyto-écologique et agronomique des prairies permanentes. Compte-rendu du séminaire de Montpellier des 25–26 Mars 1969, Doc. C.E.P.E. 56, 206 pp.

Feuillas, D., 1979. Méthodes et techniques d'estimation de la biomasse végétale épigée des formations arbustives et leurs applications au maquis corse dans la vallée du Fango. D.E.A., Orsay, 51 pp.

Lemee, G., 1966. Sur l'intérêt écologique des réserves biologiques de la forêt de Fontainebleau. Bull. Soc. Bot. Fr. 113, 5–6: 305–323.

Loisel, R., 1971. Contribution à l'étude des cistaies calcifuges de Provence. Ann. Univ. Provence Sci. 46: 63–81.

Loisel, R., 1976. La végétation de l'étage méditerranéen dans le sud-est continental français. Thèse Doct. Etat Univ. Marseille. 1 vol.

Ozenda, P., 1975. Sur les étages de végétation dans les montagnes du bassin méditerranéen. Doc. Cartographie écologique 16: 1–32.

Parde, J., 1961. Dendrométrie. Nancy. Edition de l'ecole nat. des eaux et forêts, 350 p.

Parker, K. W. & Savage, D. A., 1944. Reliability of the line interception method in measuring vegetation on the Southern Great Plains. J. Amer. Soc. Agron. 36: 97–110.

Poissonet, J. & P., 1969. Etude comparée de diverses méthodes d'analyse de la végétation des formations herbacées denses et permanentes. C.E.P.E., C.N.R.S., Doc. 50, 120 pp.

Simi, P., 1964. Le climat de la Corse. Ministère de l'Education Nationale, Comité des travaux historiques et scientifiques. Bull. Sect. Géogr. 76: 1–22.

Accepted 15.7.1981.

Development of flora, vegetation and grazing value in experimental plots of a *Quercus coccifera* garrigue*

J. Poissonet, P. Poissonet & M. Thiault
C.E.P.E. Louis Emberger, B. P. 5051, 34033 Montpellier-Cedex, France

Keywords: Dynamics, Experiment, Flora, Garrigue, Grazing value, *Quercus coccifera* L., Vegetation

Abstract

From 1969–1977 an experiment was carried out in a *Quercus coccifera* garrigue, to follow the herbaceous species development after removal of the woody vegetation. Controlled variables are cutting date and fertilization. Flora and vegetation were regularly registrated on 18 out of 63 plots in a factorial design. Two types of results are obtained:

1. Five groups of species regarding their response to the disturbance are distinguished: disappearing, strongly decreasing, constant, strongly increasing and newly appearing species. In general, the floristic composition develops towards an equilibrium.

2. The number of species and the vegetation density are increasing with fertilization, although this increase is not proportional to the quantity of fertilizers.

A method is suggested to calculate the grazing value and the corresponding sheep density, based on flora and vegetation knowledge. The results indicate that the stock can be multiplied by 3, at least, and by 7, at most, from the 3rd year after the beginning of the experiment.

Introduction

Vegetation dynamics is usually studied by synchronously comparing different succession stages which are juxtaposed in space. Hypotheses resulting from such studies need to be verified by experiments. In our case, hypotheses were concerned with the development of the *Quercus coccifera* garrigue and the respective experiments started in 1969, at Puech du Mas du Juge (commune of St.-Gély du Fesc), near Montpellier.

The term 'garrigue' indicates an evergreen ligneous vegetation, under a humid or sub-humid mediterranean climate (in the sense of Emberger, 1942) and on calcareous soil. The *Quercus coccifera* garrigue is a low and closed vegetation. It varies from ca. 30 cm – 2 m in height, but usually it is between 60 cm and 1 m high. The garrigue studied is to be classified in the *Cocciferetum brachypodietosum* Br.-Bl. 1935 (Braun-Blanquet *et al.*, 1952). Before, the area was grazed and earlier it was used as a field.

One of the experiments, starting in 1969, aimed at following the development of flora and vegetation after removal of the woody vegetation, by testing the combined influence of repeated cutting (simulated pasture) and fertilization on the development of the herbaceous vegetation. This experiment was continued till 1977, and its design has been described before (D.P.E.G., 1967; Long *et al.*, 1976, 1978; Poissonet *et al.*, 1978a).

---

* Nomenclature follows Flora Europaea

Vegetatio 46, 93–104 (1981). 0042-3106/81/0462-0093/$2.40.
© Dr W. Junk Publishers, The Hague.

Experimental design and methods

The design involves 63 plots of 10×5 m, grouped in 7 blocks. The term 'block' only refers to the spatial distribution of plots; it does not imply that the inter-block variation is higher than the intra-block one. Discussions on this subject were published by Poissonet & Thiault (1978b) and Daget & Poissonet (1978). In each block, two factors are controlled at three levels, according to a factorial design. These two factors are the fertilization and the cutting date.

With the fertilization experiment, we can check the possibilities for the development of the present herbaceous species. There are three levels of fertilization:
- no fertilization (level F1)
- 100 kg ha/yr of N, 100 kg of P_2O_5, 100 kg of K_2O) (level F2)
- 200 kg/ha/yr of each fertilizer (N, P_2O_5, K_2O (level F3)

By varying the cutting date, we can simulate pasture development under different mowing regimes. The three dates are:

- date J, when the mean vegetation is 15 cm high
- date J + 7, i.e. 7 days after date J
- J + 14, i.e. 14 days after date J

Before the above-ground vegetation was removed in 1969, the initial vegetation was analyzed, which showed that 7 of these 18 plots were very similar as to their flora and vegetation (according to the contact frequency, i.e. the number of hits per species at each point-quadrat) (Table 1). Two of these plots were given the treatment F1, J + 7, two the treatment F2, J + 7, one the treatment F2, J + 14, and two the treatment F3, J + 14. Consequently, only the factor 'fertilization' can be validly studied with: two plots F1, three plots F2 and two plots F3.

Development of flora and vegetation, 1969–1977

The following aspects were followed:
- number of species;
- contact frequency of species, including histograms and mean values per point-quadrat;
- types of species response;
- relative proportions of species and species groups.

Table 1. Species registered in the seven plots (see text) in 1969 before removal of the woody vegetation ($+ = <1$ hit/100 points).

Mean rank of the species	Names of the species	Mean number of hits/100 points
1	*Quercus coccifera* L.	366
2	*Brachypodium retusum* (Pers.) Beauv.	61
3	*Dorycnium pentaphyllum* Scop Subsp *pent.*	14
4	*Rubia peregrina* L.	10
5	*Cephalaria leucantha* (L.) R. et S.	8
6	*Cistus monspeliensis* L.	7
7	*Carex humilis* Leysser	6
8	*Carex hallerana* Asso	4
9	*Smilax aspera* L.	4
10	*Genista scorpius* (L.) D.C.	4
11	*Teucrium chamaedrys* L.	3
12	*Arrhenatherum elatius* (L.) Beauv.	3
13	*Daphne gnidium* L.	2
14	*Phyllirea angustifolia* L.	1
15	*Asparagus acutifolius* L.	1
16	*Bromus erectus* Hudson	+
17	*Festuca lemanii* Bast.	+
18	*Aphyllantes monspeliensis* L.	+
19	*Fumana ericoides* (Cav.) Gand.	+
20	*Euphorbia characias* L.	+
21	*Lonicera implexa* Aiton	+
22	*Rubus ulmifolius* Schott.	+

Number of species

The numbers of species in the three treatments, before the experiment started, were 14, 15 and 15. These numbers strongly decreased after the removal of the woody vegetation and then increased (c.f. Poissonet *et al.,* 1978a). In 1977 we found:

– 17 species in the unfertilized plots (F1);
– 24 species in the moderately fertilized plots (F2);
– 26 species in the highly fertilized plots (F3).

These values were already reached in 1973 with treatments F1 and F2, but only in 1976 with treatment F3. Thus, cutting induces a rapid but slight increase; cutting plus fertilization induces a slower, but more important increase in the number of species.

Species contact frequency

Histograms

The histograms referring to the frequency distribution of the number of hits, as shown in Figure 1, indicate if there is an equilibrium between species. In this case, each histogram has only one mode and the distribution is a log-normal one. (Poissonet, 1968; Poissonet & César, 1972). If there is no equilibrium, each histogram has several modes (Daget & Poissonet, 1970; Poissonet, 1978).

In 1969, the distribution of species frequency is very unequal, because the histograms have several modes. The first mode is due to the very important dominance of *Quercus coccifera* in the class 256–512 hits. The second mode is due to the second dominant species, *Brachypodium retusum,* dominating in the classes 32–64 or 64–128 hits. At least one or two modes appear in every histogram with the other species grouped in the classes 8–16 or 16–32 hits per species.

In 1977, the distribution over the species is more balanced than in 1969, the histograms being uni- or bi-modal. The importance of the species which was dominant in the past is not so prominent any more. So, cutting, with or without fertilization, induces a distribution of species frequency which is more balanced than before the experiment, although in 1977 the equilibrium was not yet reached.

Mean species frequency

Table 2 shows that, before removal of the woody vegetation, the number of hits per 100 point-

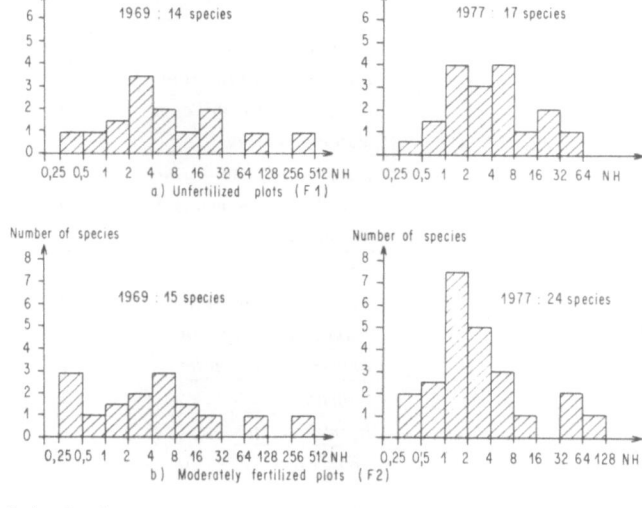

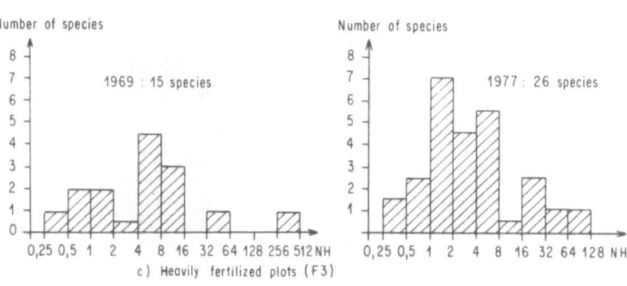

Fig. 1. Histograms of species contact frequency as number of hits (NH) per treatment, in 1969 and 1977 (logarithmic classes).

quadrats is close to 500. i.e. 5 hits per point on an average. In 1977, this number is only ca. 1.5 in the unfertilized plots, 2.3 in the moderately fertilized ones and 2.7 in the highly fertilized ones. The values obtained from 1975–1977 are lower than the ones of the preceding years, as is shown in Figure 2. In 1973 and 1974, the mean number of hits per point is 2.5–2.8 for the unfertilized plots and 4.5–5.2 for the fertilized ones.

The two decreases seem to be related to two dry periods. The first one (1975) began in October 1974 and ended in April 1975 with a rainfall of 207.2 mm (against 500–1300 mm in other years in the same period). The second decrease (1977) seems to be related to the very dry summer of 1976.

Types of species response

According to the changes in their frequency between 1969 and 1977, the species can be ranked in

Table 2. Mean contact frequency (nr. of hits per 100 point-quadrats) in 1969 and 1977 in three treatments.

1969		1977	
Quercus coccifera	375.5	*Brachypodium retusum*	49.5
Brachypodium retusum	71.5	*Quercus coccifera*	26.5
Cephalaria leucantha	20.5	*Carex hallerana*	19.5
Rubia peregrina	17	*Cephalaria leucantha*	13.5
Dorycnium pentaphyllum	14.5	*Dorycnium pentaphyllum*	7
Teucrium chamaedrys	5	*Bromus erectus*	5.5
Daphne gnidium	5	*Rubia peregrina*	4.5
Carex humilis	3.5	*Teucrium chamaedrys*	4
Arrhenatherum elatius	3	*Crepis sancta*	4
Asparagus acutifolius	2	*Sonchus oleraceus*	2.5
Phyllirea angustifolia	2	*Carex humilis*	2
Genista scorpius	2	*Thymus vulgaris*	2
Bromus erectus	0.5	*Ononis minutissima*	1.5
Smilax aspera	0.5	*Aphyllantes monspeliensis*	1.5
Sum	522.5	*Fumana ericoides*	1
		Arrhenatherum elatius	1
——— Species which have disappeared		*Galium parisiense or*	0.5
– – – Species which have strongly decreased		*Galium divaricatum*	
—— Species which have strongly increased		*Sum*	146
········ New species			

a) Unfertilized plots (F1)

1969		1977	
Quercus coccifera	338.3	*Brachypodium retusum*	79.3
Brachypodium retusum	65	*Crepis sancta*	56.7
Dorycnium pentaphyllum	19.7	*Carex hallerana*	39
Rubia peregrina	9	*Quercus coccifera*	10
Carex hallerana	8	*Cephalaria leucantha*	6.7
Cistus monspeliensis	6	*Sedum sediforme*	5.3
Carex humilis	5.7	*Lactuca virosa*	4.7
Teucrium chamaedrys	4	*Crepis vesicaria*	3.3
Genista scorpius	2.7	*Lactuca serriola*	2.7
Cephalaria leucantha	2	*Seseli elatum*	2.7
Arrhenatherum elatius	1.3	*Teucrium chamaedrys*	2.7
Rubus ulmifolius	0.7	*Carex humilis*	2
Phyllirea angustifolia	0.3	*Silene italica*	2
Aphyllantes monspeliensis	0.3	*Dorycnium pentaphyllum*	1.7
Euphorbia characias	0.3	*Festuca lemanii*	1.3
Sum	463.3	*Bromus madritensis*	1.3
		Muscari neglectum	1.3
		Lamnium amplexicaule	1.3
		Rubia peregrina	1
——— Species which have disappeared		*Bromus erectus*	1
– – – Species which have strongly decreased		*Cerastium glomeratum*	1
—— Species which have strongly increased		*Echium vulgare*	0.7
········ New species		*Muscari comosum*	0.3
		Lactuca perennis or	0.3
		Lactuca ramosissima	
		Sum	228.3

b) Moderately fertilized plots (F2)

Table 2. (Continued)

1969		1977	
Quercus coccifera	384.5	*Brachypodium retusum*	100.5
Brachypodium retusum	46	*Crepis sancta*	60.5
Cistus monspeliensis	15	*Carex hallerana*	21
Smilax aspera	11	*Taraxacum officinale*	18.5
Carex humilis	10	*Senecio vulgaris*	16
Dorycnium pentaphyllum	7.5	*Torillis nodosa*	7.5
Genista scorpius	6.5	*Cerastium glomeratum*	6.5
Rubia peregrina	5	*Rubia peregrina*	5
Carex hallerana	5	*Polygonum aviculare*	5
Arrhenatherum elatius	4	*Crepis vesicaria*	4
Festuca lemanii	1.5	*Petrorhagia prolifera*	4
Cephalaria leucantha	1	*Bromus madritensis*	4
Fumana ericoides	1	*Teucrium chamaedrys*	2.5
Lonicera implexa	0.5	*Quercus coccifera*	2
Teucrium chamaedrys	0.5	*Sonchus oleraceus*	2
Sum	499	*Lonicera implexa*	2
		Cephalaria leucantha	2
		Dorycnium pentaphyllum	1.5
		Hippocrepis comosa	1.5
		Euphorbia characias	1.5
		Tragopogon sp.	1.5
		Muscari comosum	1
		Odontites lutea	1
		Bromus erectus	0.5
		Sanguisorba minor	0.5
		Veronica arvensis	0.5
		Sum	272.5

———— Species which have disappeared
– – – – Species which have strongly decreased
——— Species which have strongly increased
············ New species

Heavily fertilized plots (F3)

five groups (Table 2).

1. Species present in 1969 and absent in 1977; most of them are ligneous species with almost no growth under repeated cutting *Genista scorpius, Cistus monspeliensis, Phillyrea angustifolia* and the grass *Arrhenartherum elatius,* essentially eliminated by repeated cuttings, while increasing in the neighbouring experiments with sheep grazing after removal of woody vegetation.

2. Species (mostly ligneous) which very strongly decrease: *Quercus coccifera, Dorycnium pentaphyllum* ssp. *pentaphyllum.*

3. Species with constant frequency: *Brachypo-* *dium retusum, Cephalaria leucantha.*

4. Species which strongly increase: *Carex hallerana, Bromus erectus.*

5. Species which appeared after 1969, and became important in 1977, particularly in the fertilized plots (especially *Compositae* and *Gramineae*).

Relative proportions of species and species groups

The point-quadrat observations not only indicate the presence and contact-frequency of species, but also the amount of bare ground, i.e. soil not covered by vegetation, and the relative proportions

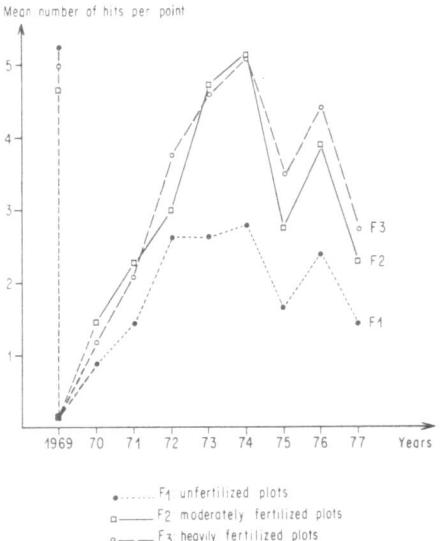

Fig. 2. Variation in the mean number of hits per point-quadrat, per treatment, from 1969–1977.

of species and species groups.

Table 3 shows that the percentage of bare ground was low (<10%) before the experiment, high (40–60%) in 1970. i.e. one year after vegetation removal and, during the following years it was decreasing. This decrease showed a hyperbolic trend and continued till 1973 in the highly fertilized plots and till 1974 in the other ones, to reach values <10% in the fertilized plots and between 15 and 25% in the unfertilized ones. Due to the observed fluctuations the year-to-year variation in amplitude was ca. 10% in the fertilized plots and 20% in the unfertilized ones; only the fertilized plots reach a pre-experiment percentage of bare ground again during the last years of the experiment. Extrapolations of the hyperbolic trend obtained from 1970 till 1977 in the unfertilized plots indicate that several more years would be necessary to reach a mean value of the bare ground percentage <10%.

Table 3 also presents the relative frequency values of species and some species groups. This frequency was called 'contact specific contribution', C.S.C. (Poissonet & Poissonet, 1969) to distinguish it from other expressions of specific frequency. In dense herbaceous formations with a near to 100% cover values are calculated as follows:

(1) CSC_i = number of hits of species $_i$ (or group $_i$ of species)/ number of hits of all species

The sum of CSC values for all species in the plot is 100. In more or less open herbaceous formations as the experimental garrigue, it is necessary to take into account the bare ground by distinguishing between a point-quadrat with bare ground and one point-quadrat with vegetation. In fact, at one point-quadrat, either one single hit of bare ground (soil, litter, pebble, to be specified eventually), or one or several hits with vegetation can be noted, so

$$CSC(p)_i = CSC(v)_i (100 - y)$$

$CSC(v)_i$ is the contact specific contribution for vegetation hits only, where y = percentage of bare ground, and $CSC(p)_i$ the resulting plot value and the sum $(y + CSC(p)_i$ is equal to 100. An alternative formula: number of hits of species i or bare ground/ total number of hits, was rejected because no correct agricultural values could be based on this.

In Table 3, species with a clear type response have been listed separately, the other ones (other *Gramineae,* other ligneous species) have been grouped. Clearly all species (species groups) are present throughout the period of observation; so, no species (or species group) has appeared or disappeared during the experiments.

Species with the highest CSC values and high bare ground values up to a total of 85% have been underlined.* The following results may be mentioned

In 1969, the main components in the unfertilized plots are bare ground, *Quercus coccifera* and *Brachypodium retusum;* in 1970 *Teucrium chamaedrys* is added and, in 1971, *Dorycnium pentaphyllum.* After 1975, *Teucrium chamaedrys* decreases and *Carex hallerana* appears between the main components. In 1976 and 1977, *Cephalaria leucantha* and *Bromus erectus* appear.

In the fertilized plots, *Brachypodium retusum* is the single species among the main components from 1969–1977. The moderately fertilized plots have an intermediate position.

In all plots, the number of main components is higher in 1977 than in 1969 and the difference is more important as the fertilization level is higher.

The similarity between any two treatments or

---

* The value of 85% may seem arbitrary. However, in equilibrated, dense, herbaceous formations, a Lorenz curve (Aitchison & Brown, 1966) is obtained for the CSC(p) values, with a characteristic point at 15–85 (Poissonet & Poissonet, 1969).

Table 3. Variation of the relative proportion of species and species groups per treatment, from 1969 till 1977. Underlined values: main components the sum of which is at least equal to 85% for the corresponding year.

Years	1969	1970	1971	1972	1973	1974	1975	1976	1977
Bare ground	6.0	60.0	41.5	23.0	17.5	14.0	29.0	11.5	25.5
Quercus coccifera	67.7	13.3	19.3	21.1	28.7	10.5	17.6	27.2	13.6
Brachypodium retusum	12.9	10.8	26.1	27.7	29.1	52.4	29.2	24.2	24.9
Teucrium chamaedrys	0.9	5.4	3.1	5.9	2.7	4.5	1.5	0.9	2.0
Dorycnium pentaphyllum	2.6	2.6	4.6	8.2	7.2	4.2	4.9	10.4	3.7
Carex hallerana		0.4	1.9	3.5	4.5	4.1	8.8	8.3	9.8
Cephalaria leucantha	3.7	4.7	2.6	4.8	3.2	3.3	1.8	7.2	6.9
Bromus erectus	0.1	0.4	0.3	0.7	1.3	3.7	1.9	3.0	2.7
Rubia peregrina	3.0	0.8	0.8	1.3	1.7	0.6	1.0	2.3	2.4
Arrhenatherum elatius	0.6	1.0	0.2	0.6	0.3	0.7			0.5
Carex humilis	0.6		0.3	1.6	1.8	0.2	2.3	2.9	1.1
Other ligneous sp.	2.1			0.2	0.4	0.2	1.1	1.1	1.5
Other herbaceous sp. (except gramineae)		0.8		1.2	1.0	0.5	0.5	0.2	1.1
Other gramineae					0.4	0.7	0.7	0.8	0.8
Compositae						0.6	0.3	0.2	3.5

a) Unfertilized plots (F1)

Years	1969	1970	1971	1972	1973	1974	1975	1976	1977
Bare ground	6.0	41.0	24.0	9.0	3.0	1.0	12.0	3.0	12.0
Quercus coccifera	68.5	10.1	11.0	5.5	4.6	4.1	3.4	4.8	4.3
Dorycnium pentaphyllum	4.1	4.5	5.2	3.2	2.4	0.5	0.6	1.1	0.6
Brachypodium retusum	13.3	26.1	41.0	54.1	53.1	47.1	57.0	50.4	30.0
Carex humilis	1.1	6.3	5.5	3.7	2.7		2.3	1.2	0.7
Arrhenatherum elatius	0.3	7.4	9.1	7.7	9.6				
Teucrium chamaedrys	0.8	1.9	1.3	3.9	4.4	6.2	4.1	5.0	1.2
Carex hallerana	1.7	0.2	1.9	8.5	7.0	4.6	5.2	3.9	15.8
Sedum sediforme					3.0	8.5	4.7	8.3	1.9
Compositae				1.6	5.4	23.4	7.3	11.1	25.8
Other herbaceous sp. (except gramineae)	0.1	1.1	0.1	0.9	0.3	2.0	1.7	4.4	3.6
Cephalaria leucantha	0.4	0.2		0.4	0.7	0.1	0.2	3.5	2.4
Other ligneous sp.	2.1	0.1	0.1	0.7	1.3	1.2	0.2	1.2	
Rubia peregrina	1.8	0.7	0.1	1.5	1.6	0.1	0.5	2.0	1.0
Other gramineae *Aphyllantes monspeliensis*	0.1	0.4			0.8	1.0	0.4	0.9	0.9
Bromus erectus				0.2	0.3		0.6	0.5	0.4

b) Moderately fertilized plots (F2)

Years	1969	1970	1971	1972	1973	1974	1975	1976	1977
Bare ground	7.5	51.0	25.0	5.5	2.0	10.0	8.5	1.5	11.0
Quercus coccifera	71.2	6.6	13.2	7.0	3.4	1.8	1.9	1.9	0.6
Brachypodium retusum	8.6	15.1	32.3	38.5	45.7	55.8	59.4	54.3	32.5
Carex humilis	1.8	6.3	9.2	12.4	11.2	3.3	2.8	1.3	
Arrhenatherum elatius	0.7	10.5	12.3	8.0	11.0	1.0	1.3	0.7	
Other gramineae	0.3		4.4	3.9	7.0	1.9	3.7	5.4	1.2
Compositae				1.2	4.0	14.3	7.0	7.4	33.7
Carex hallerana	0.9	6.2		7.8	5.4	2.5	3.7	5.7	6.8
Bromus erectus		0.2		3.6	3.9	1.3	3.7	9.1	0.2
Other herbaceous sp. (except gramineae)			0.2	0.4	1.8	1.2	1.7	3.4	9.8
Teucrium chamaedrys	0.1	0.5	0.5	5.1	2.0	4.5	2.7	3.4	0.8
Other ligneous sp.	6.4		0.9	0.9	0.7	1.1	1.1	0.3	0.6
Dorycnium pentaphyllum	1.5	2.2	1.8	2.4	0.5	0.4	1.3	0.7	0.5
Rubia peregrina	1.0	0.3	0.2	2.0	1.2	1.5	1.4	2.0	1.7
Cephalaria leucantha	0.2		0.2	1.4	0.7	0.4		2.8	0.6

c) Heavily fertilized plots (F3)

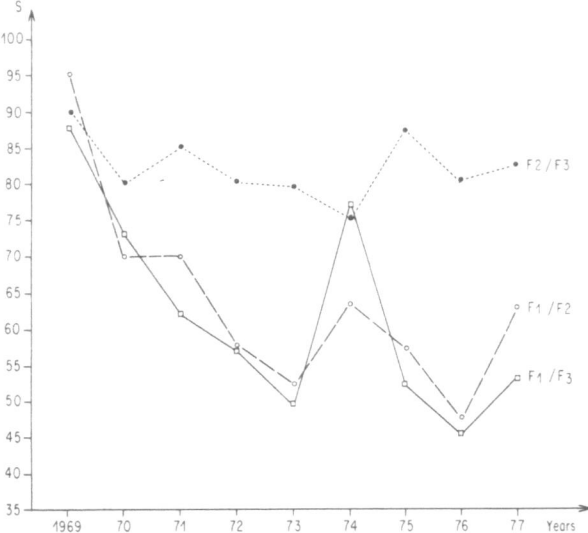

Fig. 3. Variation in the similarity (S) between treatments in pairs, from 1969-1977. F1: unfertilized plots; F2: moderately fertilized plots; F3: heavily fertilized plots.

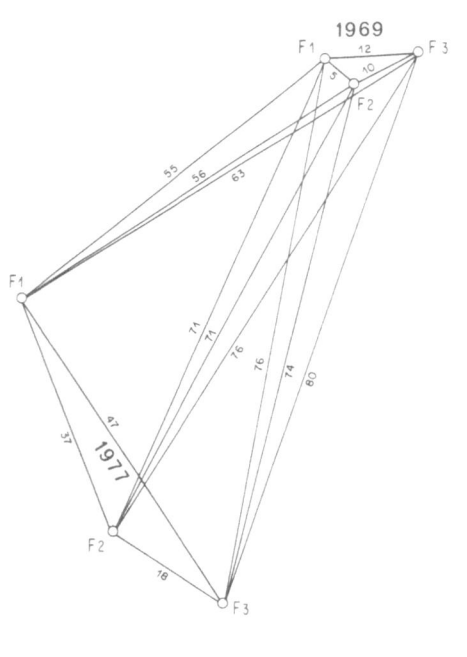

F1: unfertilized plots
F2: moderately fertilized plots
F3: heavily fertilized plots

Fig. 4. Differences (D = 100-S) between treatments, before (1969) and at the end of the experiment (1977).

groups of treatments per year may be expressed as the sums of the common CSC values per species in the two treatments S. It varies from 0-100. The results obtained are presented in Figure 3. The fertilized plots (F2 and F3) are very similar, with S = 82% ± 8%. The differences between unfertilized plots (F1) and fertilized ones (F2 and F3) increase from 1969-1973 and from 1974-1976, decrease from 1973-1974 and from 1976-1977. The general tendency however, is an increase in the differences between fertilized plots and unfertilized ones.

Figure 4 expresses the differences (as 100-S values) between treatments, before the start of the experiment and at the end of the experiment.

Development of grazing value

From the CSC values in each treatment and an agricultural value index per species (Table 4), the grazing value and the corresponding optimal sheep grazing density may be calculated (In some cases the CSC value is not known, and replaced by the 'presence specific contribution' (Poissonet & Poissonet, 1969)).

Grazing value index

De Vries (1937, 1938), de Vries *et al.* (1949), Klapp *et al.* (1953), Klapp (1954), Areu & Pidal (1961) and Daget & Poissonet (1971, 1972) give grazing quality index values ('agricultural' value, de Vries (1948)) for numerous species. These values are determined by growth rapidity, nutrient value, palatability and digestibility. They are usually expressed in a scale from 0-10 or from 0-5. Most of the species indexed in this way are temperate grassland species.

The index values in Table 4 are only a first approximation of the grazing values of Mediterranean species. Furthermore, certain, especially ligneous, species have an index value of 0 since they are not usually browsed. However, under special conditions, they can provide some additional food. For example, sheep may eat *Genista scorpius* flowers or *Quercus pubescens* sprouts in spring or, in a bad season, acornes and shrub leaves (Thiault *et al.*, 1979).

Table 4. Grazing values (0–10) of species growing in the Montpellier area.

Gramineae

Aegilops geniculata or	1	Festuca arundinacea	6
Aegilops neglecta*		Festuca lemanii	2
Aira caryophyllea	0	Festuca arvernensis	2
Dichanthium ischaemum	2	Festuca ovina	2
Arrhenatherum elatius	8	Festuca rubra	4
Avena barbata	1	Festuca paniculata	0
Avenula bromoides	3	Hordeum murinum	3
Brachypodium distachyon	1	Koeleria macrantha or	2
Brachypodium phoenicoides	1	Koeleria pyramidata*	
Brachypodium retusum	2	Koeleria vallesiana	2
Bromus erectus	4	Lolium perenne	10
Bromus madritensis	2	Melica ciliata	1
Bromus hordeaceus	2	Phleum phleoides	6
Bromus sterilis	1	Phleum pratense	8
Cynodon dactylon	4	Poa annua	6
Cynosurus echinatus	1	Poa bulbosa	4
Dactylis glomerata	10	Poa pratensis	8
Deschampsia media	1	Poa trivialis	8
Desmazeria rigida	1	Trisetum flavescens	6
		Vulpia ciliata	1

Leguminosae

Anthyllis vulneraria	5	Scorpiurus muricatus	3
Argyrolobium zanonii	1	Trifolium angustifolium	1
Astragalus monspessulanus	1	Trifolium arvense	2
Coronilla minima	4	Trifolium campestre	1
Dorycnium hirsutum	0	Trifolium fragiferum	4
Dorycnium pentaphyllum	3	Trifolium lappaceum	1
Hippocrepis comosa	4	Trifolium pratense	8
Lathyrus aphaca	2	Trifolium repens	8
Lathyrus cicera	5	Trifolium scabrum	1
Lotus corniculatus	6	Trifolium stellatum	1
Medicago lupulina	5	Trigonella monspeliaca	1
Medicago minima	5	Vicia hybrida	4
Medicago orbicularis	5	Vicia sativa	4
Medicago rigidula	5	Vicia sativa ssp nigra	4
Melilotus sulcata	1	Vicia sativa ssp amphicarpa	4
Onobrychis supina	4	Vicia tetrasperma	2
Ononis minutissima	1	Vicia tenuissima	2
Psoralea bituminosa	6		

Miscellaneous species

Cyperaceae		Compositae	
Carex divisa	1	Centaurea pectinata	2
Carex divulsa	1	Crepis sancta	2
Carex flacca	1	Crepis vesicaria	3
Carex hallerana	3	Hypochoeris radicata	1
Carex humilis	3	Lactuca sp.	2
		Picris echioides	4
Liliaceae		Picris hieracioides	4
Aphyllantes monspeliensis	2	Reichardia picroides	3
		Scorzonera hirsuta	2
Rosaceae		Scorzonera laciniata	2
Sanguisorba minor	6	Senecio vulgaris	1
		Sonchus asper	2
		Sonchus oleraceus	2

Table 4. (Continued)

Miscellaneous species			
Umbelliferae		Compositae (cont.)	
Daucus carota	2	*Taraxacum sect. erythrospermum*	1
Seseli elatum	1	*Taraxacum officinale*	4
Seseli montanum	1	*Tragopogon pratensis*	2
		Urospermum dalechampii	1
Convolvulaceae			
Convolvulus arvensis	2	Other species	0
Convolvulus cantabrica	2		
Plantaginaceae			
Plantago lanceolata	4		
Rubiaceae			
Galium parisiense			
Galium divaricatum	1		
Dipsacaceae			
Cephalaria leucantha	2		

* These flora Europaea species were not recognized separately at the time of our investigations.

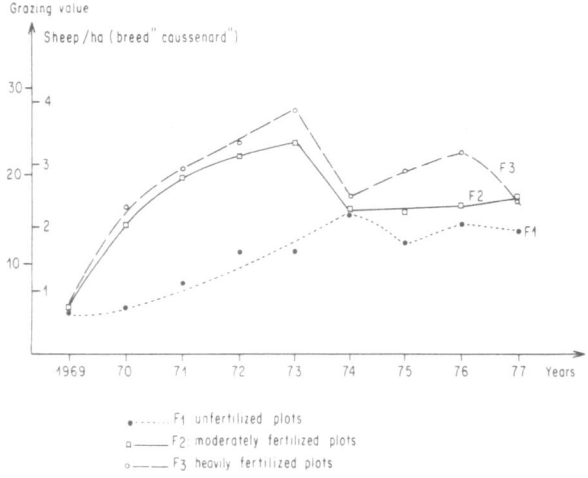

Fig. 5. Variation in grazing value per treatment, from the start (1969) till the end of the experiment (1977).

Grazing value

To estimate the grazing value (GV). The $CSC(p)_i$ are multiplied by the specific indexes (SI). While the $CSC(p)_i$ not always is known, the 'presence-specific-contribution' (PSC), currently called, 'specific contribution' (SC), is often used. PSC is linked to CSC in certain cases (Poissonet & Poissonet, 1969).

The values obtained are added and expressed as a percentage:

$$GV = a* \sum_{i=1}^{n} CSC(p)_i *SI_i$$

where n = number of species in the plot, and a = coefficient equal to 0.2 if the index values vary from 0–5, and equal to 0.1 if they vary from 0–10.

The results obtained in the experimental plots are presented in Figure 5. They show: a slow increase of the grazing value in the unfertilized plot (F1) between 1969 and 1974. This increase is followed by a relative stability between 1974 and 1977 close to the value 14/100 (i.e. about three times more than the initial grazing value).

We also observe a very quick increase in the fertilized plots (F2 and F3) between 1969 and 1973. In 4 yrs, the grazing value has been multiplied by 5 to 7. This increase is followed by a straight decrease in the spring of 1974 (after the summer drought of 1973) which reduces the grazing value of fertilized plots to the value of the unfertilized plots, with the disappearance of *Arrhenatherum elatius* and the development of *Compositae* without grazing. Then, between 1974 and 1977, the grazing value in F2 is slightly above the one in F1, whereas, in F3, after an increase from 1974-1976, another decrease in 1977 brings back the grazing value down to the one of F2 (after the summer drought of 1976).

Conclusions

In the initial garrigue community *Quercus coccifera* is the main species. After removing the above-ground vegetation and repeated cuttings, we obtain a completely different community which is more equilibrated and richer in herbaceous species than the initial one. These species indicate an artificial 'eutrophication' rather than a change into a community rich in grassland species, although the grazing value has been multiplied by 3 or more.

The changes in the composition and grazing value are quicker when the quantity of fertilization is higher. However, the differences observed are not proportional to the quantity of fertilizers: the differences between the treatments F3 (200 units of N, P_2O_5, K_2O) and F2 (100 units of the same fertilizers) are lower than the ones obtained between the treatments F2 and F1 (no fertilization).

The general dynamic tendencies are not observed in certain years. These deviations seem to be limited to 1974 and 1977, and related to the low quantities of rain in June, July and August 1973 and 1976. In fact, certain plants *(Gramineae, Cyperaceae)* can disappear altogether with persisting drought (Thiault, 1979). The high level of fertilizing makes the vegetation more susceptible to this climatic risk. The disappearance of species corresponds, first of all, to an increase of bare ground, and then to an increase of *Compositae*.

It is possible to conclude that the *Quercus coccifera* garrigue, after being destroyed and with periodical mowing develops into a dense herbaceous formation. The development is faster with a high level of fertilization and slower without fertilization. With the chosen times of cutting, fertilization makes the vegetation susceptible at drought periods and allows certain unwanted species to proliferate in the bare spots. For this reason, another experiment including sheep grazing was set up in 1974.

References

Aitchison, J. & Brown, J. A. C., 1966. The lognormal distribution with special reference to its uses in economics. Cambridge University Press, London, 176 pp.

Areu, & Pidal, 1961. Methodo de Klapp-König y Stählin para expressar el valor alimentico di un pastizal. Montes 102: 603-609.

Braun-Blanquet, J., Roussine, N. & Nègre, R., 1952. Les groupements végétaux de la France méditerranéenne. C. Nat. Rech. Sc., Paris, 297 pp.

Daget, Ph. & Poissonet, J., 1965. Contribution à l'étude des herbages des plateaux basaltiques de l'ouest du Cantal. Doc. no. 16, C.E.P.E., C.N.R.S., Montpellier, 117 pp.

Daget, Ph. & Poissonet, J., 1970. Distribution des fréquences spécifiques dans les phytocénoses herbacées en déséquilibre. 5è. Coll. Ecol., E.N.S. Paris, 25 p.

Daget, Ph. & Poissonet, J., 1971. Une méthode d'analyse phytologique des prairies. Ann. Agron. 22: 5-41.

Daget, Ph. & Poissonet, J., 1972. Un procédé d'estimation de la valeur pastorale des pâturages. Fourrages 49: 31-39.

Daget, Ph. & Poissonet, J., 1978. Un aspect méthodologique des études comparatives de carrés permanents sur prairies naturelles. Phytocoenosis 7: 151-164.

D.P.E.G. (Division de Phyto-écologie Générale), 1967. Etude du dynamisme de la végétation dans ses rapports avec le milieu. Programme de recherches envisagées sur la garrigue à Quercus coccifera de la série du Quercus ilex. C.E.P.E., C.N.R.S., Montpellier, 89 pp.

Emberger, L., 1942. Un projet de classification des climats du point de vue phytogéographique. Bull. Soc. Hist. Nat. Toulouse 77: 97-124.

Klapp, E., 1954. Wiesen und Weiden. Parey, Berlin, 519 pp.

Klapp, E., Boeker, P., Köning, F., Stählin, A., 1953. Wertzahlen der Grünlandpflanzen. Das Grünland, 5: 38-40.

Levy, E. A. & Madden, E. A., 1933. The point-method of pasture analysis. N. Zeal. J. Agric., 44: 267-279.

Long, G., Poissonet, P., Thiault, M. & Trabaud, L., 1976. Etude expérimentale et diachronique d'une phytocénose de Quercus coccifera L. Conf. MAB Médit. Montpellier, 19 pp.

Long, G. A., Etienne, M., Poissonet, P. S. & Thiault, M. M., 1978. Inventory and evaluation of range resources in 'maquis' and 'garrigues' (French Mediterranean area): Productivity levels. Proc. 1st Int. Rangeland Congr.: 505-509.

Poissonet, J., 1968. Recherche sur les lois générales d'équilibre dans la composition floristique des formations herbacées denses. Premiers résultats et hypothèses. C.E.P.E., C.N.R.S., Montpellier, 18 pp.

Poissonet, J., 1978. Equilibre et déséquilibre des phytocénoses herbacées. 10 Journées du Grenier de Theix, I.N.R.A., Versailles, 461-468.

Poissonet, J. & Poissonet, P., 1969. Essai de comparaison de méthodes d'analyses phytosociologiques et agronomiques des formations herbacées denses permanentes. C.E.P.E., C.N.R.S., Montpellier, 50, 120 pp.

Poissonet, J. & César, J., 1972. Structure spécifique de la strate herbacée dans la savane à palmier ronier de Lamto (Côte d'Ivoire). Ann. Univers. d'Abidjan, Ecol. 5: 577-601.

104

Poissonet, P., 1966. Etude méthodologique en écologie végétale à partir de photographies aériennes. Thèse de spécialité (Biologie végétale, Ecologie), Fac. Sc. Montpellier, 106 pp.

Poissonet, P., Romane, F., Thiault, M. & Trabaud, L., 1978a. Evolution d'une garrigue de Quercus coccifera. L. soumise à divers traitements: quelques résultats des cinq premières années. Vegetatio, 38: 135–142.

Poissonet, P. & Thiault, M., 1978b. Quelques problèmes posés par une série d'expériences faites dans une garrigue de Quercus coccifera L. Phytocoenosis, 7: 191–202.

Thiault, M., 1979. Présentation des parcours méditerranéens. III. Réflexions à partir de quelques aspects bioclimatiques. 10è. Journées du Grenier de Theix, I.N.R.A., Versailles: 361–373.

Thiault, M., Prud'hon, Reboul, Béchet, Molénat & Theriez, 1979. Amélioration pastorale de la garrigue. 10è Journées du Grenier de Theix, I.N.R.A., Versailles: 375–396.

Tutin, T. G., Heywood, V. H., Burges, N. A., Moore, D. M., Valentine, D. H., Walters, S. M., Webb, D. A., 1964, 1968, 1972, 1976, 1980. Flora Europaea, 5 vol., Cambridge University Press, 1, 464 pp. (1964), 2, 375 pp. (1968), 3, 370 pp. (1972), 4, 505 pp. (1976), 5, 452 pp. (1980).

Vries, D. M. de, 1937. Methods of determining the botanical composition of hay fields and pastures. Rep. IV th. Int. Grassl. Congr., Aberystwyth: 474–480.

Vries, D. M. de, 1938. The plant sociological combined specific frequency and other methods. Chronica botanica, 4: 115–117.

Vries, D. M. de, 1948. Method and Survey of the characterization of Dutch grasslands, Vegetatio 1: 57.

Vries, D. M. de, Boer, Th. de & Driven, J., 1949. Evaluation of grassland by botanical research in the Netherlands. Nations-Unies, E/CONF. 7/SEC/W 159, 8 p.

Accepted 27.7.81.

Changes in the floristic composition of a *Quercus coccifera* L. garrigue in relation to different fire regimes*

L. Trabaud[1] & J. Lepart[2]**

[1] *Département d'Ecologie générale, Centre d'Etudes phytosociologiques et écologiques L.-Emberger, B.P. 5051, 34033 Montpellier Cedex, France*
[2] *Ecothèque Méditerranéenne, B.P. 5051, 34033 Montpellier Cedex, France*

Keywords: Dynamics, Floristic composition, Garrigue, Languedoc, Mediterranean, Prescribed burning, *Quercus coccifera*

Abstract

To analyse the impact of fire on plants, an experiment has been set up in a *Quercus coccifera* L. garrigue near Montpellier. The objectives of the study were to follow the changes of the vegetation after fire in relation with different prescribed burning regimes: a fire every six years, a fire every three years and a fire every two years, lit at two different seasons: end of spring or beginning of autumn. In spite of the frequent burning, the main floristic composition remains constant on the whole. The characteristic dominant taxa remain present. The changes observed are associated both with the frequency and the season of burning. An increase of fire frequency leads to an increase of the number of taxa which cannot persist in the plots. The dates of burning essentially have an essential effect upon the number of taxa which tend to settle: the increase of floristic richness is much more pronounced for the autumn burnings. This relative stability is explained by the fact that most of the taxa present before the burnings regenerate principally by vegetative means, while taxa appearing during the year after each burn are rapidly eliminated by the plants which existed before the fires.

Introduction

Fire has often been considered as a powerful factor in the regressive succession of vegetation, and for many authors it is a main factor determining the present state of the mediterranean vegetation. According to Braun-Blanquet (1935, 1936), Kuhnholtz-Lordat (1938, 1952, 1958), Kornas (1958), burning in combination with agriculture and sheep-farming, caused the gradual change from *Quercus ilex* forests, via *Quercus coccifera* garrigues to *Brachypodium ramosum* swards.

In a diachronic study (direct approach) of the development of the vegetation and flora after wildfires in Languedoc (mediterranean Southern France), Trabaud & Lepart (1980) demonstrated that there was a quick return towards a stage close to the initial state existing prior to the fire. This result is in agreement with findings by authors working with analogous plant communities (such as Californian chaparral: Sampson, 1944; Horton & Kraebel, 1955; Sweeney, 1956, 1967; Hanes, 1971; Israelian maquis: Naveh, 1974, 1975; Australian scrub: Purdie & Slatyer, 1976). Very often, however, the initial vegetation composition is reconstructed from stubs or snags remaining on the burnt areas, and from the surrounding unburnt vegetation.

*Nomenclature follows Fournier (1961)
**We gratefully thank Dr. G. Long who spent much time in discussing the manuscript, Professors M. Godron and Ch. Sauvage for their comments of an earlier french text, M. Gautier for helping us with the translation, the technicians for their work during the burnings and observations, and V. Aussaresses for typing the manuscript.

Vegetatio 46, 105–116 (1981).   0042-3106/81/0462-0105/$2.40.

By using prescribed burning it is possible (1) to compare the post-fire floristic composition with the initial floristic composition; (2) to study the modifications induced by various fire freqencies and seasons of burning; and (3) to analyze the effect of fire on the behaviour of typical species of the *Quercus coccifera* garrigue.

Experimental design and methods

Location of the study area

The experiment has been set up on a hill called Puech-du-Mas-du-Juge, about 10 km N of Montpellier. The plant community here is a *Quercus coccifera* (kermes scrub oak) garrigue (*Cocciferetum* Br. Bl. 1924, sub-association *brachypodietosum* Br. Bl. 1935; Braun-Blanquet *et al.*, 1952). The choice of this plant community follows from the large area it occupies in Southern France: more than 100 000 ha. The locality has been the object of numerous previous studies: Long *et al.* (1961, 1967), Trabaud (1962), Poissonet (1966).

The climate of the area is Mediterranean Humid according to Emberger's classification (1942, 1971). It is characterized by two contrasting seasons. In winter there is a rainy period with low temperatures; the summer is generally hot and dry. The annual mean temperature is about 14.4 °C; the mean temperature of the coldest month (january) equals 2.4 °C, while that of the warmest month (july) is 27.5 °C; the mean annual rainfall (over a period of nine years) (1969–1977) is 1102 mm.

This garrigue was swept by many wildfires (the latest ones were in September 1943, and August 1951). Since the beginning of the experiments (1969), the studied area is completely protected from any external human action. At that time, the garrigue was a dense scrub of *Quercus coccifera*, about 1 m high, with a cover of 80 to 100%. The other dominant taxa were: in the shrub layer *Dorycnium suffruticosum, Genista scorpius, Teucrium chamaedrys*; in the herb layer *Brachypodium ramosum*, with *Rubia peregrina, Carex halleriana* and *Carex humilis*.

Burning treatments

The *burning times* were chosen in relation to some mean phenological stages of the *Quercus coccifera* population (Trabaud, 1974, 1980).

First period: kermes scrub oak has started its spring growth; the first annual shoots and the young leaves have already been developed; the flowers have appeared. *Quercus coccifera* is then in a turgid maximally photosynthetic stage. Burnings are usually set at the end of May or the beginning of June according to the meteorological conditions.

Second period: the beginning of autumn, after the lignification of the young twigs when the vegetation seems to be at rest. The burnings are set at the beginning of September.

The *burning frequencies* are: (1) 'one fire every six years', (2) 'one fire every three years' and (3) 'one fire every two years'.

The experimental design is a factorial one:
First factor: 'burning season'
   – spring fires (P)
   – autumn fires (A)
Second factor: 'burning frequencies'
   – plots burned every six years (6)
   – plots burned every three years (3)
   – plots burned every two years (2)
Control plots (T) are never burned.

Five replicates have been established for every combined treatment; thus there are 35 basic plots.

Vegetation analysis

A floristic list of all the taxa present in each plot is made every year in May, before the spring burnings, and also four months after every burning. Furthermore, a permanent 10 m long line is located in the middle of the plot in the longest direction; observations are made every 10 cm by means of needles. At each point, the presence and the number of hits per layer for each taxon are noted. In addition, the occurrence of every taxon observed in the vertical plane determined by the needles and the 10 cm segment between the two needles is recorded.

Results

Development of the floristic composition

Development of the floristic richness (Table 1)

Only the seven years (1969–1975) corresponding to the first cycle of burnings defined by the experimental design will be studied here. Four months

Table 1. Average number (and standard error) of taxa per plot, during the observations: 1969–1975 (averages are from 5 replicates).

	Number of taxa present at the beginning of the observations col. 1	Number of taxa present at the end of the observations col. 2	Number of taxa present at the beginning of observations but absent at the end col. 3	Number of taxa absent at the beginning but present at the end of observations col. 4	Number of taxa which are only present at a time during the observations col. 5	Number of taxa present at the beginning and at the end of the observations col. 6
unburnt vegetation (T)	26.6 (±1.0)	25.4 (±0.6)	3.2 (±0.9)	2.0 (±0.9)	1.6 (±0.8)	23.4 (±2.1) 77%*
Vegetation burnt every six years in spring (6 P)	26.8 (±0.9)	28.8 (±1.9)	1.4 (±0.7)	3.4 (±0.2)	2.8 (±0.6)	25.2 (±0.9) 77%*
Vegetation burnt every three years in spring (3 P)	24.2 (±1.2)	25.6 (±1.0)	2.0 (±0.7)	3.4 (±0.8)	4.8 (±1.4)	22.2 (±1.6) 69%*
Vegetation burnt every two years in spring (2 P)	27.0 (±1.2)	26.0 (±0.6)	4.2 (±0.5)	3.2 (±0.9)	9.6 (±1.3)	22.8 (±1.1) 57%*
Vegetation burnt every six years in autumn (6 A)	29.6 (±0.5)	30.8 (±0.6)	1.4 (±0.5)	2.6 (±0.9)	6.4 (±1.2)	28.2 (±1.2) 73%*
Vegetation burnt every three years in autumn (3 A)	27.6 (±0.8)	28.8 (±0.8)	2.4 (±0.7)	3.6 (±1.1)	7.0 (±2.5)	25.2 (±1.1) 66%*
Vegetation burnt every two years in autumn (2 A)	24.8 (±0.7)	23.2 (±0.7)	4.6 (±1.0)	3.0 (±0.6)	11.2 (±2.9)	20.2 (±0.2) 52%*

* Percentage of permanent taxa with regard to all taxa recorded during the observations.

after burning, whatever the season, floristic richness of every plot (number of visible living taxa) is always lower than before the fire, later the richness increases up to and sometimes beyond the total number of taxa present in the unburnt plots (Trabaud, 1980).

Generally, after more than four months, the autumn burnt plots have a higher number of taxa than the spring burnt plots (Trabaud, 1980). This difference may result from the fact that *Quercus coccifera* is in active growth in spring, producing sprouts which quickly cover the bare ground after the fire passage, leaving very little place for other taxa to establish. In autumn, *Quercus coccifera* is less active in sprouting and does not re-occupy the surface so rapidly. This will permit other taxa to establish and, even to germinate the following spring. Hence, particularly annual taxa will be more numerous.

Inspection of Table 1 leads to some additional statements:

1. The number of taxa which were present at the beginning of the observations and which then disappeared (column 3) is higher in the unburnt vegetation than in the vegetation burnt every six or three years (but lower than in the plots burnt every

two years); this may be due to the closing of the vegetation. In fact, the species involved are perennial and heliophilous. However, this loss is partially compensated by other taxa which have established (column 4). Very few taxa attempted to establish and they were not able to persist in the unburnt vegetation (1.6 taxon per plot, column 5). On the whole, there are few taxa in the unburnt vegetation which disappear: 77% of all the taxa recorded during the observations in the treatment (and 87% of the taxa present at the end of the observations) were present at the beginning.

2. The number of taxa which appear and remain (column 4) in the burnt vegetation is low; this does not depend on the season nor the frequency of burnings.

3. On the contrary, for each burning frequency the number of taxa which disappear is varying somewhat according to the burning season (column 3). However, the higher the frequency, the more this number increases; it increases from 1.4 for burning every six years to more than 4.0 for burning

every two years. In this case, this loss of taxa is badly balanced with the entrance of persistent taxa (column 4); on the other hand, numerous taxa appear which are present only temporarily (column 5). They are mainly annual, seed-reproducers and they thrive only during the period the soil remains bare.

4. The number of taxa that appear only ephemerally (column 5) is associated with frequency and date of burning: the higher the frequency, the greater this number is. Besides this number is greater in the autumn-burnt plots.

5. The number of taxa which are present at the beginning of the experiment, and which are all present seven years later (column 4), is notably similar for most treatments.

6. On the whole, the number of taxa for each plot is very little modified by repeated burnings; for example, the flora of plots burnt every two years in autumn keeps 52% of all the taxa recorded during the observations in the treatment (81% of taxa present at the beginning of the experiment and

Table 2. Mean number of taxa disappearing per plot and percentage with regard to the total number of taxa present in each frequency class (averages are from 5 replicates).

Fire regimes	Segment frequency classes	0	1	2 and 3	4 to 7	8 to 15	16 to 31	32 to 64	64 to 100
Unburnt vegetation (T)		2.2 / 21.1%	0.8 / 26.7%	0.2 / 5.6%	0	0	0	0	0
vegetation burnt in spring	every six years (6 P)	1.4 / 10.9%	0	0	0	0	0	0	0
	every three years (3 P)	1.8 / 15.0%	0.2 / 7.7%	0	0	0	0	0	0
	every two years (2 P)	3.4 / 20.2%	0.2 / 12.5%	0.4 / 20.0%	0.2 / 10%	0	0	0	0
vegetation burnt in autumn	every six years (6 A)	1.2 / 8.2%	0.2 / 9.1%	0	0	0	0	0	0
	every three years (3 A)	2.2 / 15.9%	0	0	0	0	0.2 / 12.5%	0	0
	every two years (2 A)	3.4 / 27.4%	0.2 / 11.1%	0.2 / 9.1%	0.4 / 20.0%	0.4 / 15.4%	0	0	0

permanent during the time of the observations).

In spite of successive burnings and the entrance of new taxa, the floristic richness of the *Quercus coccifera* garrigue remains rather stable, through the time of observation.

Importance of taxon frequency

Taxa which disappear, or appear, are always taxa with a low frequency at the time of observation and a low cover. Those which are highly frequent and possess high cover values remain present.

Repeated fires cause a notable decrease of the phytomass of the woody species; this is due to the shortening of the growth period and to the exhaustion of stumps and roots after repeated fires (Trabaud, 1980). For example, in May 1979 the woody phytomass in the unburnt vegetation (control, T) reached 2954 g $\cdot$ m^{-2}, in the vegetation burnt every two years in autumn (2A) only 157 g $\cdot$ m^{-2}.

Most of the dominant taxa stay in the plots and probably grow again vigorously after burning would stop.

We now consider if a point of no return has been reached, i.e. if taxa which occupied an important place in the plots have disappeared from the plots, or other taxa have appeared and play an important part now.

To investigate this, the taxa were grouped according to the importance value they have in the plots by using their presence scores in the segments.

Frequency values have been grouped in approximately $\times$ 2 classes. For each frequency class, the number of taxa wich have disappeared or appeared were calculated (Tables 2 and 3).

Ca. 10% of the taxa which have disappeared had a segment frequency higher than 1% (this group representing 50% of all the taxa in the plots). Thus, the number of taxa which disappear is nine times higher for the taxa with a frequency < 1% than for taxa with a frequency > 1%.

With only one burning there are no taxa with a segment frequency > 1% that disappear; with two burnings (3P and 3A) only one taxon of this category disappeared out of 10 plots. When the burnings are more frequent (every two years: 2P and 2A), there are 0.8 taxa of this category which disappear per plot.

Table 3. Mean number of taxa appearing per plot and percentage with regard to the total number of taxa present in each cover class.

		0	1	2 and 3	4 to 7	8 to 15	16 to 31	32 to 64	64 to 100
Unburnt vegetation (T)		1.4 14.0%	0.4 20.0%	0.2 5.5%	0	0	0	0	0
vegetation burnt in spring	every six years (6 P)	2.8 25.5%	0.4 12.5%	0.2 5.3%	0	0	0	0	0
	every three years (3 P)	3.0 22.7%	0.4 16.7%	0	0	0	0	0	0
	every two years (2 P)	2.4 17.1%	0.4 16.7%	0.4 18.2%	0	0	0	0	0
vegetation burnt in autumn	every six years (6 A)	2.0 13.3%	0	0.2 6.2%	0.2 10.0%	0.2 7.1%	0	0	0
	every three years (3 A)	2.8 21.2%	0.4 15.4%	0.4 10.5%	0	0	0	0	0
	every two years (2 A)	2.0 21.3%	0.4 12.5%	0.4 15.4%	0.2 10.0%	0	0	0	0

110

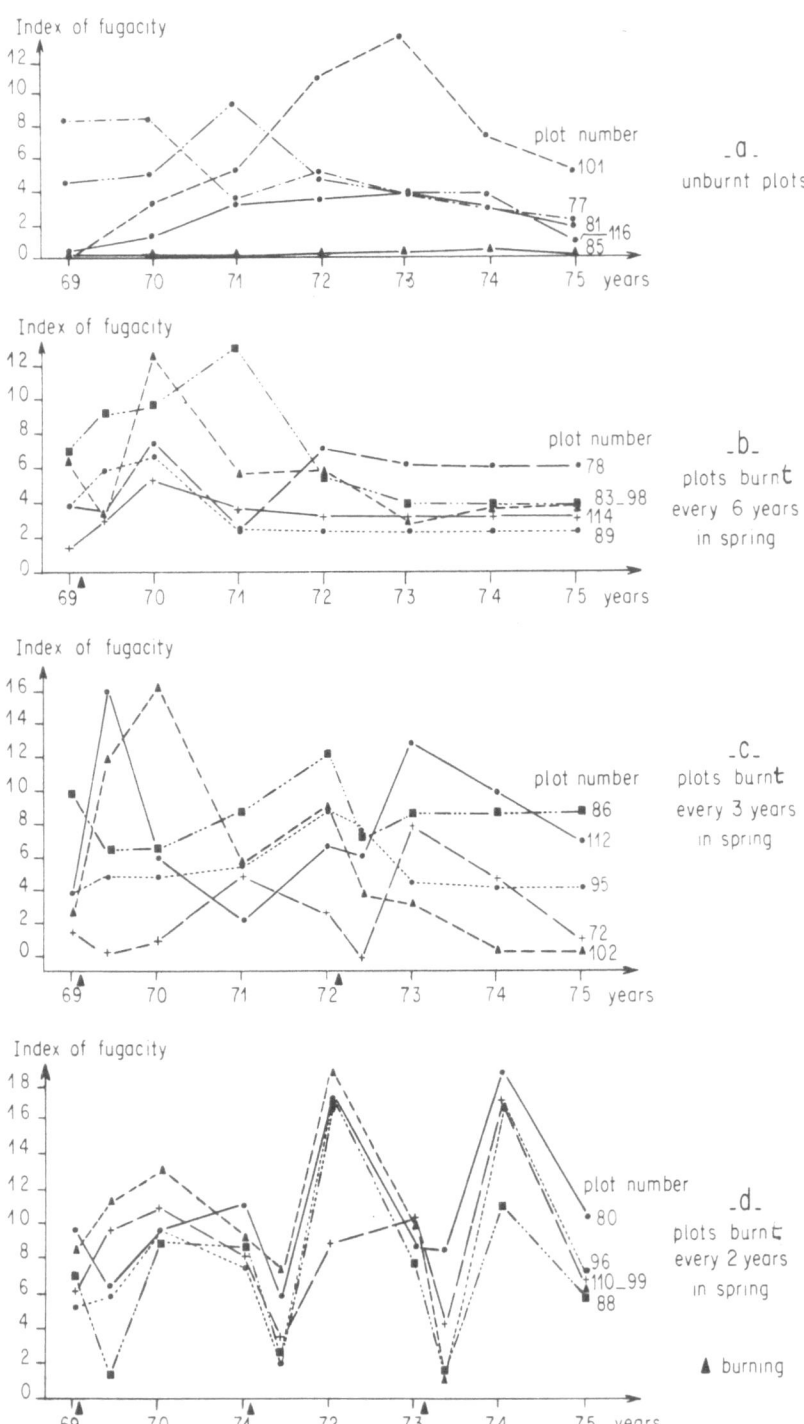

Fig. 1. Changes in the index of fugacity according to no burning and spring burnings.

One notes that the taxa which disappear are more numerous in the unburnt vegetation than in the vegetation burnt every six years (only one fire). *Sanguisorba minor* is the only taxon with a segment frequency higher than 1% which disappears; it is a perennial herb which is more frequent in more open communities than in the *Quercus coccifera* garrigue.

For the appearing taxa, the value of the frequency (associated with the segment) is apparently less closely linked to the burning frequencies (Table 3).

In conclusion, the taxa which appear or disappear are on the whole not dominant in the community.

Permanence of the flora: fugacity of the taxa

A taxon can be considered as 'fugacious' if it does not persist (in the form of alive above-ground individuals) in the plots over the period of observation. The fugacity of each taxon is simply measured by the number of observations where it is absent within the limits of the permanently adopted experimental design. We are interested in the period of establishment of the fugacious taxa and in any pattern of fugacity, i.e. a dominant model of distribution of the fugacious taxa that could be related to the burning times. An index of fugacity is calculated, which corresponds to the average fugacity of the taxa present at a given time in the plots. This fugacity can be considered as a measure of the level of instability of the floristic composition of a plot; it has already been used by Trabaud & Lepart (1980) in a slightly different form.

In the present study the index has been modified to avoid amplifications due to the high frequency of the burnings:

$$IF_i = \frac{Amax - Ai}{Amax} \times 100$$

in which:

$Amax$ = number of observations
Ai = average number of observation between, the first and the last presence of every taxon observed at time *i*.

The index is high when taxa are present for a very short time; it is low when taxa remain present in the plots for a long period.

Fugacity appears to vary with floristic richness,

which indicates that the taxa appearing for a short time are responsible for an increase in floristic richness.

In the unburnt vegetation (Fig. 1a), the index does not show a clear tendency. The maxima of fugacity appear at different times in the different plots, which means that taxa can appear in a plot, but disappear almost immediately after. The indices are generally low, and sometimes even close to zero (plot number 85); only plot 101 presents an important peak (due to the sudden appearance and disappearance of two annual taxa). The variations are progressive and the tendency is towards a weak decrease (except for plot 101), which implies a progressive disappearence of taxa which do not tolerate the closing of the vegetation cover.

In the burnt vegetation (Figs. 1b, 1c, 1d and 2a, 2b, 2c), the index of fugacity is generally very low four months after burning: the taxa which establish soon are mostly perennial. The maximum fugacity is then reached during the first and the second year after fire; then the index of fugacity decreases towards the initial level, when burnings occur every six years.

With burnings every two years, there is an increase in the variations of fugacity values but a greater regularity of the fluctuations. It seems that these two phenomena are linked with a decrease in the establishment rate of perennials which allows an easier settlement of annuals.

There is a strong nucleus of ever present taxa with some additional taxa which take advantage of the burnings to appear for a rather short time. Therefore on the whole the flora of the *Quercus coccifera* garrigue has a strong resistance to fire action and a high stability, despite the stresses brought about by the experiments.

Behaviour of some taxa

To determine the responses of some taxa to fire, we used the method of ecological profiles of corrected frequencies (Godron 1966, 1968; Guillerm 1969, 1971; Daget *et al.,* 1970); based on 305 observations over 7 yrs, i.e. systematic observations in spring every year for all plots, and for burnt plots observations four months after a burning. Only the 45 taxa present in at least 40 observations have been taken into account (Table 4).

A first interpretation was given by Trabaud

Table 4. Behaviour of some taxa according to frequency and season of burning. Data from experimental plots of the *Quercus coccifera* garrigue 1969–1975.

| | Behaviour of taxa at the beginning and the end of the considered period according to the frequency and the season of burnings. | | | | | |
| | A fire every 6 years | | A fire every 3 years | | A fire every 2 years | |
	Spring	Autumn	Spring	Autumn	Spring	Autumn
Sonchus asper (L) Hill	++/00	++/00	++/00	++/00	++/00	++/00
Euphorbia nicaeensis All.	++	+	++	++	++	++
Hippocrepis comosa L	++		++	++	++	++
Avena bromoides Gouan	0*	=	++	+	++	++
Sanguisorba minor Scop.	+	0*	+	+	+	+
Arrhenatherum elatius (L) Mert et K	+	=	=	+	=	+
Galium asperum Schreb.	+	+	+	=	=	+
Bupleurum rigidum L.	+	0	+	+	=	=
Sedum nicaeense Allioni	=	+		+		=
Cephalaria leucantha (L) Schrad	=	+	+	=	=	=
Rubus ulmifolius Schott.	=	=	=	+	=	=
Lonicera implexa Ait.	=	=	=	=	=	+
Asparagus acutifolius L.	=	=	+	=	=	=
Asphodelus cerasifer Gay	=	=	=	=	+	=
Clematis flammula L.	=	=			+	
Rosa sempervirens L.	=		+	=		=
Brachypodium ramosum (L) R et S.	=	=	=	=	=	=
Dorycnium suffruticosum Vill.	=	=	=	=	=	=
Quercus coccifera L.	=	=	=	=	=	=
Rubia peregrina L.	=	=	=	=	=	=
Teucrium chamaedrys L.	=	=	=	=	=	=
Stachys officinalis (L.) Trevisan	=	=	=	=	=	=
Daphne gnidium L.	=	=	=	=	=	=
Carex humilis Leyss.	=	=	=	=	=	=
Carex halleriana Asso.	=	=	=	=	=	=
Quercus ilex L.	=	=		=	=	
Festuca spadicea L.	=	=	=		=	=
Pistacia lentiscus L.	=			=	=	=
Centaurea pectinata L.	+	0*	=	+	=	=
Brachypodium phoenicoides R et S	=	+	0	=	+	=
Aphyllanthes monspeliensis L.	=	+	0	=	+	=
Phillyrea angustifolia L.	=	=	+	=	0	
Hieracium murorum L.	0*	0*	=	+	+	0
Euphorbia characias L.	=	+	0	=	0	0
Genista scorpius (L) Link	+	=	=	0	0	0
Bromus erectus Huds	=	=	=	=	0	=
Hieracium pilosella L.	=	=	=		0	=
Cistus salviaefolius L.	=	=	=	0	0	=
Viola scotophylla Jord.	=	=	=	=	0	0
Festuca duriuscula L.	0*			=	0	0
Smilax aspera L.	=	=	=	=	0	00
Cistus monspeliensis L.	+	=	0	0	00	0
Juniperus oxycedrus L.	=	=	=	0	00	00
Rhamnus alaternus L.	=	00	0	00	0	
Fumana coridifolia (Vill). P.F.	=	=	0	00	00	00

++ increase of the number of presences per plot, taxa not present at the beginning of the experiment,

+ increase of the number of presences per plots, taxa present at the beginning of the experiment,

= no change,

0 decrease of the number of presence per plots, taxa present at the beginning of the experiment,

00 disappearance of individuals.

* decrease in the number of presences of a taxon does not depend on burnings.

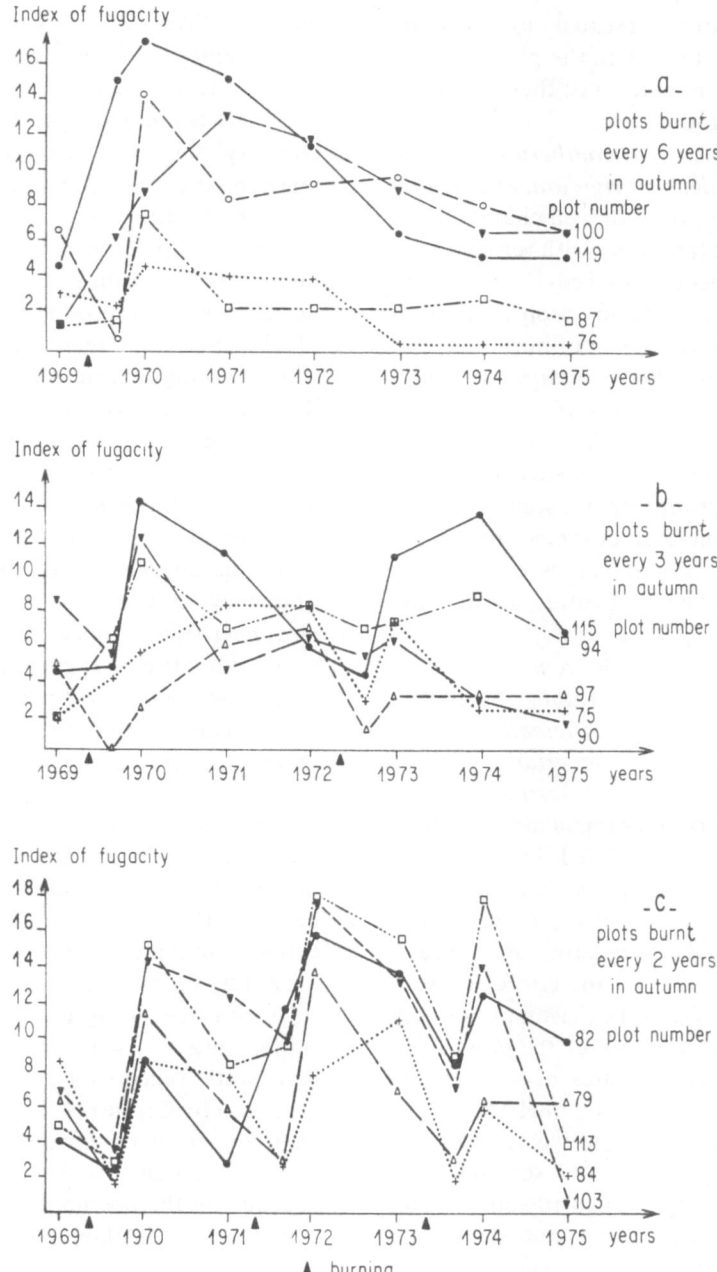

Fig. 2. Changes in the index of fugacity according to autumn burnings.

(1974). Now, we can check this behaviour with an experimental complete cycle of seven years.

Sonchus asper is the only one taxon, observed more than 40 times, which appears by the help of fire and disappears suddenly as soon as the fire action vanishes. Most of less encountered taxa which have such a behaviour are annual plants belonging to the *Compositae*, propagating by anemochorous propagules.

Euphorbia nicaeensis, Hippocrepis comosa,

Avena bromoides, are species which settle down owing to the conditions created by burning. Generally, they did not occur in the plots prior to the fires; but they do remain once they are there. Hence they are favoured by fire.

Sanguisorba minor, Arrhenatherum elatius, Galium asperum, Bupleurum rigidum, and *Sedum nicaeense* constitute a group of well characterized species. They all react favorably with regard to the fire action: their number of presences is increased at the end of the period of observations, whatever are the frequency and the season of burnings.

Cephalaria leucantha, Rubus ulmifolius, Lonicera implexa, Asparagus acutifolius, Asphodelus cerasifer. Clematis flammula, Rosa sempervirens are also species favoured by fire, but at a lower level than the previous group. Only some burning regimes do cause an increase of species presences, and then several burnings are necessary to bring about this tendency. They constitute a transition group with the following taxa.

Brachypodium ramosum, Dorycnium suffruticosum, Quercus coccifera, Rubia peregrina, Teucrium chamaedrys, Stachys officinalis, Daphne gnidium, Carex humilis, Carex halleriana, Quercus ilex, Festuca spadicea ssp. consobrina, Pistacia lentiscus form an important group of taxa; the most numerous among the considered ones. Except *Festuca spadicea ssp. consobrina,* all these taxa belong to the floristical suite of the kermes scrub oak garrigue (*Cocciferetum* Br. Bl., 1924). Apparently fire does not have any effect on them. Whatever the frequency and the season of burning, they remain present at the end of the considered period and the number of presences does not change with regard to the original one. These taxa withstand fire very well. They survive by sprouting vigorously or both sprouting and seeding. And though fire does not seem to have any effect upon them, it is not impossible that outside of the delimited plots, fire promotes their extension to the detriment of other taxa.

Centaurea pectinata, Brachypodium phoenicoides, Aphyllanthes monspeliensis, Phillyrea angustifolia, Hieracium murorum, constitute a group which do not react clearly with regard to fire. Sometimes fire seems to favour the extension of these taxa, sometimes on the contrary the number of presence decreases independently of the frequency or season of burnings.

Euphorbia characias, Genista scorpius, Bromus erectus, Hieracium pilosella, Cistus salviaefolius, Viola scotophylla, Festuca duriuscula, are taxa sensitive to fire action. Generally their number of presences is not decreased by only one fire; on the contrary, this number remains constant, or even tends to increase. But if the burnings are more frequent (a fire every three or two years), the number of presences is decreasing; the more numerous the burnings are the more important the decrease is; in this group the number of decreases is 11 when burnings are set every two years, but only 3 when burnings are lit every three years. However, these taxa differ from those of the following group because they do not completely disappear.

Smilax aspera, Cistus monspeliensis, Juniperus oxycedrus, Rhamnus alaternus, Fumana coridifolia, are very sensitive to the action of fire since they can disappear completely. The higher the frequency of burnings the more generally they disappear (the number of disappearances for this group is 6 if fire is set every two years, 2 with a fire every three years and 1 when fire is lit up every six years). The season of burning has a notable effect on the disappearance rate: the disappearances are more numerous with autumn burnings than with spring burnings: 66,6% of the cases.

The analysis of the behaviour of the taxa shows that most of them remain present during the entire period of the experiments. This permanence on the experimental plots of the garrigue species is due to the ability for most of them to possess fire survival strategies depending on the perenniality of their subterranean organs and their capability to regenerate vegetatively. Contrary to the model proposed by Grime (1977) the plants of the burned *Quercus coccifera* garrigue present a strategy to withstand a high disturbance (fire) which essentially depends on the perenniality of individuals, reproduction by seeds playing only a weak role.

Conclusion

The floristic composition of the studied *Quercus coccifera* garrigue is stable on the whole regarding the impact of fire, particularly the characteristic taxa. Only a few species appear or disappear; the more numerous the frequencies of burnings the more important the changes are, and more particularly with autumn burnings.

This relative stability proceeds from the fact that most of the taxa present in the plots before the burnings regenerate principally by vegetative means (sprouts), whereas 'invading' taxa appear during the first year after each burning only as individuals issued from seeds and they are rapidly eliminated again by sprouting plants.

In the case of such a detailed study on the development of the flora of the *Quercus coccifera* garrigue according to different fire regimes, burning causes the disappearance of some taxa; but most of these are rare and not dominant. Only a high frequency of burnings (a fire every two years) causes some taxa with a cover higher than 1% to disappear. However, the disappearance rate remains very low (less than 1 taxon of that kind per plot).

For a short period fire does not seem to be an ecological factor which can bring the *Quercus coccifera* garrigue to a point of no return. This result is doubtless linked with a selection during several centuries of taxa withstanding this disturbance. It is not certain that similar results can be obtained in *Quercus ilex* or *Pinus halepensis* communities; it is particularly unlikely for the latter because reproduction by seeds plays an important role in the regeneration of pine stands.

References

Braun-Blanquet, J., 1935. Un problème économique et forestier de la garrigue languedocienne. Comm. SIGMA 35: 11–22.

Braun-Blanquet, J., 1936. La forêt d'Yeuse languedocienne (Quercion ilicis). Monographie phytosociologique. Mém. Soc. Etud. Sci. Nat. Nimes 5: 147 p.

Braun-Blanquet, J., Roussine, N. & Negre, R., 1952. Les groupements végétaux de la France méditerranéenne. CNRS, Paris, 297 p.

Daget, P., Godron, M. & Guillerm, J. L., 1972. Profils écologiques et information mutuelle entre espèces et facteurs écologiques. In: E. van der Maarel & R. Tüxen (eds.) Grundfragen und Methoden in der Pflanzen soziologie, p. 121–149. Junk, Den Haag.

Emberger, L., 1942. Un projet de classification des climats au point de vue phytogéographique. Bull. Soc. Hist. Nat. Toulouse 77: 97–124.

Emberger, L., 1971. Considérations complémentaires au sujet des recherches bioclimatologiques et phytogéographiques-ecologiques. In: Travaux de Botanique et d'Ecologie. Masson et Cie, 291–301.

Fournier, P., 1961. Les quatre flores de France. Lechevallier Ed., Paris, 1105 p.

Godron, M., 1966. Application de la théorie de l'information à l'étude de l'homogénéité et de la structure de la végétation. Oecol. Plant. 2: 187–197.

Godron, M., 1968. Quelques applications de la notion de fréquence en écologie végétale. Oecol. Plant 3: 185–212.

Grime, J. P., 1977. Evidence for the existence of three primary strategies in plants and its relevance to ecological and evolutionary theory. Amer. Natur. 111: 1169–1194.

Guillerm, J. L., 1969. Relations entre la végétation spontanée et le milieu dans les terres cultivées du Bas-Languedoc. Thèse 3ème cycle (Ecologie), Fac. Sc. Univ. Montpellier, 165 p.

Guillerm, J. L., 1971. Calcul de l'information fournie par un profil écologique et valeur indicatrice des espèces. Oecol. Pla 6: 209–226.

Hanes, T. L., 1971. Succession after fire in the chaparral of southern California. Ecol. Monogr. 41: 27–52.

Horton, J. S. & Kraebel, C. J., 1955. Development of vegetation after fire in the chamise chaparral of southern California. Ecology 36: 244–262.

Kornas, J., 1958. Succession régressive de la végétation de garrigue sur calcaires compacts dans la montagne de la Gardiole près de Montpellier. Acta Soc. Bot. Pol. 27: 563–596.

Kuhnholtz-Lordat, G., 1938. La terre incendiée. Essai d'agronomie comparée. La Maison Carrée, Nîmes, 361 p.

Kuhnholtz-Lordat, G., 1952. Le tapis végétal dans ses rapports avec les phénomènes actuels de surface en Basse-Provence. Ed. Le Chevallier, Paris, 208 p.

Kuhnholtz-Lordat, G., 1958. L'écran vert. Mém. Mus. Nat. Hist. Nat. Série B 9: 276 p.

Long, G., Visona, L. & Rami, J., 1961. La végétation du domaine de Coulondres (Hérault). Relations avec les problèmes de mise en valeur. Boll. Inst. Bot. Univ. Catania 3: 5–52.

Long, G., Fay, F., Thiault, M. & Trabaud, L., 1967. Essais de détermination expérimentale de la productivité d'une garrigue de Quercus coccifera, CEPE, CNRS, Doc. N° 39, 28 p.

Naveh, Z., 1974. The ecology of fire in Israël. Proc. Ann. Tall Timbers Fire Ecol. Conf. 13: 131–170.

Naveh, Z., 1975. The evolutionary significance of fire in the Mediterranean region. Vegetatio 29: 199–208.

Poissonet, P., 1966. Etude méthodologique en écologie végétale a partir des photographies aériennes. Thèse 3ème cycle (Ecologie), Fac. Sci. Univ. Montpellier, 107 p.

Purdie, R. W. & Slatyer, R. O., 1976. Vegetation succession after fire in sclerophyll woodland communities in south-eastern Australia. Aust. J. Ecol. 1: 223–236.

Sampson, A. W., 1944. Plant succession on burned chaparral lands in northern California. Calif. Agr. Exp. Sta. Bull. 685, 144 p.

Sweeney, J. R., 1956. Responses of vegetation to fire. A study of the herbaceous vegetation following chaparral fires. Bot. Publ. Univ. California 28: 143–250.

Sweeney, J. R., 1967. Ecology of some 'fire-type' vegetation in northern California. Proc. Annu. Tall Timbers Fire Ecol. Conf. 7: 110–125.

Trabaud, L., 1962. Monographie phytosociologique et écologique de la région de Grabels-St Gély-du-Fesc. Thèse 3ème cycle (Ecologie), Fac. Sci. Univ. Montpellier, 131 p.

116

Trabaud, L., 1974. Experimental study of the effects of prescribed burning on a Quercus coccifera L. garrigue. Proc. Annu. Tall Timbers Fire Ecol. Conf. 13: 97–129.

Trabaud, L., 1980. Impact biologique et écologique des feux de végétation sur l'organisation, la structure et l'évolution de la végétation des zones de garrigues du Bas-Languedoc. Thèse d'état Sciences nat., Univ. Sci. Tech. Lang. Montpellier, 291 p.

Trabaud, L. & Lepart, J., 1980. Diversity and stability in garrigue ecosystems after fire. Vegetatio 43: 49–57.

Accepted 6.6.1981.

The effects of protection on steppic vegetation in the Mediterranean arid zone of Southern Tunisia*

C. Floret

Centre d'Etudes Phytosociologiques et Ecologiques L. Emberger, BP. 5051, 34033 Montpellier, France

Keywords: Arid zone, Dynamic, Protection against grazing, Regeneration, Steppe, Tunisia, Vegetation

Abstract

The results of observations and measurements on five units of steppic vegetation in the mediterranean arid zone of S. Tunisia are presented in this paper. These units were totally protected against big herbivore grazing using barbed wire fencing during seven consecutive years.

The steppes in question are composed of small shrublike chamaephytes. Annual plants appear when there is sufficient rainfall. This vegetation has been subjected to extensive overgrazing and is in general, degraded.

Our objective was to study regeneration of the natural vegetation of five steppe types in order to recommend an optimum period for protection:

1) a steppe on a gypseous crust resulting from overgrazing of the *Zygophyllum album – Anarrhinum brevifolium* association;

2) a steppe on a sandy, gypseous, rocky-surfaced colluvium resulting from overgrazing of the *Rhantherium suaveolens-Artemisia campestris* association, *Atratylis serratuloides* sub-association;

3) a very open steppe on a deep, gypseous, sandy colluvium-alluvium (sierozem) having the appearance of a post-cultivated grazed *Plantago albicans* facies of the *Rhantherium suaveolens* association;

4) a steppe in good condition on a deep, sandy-soiled plain representative of the *Rhantherium suaveolens – Artemisia campestris* association;

5) a post-cultivated facies of the *Artemisia herba-alba – Arthrophytum scoparium* association on a deep piedmont, loamy, and gypseous colluvium soil, traditionally cultivated with cereals.

The effect of protecting these steppes was mainly the increase in cover of the perennial species. The cover of the steppe vegetation on sandy soils increased more than that of the steppes on loamy or gypseous crusts; this must be taken into consideration for pastoral management. The results also show that a protection period of 7 yr is not sufficient in an arid zone for new species to appear nor for succession to reach a next stage.

Introduction

Results of observations and measurements carried out on five different types of steppic vegetation are given in this paper. They are located in the regions 'Djeffara' and 'Basses Plaines Meridionales Orien-

tales' in the eastern part of pre-Saharan S. Tunisia. These steppes were completely protected from grazing by big herbivores using barbed wire fencing during seven consecutive years. The steppes are mainly composed of small-shrublike chamaephytes with a maximum height of 20–30 cm. They are covered with annual plants in spring and again in autumn if there is sufficient rainfall.

Since these steppes have been degraded by too intensive grazing, our objective was to study the

---

* Nomenclature follows Quezel & Santa (1962) in 'Nouvelle Flore de l'Algérie et des régions désertiques méridionales'. For typical Tunisian species, the authors are cited.

Vegetatio 46, 117–129 (1981). 0042-3106/81/0462-0117/$2.60.
© Dr W. Junk Publishers, The Hague.

possibilities for regeneration (floristic composition, plant cover and litter production).

While long-term protection has been achieved and maintained in arid zones in Australia (Trumble & Woodroffe, 1954; Noble, 1977) and the United States (Buffington & Herbel, 1965; Smith & Schmutz, 1975) few quantitative and continuous observations on protection, have been made in the arid countries in North Africa and the Near East. Le Houerou (1969, 1977) presented data for protected plots in Tunisia, Libya and Iraq. Naegele (1959) has drawn conclusions on arid vegetation in Mauritania after several years of protection. Recently, Thalen (1979) compared the results obtained within and outside exclosures on different types of vegetation in Iraq.

The present investigation on the dynamics of vegetation was carried out within the framework of a large study program on the Tunisian arid zone. (Institutions involved in this program are: Departments within the Tunisian Ministry of Agriculture, Centres d'Etudes Phytosociologiques et Ecologiques Louis Emberger, Office de la Recherche Scientifique et Technique d'Outre Mer. The United Nations Development Program, F.A.O. and UNESCO.) The main results concerning the natural vegetation and its relationships to climate, soil, and use by man have already been published (Floret & Pontanier, 1978; Floret *et al.*, 1976; Floret *et al.*, 1978). All of these results, and in particular those dealing with vegetation dynamics, have served as the basis for proposals for agricultural and pastoral management of the mediterranean arid zone. In this case, the problem was to recommend an optimum

period of protection for regeneration of several types of steppic vegetation.

The average annual rainfall in the regions studied here is 180-190 mm, with considerable annual variability. The rainfall occurs during the cool season. The average minimum temperature in the coldest month is 4-7 °C. The average maximum temperature in hottest month is 32-36 °C. According to Emberger's climatic classification, these regions are situated in the mediterranean arid belt (inferior part) with a mild to temperate winter.

Methods

Five stations were chosen in a typical 'toposequence', representative of the hills in the Djeffara region, near Gabes on the versant of a hill culminating at 147 (Fig. 1). Three units of vegetation were studied there from 1972 to 1979. The associations were described in detail by le Houerou (1954). From top to bottom these stations are:

Station 1: a steppe on a gypseous crust resulting from degradation of the *Zygophyllum album* (1) – *Anarrhinum brevifolium* (Coss. and Kra) association by overgrazing;

Station 2: a steppe on a sandy, gypseous, rocky-surfaced colluvium resulting from degradation of the *Rhantherium suaveolens – Artemisia campestris* association, *Atratylis serratuloides* sub-association, by overgrazing;

Station 3: a very open steppe on a deep, gypseous, sandy colluvium-alluvium (sierozem) having the appearance of a post-cultivated grazed *Plantago*

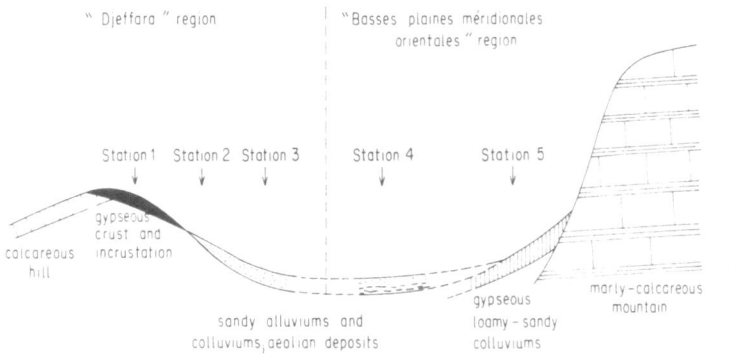

Fig. 1. A diagramatic section showing the localization of the stations studied and the main topographical and geomorphological characteristics.

albicans facies of the *Rhantherium suaveolens* association.

Farther west in the region of 'Basses Plaines Meridionales Orientales', two other stations were also protected and measurements on vegetation were made there from 1972–1979;

Station 4: a steppe in good condition on a deep, sandy-soiled plain representative of the *Rhantherium suaveolens – Artemisia campestris* association;

Station 5: a post-cultivated facies of the *Artemisia herba-alba – Arthrophytum scoparium* association on a deep piedmont, loamy, and gypseous colluvium soil, traditionally cultivated with cereals.

These five stations were chosen because they are representative of the steppic vegetation found in large areas of the region. The other main types of vegetation in these regions are found in slightly degraded depressions, covered with halophytic vegetation, and on calcareous mountain tops, which are dominated by *Stipa tenacissima*. Gaddes (1978) studied the regeneration of this type of vegetation.

At each station, the development of the plant and litter cover was followed. Permanent lines were installed in order to measure the cover by the quadrat-point method. Each line was delimited by two cemented stakes. A 20 m ribbon was stretched between the stakes at least two or three times a year (spring and autumn). A fine pin was descended to the ground every 5 cm along the ribbon. The presence of any species (living or dead) or any piece of litter hit, at least once by the pin was scored.

If n is the number of presences of a plant species on a line having N check points, the specific frequency (SF) of the plant, or its cover (c) expressed in per cent, is given by the following equation:

$$SF = c = \frac{n}{N} \times 100$$

Cover is proportional to the specific frequency (Greigsmith, 1964; Godron, 1968).

It should be noted that the sum of the specific frequencies, SSF, will be greater than the total plant cover if contact is made with several different plant species at the same check point. At each reading, 640 points were scored at stations 1, 2, 3 and 5 and 800 points at station 4. Given the number of points scored, for each species covering 20% of the surface area of the ground, there is a 95% chance of measuring an SF of 17–23% (Morice & Chartier, 1954). The number of points scored is, therefore, insufficient for species having a low cover. It should also be noted that a bias is introduced in the readings because of the thickness of the pin (Goodall, 1952) and the wind which, even when weak, causes the leaves to move and as a consequence additional contacts to be made. At station 4, in addition to estimating variations in cover by specific frequency measurements, a permanent device was installed to record the density of the main perennial species. Square plots, each measuring 4 m $\times$ 8 m and representing a total protected surface area of 768 m², were marked off at the station. Periodically, the individual perennial species were counted.

At each station, rainfall was measured during the 7 years of the study period. The pluviometric year extended from Sept. 1 of one year to Aug. 31 of the following year.

Data on annual precipitation rates are shown in Table 1. The precipitation in the first 4 years of the study was equal or greater than the average precipitation (175–185 mm) while the last 3 years were, in general, quite dry. The year 1978–1979 was a dry year at stations 4 and 5 only.

Table 1. Mean and annual precipitation (in mm) in the area of the 5 stations (rainfall period: 1 Sept. – 31 Aug.).

Stations	Rainfall period 1972–73	1973–74	1974–75	1975–76	1976–77	1977–78	1978–79	Mean
1, 2 et 3	164	260	199	534	102	136	167	221
4	210	315	186	295	85	136	88	188
5	164	371	167	357	96	156	114	203

N.B.: Gabes (nearest meteorological station): mean 1884–1977: 188 mm

Table 2. Variation of the specific frequencies (% FS) at station 1 (gypseous crust).

Species	17 Oct. 1972	27 April 1973	30 Sept. 1973	25 April 1974	10 Oct. 1974	3 May 1975	11 Oct. 1975	15 May 1976	4 Novem. 1976	5 April 1977	15 March 1978	20 April 1979
Atractylis serratuloides	2.3	2.7	2.8	3.4	1.3	3.8	0.7	4.4	6.7	4.3	6.7	5.7
Helianthemum lippi var *intricatum*	4.1	3.6	4.1	4.4	4.2	5.4	2.7	6.6	5.2	5.9	5.4	6.7
Anarrhinum brevifolium	0.2	0.2	0.3	0.8	0.2	0.6		0.5		0.3	0.7	0.7
Thymelaea microphylla	0.2	0.3	0.5		0.5	0.5	0.5	0.6	0.8		0.6	0.5
Zygophyllum album	0.2	0.2		0.8	0.5	0.9	0.8			0.8	0.4	1.2
Pituranthos tortuosus			0.6	0.3		0.6						
Helianthemum kahiricum			0.3		0.2							
Herniaria fontanesii				0.6	0.3			0.5	0.9	0.6	0.3	
Argyrolobium uniflorum				0.2	0.2	0.2		0.3				
Helianthemum crassifolium					0.3	1.2	0.6	0.2	1.7	0.8	0.3	0.8
Gymnocarpos decander						0.2		0.3			0.3	0.3
Lygeum spartum						0.9	0.3	0.5	1.4	0.2	0.3	1.1
Pituranthos chlorantus ssp. *cossonianus*								0.8	1.4	2.3	1.5	1.2
Echium pycnanthum ssp. *humile*								0.2				
Litter	0.1	0.1	0.3	0.2		1.2	0.2	1.9	3.6	1.9	1.6	4.0
Total (perennial plants)	7.0	7.1	8.9	10.7	7.7	15.5	5.8	16.8	21.7	17.1	18.1	22.2
Total minus litter	6.9	7.0	8.6	10.5	7.7	14.3	5.6	14.9	18.1	15.2	16.5	18.2

Results

Station 1 (Table 2): This station only had perennial plants, which probably established during very exceptional rainy years. Annual species were not able to germinate on this substrate. The sum of the specific frequencies varied relatively slightly from one season to the next.

Gypsophilous perennial species, although present before protection of the station, yielded too low a cover to be detected: *Helianthemum crassifolium, Lygeum spartum, Pituranthos chloranthus* ssp. *cossonianus.* The cover of all the other species also increased. This vegetation changed a little with a 18% cover increase after 7 yr (a 2.6 fold increase). The litter cover increased from nothing to 4%.

Station 2 (Table 3): After a steady increase over a 3 yr period, the *SSF* value for the perennial vegetation in the spring leveled off at 35–40% for the next 4 yr (a two fold increase). At this station as well, species such as *Helianthemum kahiricum* and *Atractylis serratuloides* appeared, but they did not develop very much.

The species with relatively the greatest development were *Rhantherium suaveolens* and *Thymelaea microphylla*. These species were present in 1972, but they were affected by overgrazing. Station 2 gradually developed the physiognomy of a pastoral steppe in good condition. The cover of *Plantago albicans,* a very good pastoral species, was expected to develop rather well, but instead, fluctuated with the annual precipitation. The *Echiochilon fruticosum* and *Argyrolobium uniflorum* cover increased regularly during the rainy period (until 1977) and

Table 3. Variation of the specific frequencies (% FS) at station 2 (truncated sierozem).

Species	17 Oct. 1972	26 April 1973	28 Sept. 1973	25 April 1974	11 Oct. 1974	3 May 1975	11 Oct. 1975	15 May 1976	4 Novem. 1976	15 April 1977	15 March 1978	20 April 1979
Helianthemum lippii var. *sessiliflorum*	1.6	1.2	0.9	1.2	0.6	2.7	2.8	1.4	2.8	3.1	1.8	1.5
Plantago albicans	3.0	6.9	2.0	5.5	0.8	7.7	0.5	6.7	6.3	6.7	4.0	4.9
Rhantherium suaveolens	3.0	3.7	4.4	5.0	4.7	6.2	4.2	7.5	8.9	10.0	12.5	10.0
Linaria aegyptiaca	0.2	1.1	0.5	0.6	0.2	1.2	0.2	0.9	0.2	–	–	–
Echiochilon fruticosum	1.7	3.6	1.9	3.2	2.3	4.0	3.0	5.3	6.1	7.1	4.8	3.4
Argyrolobium uniflorum	0.2	–	0.2	0.6	0.3	0.7		1.9	2.5	0.5	0.4	0.7
Thymelaea microphylla	1.9	1.2	0.8	2.8	1.7	3.8	2.9	4.8	4.7	3.1	5.0	4.6
Pituranthos tortuosus	0.8	0.9	0.2	3.1	1.1	4.0	1.2	1.1	0.8	1.5	0.1	4.0
Pituranthos chlorantus ssp. *cossonianus*	–	–	–	–	–	–	–	–	–	–	8.8	–
Artemisia campestris	–	0.2	–	–	0.2	–	–	6.6	6.4	0.3	–	–
Helianthemum kahiricum	–	–	–	–	0.2	0.8	–	2.0	1.7	–	2.0	2.8
Coris monspeliensis	–	–	–	–	–	–	–	0.5	0.8	0.5	–	–
Helianthemum lippii var. *intricatum*	–	–	–	–	–	–	–	1.0	1.4	0.9	–	1.7
Atractylis serratuloides	–	–	–	0.2	–	0.3	–	1.1	1.9	1.5	0.3	0.6
Total perennial plants	12.4	18.7	10.9	22.4	12.1	31.4	14.8	40.8	44.5	35.2	39.7	34.2
Total annual plants	–	–	–	1.5	–	0.8	–	9.5	0.3	–	1.1	5.6
Litter	0.3	2.5	3.6	2.5	2.2	5.2	1.0	3.8	9.8	2.1	9.5	6.5
Total	12.7	21.2	14.5	26.4	14.3	37.4	15.8	54.1	54.6	37.3	51.1	46.3
Total minus litter	12.4	18.7	10.9	23.9	12.1	32.2	14.8	50.3	44.8	35.2	41.6	39.8

Table 4. Variation of the specific frequencies (% FS) at station 3 (sierozem).

Species	5 Oct. 1972	24 April 1973	27 Sept. 1973	25 April 1974	10 Oct. 1974	3 May 1975	11 Oct. 1975	15 May 1976	4 Nov. 1976	4 April 1977	14 March 1978	20 April 1979
Helianthemum lippi var sessiliflorum	0.5	2.3	1.6	3.2	2.0	4.5	2.9	9.8	7.7	12.2	8.7	8.4
Plantago albicans	5.2	19.8	8.7	13.3	4.8	20.3	4.8	23.7	14.4	13.8	19.0	20.9
Echiochilon fruticosum	2.8	3.0	3.4	2.3	1.6	4.7	2.7	5.0	5.2	7.2	6.5	7.9
Argyrolobium uniflorum	0.6	2.3	0.5	1.5	0.5	4.3	0.9	6.8	3.0	2.5	1.1	0.3
Thymelaea microphylla	0.2	0.8	0.5	0.9	–	1.6	1.5	1.3	2.3	3.2	3.0	2.1
Linaria aegyptiaca	–	0.6	0.8	1.5	–	0.5	0.2	1.3	1.4	1.1	1.2	3.1
Astragalus armatus spp. tragacanthoides	0.5	0.8	0.9	0.9	0.2	1.4	1.1	1.6	2.0	1.7	1.4	0.8
Pituranthos tortuosus	0.5	1.2	1.6	5.0	0.5	4.0	2.2	–	2.4	2.2	1.1	3.8
Rhantherium suaveolens	0.8	1.9	1.7	2.7	2.0	3.4	3.3	5.0	3.6	3.8	3.9	4.9
Teucrium polium	0.2	0.6	0.3	0.3	–	0.6	–	–	–	0.4	0.7	–
Artemisia campestris	–	–	–	0.2	–	–	–	0.3	0.5	0.3	0.2	0.1
Pituranthos chloranthus ssp. cossonianus	–	–	–	–	–	–	–	5.5	7.8	5.6	5.9	8.7
Total perennial plants	11.5	33.3	20.0	31.8	11.6	45.3	19.6	60.3	50.3	54.0	52.7	61.0
Total annual plants	1.2	2.4	–	8.1	0.5	7.7	0.2	24.4	0.3	0.9	9.2	16.1
Litter	3.0	5.6	6.7	2.5	5.8	5.9	5.4	51.1	19.8	23.6	12.6	8.0
Total	15.7	41.3	26.7	42.4	17.9	58.9	25.2	135.8	70.4	78.5	74.5	85.1
Total minus litter	12.7	35.7	20.0	29.9	12.1	53.0	19.8	84.7	50.6	54.9	61.9	77.1

Table 5. Variation of the specific frequencies (% FS) at station 4 (sierozem).

Species	1972	1973			1974			1975			1976			1977		1978	1978	1979
	19 Nov.	19 March	30 May	16 Nov.	14 March	11 Jun.	11 Nov.	6 May	21 Jul.	10 Oct.	16 Feb.	21 May	8 Nov.	23 Feb.	24 May	6 March	29 May	17 April
Rhantherium suaveolens	31.7	35.5	41.0	32.7	31.7	46.1	45.7	47.9	50.0	51.1	51.1	51.4	48.5	48.4	42.0	40.0	44.6	41.1
Arthrophytum schmittianum	0.1	0.4	0.8	0.2	0.5	0.7	2.2	3.1	3.2	4.0	3.0	4.5	4.5	4.1	4.0	4.0	4.0	4.3
Plantago albicans	5.7	11.2	9.6	6.7	8.6	14.0	19.6	25.6	19.9	14.0	27.0	22.6	16.4	37.2	22.2	21.1	18.9	4.6
Argyrolobium uniflorum	0.1	0.7	1.0	0.4	0.5	1.2	0.7	1.9	0.7	1.6	2.0	4.1	1.7	0.7	0.7	0.5	–	0.5
Pituranthos tortuosus	–	–	0.1	0.1	1.2	1.3	1.1	1.2	1.4	1.5	1.0	1.5	1.4	1.6	1.5	1.6	1.6	1.6
Helianthemum lippii var sessiliflorum	–	0.5	0.4	0.1	1.5	4.2	2.4	6.7	6.2	6.0	5.6	9.2	8.4	9.1	9.0	8.6	8.6	9.0
Stipa lagascae	0.6	0.6	1.6	1.2	1.0	1.4	1.1	2.6	2.0	1.9	2.7	4.6	4.7	4.7	6.1	6.0	10.8	6.5
Teucrium polium	–	0.2	0.4	–	–	–	–	0.4	0.2	–	0.2	0.4	–	–	–	0.3	0.3	0.3
Total perennial plants	38.4	49.1	54.9	41.4	45.0	68.9	72.6	89.4	83.6	80.1	92.6	98.3	85.6	105.8	85.5	82.1	88.8	67.9
Total annual plants	8.3	47.7	27.6	0.8	14.6	14.6	13.6	32.0	11.5	11.2	34.2	26.7	1.3	5.0	0.0	28.2	0.0	1.1
Litter	9.0	5.5	8.3	24.1	8.9	15.6	12.1	22.8	11.5	19.5	15.2	13.8	18.3	18.0	23.7	14.3	27.3	22.0
Total	55.7	102.3	90.8	66.3	68.5	99.1	98.3	144.2	106.6	110.8	142.0	138.8	105.2	128.8	109.2	124.6	116.1	91.0
Total minus litter	46.7	96.8	82.5	42.2	59.6	83.5	86.2	121.4	95.1	91.3	126.8	125.0	86.9	110.8	85.5	110.3	88.8	69.0

124

Table 6. Variation of the density of tufts per hectare of the different plant species at station 4.

Species	Tufts	31/5/72	13/6/73	20/10/73	2/10/77	% Increase
Rhantherium suaveolens	live	26 549	27 981	28 333	30 142	13,5
	dead	445	664	1 731	2 526	467,6
Pituranthos tortuosus	live	273	260	325	466	70,7
Stipa lagascae	live	1 783	2 850	3 307	4 088	129,3
Arthrophytum schmittianum	live	1 158	1 522	1 587	1 717	48,3
Salsola vermiculata	live	195	425	481	481	146,7
Helianthemum sessiliflorum	live	263	1 275	1 940	2 669	914,8
Linaria aegyptiaca	live		685	1 067	1 535	
Teucrium polium	live	247	351	455	676	173,7
Aristida ciliata	live		26	39		
Artemisia campestris	live	78	182	247	286	266,7
Lygeum spartum	live		39	156	145	
Astragalus armatus spp. *tragacanthoides*	live	26	65	104	112	330,8
Argyrolobium uniflorum	live		2 135	2 669	2 747	
Marrubium deserti	live	143	130	169	208	45,5
Echichilon fruticosum	live		130	156	118	
Astragalus caprinus lanigerus	live		26	39		
Atractylis serratuloides	live		26	78	65	
Total living plants	live		38 108	41 152	45 455	

then decreased during the dry period.

The litter, which initially showed a *SF* of 0.3% reached 10% during certain periods.

Station 3 (Table 4): The vegetation at this station, situated on deep, sandy soil and subjected to overgrazing, benefited from the protective measures. The SSF of the perennial plants progressed from 33 to 61% in the spring over the 7 yr period (a 1.8 fold increase). This increase was mainly due to the development of *Helianthemum lippii* var. *sessiliflorum*, *Echiochilon fruticosum* and *Rhantherium suaveolens*. The latter species which had been practically eliminated by cultivation slowly reappeared. *Pituranthos chloranthus* ssp. *cossonianus* appeared and proliferated extensively during the last 3 dry years. *Argyrolobium uniflorum* developed, as at station 2, during the rainy period.

Plantago albicans developed only slightly. Figure 2 shows the wide seasonal variation in the total sum of the specific frequencies which correlated with the development of annual species in the spring. Among the annual species, *Hedysarum spinosissimum* seems to be the species that benefited most from the protective measures.

The litter increased considerably, mainly as from the beginning of the dry period.

Station 4 (Tables 5 and 6): On this deep sandy soil, the steppic vegetation was already in a rather good condition before protection. The *SSF* of the perennial species progressed in the spring from 50% to about 100% during the succession of rainy years. This progression was due, in the first place, to *Rhantherium suaveolens* which developed from 35 to 51% to return finally to 41%. But the cover of

Table 7. Variation of the specific frequencies (% FS) at station 5 (loamy piedmont).

Species	1972	1973		1974			1975		1976		1977	1978	1979
	11 Oct.	19 March	14 Nov.	15 March	23 April	10 Dec.	7 May	30 Oct.	5 March	27 Oct.	4 April	28 Feb.	16 April
Arthrophytum scoparium	3.1	5.3	6.6	6.7	6.3	6.5	9.9	8.3	8.9	13.3	15.9	13.0	10.4
Moricandia arvensis ssp. *suffruticosa*	0.5	2.7	1.6	3.1	3.2	1.2	3.3	1.3	4.4	5.3	4.4	5.3	5.9
Cynodon dactylon													1.2
Total perennial plants	3.6	8.0	8.2	9.8	9.5	7.7	13.2	9.6	13.3	18.6	20.3	18.3	17.5
Total annual and biennal plants	0.0	18.2	0.2	16.9	12.6	0.0	2.9	0.2	13.7	2.4	13.7	43.0	28.9
Litter and dead plants	0.5	4.0	11.3	5.3	3.3	4.7	4.6	2.8	1.8	76.7	59.7	42.7	21.5
Total	4.1	30.2	19.7	32.0	25.4	12.4	20.7	12.6	28.8	97.7	93.7	104.1	67.9
Total minus litter and annual dry plants	3.6	26.2	8.4	26.7	22.0	7.7	16.1	9.8	27.0	21.0	34.0	61.3	46.4

126

numerous other perennial plants increased: *Helianthemum lippii* var *sessiliflorum* and *Stipa lagascae* went from 0 to about 10%. *Plantago albicans* showed an *SF* of greater than 20% quite regularly in the spring.

It is interesting to compare the development of the cover of the perennial plants (Table 5) with that of the densities of the same species (Table 6). It appears that the number of *Rhantherium suaveolens* tufts increased only to a small extent and that the increase in cover was mainly due to the increase in volume of these tufts.

It is important to note the big increase in the number of dead tufts as the density of *Rhantherium* increased. In all, in 1977, the density was 3 tufts of *Rhantherium* per m² and 4.5 tufts of perennials per m² (except *Plantago albicans*).

Pituranthos tortuosus and *Arthrophytum scoparium* produced few new tufts as well. On the contrary the number of tufts of *Stipa lagascae*, of *Helianthemum lippii* var *sessiliflorum*, and of other species increased considerably, and it is these new tufts that accounted for the increase of their *SF*.

The *SSF* of the annual plants varied greatly as a function of the annual precipitation. A value of 48% was reached the first year after protection began because of consistent rainfall. There did not seem to be a tendency for particular species to develop following protection. *Medicago truncatula*, *Picris coronopifolia*, *Paronychia arabica* and *Hippocrepis bicontorta* were the species that, periodically, covered most of the surface area of the ground, even though there was a greater number of individual *Poaceae* present such as *Schismus barbatus* ssp. *calcycinus* and *Cutandia dichotoma*.

The litter cover increased threefold over the 7 year study period.

Station 5 (Table 7): On this post-cultivated field (cereal), on a loamy soil, development of the few perennial species that had resisted ploughing was very slow (*Arthophytum scoparium*, *Moricandia arvensis* ssp. *suffruticosa*). The *SSF* increased from 8 to 20% (a 2.5 fold increase) in 5 years, then regressed after 2 dry years.

The cover of the annual and biennial species was excellent during the wet years (more than 70% in 1977). *Stipa retorta* reached the greatest cover.

Artemisia herba-alba, which originally accounted for the physiognomy of the steppe, did not reappear in spite of 7 years of protection; cultivation had totally eliminated it. Measurements made at a nearby station, which was protected as well, but where *Artemisia herba-alba* had a cover of 9.6% at the beginning of protection, showed that this species had doubled its cover in two years, not only by an increase in the volume of the tufts but also by the appearance of new individual plants.

Discussion

The data presented here illustrate the great variability in plant cover according to the seasons and the years in the arid zone (Figs. 2 and 3). This variability was mainly due to annual species, but frequently the cover of perennial species varied by as much as 25%. During the dry periods, often only the woody parts of the plants remained standing; and the litter, though generally redistributed by the wind, was abundant on the ground. Noy-Meir (1974) has drawn attention to this weak 'persistence' of ecosystems in arid zones and their strong 'resilience' resulting from their adaptation to regeneration when conditions become favorable again.

The increase of the cover of annual species, following years of protection which should favor reconstitution of seed stocks, was not obvious because this cover is so much affected by the amount of rainfall and, even more so, by its seasonal distribution. The annual species in this arid zone have seeds which can remain dormant in the ground for long periods and then germinate and

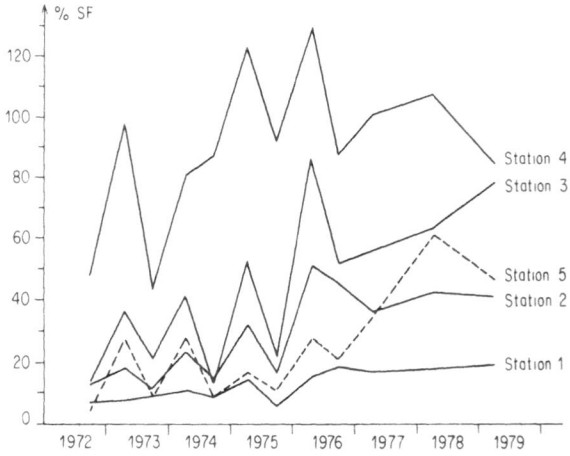

Fig. 2. Variation of the sum of the specific frequencies of the perennial and the annual plants at the five stations.

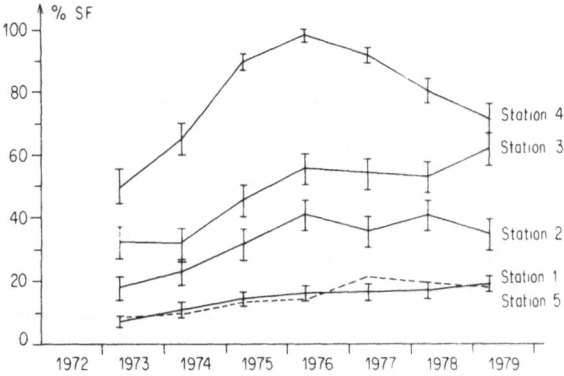

Fig. 3. Variation of the sum of the specific frequencies of the perennial plants in spring at the five stations.

develop very quickly when conditions are favorable (Evenari *et al.,* 1971).

The effect of protection was evident but different for each type of steppic vegetation studied. The differences in plant cover on these steppes were small at the beginning of the study because they had been subjected to overgrazing (except at Station 4 where the vegetation was in rather good condition) and increased progressively over the course of the years (Figs. 2 and 3). The differences were even more important in the spring and during the years of heavy rainfall.

The steppes in the deep, sandy zones responded best to protection. The pastoral perennial plants, subjected to overgrazing under the usual conditions, which benefited most from this protection are *Helianthemum lippii* ssp. *intricatum, Stipa lagascae. Rhantherium suaveolans. Argyrolobium uniflorum* and *Echiochilon fruticosum* developed well in rainy periods but regressed significantly in dry periods. *Plantago albicans,* a plant sought after by the animals but which seems to adapt well to frequent rejuvenation under the effect of grazing, did not develop very much.

It must be noted that an increase in the steppic plant cover does not mean a proportional increase in its pastoral value. *Rhantherium suaveolens,* for example, showed a progressive increase in shrub volume but the ratio new shoots/woody parts gradually became smaller. On the contrary other species such as *Stipa lagascae,* multiplied with the appearance of new tufts which the animals appreciate. For this same *Rhantherium* species, after the important increase of its cover during the rainy

period, we observed a high mortality during the dry period which followed (Table 6). Because of this mortality, new young tufts of this species or of other perennial plants developed in spots.

Protection of the gypseous crusted steppes for pastoral reasons turned out to be of little interest. Indeed, these crusts had been formed as a result of ablation of the surface horizon and stripping of the deep seated gypseous geological layers which hardened rapidly once exposed to the surface. The presence of this crust prevented the return of the steppe to its initial state before degradation. Even if the cover increases, good pastoral species are not able to develop again on these steppes.

In the case of the loamy steppes, regeneration was also very slow. Repeated mechanical ploughing led to the disappearance of most of the original steppic species, mainly *Artemisia herba-alba,* over a large area. The absence of seeds from this species prevented re-establishment in spite of its protection. It can be noted that if a small surface area is cleared for cultivation in the middle of an *Artemisia* steppe, *Artemisia* can take over again rather easily. On the other hand, after protection, a rain-beat seal may develop on the surface. This is particularly important on loamy soils. The origin of this rain-beat seal seems to be the development of algae, fungi and microcrystals of salts present in the soil. The ground 'closes up' and, in the absence of ploughing and animal trampling, this film makes germination difficult. Noble (1977) observed the same phenomenon in a protected area in the Australian arid zone.

On the gypseous steppes and the loamy fallow lands, simple protection did not lead to a return of the steppe to its original state before degradation, at least not within a reasonable lapse of time. These steppes had gone beyong the point where degradation phenomena could be reversed. According to Godron (1979), these steppes exceeded their 'elasticity limit'.

The stability of many ecosystems in the arid zone is endangered because of their vulnerability to erosion and to modification of the surface horizon of the soil. A very strong human pressure (ploughing and overgrazing) can thus lead to new types of vegetation in equilibrium with this pressure but with much lower productivity (Le Houerou, 1974). The term 'desertification' (United Nations, 1977) has been used to refer to this loss of the environment's potential.

However, in the case of the pre-Saharan region studied, the very important development of the perennial plants following protection acted indirectly on the environment and rather rapidly caused important modifications. In fact, the above-ground biomass of the plants serves as an obstacle to sand frequently carried by the wind and accumulating at the foot of shrubs. As an example, 225 tons of accumulated sand per hectare were found at Station 4 after three years of protection. The presence of this trapped sand permits better germination of the annual plants by covering the crusts of the surface, favors infiltration of rain water, and replaces the mulch leading to a better water balance; such a situation particularly favors a rapid and progressive development of vegetation.

The perennial plant cover, especially of the tall plants, increased. This increase affected the environment mainly because these species trapped the sand carried by the wind, but the absence of appearance or disappearance of plant species made it difficult to characterize different stages of succession in such a short time. Many authors have noted that progressive development in arid zones is slow (Chew & Chew, 1965; Kassas, 1966; Wagner, 1976; Le Houerou, 1977; Thalen, 1979).

On the basis of these results, it is difficult to recommend an optimal duration of protection for regeneration of the land in arid zones with the objective of setting up a pastoral management program. This duration depends very much upon the amount of rainfall following protection and upon local conditions, such as the initial state of the vegetation, in particular. The proximity of a 'sand source' (i.e., a cultivated zone) and the relative area of the degraded zone in relation to the surrounding steppe in good condition (distance from seed bearer) are aspects that must also be taken into consideration. Protection seems to be useless in those areas where a gypseous crust has already developed, and pastoral regeneration of post cultivated fallow lands in loamy zones is really a matter of agronomy (sowing in a favorable year). In sandy zones, 2 to 3 years of protection seem to be the maximum not to be exceeded. Generally, few grazing periods in rotation with rest periods for the vegetation is better than total protection; this grazing stimulates growth of young perennial plant shoots and favors germination of annual plants.

References

Buffington, L. C. & Herbel, C. H., 1965. Vegetational change on a semidesert grassland range from 1858 to 1963. Ecol. Monogr. 35: 135–64.

Chew, R. M. & Chew, A. E., 1965. The primary productivity of a desert shrub (Larrea tridentata) community. Ecol. Monogr. 40: 1–21.

Daget, Ph. & Poissonet, J., 1971. Une méthode d'analyse phytologique des prairies. Critères d'application. Ann. Agron. 22: 5–41.

Evenari, M., Shanan, L. & Tadmor, N., 1971. The Neguev. The challenge of a desert. Havard University Press. Cambridge, Massachusetts, 345 pp.

Floret, C., Le Floc'h, E. & Pontanier, R., 1976. Carte de la sensibilité à la désertisation en Tunisie centrale et méridionale (Processus de dégradation en cours des sols de la végétation). Echelle 1/1 000 000. Sols de Tunisie, n° 8, 68 pp., 1 carte couleur h.t.

Floret, C. & Pontanier, R., 1978. Relations climat-sol-végétation dans quelques formations végétales naturelles du Sud Tunisie (Production-bilan hydrique des sols). Inst. Rég. Arides, Médenine, Dir. Ress. Eau et Sols, Tunis, C.E.P.E./ C.N.R.S. Montpellier, O.R.S.T.O.M., Doc. Technique n° 1, Paris, 96 pp.

Floret, C., Le Floc'h, E., Pontanier, R. & Romane, F., 1978. Modèle écologique régional en vue de la planification et de l'aménagement des parcours des régions arides. Applications à la région de Zougrata. Inst. Rég. Arides. Médenine, Dir. Ress. Eau et Sols, Tunis, C.E.P.E./C.N.R.S., Montpellier, O.R.S.T.O.M., Paris, Doc. Technique n° 2, Paris 73 pp., 1 carte h.t.

Gaddes, N., 1978. Etudes des relations végétation-milieu et effet biologique de la mise en défens notamment sur l'Alfa (Stipa tenacissima L.) dans le bassin versant de l'Oued Gabès. Thèse Ecologie générale, U.S.T.L., Montpellier, 130 p.

Godron, M., 1968. Quelques applications de la notion de fréquence en écologie végétale. Oecol. Plant. 3: 185–212.

Godron, M., 1979. Eléments d'écologie des végétaux terrestres. Université Montpellier II, 66 p.

Goodall, D. W., 1952. Some considerations in the use of points quadrats for the analysis of vegetation. Aust. J. Sc. Res. Ser. B, 5: 1–41.

Greig-Smith, P., 1964. Quantitative plant ecology. Butterworths, London, 256 pp.

Kassas, M., 1966. Plant life in desert p. 145–180 in E. S. Hills (ed) Arid lands. A geographical appraisal Methness Co. Ltd. London and Unesco, Paris, 461 pp.

Le Houerou, H. N., 1959. Recherches écologiques et floristiques sur la végétation de la Tunisie méridionale. Inst. de Rech. Sah. Alger. Mém. H.S. 510 p.

Le Houerou, H. N., 1969. La végétation de la Tunisie steppique (avec référence aux végétations analogues d'Algérie, de Lybie et du Maroc). Ann. Inst. Nat. Rech. Agron. Tunisie, 42. 5.624 p. et 1 carte couleur 1/500 000.

Le Houerou, H. N., 1974. Deterioration of ecological equilibrium in the arid zones of North Africa, Spec. Publ. Agron. Res. Organ. Volcani Centre Bet. Dagan 39: 54–57.

Le Houerou, H. N., 1977. Biological recovery versus desertization Econ. Geogr. Vol. 53: 413–420.

Morice, E. & Chartier, F., 1954. Méthode statistique, 1. Elaboration des statistiques 187 pp., 2. Analyse statistique. Imprimerie Nationale. Paris.

Naegele, A., 1959. La végétation de la zone aride: Les parcelles protégées d'Atar. La Nature 72–76.

Noble, I. R., 1977. Long term biomass dynamics in an arid chenopod schrub community at Koonamore South Australia. Aust. J. Bot. 1977. 25: 639–653.

Noy-Meir, I., 1974. Stability in arid ecosystems and the effects of man on it. Proceedings of the first Intern Congress of Ecology. The Hague. The Netherlands, Sept. 8–14, 220–225.

Quezel, P. & Santa, S., 1962. Nouvelle flore de l'Algérie et des régions désertiques méridionales. C.N.R.S. Paris, 2 vol. 1170 pp.

Smith, D. A. & Schmutz, E. M., 1975. Vegetative changes on proteted versus grazed desert grassland ranges in Arizona. J. Range Manage. 26 (6): 453–58.

Thalen, D. C. P., 1979. Ecology and utilization of desert shrub rangelands in Iraq. Dr. W. Junk B.V., Publishers. The Hague, 448 pp.

Trumble, H. C. & Woodroffe, K., 1954. Influence of climatic factors in the reaction of desert shrubs to grazing by sheep. In: 'The biology of deserts' Inst. Biology. London Symp. 3: 129–147.

United Nations, 1977. Etude de cas sur la désertification, Région d'Oglat Merteba. Conférence des Nations Unies sur la désertification, Nairobi, 1977. Doc. A/Conf. 74/12, 143 pp. 1 carte couleur (anglais et français).

Wagner, F. H., 1976. Integrating and control mechanisms in arid and semi-arid ecosystems. Considerations for impact assessment. Proceeding of a Symposium on evaluation of environmental impact. 27th Annual Meeting of American Institute of Biological Sciences, New Orleans, Louisiana (In Press).

Accepted 25.6.1981.

Vegetation dynamics in a coastal grassland of Hawaii**

D. Mueller-Dombois
Department of Botany, University of Hawaii at Manoa, 3190 Maile Way, Honolulu, Hawaii 96822, USA

Keywords: Arrested succesion, Exclosures, Feral goats, Life-cycle dependent death, Mediterranean rainfall pattern, Recovery, Tropical grassland

Abstract

Vegetation development following goat removal in a tropical grassland with a Mediterranean rainfall seasonality on leeward Hawaii is reported from two permanent monitoring sites. The two sites are equipped with experimental exclosures against feral goats. The sites were monitored annually for a decade (1971–80) for changes in cover by species. The vegetation changes reported relate to both inside- and outside-exclosure developments.

Four dynamic categories were recognized among the grassland species following goat removal. These are the decreasers, the increasers and the persistent species. The latter were divided into stable and oscillating persisters. Attention was drawn to the oscillating persisters, as exemplified by the endemic vine *Canavalia kauensis,* and to the absence of any correlation of its oscillation pattern to the variation in year-to-year rainfall. It was concluded that the oscillation pattern resides in the population itself and can be attributed to its life-cycle phases. It is suggested that synchronized life-cycle dependent death in local populations is an important dynamic phenomenon in certain plant communities. This phenomenon may be considered as separate from phenology and succession, but of similar significance in understanding the processes of vegetation dynamics.

Introduction

This paper reports a 10-year, annually repeated, monitoring study of a tropical grassland on the leeward side of the island Hawaii, where feral goats (*Capra hircus*) used to be the dominant stress factor at the beginning of the study. The monitoring sites are two goat exclosures with their surrounding areas, and both are within the confines of Hawaii Volcanoes National Park (Fig. 1).

Two papers were published earlier on this study. The first one (Mueller-Dombois and Spatz, 1975) reported on the rapid vegetation development and the life-form sequence that was observed in the first few years following experimental goat exclusion. Of particular interest was the early dominance of an endemic vine, *Canavalia kauensis,* which nearly filled out the surface area of one of the exclosures after two years. The appearance and sudden dominance of this vine was especially remarkable, because native plants had been reduced to extreme rarity during the one-and-a-half century-long goat occupation of this territory. Moreover, this particular vine had never been discovered before among the native flora of Hawaii (St. John, 1972). The second paper (Mueller-Dombois, 1979) presented the vegetation recovery data through four additional years, till 1976. It was noted that recovery had become more heterogeneous than expected earlier and that the most advanced development seen as a mixed chamaephyte-vine-bunchgrass stage had become 'arrested' because of a delay

* Contribution No. 007/12 CPSU University of Hawaii, National Park Service Contract Number CX8000 6 0031.
** Nomenclature of plant names follows St. John (1973).

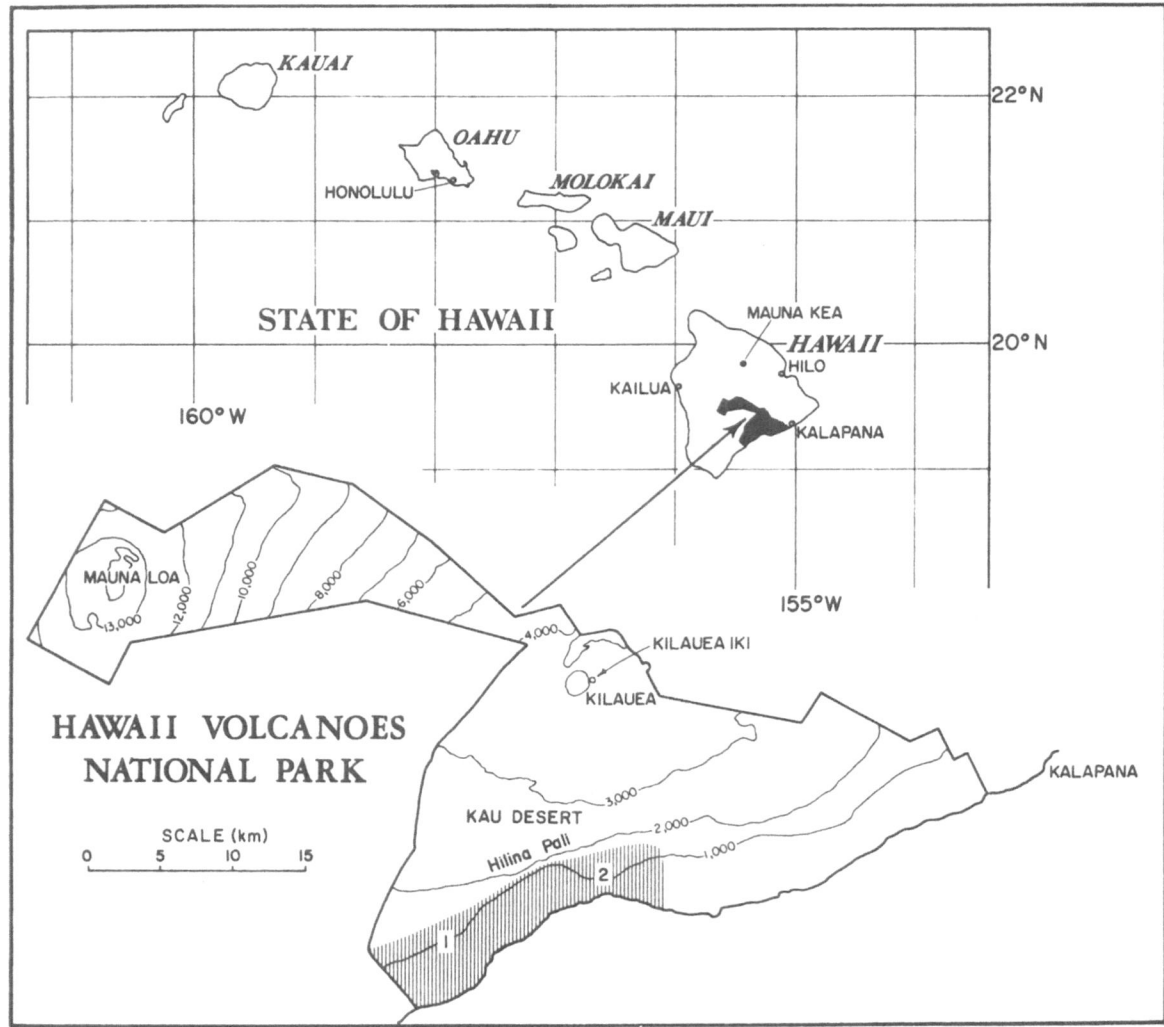

Fig. 1. Map of Hawaiian Islands with study area in the western coastal lowland of Hawaii Volcanoes National Park. 1 = Kukalau'ula exclosure. 2 = Puu Kaone exclosure. Shaded area shows approximate extent of former *Eragrostis tenella* grassland. Contour lines in feet (1000ft = 305 m).

in phanerophyte invasion. Two weed tree species, *Leucaena latisiliqua* and *Ricinus communis,* with a potential of forming the next stage in vegetation development were deliberately prevented from spreading by the National Park Service. The seed supply of indigenous phanerophytes was considered as 'exhausted' for the area, because the prolonged occupation by feral goats had eliminated the potential tree seed sources from this territory. This paper adds a new dimension to the process of vegetation dynamics in this area by the discovery of life-cycle dependent death in *Canavalia.*

General description and history of the area

The coastal lowland in the Park extends 45 km along the coast and 5 km inland to the 300 m elevation line. It thus includes a territory of about 22 500 ha (225 km²).

The climate has a typical Mediterranean rainfall pattern, i.e. a winter-wet and summer-dry season. In contrast, the month-to-month mean air temperature is distinctly tropical (24 °C mean annual) showing less than 6 °C variation between summer and winter. A strong, desiccating wind (25–50 km/hour) blows with considerable constancy during the dry summer months (June through September). The annual rainfall at the study sites varies normally between 700 and 900 mm.

The substrate in the coastal lowland is from volcanic materials and consists mostly of basaltic lava rock outcrop with pockets of soil from ash of differing depths and sizes. More detailed habitat descriptions are given in Doty & Mueller-Dombois (1966) and Mueller-Dombois (1980).

The current vegetation cover consists of woodland, shrubland, grassland with shrubs and widely scattered trees and of sparse grassland and barren areas (recent lava flows). These physiognomic types form a patchy mosaic, with the woodland and shrubland types more prevalent in the eastern coastal lowland and the grassland types more prevalent in the western part (Mueller-Dombois, 1976, 1980). There are a few endemic species, such as the trees *Metrosideros polymorpha, Diospyros ferrea, Canthium odoratum, Wikstroemia phillyreifolia* and *Erythrina sandwicensis.* Endemic shrubs include *Dodonea viscosa* and *Osteomeles anthyllidifolia.* There are no quantitatively important endemic grasses. Most of the now dominant species are introduced. This includes the most important animals, the feral goats (*Capra hircus*) and the Indian mongoose (*Herpestes auropunctatus*).

This coastal area was quite populated by native Hawaiians at the time the Europeans arrived in the latter part of the 18th century (Ellis, 1825). Hawaiians tended to live at the coast rather than in the mountains: they derived most of their protein from the sea. They cultivated sweet potatoes (*Ipomoea batatas*) and grass cover of the SE Asian bunchgrass, *Heteropogon contortus* which was apparently introduced by Hawaiians, and used for thatching their roofs (Fosberg, 1972). There is some evidence from other areas in the coastal lowland that the grass cover was maintained by periodic burning (*fide* Vogl, 1969). This practice must have destroyed much of the woody vegetation already before arrival of the Europeans. In 1778 Captain Cook introduced goats to Hawaii (Tomich, 1969). During the following decades the Hawaiians left this coastal region, and it turned into a concentration center for the feral goats. These then formed large herds during a period of at least 150 years (Marques, 1905).

When the area became US National Park in 1916,

goats were decimated by annual goat drives, and as much as 5000 were killed in certain years (Gerdes, 1964). However, they kept multiplying. When I mapped the Park's vegetation in 1965 (Mueller-Dombois, 1966; Mueller-Dombois & Fosberg, 1974), much of the coastal lowland area in its western half was reduced to an annual grassland with the pantropical love grass, *Eragrostis tenella,* forming a dominant cover together with hardy perennial creeping grasses, e.g. *Cynodon dactylon* and *Chrysopogon aciculatus.* I then suggested to the Park Service the construction of some experimental exclosures to find out what effect goat browsing had on this ecosystem.

Based on an immediate and rapid vegetation recovery, which included a significant proportion of native species as recorded in an experimental exclosure (Mueller-Dombois & Spatz, 1975), the Park Service changed its goat control strategy in 1971. The lowland was partitioned by many kilometers of fences. The feral goats were systematically eliminated from the fenced sections. Since 1973 only very few goats have been seen in the fenced areas. The control method became generally effective in the vegetation around 1975, so much that a new vegetation map (Mueller-Dombois, 1980) was prepared.

Vegetation development over a 10-year period

The Park Service built two experimental exclosures in the grass-dominated area of the western coastal lowland, one (7 m × 100 m) in 1968 at a place called Kukalau'ula and the other (10 m × 100 m) in 1971 at Puu Kaone. Here, I will call these for simplification Sites 1 and 2, resp. A third, one hectare, exclosure built at Site 2 was used for management purposes so that recovery monitoring had to be discontinued in 1976.

Monitoring was done for % shoot cover by the point-frequency method (Mueller-Dombois & Ellenberg, 1974) along 10 permanently located transects running at 10 m intervals perpendicular to the long side of each exclosure. Intercepting points were spaced 20 cm apart, i.e. five per meter. The exclosure at Site 1 was monitored with 350 points inside and 250 outside, the exclosure at Site 2 with each 500 points inside and outside. Data were obtained till 1976 by semi-annual monitoring, once each during

134

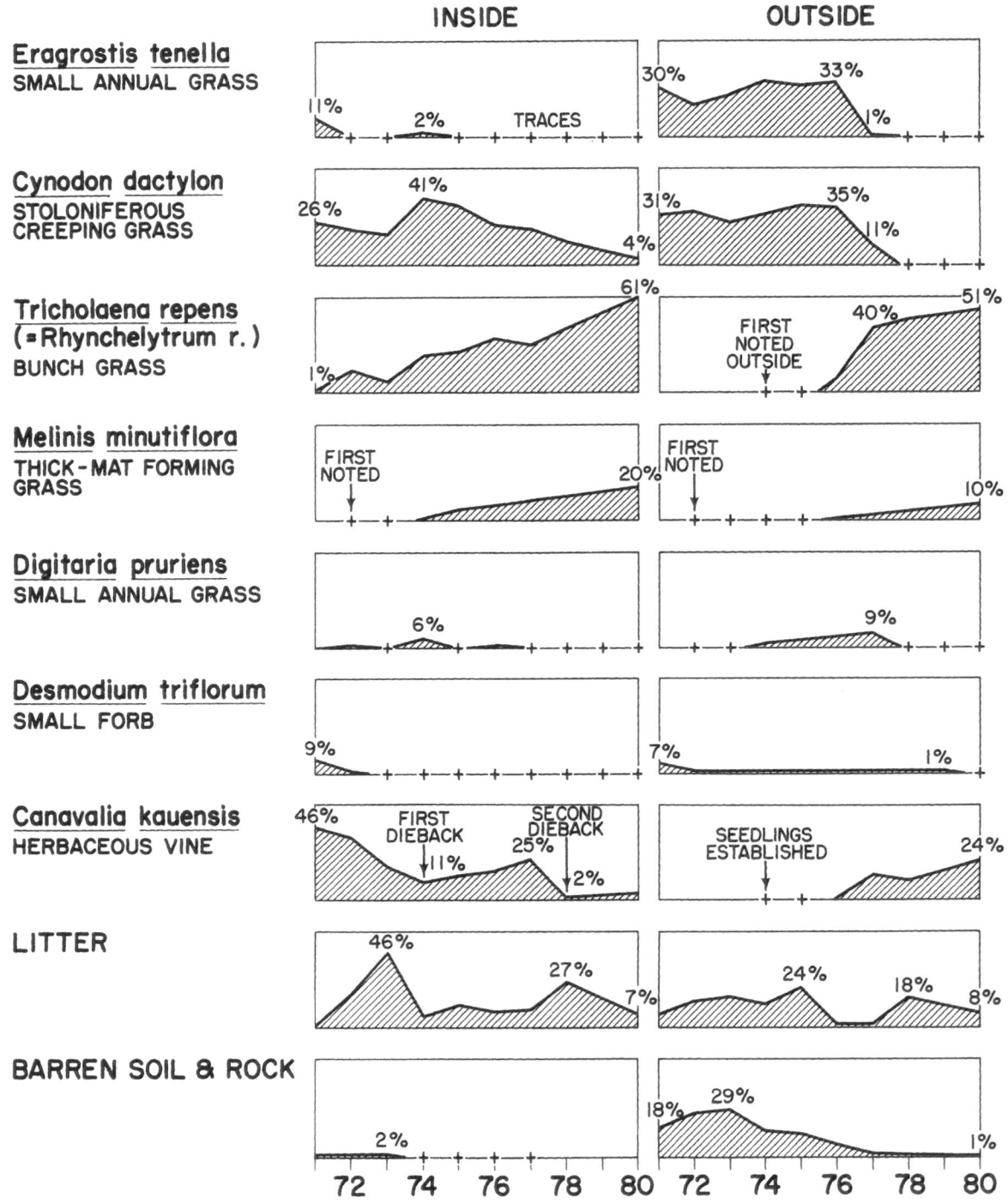

Fig. 2. Ten-year development of vegetation cover inside and outside exclosure 1 at Kukalau'ula.

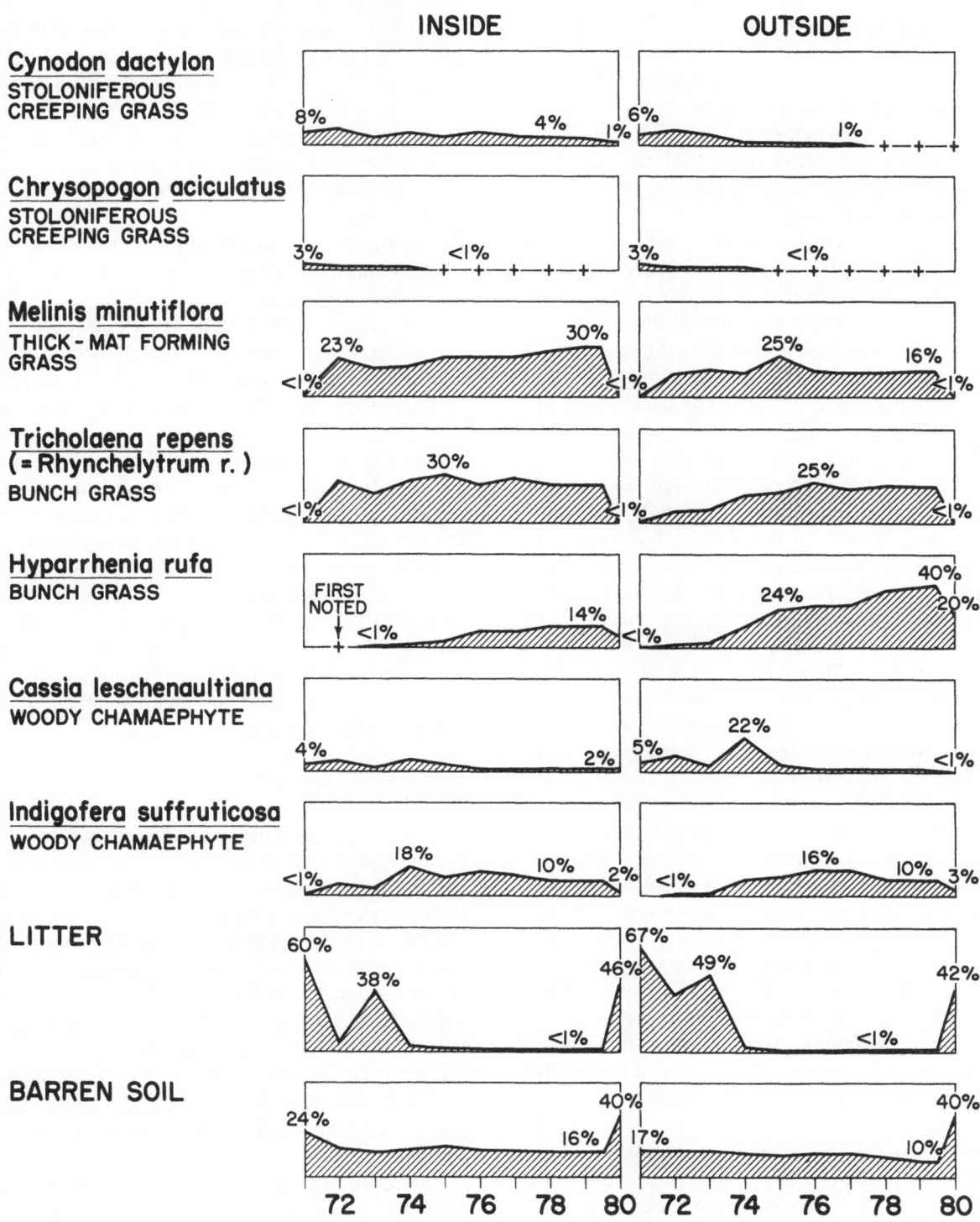

INSIDE OUTSIDE

Cynodon dactylon
STOLONIFEROUS
CREEPING GRASS

Chrysopogon aciculatus
STOLONIFEROUS
CREEPING GRASS

Melinis minutiflora
THICK - MAT FORMING
GRASS

Tricholaena repens
(= Rhynchelytrum r.)
BUNCH GRASS

Hyparrhenia rufa
BUNCH GRASS

Cassia leschenaultiana
WOODY CHAMAEPHYTE

Indigofera suffruticosa
WOODY CHAMAEPHYTE

LITTER

BARREN SOIL

Fig. 3. Ten-year development of vegetation cover inside and outside exclosure 2 at Puu Kaone.

the dry and wet seasons. Since then monitoring was repeated once a year during the dry season.

Development at Site 1

Only species with quantitatively important foliage cover are shown in Figure 2*. Their development inside the exclosure is contrasted with that outside for the period from 1971 through 1980. Data presented are from annual monitoring during the dry season of each year.

The pantropical love grass, *Eragrostis tenella*, was reduced to 11% inside the exclosure in 1971 and from then on it played a minor role to the present, while outside it was dominant with 30 to 33% cover until 1976 when the effect of goat elimination from this area became very evident in the vegetation surrounding this exclosure. The decrease of *Eragrostis tenella* is the result of competitive displacement by taller-growing plants, which established first inside the exclosure and later outside. These competitive displacers include here particularly the perennial bunchgrass *Tricholaena repens* (Natal redtop) and the herbaceous vine *Canavalia kauensis*. The latter was a great surprise, because it turned out to be a new endemic vine which had not been discovered before in the native Hwaiian flora. After three years of exclusion from goat grazing it covered almost half of the exclosure surface making the exclosure appear like an 'oasis' in a parched semi-desert of dead annual grass during the dry season.

At that time it was assumed that the vine may expand further through the entire exclosure as a layer over the perennial bunchgrasses. Moreover, when goats were eliminated, as was done through the systematic fencing program, it was thought that *Canavalia* may become a dominant cover throughout this area of the coastal lowland. It was further expected that certain woody chamaephytes, such as *Indigofera suffruticosa*, *Waltheria americana* and *Cassia leschenaultiana* (which are not shown on the diagram) may densify. It was expected further that certain phanerophytes may become established eventually. This was expected because the site potential of the area, with a mean annual rainfall of 800 mm, should indeed result in the establishment of a woodland vegetation, given enough time.

* See further Mueller-Dombois & Spatz (1975).

However, the development turned out to be different: as can be seen in Figure 2, the endemic vine *Canavalia kauensis* did not increase in cover beyond 46%. Instead it declined to a low of 11% in 1974. Thereafter it increased again but only to 25% cover in 1977, and then it died back a second time to a low of only 2% cover in 1978.

The species shown on this diagram can be categorized into three dynamic groups:
1. 'decreasers', i.e. *Eragrostis tenella* and *Cynodon dactylon*
2. 'increasers', i.e. the bunchgrass *Tricholaena repens* and the thick-mat forming molasses grass *Melinis minutiflora*
3. 'persisters', which include the three remaining species, *Digitaria pruriens*, a small native annual grass, *Desmodium triflorum*, a small leguminous forb and the endemic vine *Canavalia kauensis*.

The two additional surface covers, litter and barren soil or rock, which here were monitored similarly as the species, can also be grouped into these categories: barren soil and rock as decreasing and litter (consisting mostly of the dead or dried up grass foliage and *Canavalia* leaves) as persisting. I may add that among the 'persisters' it is possible to distinguish 'stable' or 'constant' persisters, such as *Desmodium triflorum* and 'oscillating' persisters, such as *Canavalia* and to some extent *Digitaria pruriens* and also the litter.

Development at Site 2

Figure 3 presents the year-to-year development at Site 2, the Puu Kaone exclosure. As soon as this exclosure was built for experimental purposes, the Park Service eliminated the goats from this area. Therefore, the development inside and outside is almost the same. Site 2 is on deep volcanic ash soil of loess-loam characteristics. There is no rock outcrop and *Eragrostis tenella* was never important here. Instead the two stoloniferous creeping grasses, *Cynodon dactylon* and *Chrysopogon aciculatus*, were the two dominants under high goat grazing pressure. Other life-forms were almost absent.

Again, one can classify the species into three dynamic groups, into decreasers, increasers and persisters. Here the decreasers are the two stoloniferous creeping grasses *Cynodon* and *Chrysopo-*

gon. Decreasing was also the litter at this site from 60% or more in 1971 to less than 1% in 1978. A sudden increase (from <1% to over 40%) is shown at the end of the study period, when a fire went through the area in January 1980. This also increased the area of barren soil, which over the study period had decreased slowly. The slow decrease was due to an eroded soil slope at the seaward end of the exclosure which was under the influence of constantly blowing, particularly strong desiccating winds. A typical increaser at this site is only the bunchgrass *Hyparrhenia rufa,* while the two other important grasses, *Melinis minutiflora* and *Tricholaena repens,* increased rapidly in the first few years after goat control. Thereafter they became persisters, with *Melinis* showing some oscillation between 16%, 25% and 16% cover over the study period outside the exclosure and *Tricholaena* exhibiting a somewhat more constant cover from year to year, inside and outside the exclosure. The two woody chamaephytes, *Cassia leschenaultiana* and *Indigofera suffruticosa,* can also be described as persisters, with *Cassia* showing more oscillation than *Indigofera.* The dominance of persisters indicates further that vegetation development was more or less arrested here at Site 2 before the fire.

Correlation with year-to-year rainfall

Apart from fire, which occurred so far only once in January 1980, and only at Site 2, the most important control-variable after goat removal may be considered to be the annual variation in rainfall.

This variation is shown in Figure 4 for Site 1. The normal rainfall at this site is 800 ± 100 mm/yr; at Site 2 it is slightly less (750 ± 100 mm). Considerable year-to-year rainfall variation is indicated. Normal rainfall between 700 to 900 mm/yr occurred five times out of the nine study years. Almost 50% of the study years had abnormal rainfall. Two years were very wet (1971 and 1979) and two years were very dry (1973 and 1977).

The most peculiar of the dynamic trends described before is that of the oscillating persisters, particularly as exemplified by the endemic legume vine, *Canavalia kauensis,* at Site 1.

When I plotted the % cover of *Canavalia* over the annual rainfall at Site 1, I found no correlation at all (Fig. 5). In 1971, a wet year, *Canavalia* had reached its maximum cover with 46%. It then declined from year to year with decreasing annual rainfall till 1973, the driest year. The *Canavalia* cover declined further to 11% in the following year, 1974, when the annual rainfall had increased to a high normal (with 888 mm). From then on the *Canavalia* cover increased steadily with decreasing rainfall to its second peak of 25% in 1977, when the annual rainfall was again abnormally low (with 605 mm). Following that, the *Canavalia* cover declined a second time in spite of increasing year-to-year rainfall.

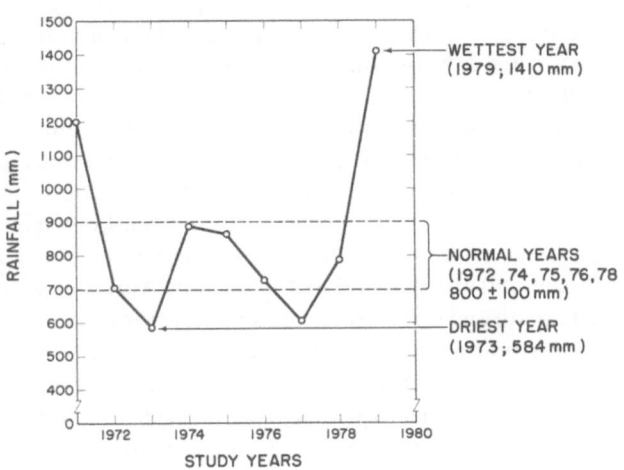

Fig. 4. Year-to-year rainfall variation at Site 1, Kukalau'ula exclosure.

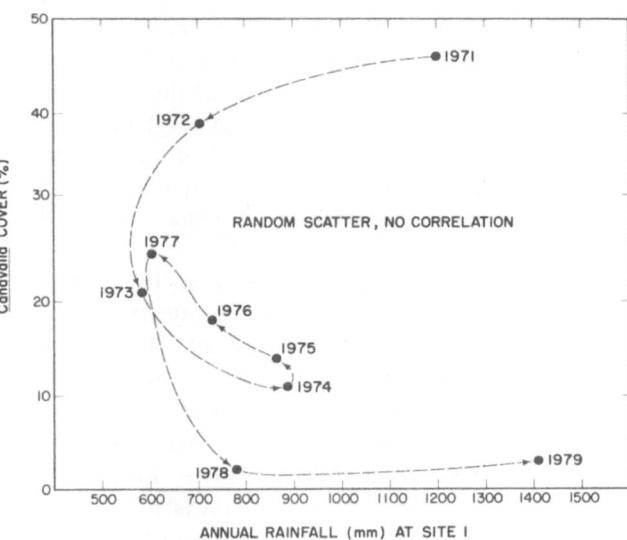

Fig. 5. Year-to-year cover variation of *Canavalia kauensis* inside exclosure plotted over annual rainfall at Site 1.

The other oscillating persister at Site 1, *Digitaria pruriens,* likewise showed no real correlation to annual rainfall. It was not recorded in 1971, a high rainfall year. It occurred with unmeasurable cover in the driest year, 1973. Its peak cover of 6% inside the exclosure occurred in a normal rainfall year in 1974, and its peak cover of 9% outside occurred in an abnormally dry year, 1977.

Discussion and conclusions

The non-correlation of the *Digitaria pruriens* cover with annual rainfall can be explained perhaps as a relationship to shifting patterns of cover of taller-growing species, which often overtop and locally shade out this low-growing and small annual grass. Its presence during the dry season of dry years, however, puts doubt on its consideration as an annual. Its behavior is more like that of a short-lived perennial.

Decline from shading of overtopping plants cannot be considered a reason for the periodic dieback observed in the vine *Canavalia.* Since its dieback is also not relatable to the variation in annual rainfall, competition for available water in the root zone is not a likely cause of its periodic dieback and recovery. Another factor is probably at work.

The exotic black and gold stink-bug (*Coptosoma xanthogramma*) has been observed to infest dying *Canavalia* plants. But the *Coptosoma* populations tend to be more abundant on dead plants than on those of low vigor. The *Coptosoma* infestation thus appears to be secondary to a primary cause responsible for initiating the dieback of local populations of *Canavalia.* It is suggested here that the primary cause of the observed dieback lies in the plant population itself.

Groups of individuals of the same life-stage, i.e. cohorts, appear to have reached maturity together. They then progressed to a stage of senility, when synchronized dieback occurred primarily as a response to the termination of their life cycle.

Since writing this paper, I was made aware of a *Canavalia kauensis* individual that was planted from seed in a forest nursery on the island of Maui. The individual developed flowers and fruits in the first year as well as in the second at the end of which it died without an apparent cause. There was no competition from other plants, no lack of water or nutrients (Robert Hobdy, pers. comm.). Similarly short life-spans have been observed in other endemic *Canavalia* species grown in the same nursery. Therefore, there is no doubt that these native *Canavalia* vines are short-lived perennials, but they are not typical biennials as known from temperate environments and also they are not monocarpic species (hapaxants), which are known to die after they have set seed once.

Synchronized death in non-monocarpic plants may occur when individuals of local populations germinate, grow up and reach maturity and senility at the same time, particularly if the individuals of such cohorts are of comparable vigor. Synchrony in death may, moreover, be triggered by secondary factors, such as an insect infestation at the stage of senility, or a minor environmental perturbation, such as a strong, drying wind.

During the ten years of monitoring the coastal lowland ecosystem, I have observed 'out-of-season' death of smaller and larger local populations in a number of grass species, particularly in *Melinis minutiflora, Andropogon virginicus,* but also in *Tricholaena repens* and in *Heteropogon contortus.* These grasses seem to live 'indefinitely' when they are periodically disturbed by mechanical breakage such as resulting from trampling, grazing or fire. Yet when they are allowed to outlive their life-span without any mechanical disturbance, they seem to die as individuals or groups of individuals in relation to their vigor, which is determined by both age and type of habitat. The same species may not become as old when growing on a marginal habitat than when it grows on a physiologically favorable habitat.

My thesis here is that life-cycle dependent and locally synchronized death of non-monocarpic perennials is a dynamic phenomenon in vegetation, equal in rank and importance to the well-known dynamic concept of phenology and succession. Phenology is usually related to seasonal variations in climate and soil factors, and succession is usually understood as a directional change in vegetation that comes about through the decrease and increase of different species, primarily as a result of their competitive interactions or through changing environmental factors.

Synchronized, life-cycle dependent death of cohorts, on the other hand, may occur without competition or environmental change; it may occur without any relation to seasonality. Neither Harper (1977) nor Grime (1979) in their more recent and detailed treatments of dynamic processes of plant populations in natural communities emphasize the phenomenon of synchronized life-cycle dependent death. I believe one reason why this third phenomenon has not yet received the attention it deserves is that most investigators did not have the chance to observe the same vegetation, year after year, for a long enough time. Locally dead populations, when seen in the vegetation, are often considered as part of a successional or phenological process, or as an outcome of a disease.

Little is known about the life-span and its behavioral manifestations in perennial species when growing in natural communities, and for the reason just given, it is easier to detect life-cycle dependent death in short-lived perennials, i.e. populations with life-spans from 2–5 yr. Among the species considered here in the 10-yr monitoring sequence as 'constant' or 'stable persisters' may be those (e.g. *Melenis* and *Tricholaena* at Site 2) with somewhat longer life-spans, lasting perhaps a decade or more. Undoubtedly, disturbance such as grazing and fire will change the life-span behavior manifestations, perhaps commonly into vegetative rejuvenations. Periodic disturbances, which are normal for unprotected communities, can be considered another reason why synchronized life-cycle dependent death has not yet received the attention it deserves. A third reason could be that this phenomenon is more difficult to detect in species-rich community mosaics. In contrast, species-poor community mosaics are more likely to lack aggressive competitive displacers in certain successional situations. They also allow more typically for certain species-populations to assume spatial dominance in succession, thus rendering synchrony in death of such populations a more dramatic or more easily detectable phenomenon. Moreover, it takes an approach of combining plant population and community studies, but the two areas still suffer under an artificial disciplinary barrier.

It will be of interest to follow the readjustment of the vegetation cover to the local break down of cohorts. At this time, it is too early to make any predictions. The break down of *Canavalia* cohorts may indicate a point in succession favorable for the invasion of phanerophytes, because of locally improved seed bed conditions by enrichment with organic matter, sheltering from wind and intensive radiation and by the absence of any root competition. However, in the absence of an effective seed source of phanerophytes, almost any sort of replacement pattern among the currently existing plant species can be expected, i.e. *Canavalia* death followed by *Canavalia* rejuvenation from seedlings, or followed by a densification of woody chamaephytes, or by reemergence of the more shade-resistent grass mats of *Cynodon dactylon*. The latter has occurred in some patches, while *Canavalia* has moved out of the exclosure and now oscillates spatially and temporally over a much larger area covering more than a hectare. Continued monitoring of the coastal grassland should make it possible to test Watt's (1947) thesis that a more or less complete spatial rotation may occur among the species that make up a species-constant or species-saturated community.

References

Doty, M. S. & Mueller-Dombois, D., 1966. Atlas for bioecology studies in Hawaii Volcanoes National Park. Hawaii Bot. Sc. paper 2, 507 pp. (Republ. 1970 as College of Trop. Agric., Hawaii Agric. Expt. Sta. Miscell. Public. 89).

Ellis, W., 1825. A journal of a tour around Hawaii, the largest of the Hawaiian Islands. Boston. (Repr. in part in Am. J. Science 11(1): 7–36, 1926).

Fosberg, F. R., 1972. Guide to excursion III, Tenth Pacific Science Congress. Rev. ed. Publ. by Univ. of Hawaii, Honolulu, 249 pp.

Gerdes, R. G., 1964. History of the feral goat control program in Hawaii Volcanoes National Park. Mimeogr. report, Hawaii Volcanoes National Park, Headquarters. 17 pp.

Grime, J. P., 1979. Plant strategies and vegetative processes. John Wiley & Sons, New York. 222 pp.

Harper, J. L., 1977. Population biology of plants. Academic Press, London, N.Y., San Francisco. 892 pp.

Marques, A., 1905. Goats in Hawaii. Thrum's Haw. Annual for 1906: 48–55. Honolulu.

Mueller-Dombois, D., 1966. The vegetation map and vegetation profiles. In: M. S. Doty & D. Mueller-Dombois. Atlas for bioecology studies in Hawaii Volcanoes National Park pp. 391–441. Hawaii Botanical Science Paper (Republ. 1970 as College of Trop. Agric., Hawaii Agric. Expt. Sta. Miscell. Public. 89).

Mueller-Dombois, D., 1976. The major vegetation types and ecological zones in Hawaii Volcanoes National Park and their application to park management and research. In: C. W. Smith, ed. Proceed. First Confer. in Natural Sciences, Hawaii Volcanoes National Park. pp. 149–161, Coop. Nat. Parks Stud. Unit, Dept. of Botany, Univ. of Hawaii, Honolulu. 243 pp.

140

Mueller-Dombois, D., 1979. Succession following goat removal in Hawaii Volcanoes National Park. Proceed., First Confer. on Scientific Research in the National Parks, II: 1149–1154. (Superintendent of Documents, U.S. Gov. Printing Office, Washington, D.C. 20 402).

Mueller-Dombois, D., 1980. Spatial variation and vegetation dynamics in the coastal lowland ecosystem, Hawaii Volcanoes National Park. (In press). In: C. W. Smith, ed. Proceed. Third Confer. in Natural Sciences, Hawaii Volcanoes National Park Studies Unit, Dept. of Botany, Univ. of Hawaii, Honolulu.

Mueller-Dombois, D. & Ellenberg, H., 1974. Aims and methods of vegetation ecology. John Wiley & Sons, New York, London, Sydney. 547 pp.

Mueller-Dombois, D. & Fosberg, F. R., 1974. Vegetation map of Hawaii Volcanoes National Park at 1: 52 000. Cooper. Nat. Parks Studies Unit, Bot. Dept., Univers. of Hawaii, Tech. Report. 4, 44 pp.

Mueller-Dombois, D. & Spatz, G., 1975. The influence of feral goats on the lowland vegetation in Hawaii Volcanoes National Park. Phytocoenologia 3: 1–29.

St. John, H., 1972. Canavalia kauensis (Leguminosae) a new species from the Island of Hawaii. Hawaiian Plant Stud. 39. Pacific Science 26: 409–414.

St. John, H., 1973. List and summary of the flowering plants in the Hawaiian Islands. Pac. Trop. Bot. Garden Memoir 1, 519 pp. (Lawai, Kauai, Hawaii).

Tomich, P. Q., 1969. Mammals in Hawaii. B. P. Bishop Museum Special Public. 57. Bishop Museum Press, Honolulu, Hawaii, 238 pp.

Vogl, R. G., 1969. The role of fire in the evolution of the Hawaiian flora and vegetation. Proceed. Annual Tall Timbers Fire Ecology Confer.: 5–60.

Watt, A. S., 1947. Pattern and process in the plant community. J. Ecol. 35: 1–22.

Accepted 23.3.81.

Dynamics of some Western Australian ligneous formations with special reference to the invasion of exotic species*

P. B. Bridgewater & D. J. Backshall*
School of Environmental and Life Sciences, Murdoch University, Murdoch, W.A. 6150, Australia

Keywords: Exotic species, Fire, Management, Shrublands, W. Australia

Abstract

Invasion of natural vegetation in Australia by exotic species has become pronounced in the last 150 years of European settlement. Changes in the frequency and type of fire have been a major disturbance to the dynamics of natural vegetation, and this disturbance has allowed rapid spread of exotics, particularly in urban and semi-rural regions. Less obvious disturbances include changes in nutrient levels and hydrological regimes, which also allow exotic species to establish and spread, at the expense of native species. Permanent plots established in various vegetation types show a decline in diversity which is associated with an increase in invasion by exotic species. Possible management strategies, designed to maintain the natural vegetation, which can be postulated include control and change of fire regime, removal of exotic species before seed-set, and the active introduction of native plants adapted to disturbance.

Introduction

Much of Western Australia included as climate type IV by Walter & Lieth (1960) is covered by ligneous formations, ranging from low schlerophyllous species-rich vegetation (Marchant *et al.,* 1980) to *Eucalyptus diversicolor* tall open forest (Specht *et al.,* 1974). This paper deals with the structure and dynamics of some components of these ligneous formations in the Swan coastal plain of the central South West Australian coast. (Fig. 1). In this region ligneous formations range from species-rich heathlands developed over limestone (Marchant *et al.,* 1980), through woodland (Specht, 1970) dominated by *Banksia menziesii* and *B. attenuata* to open-forest (Specht, 1970) dominated by *Eucalyptus gomphocephala. Banksia* woodland

often has a species-rich shrub understorey, sharing many species with the limestome heathland. In contrast, the *E. gomphocephala* open-forest has a rather open understorey, with a greater number of grasses and forbs and fewer shrub species. Prior to land settlement by European man in 1829, the main environmental control of these formations would have been fire. Both natural fire from storms and deliberate fire through the intervention of aboriginal man were important in these formations. Many of the component species of this vegetation are clearly fire adapted, requiring fire to aid seed release, seed germination and rejuvenation after senescence. Many of the shrub species reach senescence after 25–30 yr of growth, although some species may survive for much longer periods. At an ecosystem level, fire would have helped nutrient cycling in the nutrient poor soils, and promoted a mosaic of regeneration phases which supported the widest range of animals. One study which illustrates this well from a similar region in south eastern Australia is that of Cockburn (1978).

* We gratefully acknowledge the assistance of R. Hume for cartography and Doreen Jones for typing. One of us (DJB) was in receipt of a Commonwealth of Australia postgraduate award during the course of this work.

Vegetatio 46, 141–148 (1981). 0042-3106/81/0463-0141/$1.60.
© Dr W. Junk Publishers, The Hague.

142

During the last 150 years of European settlement however, there has been a change in frequency and intensity of fire in the landscape, additional inputs of nutrients into ecosystems via the groundwater and aerial dust, and deliberate and also accidental introduction of many exotic species from mediterranean Europe and South Africa. There is now considerable evidence accumulating to suggest that these factors are combining in various ways to change the structure and floristics of native communities. This paper reports the first results of some studies set up to monitor the proportion of exotic species in relation to degree of disturbance in the vegetation.

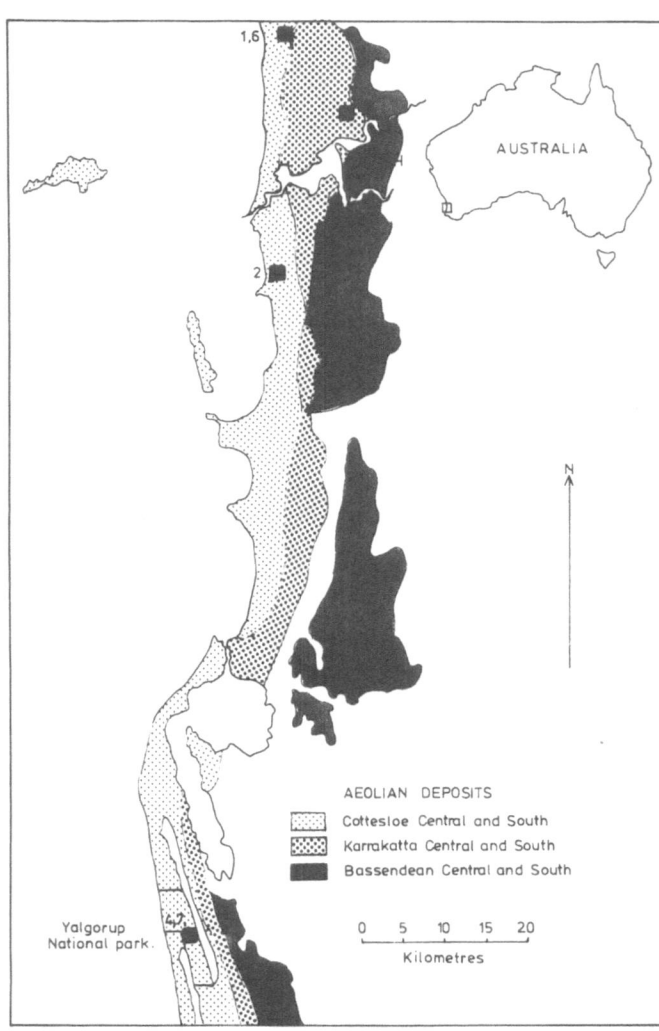

Fig. 1. Location map of sampling sites.

Structure and floristics of the vegetation

Most published descriptions of the vegetation have used primarily structural concepts, with some minor floristic data added. Thus Seddon (1972) describes the vegetation as closed heath, and Tuart-Jarrah-*Banksia* formation. Specht *et al.* (1974), using a slightly different approach, has the categories named as closed-heath of the *Proteaceae-Leguminosae-Myrtaceae* alliance; *Banksia menziessi-B. attenuata-Casuarina fraserana-E. todtiana* low woodland, *Eucalyptus gomphocephala* woodland. Bridgewater & Zammit (1979), looking solely at the limestome heath, described two associations from this region – the *Dryandra – Calothamnetum quadrifidae* and the *Dryandra – Acacietum cuneatae*.

Floristic data for the *Banksia* woodland and *E. gomphocephala* forest gathered in the course of this work will be published elsewhere.

Vegetation and soils

McArthur & Bettenay (1960) described the Swan coastal plain as consisting of quaternary units formed almost entirely from depositional material of either fluviatile, marine or aeolian activity. An area of Pleistocene aeolianite known as the Coastal Limestone (Geological Survey of W.A., 1975) parallels the present coastline of the Swan coastal plain as a series of dune systems ranging from recent (Quindalup Dune System), mid (Spearwood Dune System), to early Pleistocene (Bassendean Dune System). The Spearwood dune system consists of two principal soil associations, the Karrakatta and the Cottesloe series. The distribution of these soils is shown in Figure 1. Limestone heath occurs on areas of exposed dunal limestone, *Banksia* woodland on areas of deeper sand cover in the dunal depressions, and *E. gomphocephala* woodland in areas of deeper sand over limestone on the dune coasts and slopes. These three vegetation types are found thus intermixed on the Spearwood dune system, with *Banksia* woodland extending to the Bassendean dunes. Because of their proximity to urban and semi-rural surroundings of Perth, these vegetation types are particularly subject to disturbance, and often have large numbers of exotic species associated with them. Many of these exotics

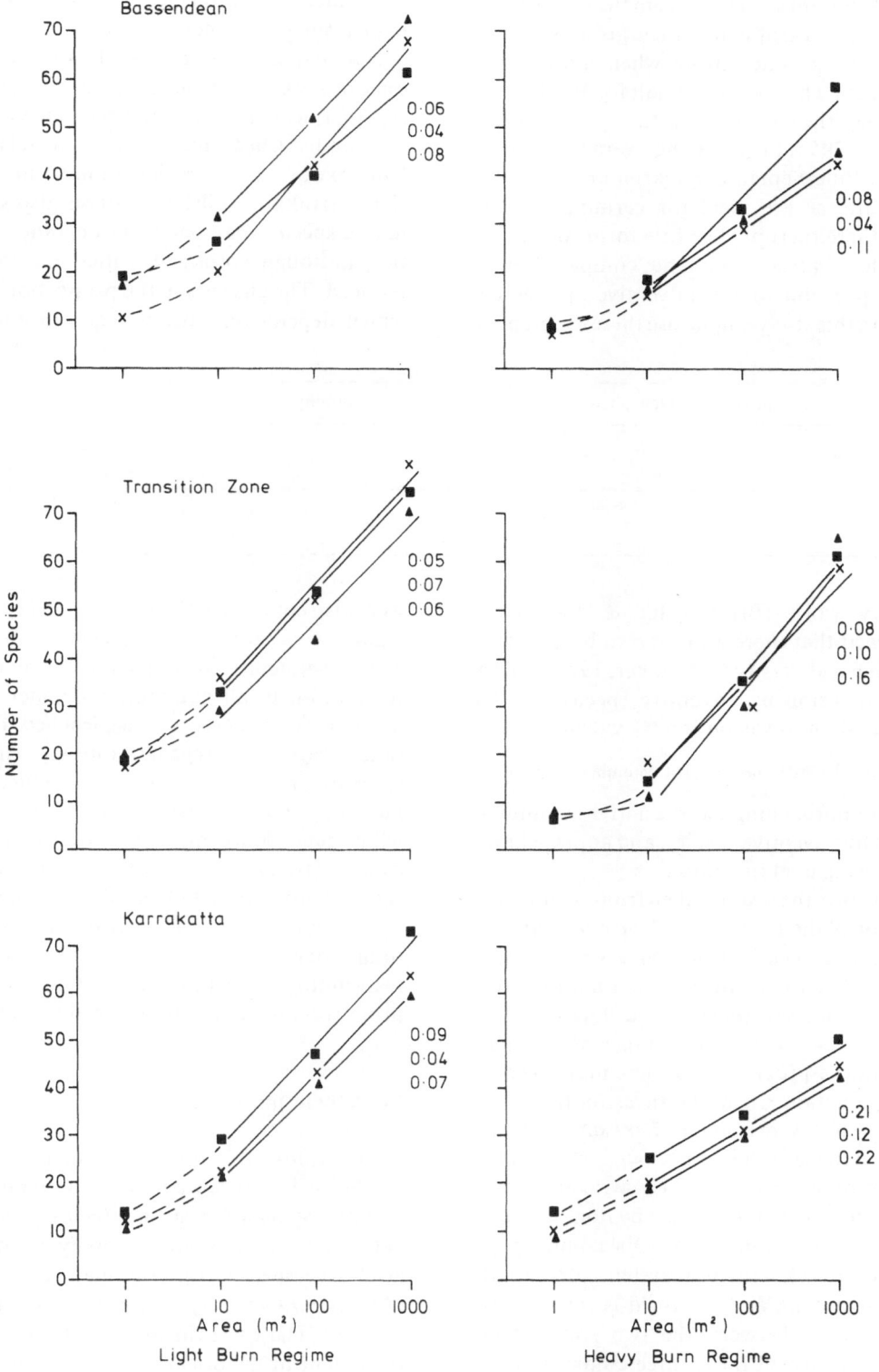

Fig. 2. Data from 18 nested diversity quadrats sampled in communities from Karrakatta, Bassendean soils and a transitional region between Karrakatta and Bassendean are shown for fire disturbed and near-natural areas. The invasion indices associated with these curves are also shown, emphasising that soil type and disturbance are associated factors in the invasion process.

have quite different strategies from the native flora. Being typically therophytes, they grow primarily during the winter and spring when most of the rainfall occurs. The annual rainfall for Perth is 883 mm, whereas the average for the quarter June-September is 395 mm, while the average temperature for the June-September quarter arise 17.2 °C. These conditions are ideal for germination and growth of these therophytes. Life form spectra for an uninvaded *Banksia* woodland, compared to the life form spectrum of the adventive species encountered in this study emphasise these differences:

	Phanerophyte	Chamaephyte	Hemicryptophyte	Cryptophyte	Therophyte
Native Community	9	55	31	3	2
Adventive Species	–		20	10	70

Preliminary work (Bridgewater & Kaeshagen, 1979) showed that there appeared to be a correlation between α-diversity (Whittaker, 1975) and the degree of invasion by adventive species. In that study we used an 'Invasion Index', calculated as –

I = Number of adventive species Total number of species

with values approaching 1 as the native community is replaced by adventive species, and approaching 0 with decreasing contamination.

Table 1 shows the results taken from six quadrats in vegetation of the *Dryandra - Acacietum cuneatae* near Fremantle. The first two relevés were from a relatively isolated undisturbed site on a limestone ridge, the second pair on the ridge slopes, and the third at the base of the ridge, adjacent to a busy road. There is a progressive decline in α-diversity, with a concomitant rise in the Invasion Index.

Similar trends were found in *Banksia* dominated woodland in the same localities. *Banksia* woodland on elevated dunes spanning the Karrakatta - Bassendean systems can be classified by floristic means into three distinct communities. One community is confined to the Karrakatta system, one to the Bassendean system, and the third is found in the transitional zone between the two soil systems. Sites were selected in these three communities to enable a comparison of undisturbed, rarely burnt vegetation with areas which, because of their closer

proximity to settlement, had been subjected to higher burn frequency.

Native species – area curves from nested 1 000 m² quadrats taken in these areas are shown in Figure 2, together with invasion indices. These emphasize that undisturbed communities are relatively free from exotics, even on the younger, more nutrient rich Karrakatta soils. The curves also show many native species may persist in degraded communities, although species densities are considerably reduced. The changes in the proportions of species which depend on various regeneration strategies were also compared. Species that regenerate from a viable seed-bank (e.g. *Asteraceae, Mimosaceae, Fabaceae*) significantly increased proportional representation in all communities under the higher burning frequencies. In the leached Bassendean sands vegetative regeneration via rhizomes or spreading roots *(Cyperaceae, Restionaceae)* is significantly more resistant than other strategies, while dependence on resprouting from a woody caudex, rootstock or corky sub-surface roots is significantly more 'vulnerable. Geophyte species *(Orchidaceae, Liliaceae)* were reduced in the transitional zone sites, while those species which combine resprouting and seed-bank replacement retained proportional representation throughout.

Permanent plots

To explore this further, permanent plots were established in seven examples or remnants of *E. gomphocephala* forest and *Banksia* woodland to determine biomass and diversity changes in the adventive species over the main growing season (August to October). Site locations are shown in Figure 1. These were located to reflect a variation in the extent of historical disturbance received since settlement. At each location four 0.25 m² plots were harvested each month, each plot located so as to be

Table 1. Species composition and Invasion Index value in six quadrats of the *Dryandra - Acacietum cuneatae*

Relevé number:	3	4	9	10	11	12
Invasion Index:	29	27	36	36	42	48
Native Species						
Hibbertia hypericoides	3	3	3	2	+	1
Dryandra nivea	2	+	3	2	2	2
Grevillea thelemanniana	2	2	1	+	2	+
Schoenus grandiflorus	1	+		+		+
Lomandra suaveolens	2	2	+	+	1	2
Acacia truncata	1	1	+	+	-	+
Loxocarya flexuosa	2	1	2	+		
Phyllanthus calycinus	2	1	2	+	+	
Melaleuca acerosa	1	1	3	2	+	+
Conostylis candicans	2	2	2	+	2	+
Crassula colorata	+	+			1	
Trachymene pilosa	+	-		2		+
Xanthorrhoea preissii	2	1	4			5
Thysanotus patersonii	+	+				+
Lepidosperma gracile	2	2	2			
Hybanthus calycinus	+	1			1	
Borya sp.	1	2				+
Cryptandra mutila	+	+				
Senecio lautus	+	1				
Stipa elegantissima	+	+		+		
Pimelea rosea	-	-	1	+	+	
Acanthocarpus preissii						
Templetonia retusa						
Acacia cochlearis			1	2		
Acacia pulchella	+	-	1			
Drosera macrantha		+		+		
Opercularia vaginata	+	-	+			
Calothamnus quadrifidus	-	-	+	+		
Scaevola canescens	2	-		+		
Melaleuca heuglii	-	2				
Anigozanthos humilis	-	1			+	
Dryandra sessilis				+		
Kennedia prostrata				+		
Cynodon dactylon				2	2	
Adventive Species						
Hypochaeris glabra	2	2	2	2	3	2
Anagallis arvensis	+	1	2	2	2	1
Petrorhagia prolifera	1	1		2	1	1
Briza maxima	2	2	1	3	2	-
Erharta longiflora	-	1	+	2	1	1
Arctotheca calendula	+	+		2	2	2
Carpobrotus edulis	-	-	+	+	+	-
Brassica tournefortii	+	+	+	+	-	+
Pelargonium capitatum	1	+		3		+
Lagarus ovatus	+	1		+	-	-
Trifolium campestre			+			+
Erodium moschatum	-	-		2		2
Minute caryophyllaceae	-	1	1			1
Lupinus consentinii	-	-		1		-
Asphodelus fistulosus						
Trachyandra divaricata						
Asteraceae sp. indet.						
Sonchus oleraceus	+	-			+	
Vulpia sp.	-	-			1	+
Stellaria media	1	-			+	

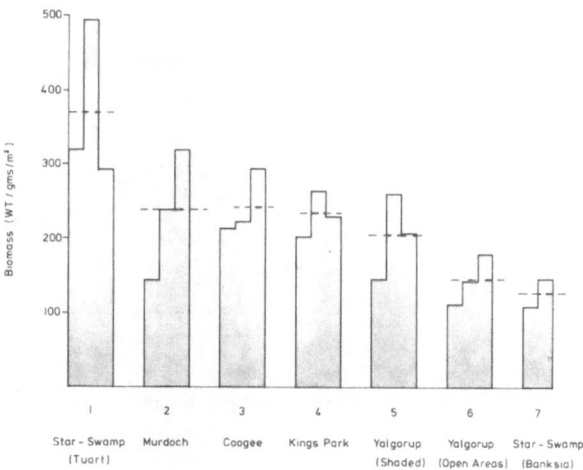

Fig. 3. A summary of mean (with standard error) biomass for 4 replicates/site, in 7 sites for each sampling period. The October value for site 6 is a mean of 3 replicates only.

as free as possible from the influence of ligneous shrub and overstorey components. All herbaceous species were cut at ground level, counted and weighed in the field.

Figure 3 shows biomass from three harvests for the seven sites. Plot biomass for each siteperiod, and the relative biomass of dominant species are respectively related to invasion index in Figure 4 and number of species in Figure 5. With the exception of site 7, which contains relatively few exotic species, a general biomass response to soil fertility is shown. Although there is no clear relationship between invasion index and total biomass of species, the role of dominant exotic species within a changing community structure is indicated, these changes are reflected in a plot of invasion index and site species richness (Fig. 6). Because of the very short life-span of many of the therophytes, each site can vary considerably in diversity through the spring growth period. This variation is shown in

Additional Species

3 - *Podolepis* sp. 1.
4 - *Astroloma pallidum* +;
    Eryngium pinnatifidum +;
    Hakea prostrata +;
    Mesomelaena stygia +.
9 - *Ursinia anthemoides* +.
10 - *Leucopogon propinquus* +;
    Lechenaultia linarioides +;
    Helichrysum cordatum +.
11 - *Grevillea vestita* 2.
12 - *Hardenbergia comptoniana* 2.
* - indicates an adventive species.

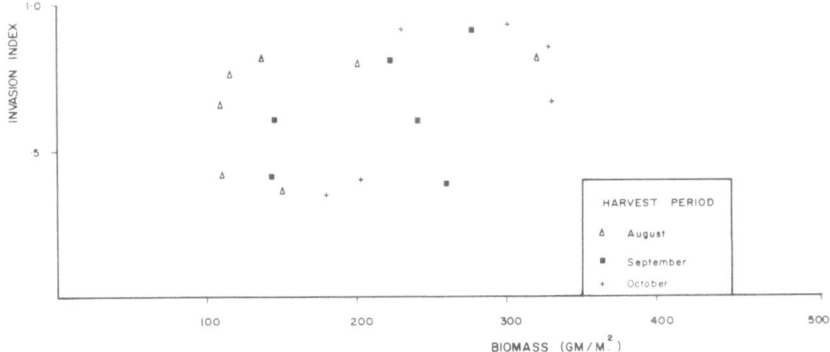

Fig. 4. The relationship between total plot biomass and the invasion index for all site periods.

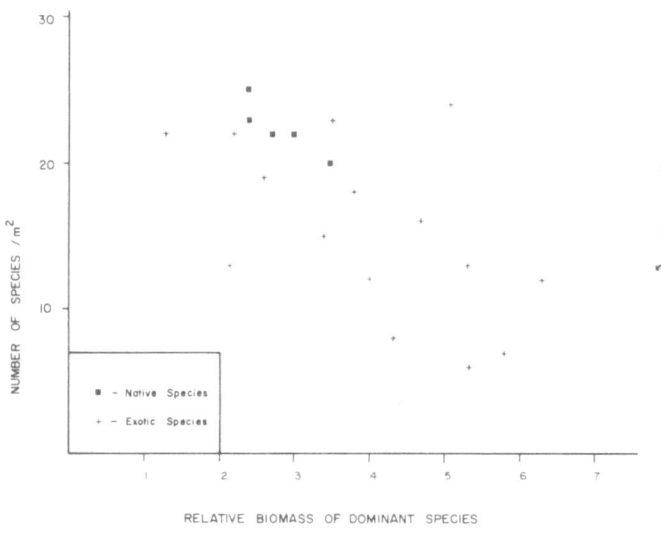

Fig. 5. The relationship between relative biomass of dominant species and number of species per m² for each site. Dominance by exotic or native species is indicated.

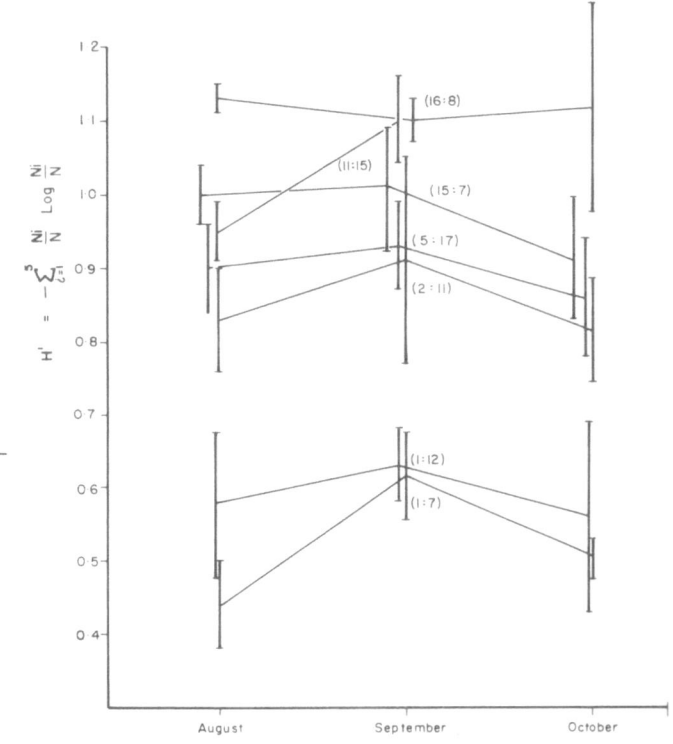

Fig. 6. The changing diversity of the sites through time.

Plot DIVERSITY for each SITE-PERIOD, values are mean of samples with S.E. The PROPORTION of NATIVE SPECIES: EXOTIC SPECIES for the PEAK PERIOD in SEPTEMBER is shown in BRACKETS.

Figure 7, emphasising synchrony in species diversity during September. Species that have established in the least disturbed *E. gomphocephala* forest are generally functionally similar to the native species, being small ephemerals. Those that have dominated more disturbed areas are at present sparsely represented in less disturbed areas (e.g. *Hypochoeris* spec., *Briza maxima*). In *Banksia* woodland and limestone heath a similar situation exists with the least disturbed sites carrying only ephemeral species, but more severely disturbed and degraded sites often carry species with life-forms competitive to the native species.

Discussion

All the data gathered from phytosociological and dynamic studies demonstrate a reduction in site diversity when there is an increase in exotic species.

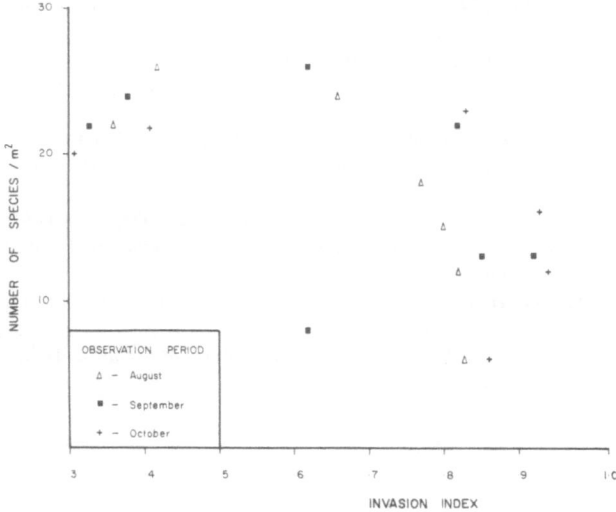

Fig. 7. The relationship between species richness (number/m²) and invasion index for each site period.

A point of debate arises concerning the effects of exotic species on the native plant communities – are these species responsible for community breakdown, or do they emerge as dominants as a consequence of other effects? Some adventive species, chiefly grasses from South Africa (eg. *Eragrostis curvula, Erharta calycina*) are hemicryptophytes. In the case of these species it seems very probable there is a strong competitive effect with native chamaephytes, especially in the face of increased fire frequency and general disturbance. Additionally, in the case of *E. calycina* strong seed predation by herbivorous rodents occurs in South Africa (S. Milton pers. comm). Where this plant species is expanding in Western Australia the native mammals are few in number, and introduced rodents (*Mus musculus, Rattus rattus*) are omnivorous or insectivorous. There is thus a high volume of seed set and release. Nevertheless, the distribution of *E. calycina* suggests it may be an important invasive species only in the Karrakatta and Transitional soils, and that in other areas it only becomes invasive as a consequence of gross physical disturbance. In contrast, many of the adventives occur widely through the vegetation of Western Australia. With regard to those species, it appears that the increasing frequency of fires, particularly in semi-urban areas, linked with an increase in nutrient-rich dust, is responsible for an initial debilitation of the native plant community,

causing death of some components. Gaps thus caused are filled by adventive therophytes, in addition to some seedlings of native species. These areas of therophytes can become small reservoirs of nutrients, cycling them in the top few centimetres of soil each year which may be a previously unexploited niche. As fires occur in late spring to autumn, when therophytes are dormant, they are usually unaffected – except to benefit from the extra input of nutrients from the burnt portions of the perennial species. This process is thus one of positive feedback, leading to gradual depletion of the native chamaephytes and hemicryptophytes. Particularly badly affected areas can carry an overstorey of *Banksia* or *Eucalyptus* species, with virtually no native species in the understorey. The prognosis for eventual regeneration of these overstorey species is clearly poor, as the role of exotic species in successions is unlikely to be the same as those of native species. Parts of Western Australia may face a future in which the native communities are replaced by synthetic communities of adventives with a native overstorey. Worse, the destruction or death of the overstorey may well result in the replacement of a perennial ligneous vegetation type by an annual therophyte/geophyte vegetation type. As this community would be largely dormant in the summer period, the possibilities for landscape degradation through wind erosion in the dry season appear high. It is possible, however, to see management solutions which could ameliorate this prognosis.

Clearly, a major priority would be to change the fire regime to one closer to pre-European frequency. Stabilisation of native communities which are being invaded by adventives, by the removal of these species before seed set, and replacement with seedling of native species components is an expensive but desirable option. Attempts at such work have been made in Sydney (Bradley, 1971) with considerable success. Such stabilisation is even more imperative where the adventive species are annual. Finally, it is of utmost importance to locate native species whose life strategy best fits disturbance, of any kind, to plant along the edges of, and paths through, native reserves and other areas of natural bushland left for posterity. In this way the native communities may be managed in their continually altered environment, leaving the landscape with the typical Australian physiognomy, and the advantage of being undegraded.

148

References

Bradley, J., 1971. Bushland Regeneration. Mosman Park Ass., Sydney.

Bridgewater, P. & Kaeshagen, D., 1979. Changes induced by adventive species in Australian plant communities. In: Wilmanns, O. & Tüxen, R. (eds.) Werden und Vergehen von Pflanzengesellschaften, p. 561–579, J. Cramer Vaduz.

Bridgewater, P. B. & Zammit, C. A., 1979. Phytosociology of SW Australian limestone heaths. Phytocoenologia 6: 327–343.

Cockburn, A., 1978. The distribution of Pseudomys shortridgei (Muridae: Rodentia) and its relevance to that of other heathland. Pseudomys. Aust. Wildl. Res. 5: 21379.

Geological Survey of W.A., 1975. The geology of Western Australia Memoir 2 Perth.

George, A. S., Marchant, N. G. & Hopkins, A. J. M., 1980. The heathlands of Western Australia. In: Specht, R. L. (ed.), Ecosystems of the world 9A. Heathlands and related shrublands. Elsevier, Amsterdam.

McArthur, W. M. & Bettenay, E., 1960. The development and distribution of the soils of the Swan Coastal Plain, Western Australia. Soil Publ. 16, CSIRO, Melbourne.

Seddon, G., 1972. Sense of place - university of W.A. Press, Perth, 274 pp.

Specht, R. L., 1970. Vegetation – In: Leeper, G. W. (ed.), The Australian environment, 4th ed., CSIRO, Melbourne, 162 pp.

Specht, R. L., Roe, E. M. & Boughton, V. M., 1974. Conservation of major plant communities in Australia and Papua New Guinea – Aust. J. Bot. Suppl. Ser. 7.

Walter, H. & Lieth, H., 1960. Klimiadiagramm – Weltatlas. Jena.

Whittaker, R. H., 1975. Communities and ecosystems, 2nd ed., Macmillan, New York.

Accepted 20.7.1981.

Community dynamics in relation to management of heathland vegetation in Scotland*

C. H. Gimingham, R. J. Hobbs & A. U. Mallik

Department of Botany, University of Aberdeen, St. Machar Drive, Aberdeen, AB9 2UD, Scotland

Keywords: *Calluna vulgaris,* Fire, Heathland, Markovian model, Permanent plots, Strategies, Succession, Vegetation dynamics

Abstract

The paper describes studies of post-fire succession in heathland vegetation in N.E. Scotland, dominated by *Calluna vulgaris.* A preliminary model (Legg, 1978) suggested good agreement between simulation of succession on the basis of a Markov chain and observations of stands at different stages of development after burning, at least in the earlier stages. Vegetation transitions are currently being recorded in permanent plots on burnt areas. First results confirm the view that (a) the post-fire succession has the properties of a Markov process, (b) this type of model remains valid when constructed from records of actual transitions, rather than data obtained by inference from evidence of transition. Comparing successional events in stands where, at the time of burning, the *Calluna* population was in pioneer-, building-, mature- and degenerate phases, shows that transition matrices generally agree with the Markov hypothesis, but not in the case of stands where *Calluna* was degenerate when burnt. The composition of establishing vegetation 1 year after fire is not confined to species normally associated with the early stages of succession, but reflects the composition of the stand before burning. Redevelopment after fire is described in terms of an initial floristic composition of species with strategies permitting early re-establishment, selected by the recurrence of the fire factor. Subsequent transitions represent changes in their relative abundance due to differing growth properties and competitive interactions. This interpretation applies only under conditions of recurrent incidence of fire (normally once in 10–15 yr). If fire does not recur, *Calluna* stands pass into the degenerate phase, where changes in the nature of relay floristics may come into play (e.g. with tree colonization).

Introduction

In many parts of Western Europe the extent of heathland vegetation has been much reduced during the past 60–100 years. In Northern Britain, however, this is not the case, and heathlands are still widespread on acid soils. Ericaceous low shrubs are the structural dominants of the vegetation (notably *Calluna vulgaris* on the more freely-drained substrata); trees and tall shrubs are sparse or absent. At altitudes below the natural timberline, such vegetation has in most cases been derived at various times in the past from formerly extensive forest, by human activity. It survives today because the lands are still used for the production of hardy breeds of sheep (e.g. Blackface sheep) and for sporting purposes (shooting, especially of the game-bird red grouse, *Lagopus l. scoticus,* and the red deer, *Cervus elaphus*), either separately or in combination. *Calluna* provides a significant proportion of the diet of the herbivores mentioned; therefore the heathlands are managed with the aim of encouraging production of the edible green shoots of *Calluna*. This form of land-use is prevalent

*Nomenclature follows Clapham, Tutin & Warburg (1962) for vascular plants; Smith (1978) for bryophytes.

Vegetatio 46, 149–155 (1981). 0042-3106/81/0463-0149/$1.40.

especially in the central and eastern uplands of Scotland and North-east England (Gimingham, 1972).

Calluna is a perennial, woody low-shrub. As stands of *Calluna* increase in age they pass through various phases which, though not sharply separated, are quite easily recognised (Watt, 1955; Barclay-Estrup & Gimingham, 1969):-

1. the *pioneer* phase: young individuals colonizing an area; cover incomplete; plants small with little lignification.
2. the *building* phase: individuals enlarging, normally establishing complete cover; canopy dense; production of green shoots and flowers reaching a peak.
3. the *mature* phase: production levels largely maintained but with a tendency for height growth to decline and for the canopy to become more diffuse.
4. the *degenerate* phase: main branches progressively collapsing and dying back; expanding gaps in the canopy.

Throughout this sequence, which in Scotland normally lasts for upwards of 40 years, a proportion of the annual production is partitioned to the woody branches, which extend in length and diameter: hence the ratio of edible green shoots to woody material declines. In the later stages, the production of green shoots per unit area p. yr may also fall, and they become increasingly inaccessible to sheep or grouse (Miller & Miles, 1969). The grazing value of a stand of *Calluna* therefore drops rapidly from about the middle of the building phase (i.e. when plants exceed ca. 15-20 cm in height, normally at an age of between 10 and 15 yr). The objective of management is therefore to destroy the above-ground parts at this stage, while promoting, so far as possible, rapid regeneration of the stand. This is usually achieved by burning, because a well-controlled fire does not kill the stem base from which vigorous vegetative regeneration may occur, leading to renewal of the stand and restoration of nearly complete cover in about 3 yr under favourable conditions.

Although under appropriate management of this kind, *Calluna* may assume dominance again in a relatively short time, a brief post-fire plant succession occurs in which lichens and bryophytes may be first to appear, and various other species become relatively prominent for a time alongside the regenerating *Calluna* (e.g. *Deschampsia flexuosa, Erica cinerea*). Where, however, fires are poorly controlled or stands have been allowed to become too old before burning, temperatures may rise to levels which cause failure of vegetative regeneration in *Calluna*, e.g. >400 °C at ground level (Whittaker, 1960). Longer and more complex post-fire successions then follow. Observations have suggested considerable variability in the details of these successions, and in the species taking part, which may be related to the age and composition of the stand before burning.

Studies are therefore in progress to investigate the nature of the vegetation changes involved, with three main questions in mind:-

1. What kind of model can best express the vegetation dynamics involved?
2. Is the idea of 'relay floristics' supported, or are the observations better interpreted on the basis of an 'initial floristic composition' with subsequent adjustment in the roles and quantities of the various species?
3. What are the characteristic strategies of species which habitually play a part in the post-fire succession?

Modelling post-fire succession

It was decided first to test whether or not the post-fire succession could usefully be modelled as a Markov-type process. If the system is Markovian, this implies that a change from one vegetation state to another depends only on the outcome of the immediately preceding change(s), rather than upon the whole history of earlier events. Put another way, future states depend only on the present, rather than upon the past (van Hulst, 1979).

To save time in setting up a preliminary model for purposes of this test, advantage was taken of the fact that burning management produces a patchwork of small stands each of a different age (time since burning), all side by side and therefore in almost identical habitats. A number of such stands in a *Calluna-Arctostaphylos uva-ursi* heath at Dinnet Moor, 56 km west of Aberdeen, NE Scotland, was sampled by Legg (1980). (*Calluna-Arctostaphylos* heath is a relatively species-rich community-type, presenting a more complex pattern of development than the *Calluna-Vacci-*

nium communities typical of many N British heathlands). Using numerous 10 cm × 10 cm quadrats on one sampling date only (autumn, 1976), Legg recorded the 'vegetation state' displayed by each quadrat at the time of sampling, together with evidence, where present, that this state had recently been derived from some other vegetation state or that change was currently occurring. ('Vegetation states' were defined in terms of occupancy of the quadrats by a given species, or sometimes a pair or small group of species in intimate mixture. Recent derivation from another state was evidenced by one species clearly having overgrown another, or the present state encompassing the remains of a previous one).

The numbers of instances of any one state changing into another were combined in a matrix, all elements of which were then converted into probabilities. The resulting matrix of probabilities was used to generate a model of successional changes on the basis of a Markov chain (Fig. 1.). The starting point of a post-fire succession is a total absence of above-ground vegetation (Legg, 1978). The first state probability vector representing the status of the vegetation at the beginning therefore contains the value 1.0 for the state 'bare ground'

and 0 for all other states. This vector is then multiplied by the transition probability matrix described here, so generating a new vector which in turn is multiplied by the transition probability matrix, and so on. Figure 1 is drawn from the results of 20 successive multiplications. The predicted sequence of 'states' (species or species-groups) rising into prominence or declining parallelled closely the observed successional trends (at least as regards the earlier stages), viz. as bare ground disappears, first grasses, then *Erica cinerea* and finally *Calluna vulgaris* rise to prominence, with subordinate peaks of lichens (amongst pioneer *Calluna*), *Hypnum cupressiforme* (under building *Calluna*) and *Pleurozium schreberi* (in the later stages), with *Arctostaphylos uva-ursi* increasing slowly to a relatively stable level in the later stages.

However, despite this degree of agreement between predictions from the model and the observed sequence, there are clear limitations to this approach. For example, the first model did not include changes consequent upon the *Calluna* passing into its degenerate phase (though these were incorporated later). More important, the transitions were inferred from such evidence as was available, and were not recorded as observed

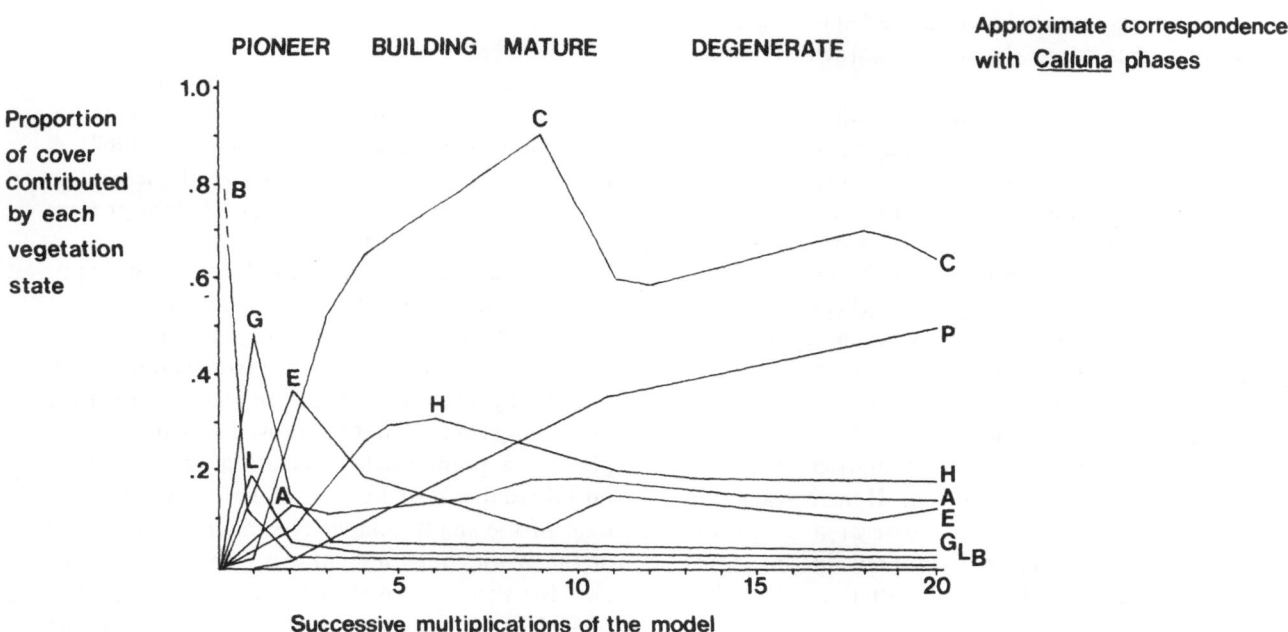

Fig. 1. Prediction of post-fire successional changes by the simple model (Markov chain). 'Vegetation states':- *B*, Bare ground; *G*, grasses; *L*, Lichens; *E*, *Erica cinerea*; *A*, *Arctostaphylos uva-ursi*; *C*, *Calluna vulgaris*; *H*, *Hypnum cupressiforme*; *P*, *Pleurozium schreberi*.

events. Moreover, dependence upon stands of different age, even though adjacent, has the disadvantage of using a spatial series to construct an assumed succession in time.

Subsequent studies, still in progress, are related to the two following questions:-

a. is the model upheld if actual changes with the passage of time are used, instead of inferred transitions?

b. is the process really Markovian in character – that is to say is it really the case that each transition depends only on the immediately preceding states, rather than representing a stage in a fixed sequence, each step of which depends on the accumulated series of previous steps (Collins, 1975)?

It also becomes possible to approach the problem of the extent to which the succession varies in relation to varying characteristics of the stands at the time of burning.

To answer these questions, permanent plots have been set up in areas burnt in 1979 and 1978 and some earlier years. In each of these areas the fires had burnt through stands having different characteristics (i.e. stands in which at the time of burning the *Calluna* population was in pioneer, building, mature or degenerate condition). The vegetation state in permanently-marked 10 cm × 10 cm quadrats was recorded at the beginning of the study and is currently being recorded at annual intervals.

Whereas for the preliminary model there was a degree of subjectivity in the recognition of vegetation states, these were now based more objectively on the cover-abundance values of the species present. Most quadrats fell readily into one of the vegetation states delimited. Actual transitions, from one vegetation state to another, where these have occurred, have therefore been available for use in transition matrices for these areas. On the one hand, all results to date have been combined in a comprehensive transition matrix (Fig. 2.) and, on the other, separate transition matrices for each stand have been constructed. Hence, transition matrices are available from several stages in the post-fire succession (times since burning) in stands which, at the time of burning, differed in regard to phase.

To test whether or not these transition matrices show the Markovian property, rather than

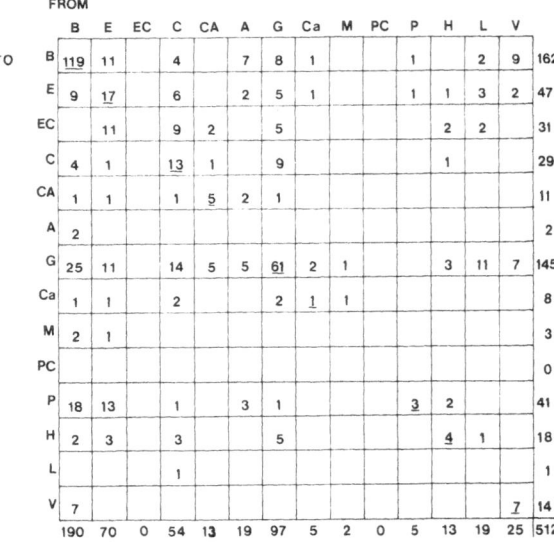

TO \ FROM	B	E	EC	C	CA	A	G	Ca	M	PC	P	H	L	V	
B	119	11		4			7	8	1		1		2	9	162
E	9	17		6		2	5	1			1	1	3	2	47
EC		11		9	2		5					2	2		31
C	4	1		13	1		9					1			29
CA	1	1		1	5	2	1								11
A	2														2
G	25	11		14	5	5	61	2	1			3	11	7	145
Ca	1	1		2			2	1	1						8
M	2	1													3
PC															0
P	18	13		1			3	1			3	2			41
H	2	3		3			5					4	1		18
L				1											1
V	7													7	14
	190	70	0	54	13	19	97	5	2	0	5	13	19	25	512

Fig. 2. Transition matrix for plots burned in April 1978, Dinnet Moor, N.E. Scotland. Figures indicate numbers of recorded transitions *from* the vegetation state shown along the top of the Figure, *to* the state shown on the left, within the period September 1978–September 1979. (Entries on the leading diagonal (underlined) show the numbers of observations of vegetation states which remained unchanged.) 'Vegetation states':- *B*, bare ground; *E, Erica cinerea; C, Calluna vulgaris; A, Arctostaphylos uva-ursi, G,* grasses; *C, Carex pilulifera; M,* Mosses (acrocarpous); *P,* Pleurocarpous mosses; *H,* Herbs; *L,* Lichens; *V, Vaccinium myrtillus.*

representing a number of statistically independent changes, the 'maximum likelihood ratio criterion' test is used (Kullback, Kupperman & Ku, 1962). A result significantly greater than χ^2 at the appropriate degrees of freedom indicates that the process is Markovian in character.

Significant agreement with the Markov hypothesis has been obtained (Table 1) for the matrix in which all results to date are incorporated, and also, separately, for matrices derived from stands which were in the building and mature phases at the time of burning (though not for those in the degenerate phase – a point further discussed below). These initial results must be viewed with caution in the light of the small number of samples at this stage in the work, but they confirm the view that the post-fire succession on heathland has the properties of a Markovian process, and show that the model remains valid when the matrix is constructed from records of actual transitions. On the strength of

Table 1. Maximum likelihood criterion test showing agreement between transition matrices and the first-order Markov chain. (The matrices are composed of transitions within the yr September 1978–September 1979 in stands at Dinnet, N.E. Scotland, which at the time of burning were in the phase indicated on the left.)

	$-2\log_e\lambda$	d.f.	χ^2	Significance
All plots combined[†]	438.25	144	179	***
Pioneer	72.37	81	103	NS
Building	105.36	64	84	**
Mature	88.82	36	51	***
Degenerate	14.20	49	66	NS
Significance levels:	*** $P = .001$			
	** $P = .01$			

[†]The transition matrix for this test is illustrated in Figure 2.

this, a more comprehensive and detailed Markov model of post-fire changes has been constructed and is currently being explored.

The 'maximum likelihood criterion test', while establishing the Markovity of the process, does not give any information on the order of the chain, i.e. whether transitions depend only upon the immediately previous state (1st order chain), or upon the immediately previous state and the one before that (2nd order chain), etc. The null hypothesis of a 1st order Markov chain can be tested (Kullback, Kupperman & Ku, 1962), but this test requires transition matrices from consecutive time intervals. At the time of writing, sufficient data are not available to carry out this test, although it is obviously important to do so and the analysis will be completed at a future date.

It is also possible to test whether, for example, the changes evident in stands burnt in 1978 and 1979 form part of the same process as those in stands burnt in 1972. The transition matrices are tested for homogeneity and the results indicate that the matrices from different stages in postfire development indeed fit into the same process of change (Table 2).

Relay floristics or initial floristic composition

Turning now to the question as to whether or not these changes represent 'relay floristics', Legg (1978) devised a system of scoring the various species concerned according to their relative abundance in stands of different age since burning, to obtain an index of their apparent position in the succession: e.g. ca. 1 for pioneer mosses; ca. 3 for certain herbs and grasses, 3–4 for ericaceous shrubs and 4–5 for some pleurocarpous mosses. The flora recorded before burning in the various stands (pioneer, building, mature and degenerate) used in the present study has been scored in this way. When such stands are burnt, if the 'relay floristics' hypothesis holds good a time sequence would be expected with all stands alike showing initial colonization only by species with low scores, which would progressively be replaced by those with higher scores. On the contrary, however, we find

Table 2. Results of significance tests of conditional homogeneity between transition matrices representing different stages in the post-fire succession, the stands being (a) in building phase (b) in mature phase at time of burning.

Phase of Calluna stand at time of burning	Comparison	Conditional Homogeneity[§]	d.f.	χ^2	Significance
(a) Building	Transition matrix for year 1–2*, and transition matrix for year 6–7[†]	110.52	182	212	NS
(b) Mature	Transition matrix for year 1–2*, and transition matrix for year 6–7[†]	119.15	132	157	NS

*Stand burnt April 1978, recorded September 1978 and September 1979.
[†]Stand burnt June 1972, recorded September 1978 and September 1979.
[§]See Kullback, Kupperman & Ku (1962).
Note: The result indicates that the transition matrices compared fit into the same process of change, since they do not show significant differences.

that 1 year after burning the flora of these stands is not uniform. In each case it reflects rather closely the composition before the fire, with some species characteristic of the later stages of succession present in all stands, amongst those typically associated with the earlier stages. It seems that most of the species represented in the successions can establish rapidly after fire, and that the successions consist of rearrangements of species abundances rather than a replacement series or 'relay'. The rule here seems to be that what re-appears after the fire depends largely on what was there before.

Attention may be drawn to the poverty in species of the recolonization stages of areas previously occupied by degenerate stands. As mentioned earlier, the transition matrix from these failed to show significant agreement with the Markov hypothesis, and it may be that such stands, because of age coupled with higher fire temperatures, are to some degree deprived both of propagules and capacity for vegetative regeneration. The succession might then be more akin to relay floristics.

Species strategies

This leads to consideration of the extent to which the nature of the successions is determined by the strategies of the species concerned. These can conveniently be classified as follows:-

A. Strategies conferring ability to survive fire

1. Hemicryptophytes with apices insulated by leaf bases e.g. rosette species such as *Pyrola media* and *Succisa pratensis;* densely tillering species such as *Carex pilulifera* and some grasses.
2. Low woody species with resistant stem-base from which new sprouts emerge at or a little above or below ground level:- e.g. *Calluna vulgaris, Arctostaphylos uva-ursi, Genista anglica, Vaccinium* spp.
3. Geophytes with rhizomes, tubers etc. unaffected by fire:- e.g. *Lotus corniculatus, Potentilla erecta, Anemone nemorosa, Lathyrus montanus.*

B. Strategies associated with rapid post-fire recolonization

1. Species maintaining a seed-bank in the soil:- e.g.

Calluna vulgaris, Erica cinerea, Arctostaphylos uva-ursi, Carex pilulifera.
2. Species re-establishing from fragments remaining in or on the soil surfaces:- e.g. certain bryophytes, lichens.
3. Species which are wind-dispersed into the area from neighbouring populations:- e.g. *Senecio sylvatica, Betula* spp.

All species in the stands examined possess at least one of these strategies; many of them combine one of the A group with one of the B group. The categories described correspond fairly closely with Noble & Slatyer's (1981) 'vital attributes' of species which occupy habitats subject to recurrent disturbances. These authors recognise, for example, species having vegetative mechanisms of persistence, with the adults either persisting as mature individuals (A1, above) or recovering by regrowth (A2 and A3). Their classification also includes groups of species with relatively long-lived seed stored in the soil (B1) and species with seeds widely dispersed and available in sufficient numbers for restocking (B3).

The effectiveness of these strategies in relation to re-colonization and re-establishment of the heath community is currently being investigated by following the details of post-fire succession in permanent 1 m^2 plots, mapped at frequent intervals. This, additionally, contributes information on whether all (or most) species of the community re-establish rapidly and concurrently after a fire, or whether there is a replacement series spread over a period of time. A summary of the results obtained during the first and second growing seasons after a fire in March 1979 is given in Table 3. Most of the species concerned are in fact established early in the succession. First to appear are plants which can regenerate vegetatively from points at or near ground level, and these are closely followed by species regenerating from below the surface. Re-establishment from seed or fragments is a little slower, and is subject to the hazards of a winter season, but none the less the plants displaying this type of strategy are also early colonists. Together, these species make up an initial floristic composition and succession consists in changes in their relative abundance due in large measure to differing growth properties and competitive interactions.

Table 3. Numbers of species representing various strategies, recorded at intervals following a fire in March 1978.

Time since burning:-	1st Season			2nd Season
	2 months	4 months	6 months	12 months
Strategy				
A(1) Hemicryptophytes with protected apices	2	6	8	8
A(2) Stem-base sprouts	3	4	4	4
A(3) Geophytes	0	4	5	6
B(1) Seed bank	0	0	4*	4
B(2) Fragments	0	0	2	5
B(3) Wind-dispersed into area	Not represented in the samples. Three species in this category occur patchily, from the 1st season.			

* Most of the seedlings of these 4 species died during the first winter, but new ones appeared in the following year.

Conclusions

The redevelopment of heathland vegetation after burning is clearly a secondary succession, but the results of this study indicate that it does not conform to the Clementsian idea of a fixed sequence of relay floristics similar to that of a 'prisere' but deprived of its earlier stages. Instead, a group of species having strategies which permit early re-establishment after fire has been selected by the recurrence of the fire factor in the course of management. These species, either because they survive the fire and regenerate vegetatively, or because their propagules are banked or rapidly dispersed into the area, set up an initial floristic composition. Subsequent transitions, which conform to a Markovian pattern, together amount to a chain in which different species increase or decline in prominence according to their growth rates, morphology, longevity, and competitive ability.

This interpretation can be sustained only so long as the incidence of fire recurs at the standard intervals of about 10–15 yr. If the succession is allowed to proceed beyond this time-span, changes in community structure permit the entry of species which are not normally present in the early stages (e.g. tall shrub or tree species). Increasingly, from the degenerate phase of the *Calluna* stand onwards, an element of relay floristics comes into play – for example, successions may proceed to stands of *Betula* spp. or *Pinus sylvestris*. This must be the subject of further studies.

References

Barclay-Estrup. P. & Gimingham, C. H., 1969. The description and interpretation of cyclical processes in a heath community. I. Vegetational change in relation to the Calluna cycle. J. Ecol. 57: 737–758.

Clapham, A. R., Tutin, T. G. & Warburg, E. F., 1962. Flora of the British Isles. Cambridge University Press. 1269 pp.

Collins, L., 1975. An introduction to Markov chain analysis. Concepts and techniques in modern Geography, 1. University of East Anglia, Norwich. 36 pp.

Gimingham. C. H., 1972. Ecology of heathlands. Chapman & Hall, London. 266 pp.

Hulst, R. van, 1979. On the dynamics of vegetation: Markov chains as models of succession. Vegetatio 40: 3–14.

Kullback, S., Kupperman, M. & Ku, H. H., 1962. Tests for contingency tables and Markov chains. Technometrics 4: 572–608.

Legg. C. J., 1978. Succession and homeostasis in heathland vegetation. Unpubl. thesis, University of Aberdeen, Scotland. 219 pp.

Legg, C. J., 1980. A Markovian approach to the study of heath vegetation dynamics. Bulletin d'Écologie (in press).

Miller, G. R. & Miles, A. M., 1969. Productivity and management of heather, in: Grouse research in Scotland, 13th Progress Report. pp. 31–45. The Nature Conservancy, Edinburgh.

Noble, I. R. & Slatyer, R. O., 1981. The use of vital attributes to predict successional changes in plant communities subject to recurrent disturbances. Vegetatio 43: 5–21.

Smith, A. J. E., 1978. The moss flora of Britain and Ireland. Cambridge University Press. 706 pp.

Watt, A. S., 1955. Bracken versus heather: a study in plant sociology. J. Ecol. 43: 490–506.

Whittaker, E., 1960. Ecological effects of moor burning. Unpubl. thesis, University of Aberdeen, Scotland. 230 pp.

Accepted 16.2.1981.

Vegetation dynamics in Brittany heathlands after fire*

B. Clément & J. Touffet**

Laboratoire d'Ecologie végétale, Université de Rennes, Complexe scientifique de Beaulieu, 35042 Rennes Cedex, France

Keywords: Brittany, Fire, Heathland, Line intercept, Permanent plot, Succession

Abstract

The vegetation dynamics of heathlands in Brittany have been followed for three years in areas subject to fire in August 1976. The pre-fire vegetation had been analysed and mapped before its destruction. The structure of the community and the processes taking place in it (in terms of biomass, primary productivity, phenology, mineral nutrition and food value) had been examined. The redevelopment of the canopy was studied by the point-contact method along permanent line transects. This semi-quantitative study permits calculation of the relative frequency of each species and, from this, its cover. The growth form of each species and the stratification of the community are also indicated by this method. Permanent plots were also used to record changes in the vegetation, by means of a census of individuals and records of the development and growth strategy of each species. The plots were located in homogeneous areas, or on bare soil around seed parents in order to examine seed dispersal and seedling establishment. These two methods yielded detailed information on the nature of the secondary successions following fire in the heathlands of Brittany.

Introduction

In the summer of 1976, many of the heathlands of Brittany were destroyed by huge fires. Summer fires of this kind occur only occasionally but profoundly affect the biocoenosis and its habitat; whereas the more frequent spring fires, which prevent the natural succession of heathland to thicket and forest, are less intense because of the higher moisture content in soil and vegetation. After spring fires, plants of the 'pyrophyte group' of heathland species, start to grow again in the first phase of reconstitution of the vegetation following a fire, and the floristic composition and primary productivity of the community are affected only temporarily.

In summer 1976, the drought magnified the extent of the fire and of its effects upon the heathland ecosystem. In addition to consuming the vegetation, the fire burnt the top layers of soil (litter, humus, and A_1 horizons), destroying nearly all the diaspores. As the fires occurred towards the end of the growing season, immediate recolonization could not take place and autumn rains swept the ashes away, leading to erosion of as much as 10–15 cm of soil in a few sites.

This erosion was especially significant on the tops and slopes of hills. Furthermore, it was correlated with the type of vegetation present before the fire, being greatest where there was abundant fuel in the form of vegetation and humus. By contrast in young heaths, where the vegetation was short after mowing and the litter layer was thin, the fire passed over rapidly without destroying the potentiality of the plants to regenerate vegetatively.

The object of this paper is to describe the

* Nomenclature follows Augier (1966) for Bryophytes, and Flora Europaea for Phanerogams.
** The considerable help of Professor Ch. Gimingham in preparing this English version of the original French manuscript is gratefully acknowledged.

regeneration of the vegetation in sites which were severely denuded as a result of burning. The observations were made in the heathlands of the Monts d'Arrée (Finistère, France), and are compared with results from Paimpont (Ille-et-Vilaine and Morbihan) obtained by Forgeard & Touffet (1979a, b).

Methods

The recolonizing vegetation was analysed using (a) permanent line transects and (b) permanent plots.

(a) Having chosen the sites, a line 10 m in length was marked out, at random, in an apparently homogeneous area. Point samples were then recorded at 10 cm intervals (cf. Daget & Poissonet, 1971). The distance between points and their number was proposed by Forgeard & Touffet (1979a) to allow for subsequent development of the vegetation, since the method must remain appropriate throughout the various stages of succession and be applicable to vegetation in which composition and structure are likely to be heterogeneous (since it consist of a mixture of cryptogams, herbs, erect and prostrate ligneous species, etc.).

The point-contact method yields quantitative data on the relative frequency of occurrence of each species at the points. If the number of samples is large enough, this can be expressed as a probability of the presence of a species. Assuming that the point where readings are made are so small as to have no diameter, the probability of the presence of a species is equivalent to its cover (Greig-Smith, 1964; Godron, 1968; cf. P. & J. Poissonet, 1969). The cover contribution of the several species can be converted to relative cover values for the community concerned.

This method can also be used to display the stratification of the vegetation by recording hits, or absence of hits, at various heightintervals on the wires (Forgeard & Touffet, 1979b).

(b) The development of each individual or group of individuals was also followed in permanent plots. These were 1 m² in area, a size convenient for recording without disturbing the canopy. Records were made using a detachable gridded square, suspended by adjustable chains, just above canopy level. This was divided into 16 grid-squares, each 0.25 × 0.25 m.

Plots were laid out to serve two main objectives. The first was to observe successional change in apparently homogeneous areas. Detailed mapping incorporated information on the number and height of the individuals, their cover and dispersion. The second was to observe the development and spread of new generations established from a seed-parent, and for this purpose the plots were located where an isolated plant, bearing seeds, was present in a space believed to be large enough to preclude seed input

Table 1. Description of sites.

Locality	Site with line numbers	Topography	Inclination in %	Orientation of line	Rock-type	Soil type	Community type
Tuchenn Gador E side	1	Middle-slope ⊥ slope	30	E-W	sandstone	podsol	*Erica ciliaris* and *Molinia caerulea* mesophilous heathland
Tuchenn Gador	2	summit	0	NE-SW	sandstone	crypto-podsolic ranker	*Erica cinerea* and *Molinia caerulea* meso-xerophilous heathland
Tuchenn Gador E side	3-4	middle-slope ∥ slope	20	E-W	sandstone	podsol	*Erica ciliaris* and *Molinia caerulea* mesophilous heathland
Roc'h Trévezel	5-6	upland	0	NNW-SSE	schists and quartzites	podsolic soil	*Erica ciliaris* and *Agrostis setacea* mesophilous heathland
Roc'h Trévezel S-E side	7	middle-slope ⊥ slope	15	NE-SW	schists and quartzites	brown crypto-podsolic soil	*Erica cinerea* and *Agrostis setacea* xero-mesophilous heathland
Roc'h Trévezel	8	depression	0	N-S	schists and quartzites	peat soil with gley	*Erica tetralix* and *Sphagnum compactum* moorland

from other individuals at the start of the experiment.

The sites

Six sites were selected with regard to their pre-fire vegetation, as analysed and mapped before it was destroyed (Clément, 1978), representing the main biotopes and associations of the heathland of this country. Their most important characteristics are listed in Table 1. In sites 1, 2, 5 and 6 one line was placed. Lines 5 and 6 were parallel, 0.5 m apart, in a heath (site 4) which consisted of a series of ridges and furrows due to ploughing, with line 5 in a furrow and line 6 on a ridge. Line 3 and 4 (in site 3) are also parallel, but separated by a part of the heath which had escaped burning.

The permanent plots were also set out in these sites.

Results

Succession as indicated by the line transects

Stages in the development of cover
Figure 1 shows the changes in total cover with time since burning at the different sites. Two years (July 1978) after burning (August 1976), cover had

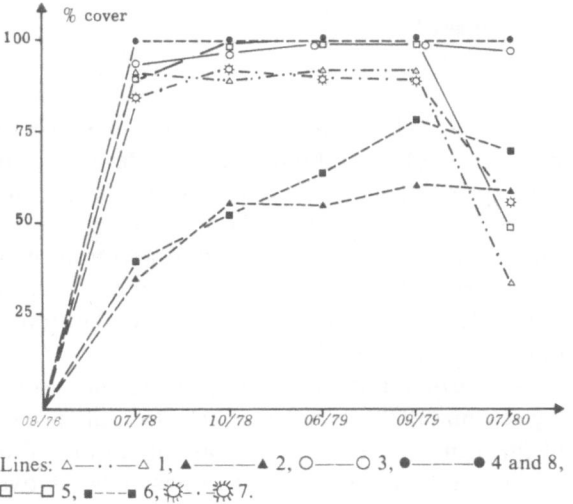

Lines: △—··—△ 1, ▲————▲ 2, ○——○ 3, ●————● 4 and 8, □——□ 5, ■·——■ 6, ☼·· ☼ 7.

Fig. 1. Total cover of the vegetation on the different lines.

reached 100% in two sites only (lines 4 and 8); three years later only four sites have complete cover. On lines 2 and 6, cover reached only about 60% and 70% resp. rain and violent winds sweeping over the hill tops were the chief causes of intensive erosion on these sites.

The low cover values shown for lines 1, 5 and 7 in July 1980 were due mainly to the death of bryophytes in the winter 1979–1980.

The total cover of the vegetation appears to depend on two principal factors: the severity of erosion, and the exposure of large stones during the winter following the fire.

Dynamics of new colonization
In Figures 2 and 3 the succession of all species in the six sites may be compared. The plants concerned were almost the same in all sites except the site 6 (which is situated in a moorland). The bryophytes were *Funaria hygrometrica, Ceratodon purpureus,* with different species of *Polytrichum* according to the water regime: *P. piliferum, P. juniperinum* and *P. formosum* on xerophilous and mesophilous heathland, and *P. commune* on moorland. *Campylopus introflexus* had a scattered occurrence. *Agrostis setacea* was the only phanerogam which was almost constant, and became dominant in areas where it was not previously characteristic (e.g. site 1), because of superior colonizing ability in comparison with other species. *Molinia caerulea,* for example, appeared to be capable of establishing seedlings only in damp sites. Some species which were favoured by the fire appeared, such as *Epilobium montanum, E. tetragonum, E. palustre* and *Senecio sylvaticus.*

The species previously most characteristic of the heathland communities *(Erica* spp., *Calluna vulgaris, Ulex* spp.) seldom reappeared; both their germination and seedling establishment seemed to be suppressed.

The first species to become visible was generally *Funaria hygrometrica,* which reached the peak of its development in 1977, retaining its importance in 1978 in those sites in which competition with *Ceratodon purpureus* was low. *Funaria hygrometrica* is an annual species growing on bare ground, whereas *Ceratodon purpureus* is perennial and after colonization tends to reduce the area of bare ground. Because of erosion in site 4, *Funaria hygrometrica* was still the most important species in

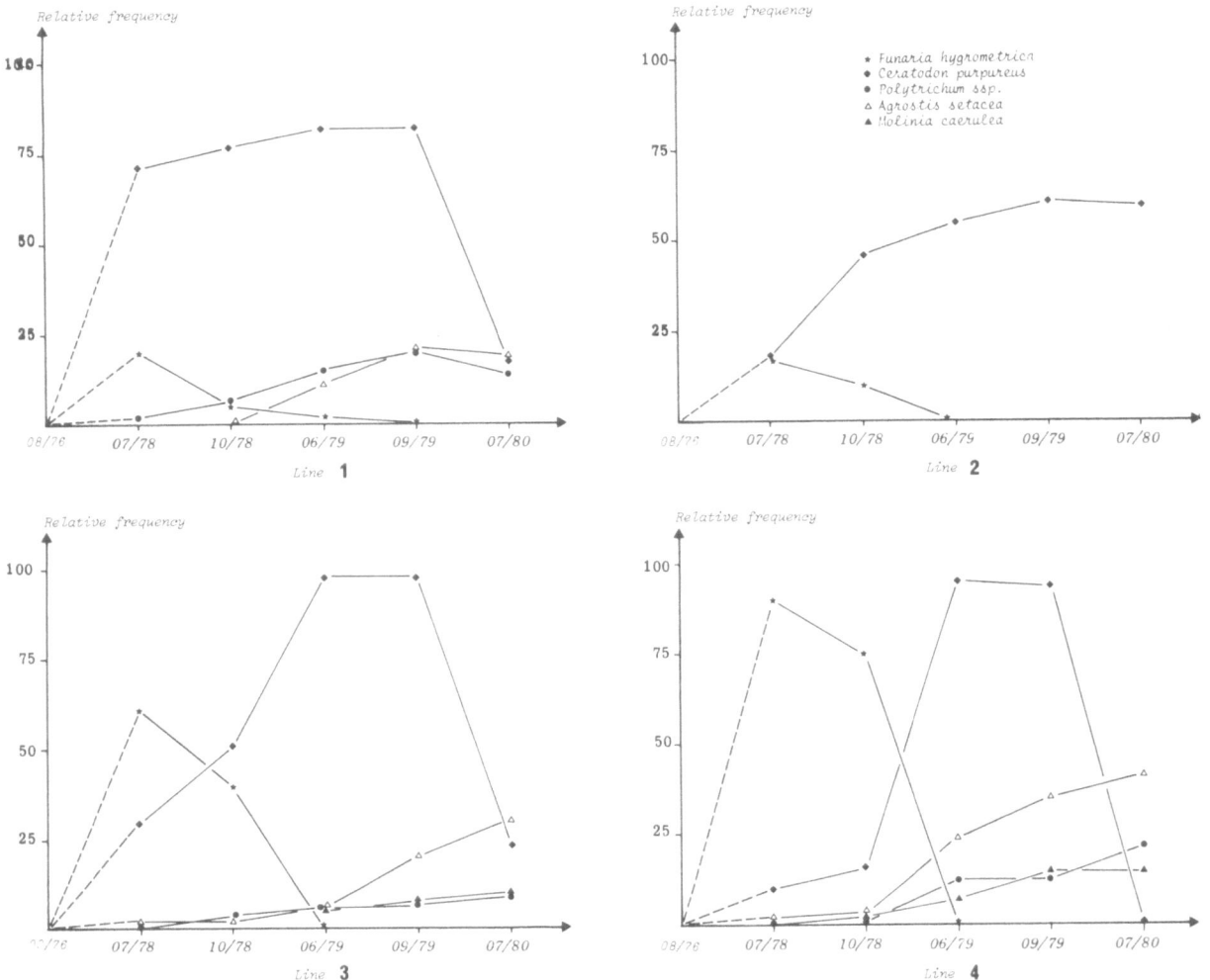

Fig. 2. Relative frequency (cover %) of the principal species on the lines 1 to 4 from 1976–1980.'

1978, but vigorous growth of *Ceratodon* in 1979 brought about its complete disappearance.

Ceratodon purpureus was the most important species in the majority of sites until autumn 1979, and was most abundant in 1978 and 1979. Weather conditions in the 1979-80 winter caused high mortality in this species, except on lines 2 and 6. It is noteworthy, however, that survival of mosses was best in sites with most severe erosion.

This may be explained by differences in the stage of development in different populations. In these two sites, the height of the vegetation did not exceed 1 or 2 cm, but it reached 5-7 cm in the remainder. Rain and frost caused depression of the patches of

Ceratodon, which desintegrated into a powdery litter in July 1980, inhibiting further development next spring. This deterioration would affect the invertebrate fauna living in the moss canopy. Birds, rabbits and roe deer also caused disappearance of *Ceratodon* in places by scraping. Disappearance of *Ceratodon* may be expected to allow the development of other species.

Polytrichum appeared during the second growing season (1978), but the contribution of species of this genus became substantial only in the third year (1979). Their spread was rather slow, relative frequencies not exceeding 25 along lines 1-6 in July 1980. Line 7 was more favourable, and on the

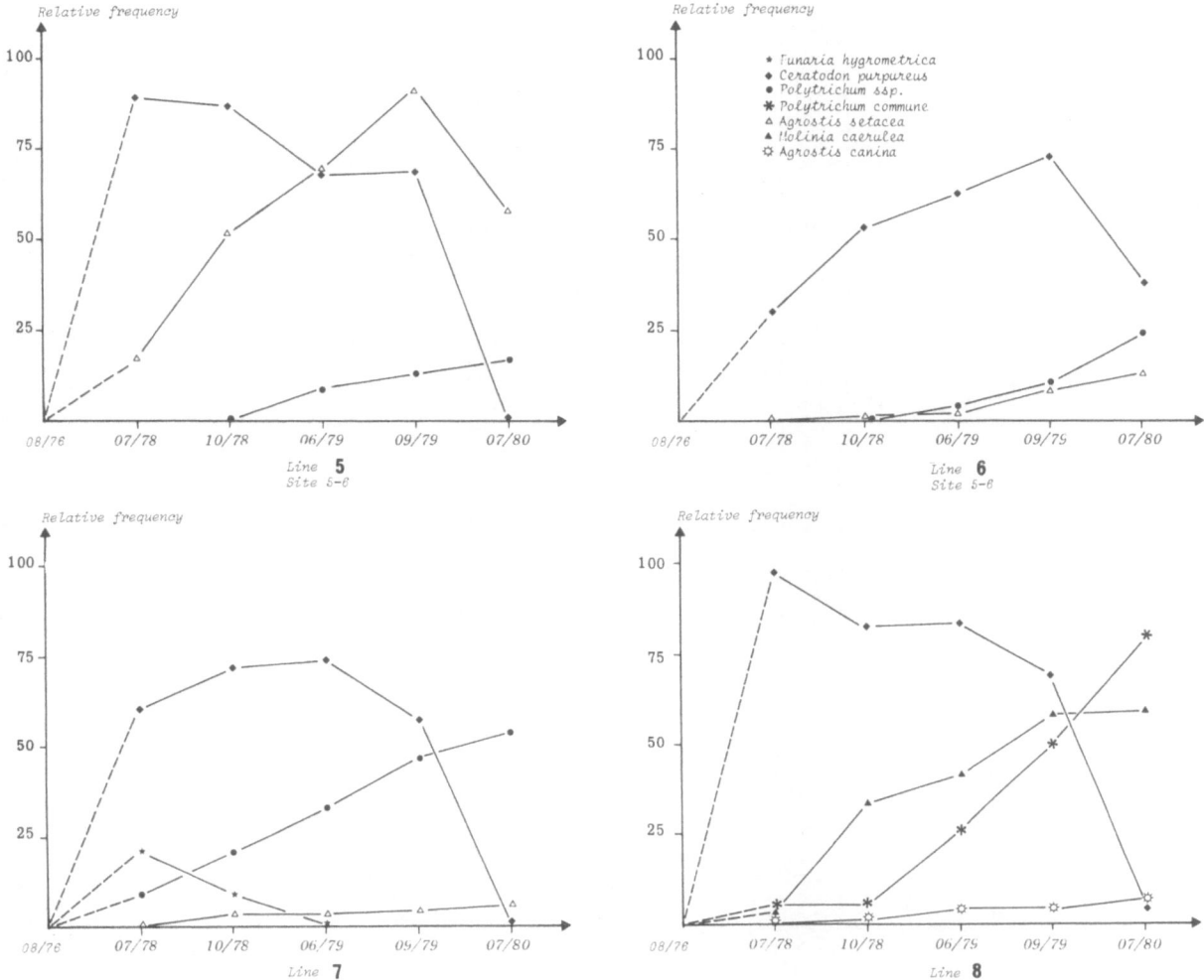

Fig. 3. Relative frequency of the principal species on the lines 5 to 8.

moorland *P. commune* was more prominent because of the high humidity during the spring growth period.

The development of *Agrostis setacea* was parallel to that of the *Polytrichum* communities. The first populations derived from seed germination were recorded in 1978, but significant numbers did not appear until 1980. However, there were few or no individuals in 1980 where erosion was severe (lines 2 and 6). Comparing lines 5 and 6, side by side in the same site, the cover of *Agrostis setacea* differed markedly, reaching 92% at the end of 1979 on line 5, but only 8% on line 6. This may be associated with the nature of the substratum. In the furrow, fine mineral grains and ash had accumulated, pro-

moting rapid growth of the *Agrostis* population. On the ridge, the removal of fine grains by water and wind restricted the establishment of *Agrostis*. The reduction of its relative frequency to 57% in July 1980 on line 5 was due to the death of a large number of individuals.

Molinia caerulea establishment took place only along lines 3 and 4 although the presence of seeds was observed in other sites. The moisture content of the substratum was perhaps the main factor controlling its spread, because lines 3 and 4 are situated on a slope and in a channel through which water drains from above. The significant development of *Molinia* in the moorland (of line 8) confirms this conclusion.

162

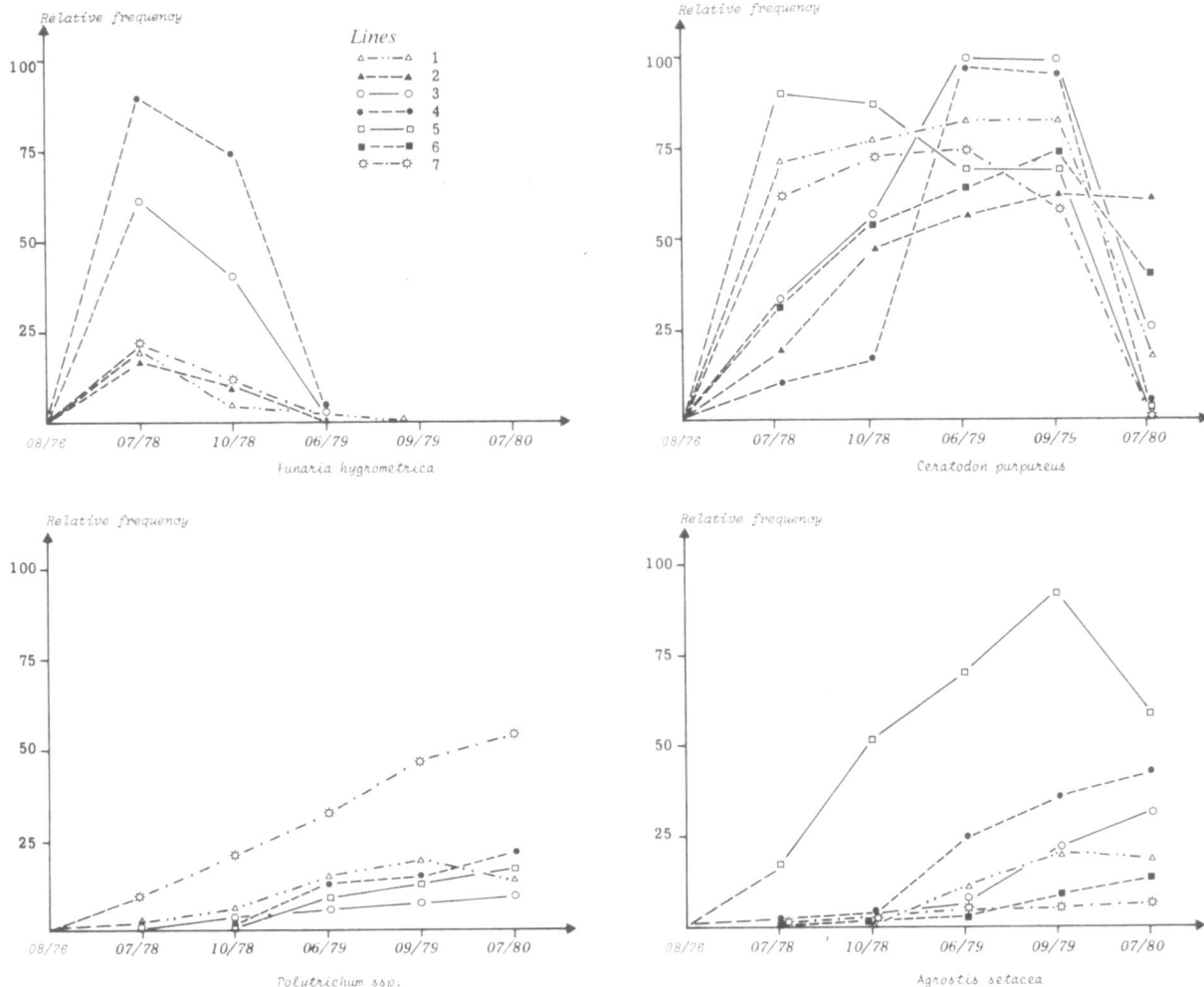

Fig. 4. Comparison of relative frequency of four species on the different lines.

The development of the four most important species in the different sites is compared in Figure 4. The species develop in a similar manner irrespective of site differences. There is therefore little influence of the type of heath community present at a site before the fire. For each species studied, the only differences are in their quantitative proportions which may change with time.

Succession of vegetation in the permanent plots

Dynamics in a homogeneous area

The succession is illustrated in Figure 5, which shows stages of development in site 7 near the line transect. Comparison of the 3 maps shows the development of moss cover, and especially the growth of *Polytrichum* in contrast to *Ceratodon purpureus*. Although the former covered about 80% of the plot in autumn 1978, its cover was only 20% in autumn 1979 and it disappeared completely in 1980. The plants present in 1978 gave rise to patches in 1979 and 1980. *Polytrichum piliferum* and *P. juniperinum* develop more slowly than *P. formosum;* the latter is more vigorous because it not only invades open spaces but also excludes the *P. piliferum* or *P. juniperinum* stands (Fig. 5, plot III).

These plots also illustrate the very limited colonization by phanerogams. Species once established grow normally, but establishment of new individuals is inhibited by the dense *Polytrichum* stands.

Dynamics from a seed parent

Figure 6 shows the development of *Agrostis setacea* from seeds produced by one individual in 1977 and 1978. Comparison of the three maps shows the growth of *Agrostis setacea* individuals, for each of which changes in the diameter of the tuft, and the height and number of inflorescences with increasing age, may be followed. This permits analysis of the reproductive activity and development of the species, while the establishment or disappearance of cohorts of individuals indicates the colonizing and competitive ability of the species, together with the effects of external factors such as predation: Plot III (Fig. 6) shows the death of many individuals due to the destruction of roots by *Tipula* larvae. However, scraping by rabbits or roe-deer in a part of this plot caused disappearance of the bryophytes, and recolonization in this disturbed part should elucidate the effects of the moss canopy on the establishment of phanerogams. The growth of the latter is in fact much faster in similar systems where cryptogams are few (Forgeard & Touffet, 1979a).

Discussion

The comparison of succession on several sites and the combination of line transects and permanent plots revealed a number of facts concerning the different steps and the speed of colonization in the burnt heathlands. Although the various stages of colonization have been mentioned in the literature (Lemée 1937; Géhu, 1960, Gimingham 1972), their importance and the parts played by different communities were unknown.

In the heathlands, the vegetation dynamics seem to depend upon interactions between the attributes of the species and the climatic conditions of the region. Variability of the substratum related to local topography merely modulates the rate of the biocoenosis succession. The first stage of the new colonization in mesophilous heathland is virtually the same as in xerophilous heathland, but the succession on moorland differs markedly from that on the xero-mesophilous heathlands. The former

may be compared to that described by Froment (1975) in the Hautes-Fagnes of Belgium. The vital role and prime importance of the bryophytes in the first steps both of plant and animal succession after fire must be emphasized.

The preponderance of *Funaria hygrometrica* and *Ceratodon purpureus* leads to the re-building of a large accumulation of organic matter (net productivity: 4–5 t. ha^{-1} yr^{-1}, Clément *et al.,* 1979). These two species create a biotope favouring the development of an invertebrate fauna.

The establishment and prevalence of *Polytrichum* spp. demonstrates clearly the strong competitive ability of these species in comparison with higher plants. The high density and greater growth rate of *Polytrichum formosum* or *P. commune* (7 to 8 t. ha^{-1} yr^{-1}, Clément *et al.,* 1979) prevents not only the establishment of new species but also the growth of any individuals of other plants already in place. These facts partly explain the continuing dynamics of the burnt heaths at the present time. New colonization by the characteristic heathland species will take a very long time if the moss communities continue to flourish. This balance could only be destroyed by the intervention of animals, but their activity is spasmodic. Moreover, the primary consumers tend to be responsible for killing seedlings of *Ericaceae* or *Fabaceae*.

The study of the plant succession on burnt heathland gives basic information on the role of the first colonists in the reconstruction of an ecosystem, and so leads to a better understanding of the status and stability of a large number of the heathlands of Brittany.

References

Augier, J., 1966. Flore des Bryophytes. Ed. Lechevalier, Paris, 702 pp.

Clément, B., 1978. Contribution à l'étude phyto-écologique des Monts d'Arrée. Organisation et cartographie des biocenoses; évolution et productivité des landes. Thèse 3ème cycle Rennes 260 pp.

Clément, B., Forgeard, F. & Touffet, J., 1980. Importance de la végétation muscinale dans les premiers stades de recolonisation des landes après incendie. Colloque 'Ecologie des landes', Rennes 1979, Bull. Ecol. 11: 359–364.

Daget Ph. & Poissonet, J., 1971. Principes d'une technique d'analyse quantitative de la végétation des formations herbacées. Document C.E.P.E.-C.N.R.S., 56 Montpellier: 85–100.

164

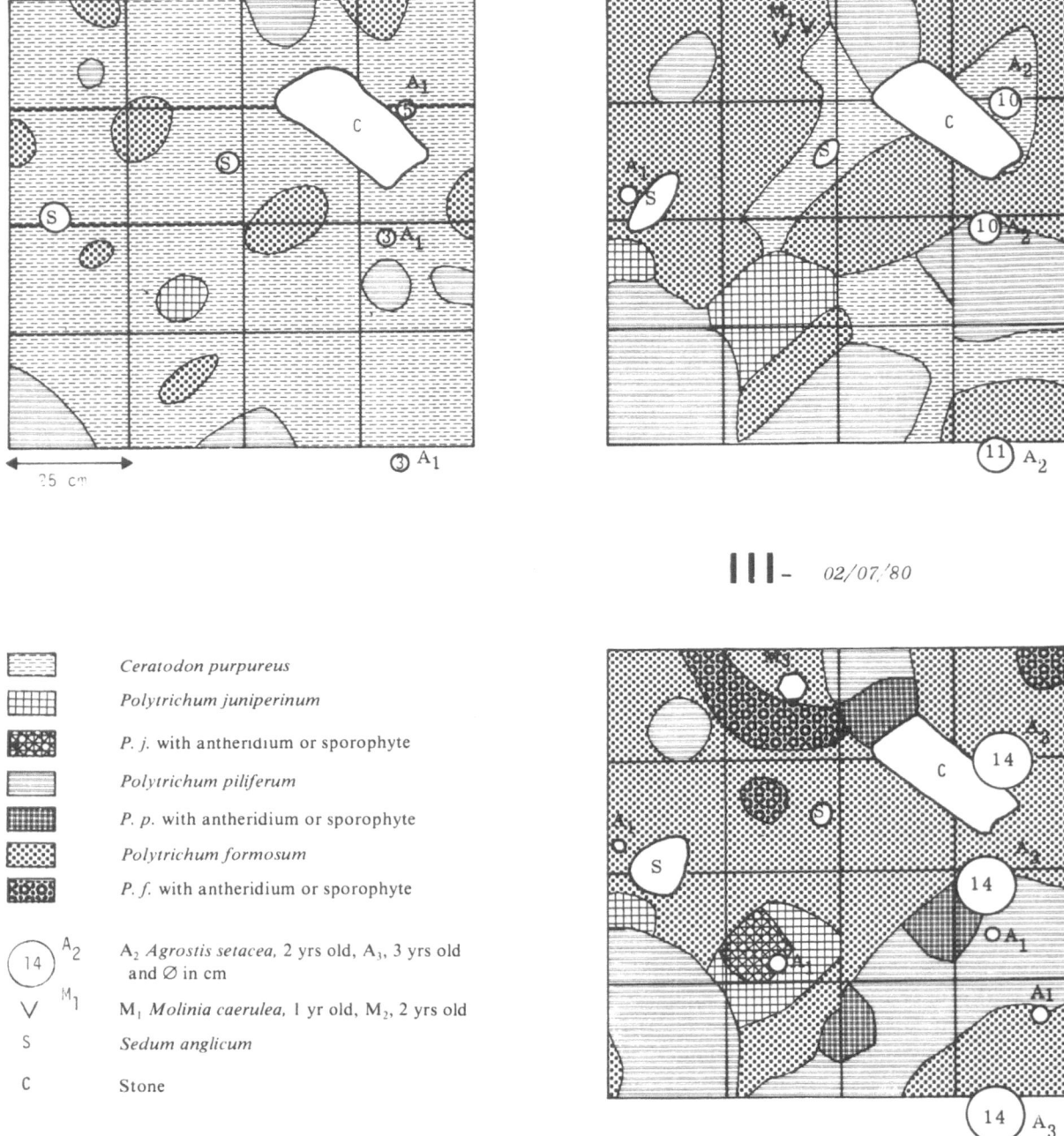

I - *12/11/78*

25 cm

③ A₁

II - *23/10/79*

A₂

⑩

A₁

S

⑩ A₂

⑪ A₂

III - *02/07/'80*

A₃

⑭

A₁

S

A₂

⑭

○ A₁

A₁

○

⑭ A₃

	Ceratodon purpureus
	Polytrichum juniperinum
	P. j. with antheridium or sporophyte
	Polytrichum piliferum
	P. p. with antheridium or sporophyte
	Polytrichum formosum
	P. f. with antheridium or sporophyte

⑭ A₂ A₂ *Agrostis setacea*, 2 yrs old, A₃, 3 yrs old
and Ø in cm

V M₁ M₁ *Molinia caerulea*, 1 yr old, M₂, 2 yrs old

S *Sedum anglicum*

C Stone

Fig. 5. Succession of the vegetation in a permanent plot (1 m²) located in a homogeneous area.

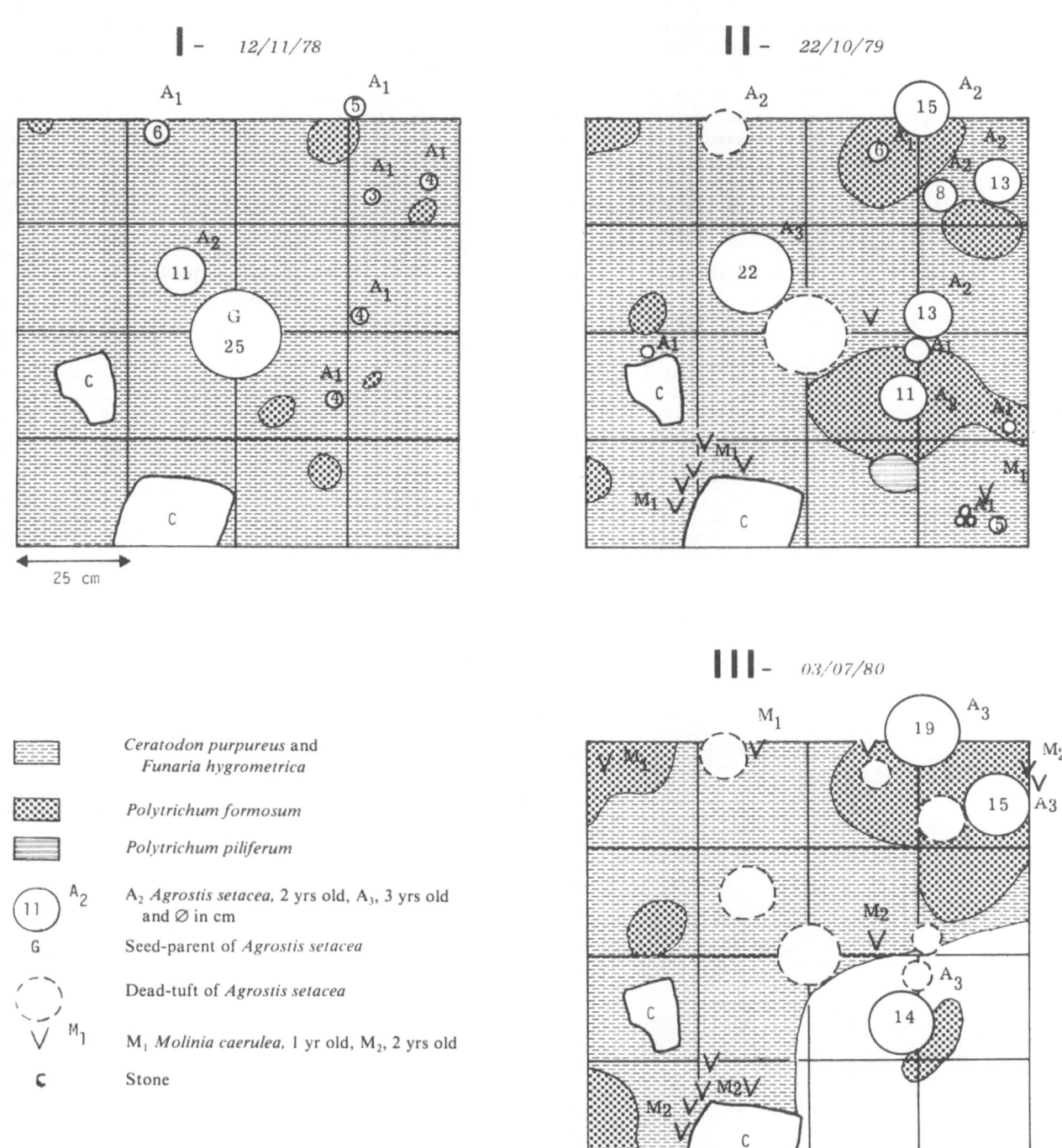

I - *12/11/78*

II - *22/10/79*

III - *03/07/80*

25 cm

Ceratodon purpureus and
Funaria hygrometrica

Polytrichum formosum

Polytrichum piliferum

A₂ Agrostis setacea, 2 yrs old, A₃, 3 yrs old
and ∅ in cm

G Seed-parent of Agrostis setacea

Dead-tuft of Agrostis setacea

M₁ Molinia caerulea, 1 yr old, M₂, 2 yrs old

C Stone

Fig. 6. Succession of the vegetation in a permanent plot (1 m²) with a seed-parent in the middle of the square.

166

Forgeard, F. & Touffet, J., 1979a. Les premières phases de recolonisation végétale après incendie dans les pelouses et les landes de la région de Paimpont (Ille-et-Vilaine). Bull. Soc. Bot. Fr., Lettres Bot. 126: 473–485.

Forgeard F. & Touffet J., 1980. La recolonisation des landes et des pelouses dans la région de Paimpont, Evolution de la végétation au cours des trois années suivant l'incendie. Colloque 'Ecologie des landes', Rennes 1979, Bull. Ecol. 11: 349–358.

Froment, A., 1975. Les premiers stades de la succession végétale après incendie de tourbe dans la réserve naturelle des Hautes Fagnes. Vegetatio 29: 209–214.

Géhu, J. M., 1960. Les incendies de 'Bruyères'. Bull. Soc. Bot. Nord Fr. 13: 63–76.

Gimingham, C. H., 1972. Ecology of heathlands. Chapman & Hall, London 266 pp.

Godron, M., 1968. Quelques applications de la notion de fréquence en écologie végétale. Oecol. Plant. 3: 185–212.

Greig-Smith, P., 1964. Quantitative plant ecology, Butterworks, London 256 pp.

Lemée, G., 1937. Recherches écologiques sur la végétation du Perche. Thèse, Paris 388 pp.

Poissonet, P. & Poissonet, J., 1969. Etude comparée de diverses méthodes d'analyse de la végétation des formations herbacées denses et permanentes. Conséquences pour les applications agronomiques. Document C.E.P.E.-C.N.R.S., 50, Montpellier 119 pp.

Tutin, T. G., Heywood, V. H., Burges, N. A., Valentine, D. H., Moore, D. M., Walters, S. M. & Webb, D. A., 1964–1980. Flora europaea, Cambridge University Press, 5 vol.

Accepted 3.6.1981.

Pattern development of the vegetation during colonization of a burnt heathland in Brittany (France)*

J. C. Gloaguen[1] & N. Gautier[2]**

[1] University of Rennes I, Laboratory of Vegetal Ecology, Scientific Complex of Beaulieu, 35042 Rennes Cedex, France
[2] University of Lyon I, Laboratory of Biometry, Boulevard du 11 Novembre 1918, 69622 Villeurbanne, France

Keywords: Brittany, Fire, Heathland, Pattern, Permanent plots, Spatial autocorrelation index, Vegetation dynamism

Abstract

The main part of the heathlands in the Paimpont area (Morbihan, and Ille-et-Vilaine, France) was burnt during spring and summer 1976. The pattern development of the vegetation was followed using permanent plots. The results presented concern the three first years: 1977, 1978 and 1979. One year after burning the area was sampled along a 128×1 m transect (samples in 50×50 cm contiguous quadrats) to situate this area in relation to its general vegetational environment, and sources of recolonizing diaspores. Permanent plots were established on a 8×16 grid, with 1 m squares. In order to choose the site of these permanent plots, a 5×5 grid was situated with 10×10 m contiguous quadrats.

The spatial autocorrelation index calculated from the proximity relation by block (Chessel & Gautier, 1979) was used to describe the monospecific patterns. The interspecific analysis used either a Principal Component Analysis when most data were percentage cover, or Correspondence analysis when the data were individual counts, frequencies or abundance-dominance coefficients. Analysis of vegetational development allows a distinction between two species groups:

– *Ceratodon purpureus* and *Polytrichum piliferum* have complementary densities the first year, the two species jointly occupy the whole grid in the second year, then regress in the third year, *Polytrichum piliferum* in small patches, *Ceratodon purpureus* in extensive patches.

– *Polytrichum juniperinum, Polytrichum formosum, Agrostis setacea* and *Ulex minor* establish regularly but very slowly. The first species is unstructured, the three others have a disjointed aggregate pattern whose diameter grew over the three years of the analysis.

One year after burning the pattern was very strong and essentially due to the heathland species (*Ulex minor* and *Agrostis setacea*) which developed on moss in which *Polytrichum formosum* showed the strongest pattern. This scheme tended to be disturbed in the second and third years.

* Plant nomenclature follows Flora Europaea (Tutin *et al.*, 1964–68–72–76) for Dicotyledons and Pteridophytes, des Abbayes *et al.* (1971) for Monocotyledons, Augier (1966) for Bryophytes and Ozenda & Clauzade (1970) for Lichens.
** We thank D. Chessel (Laboratory of Biometry, Villeurbanne) who has actively participated in the data treatment, kindly commented on the manuscript, and wrote an appendix on the spatial autocorrelation matrix.

Introduction

The larger part of the heathlands in the area of Paimpont (Morbihan, and Ille-et-Vilaine, Brittany, France) was burnt in spring and summer 1976. The period and the intensity of the fire varied considerably according to the station, and had important consequences for subsequent colonization. The

Vegetatio 46, 167–176 (1981). 0042-3106/81/0463-0167/$2.00.
© Dr W. Junk Publishers, The Hague.

168

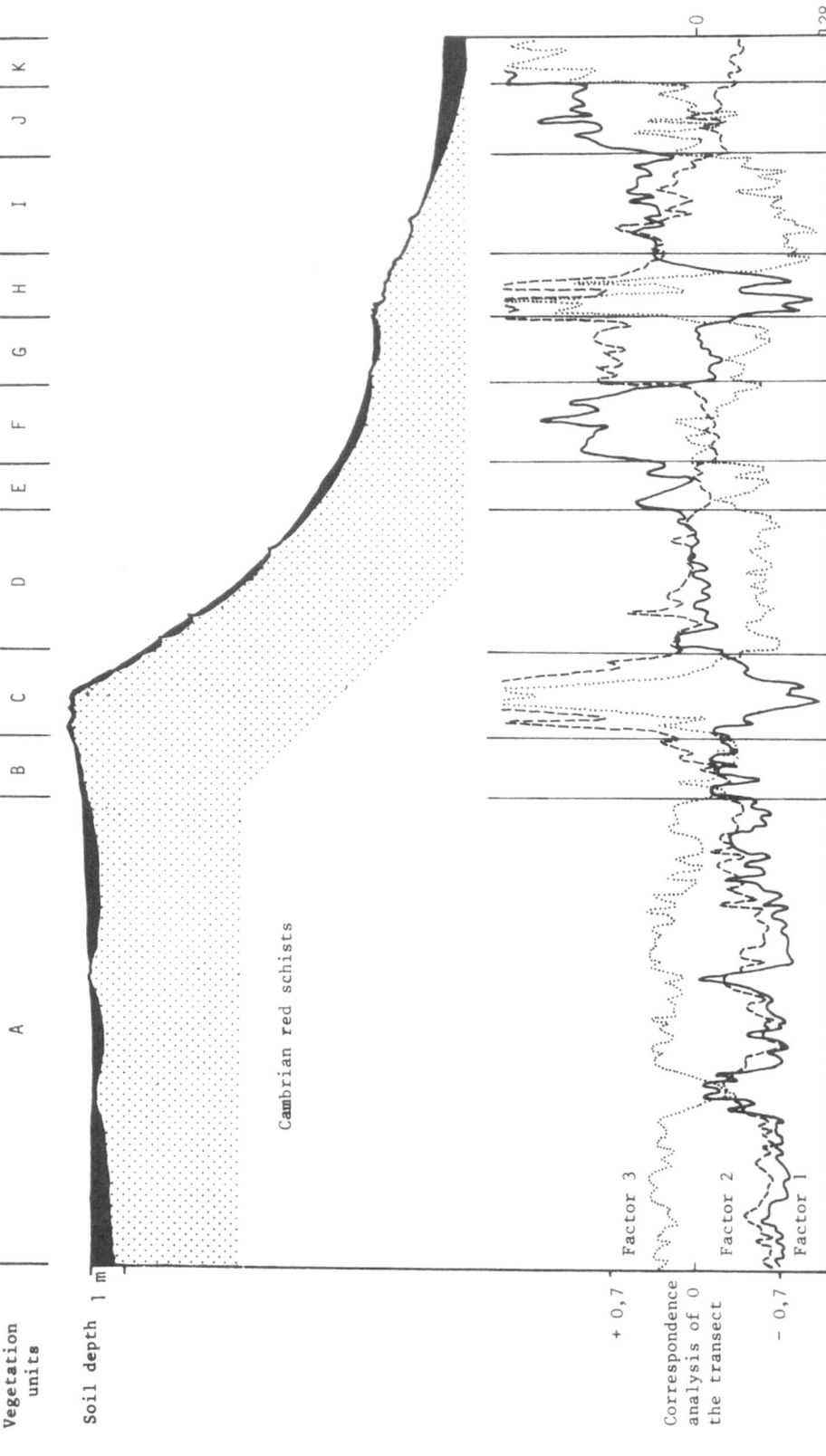

Fig. 1. Vegetation transect in the heathland, one year after burning.

area studied burnt violently in the middle of July, relatively late in the growing period. For 1976, this prevented practically all subsequent development of the vegetation which only began to establish in February 1977.

The pattern development of the vegetation was followed by using permanent plots. The presented results concern the three first years: 1977, 1978 and 1979.

Material and methods

Experimental method

The study area was situated at 'Trécesson', in the commune of Campénéac, on cambrian red schists. The previous vegetation (Forgeard, 1977) was a high mesophilous heathland, i.e. vegetation ±2 m in height, on ochreous brown soil, 30 to 60 cm deep. The following sample (Forgeard & Touffet, 1979) gives an idea of the floristic composition (Braun-Blanquet combined estimation values): *Ulex europaeus* 4, *Ulex minor* 3, *Calluna vulgaris* 2, *Erica cinerea* 1, *Erica ciliaris* 1, *Cytisus scoparius* +, *Agrostis tenuis* +, *Festuca ovina trachyphylla* +, *Asphodelus albus* +, *Cladonia* gr. *impexa* 2, *Hypnum cupressiforme ericetorum* 1, *Dicranum scoparium* +.

This vegetation was almost totally destroyed by the 1976 fire, subsequent colonization was almost exclusively from spores and seeds, except for occasional *Agrostis setacea* tussocks and *Ulex europaeus* and *Ulex minor* stumps which produced new shoots.

The permanent plots were located on a 'plateau' with a slight (2%) NE slope (Fig. 1) bordered on the South and the West by an abrupt slope, on the North by fields and on the East by a narrow road behind which was an unburnt high mesophilous heathland. A scale sampling plan, with three scales, one transect and two grids, was used (Debouzie *et al.*, 1975; Chessel & Donadieu, 1977).

One year after burning the area was sampled using a 128 m long transect, 1 m wide, with samples in contiguous quadrats (50 cm × 50 cm) (Fig. 1). This allowed an appreciation of the overall plant environment as a source of colonization by diaspores. In each square, cover % values, directly estimated, were then transformed to the usual

phytosociological abundance-dominance Braun-Blanquet values. The permanent plots on the study site were selected using a 8 × 16 grid with 1 m squares. The 8 lines were separated by 1 m strips to allow circulation (these are not shown in the figures). A 5 × 5 grid was laid out on the 'plateau' just beside the transect with contiguous 10 m squares in order to choose the site of these permanent plots. In this 5 × 5 grid all species were counted except mosses whose % cover alone was estimated. This grid was read again 2 yr after burning, but counts were replaced by estimated % cover for *Agrostis setacea*, *Agrostis canina* and *Pteridium aquilinum* which were very abundant in some squares. The 8 × 16 grid was placed on the quadrat which seemed to be the most 'homogeneous' and the most representative of the 'plateau' vegetation.

The 8 × 16 grid was read using the following protocol. The first year, the species were counted and their cover % values were directly estimated (for mosses, only estimated cover was noted). Afterwards, in order to avoid errors in cover evaluation, the 1 × 1 m quadrats were subdivided into 20 × 20 cm squares, readings (3840 presence-absence) being effectuated at intersections in 1 × 1 cm squares. The cover % values were then calculated for each species: number of presences/3840 × 100.

Readings were made at one year intervals at the end of winter or beginning of spring (end of March-beginning of April), before the vegetation season started.

Statistical analysis

The data were arranged in sample-species tables. The i-th line indicates the i-th square of the grid or of the transect, the j-th column indicates the j-th species. At the intersection of the i-th line and the j-th column a quantitative index for this species in this square is found. The main difficulty results from the use of different variables: individual counts, estimated cover, frequency on a given number of quadrat points. To ensure a uniform treatment, patterns were described using the spatial autocorrelation index calculated from the proximity relationship by block (Chessel & Gautier, 1979). A brief description of the significance and calculation of the spatial autocorrelation index is provided in the appendix.

Interspecific analysis was performed using either Principal Components Analysis when most data were estimated cover (8 × 16 grid, first year) or Correspondence Analysis (for all other grids and the transect) when the data were individual counts, frequencies, or abundance-dominance coefficients, jointly re-estimated on the individual counts and percentage cover when it was necessary to treat them together. Functional representation by transect (Estève, 1978) or cartographical representations of the factor values (Bachacou & Chessel, 1979) and spatial structure analysis of the same factors by the spatial autocorrelation index according to the Chessel & Gautier (1979) method, were also used.

Results

The small scale pattern: the transect

The different vegetation units present on the 'Trécesson' heathlands one year after burning, specified by Correspondence Analysis, are shown in Figure 1. It is possible to recognize:

1. The plant groups of the rock outcrops (especially lichens) and of the fissures (C and H), notably with *Festuca ovina trachyphylla, Agrostis setacea, Sedum anglicum, Hypericum linarifolium,* mosses such as *Campylopus polytrichoides* and *Polytrichum piliferum* and lichens like *Cladonia* gr. *impexa;*

2. On both sides of these rock outcrops, swards (B, D, G and I) are established on a shallow skeletal soil (3 to 15 cm), a little evolved ranker or acid brown soil. The most frequent species are *Agrostis setacea* (very abundant), *Sedum anglicum, Erica cinerea, Ulex europaeus, Polytrichum piliferum* and also *Polytrichum formosum;*

3. When the soil is deeper (acid and more evolved brown soil, or ochreous brown soil down to a depth of 60 cm) heath vegetation can grow. It was either dry heathland if there was a steep slope, or mesophilous if it was shallow. In the first case (E and F), the main species were *Erica cinerea, Ulex europaeus* and *Agrostis setacea.* In F a mesophilous tendency appeared with the presence of *Ulex minor* and *Molinia caerulea.* In the second case, the consequences of burning varied with the topographical position. In the hollows, the low intensity of

burning allowed development of numerous new shoots, especially of *Ulex minor* and *Erica ciliaris:* in J, *Ulex minor* clearly dominated whereas in K *Erica ciliaris* became co-dominant and *Molinia caerulea* more abundant than in J. Inversely, on the 'plateau' (A), the very intensive fire destroyed all the heathland and new shoots were rare. After one year, vegetation was practically limited to a moss carpet dominated by *Ceratodon purpureus* and *Polytrichum piliferum* but also with *Funaria hygrometrica, Polytrichum juniperinum* and *Polytrichum formosum.* Some tussocks of *Agrostis setacea* resisted, and some new shoots of *Ulex europaeus* and *Ulex minor* were also present. However most of the colonization was due to germination, especially of *Agrostis setacea, Ulex europaeus* and *Ulex minor, Erica ciliaris* and *Erica cinerea* and also of trees, such as *Pinus sylvestris, Betula pendula, Betula pubescens* and *Salix atrocinerea.*

The two 5 × 5 and 8 × 16 permanent grids were situated in this vegetation. The 'perturbations' in the Correspondence Analysis were due to a greater abundance of *Agrostis setacea, Polytrichum formosum* and also *Erica cinerea,* and correspond with more shallow soils where rock outcrops can occur.

The medium scale pattern: the 5 × 5 grid

Monospecific analysis

The pattern for each species, one and two yr after burning is shown in Table 1 (for interpretation see the corresponding monospecific analysis paragraph for the 8 × 16 grid and Chessel, 1978). In the first year all the species showed a very strong pattern. In the present example, the seeds which germinated were present in the soil before the fire (except

Table 1. Pattern development of the main species in the 5 × 5 grid.

	One year after burning	Two years after burning
Ulex minor	strongly aggregative	aggregative
Erica cinerea	strongly aggregative	aggregative
Erica ciliaris	strongly aggregative	no pattern
Agrostis setacea	intensive patches	no pattern
Pteridium aquilinum	intensive patches	intensive patches
Salix atrocinerea	patch	zone
Betula pubescens	patch with nucleus	zone

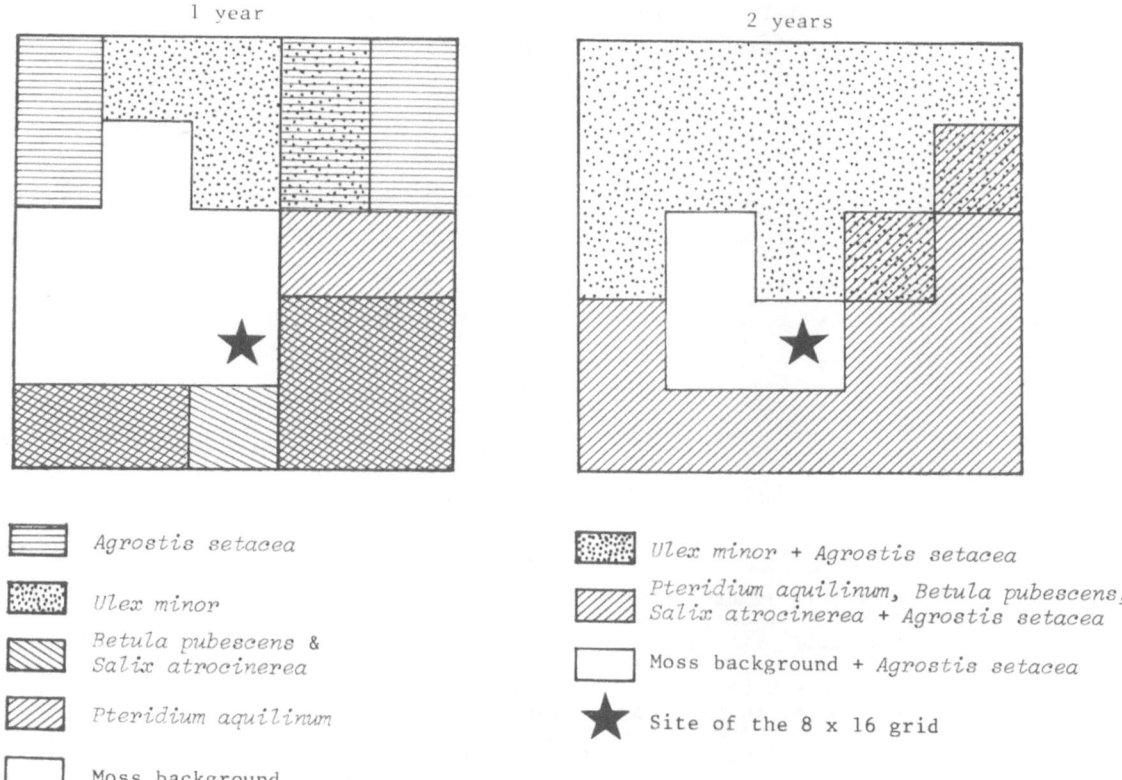

1 year

2 years

Agrostis setacea

Ulex minor

Betula pubescens &
Salix atrocinerea

Pteridium aquilinum

Moss background

Ulex minor + Agrostis setacea

Pteridium aquilinum, Betula pubescens,
Salix atrocinerea + Agrostis setacea

Moss background + Agrostis setacea

Site of the 8 x 16 grid

Fig. 2. Generalized pattern in the 5 × 5 grid.

perhaps some *Salix atrocinerea* and *Betula pubescens* seeds which could have been of external origin), because the heathland species *Ulex minor, Erica cinerea, Erica ciliaris* and *Agrostis setacea* have small dispersal potential. *Agrostis setacea* which is the must successful colonizing plant spreads its seeds up to about one or two m from a tussock (Clément *et al.,* 1980). The observed pattern therefore indicates, either the existence of the same pattern in the soil seeds, or alternatively, despite an apparent consistence in the fire intensity, some irregular action of fire on the stock of soil seeds by destruction of a certain number of seeds, or by favouring the suppression of dormancy in some. The first assumption appears the most probable because the fire seems to have burnt all the 'plateau' vegetation with the same intensity, but it is possible that soil and vegetational heterogeneity modified the quality of the fire with some consequences for germination of the seeds.

In the second year (Table 1), *Salix atrocinerea* and *Betula pubescens* had colonized a greater part of the grid whereas *Pteridium aquilinum* always appeared in dense masses where the rhizomes were undamaged. *Ulex minor* and *Erica cinerea* still had an aggregate pattern but the spatial autocorrelation index was lower than in the first year. Finally, no pattern could be shown for *Erica ciliaris* and *Agrostis setacea.* In conclusion, in the 5 × 5 grid (10 × 10 m squares), the vegetation dynamics were characterized by a progressive confounding of pattern on the heathland species *Ulex minor, Erica cinerea, Erica ciliaris* and *Agrostis setacea.*

Interspecific analysis

Figure 2 summarizes the results from the 5 × 5 grids at one and two years. The construction of the model is explained for the 8 × 16 grid in the corresponding section. Using the Correspondence Analysis, knowing the species that best define

Raw data : number of presences/m2
(for 30 reading points)

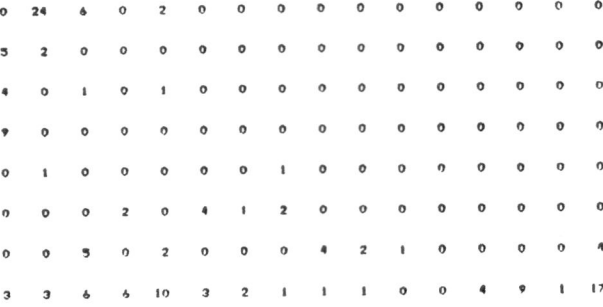

```
20   24   6    0    2    0    0    0    0    0    0    0    0    0    0    0
13   2    0    0    0    0    0    0    0    0    0    0    0    0    0    0
4    0    1    0    1    0    0    0    0    0    0    0    0    0    0    0
9    0    0    0    0    0    0    0    0    0    0    0    0    0    0    0
0    1    0    0    0    0    0    1    0    0    0    0    0    0    0    0
0    0    0    2    0    4    1    2    0    0    0    0    0    0    0    0
0    0    5    0    2    0    0    0    4    2    1    0    0    0    0    4
3    3    6    6    10   3    2    1    1    1    0    0    4    9    1    17
```

Spatial autocorrelation matrix

```
NOMBRE DE MESURES 128
MOYENNE:  1. 40625
VARIANCE: 13. 9911
        1 *     2 *     4 *     8 *     16 *
+ 0. 00*+ 5. 17*+ 5. 23*+ 3. 59*+ 3. 19*    1   *
+ 3. 80*+ 7. 07*+ 6. 06*+ 4. 14*+ 2. 89*    2   *
+ 3. 66*+ 6. 35*+ 4. 81*+ 2. 26*- 0. 71*    4   *
+ 1. 99*+ 3. 82*+ 4. 16*+ 2. 87*+ 0. 00*    8   *
```

Species map

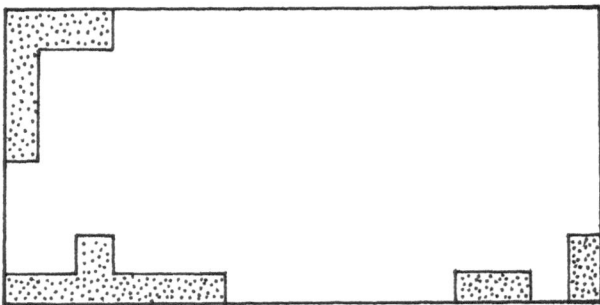

Fig. 3. Monospecific analysis: *Ulex minor.*

the prevailing conditions and operational factors in such heathland, the vegetational pattern shown by the 5 × 5 grid at one year can be summarized as follows. On a moss background the pattern is attributable to *Agrostis setacea, Ulex minor, Pteridium aquilinum, Betula pubescens* and *Salix atrocinerea.* After two years, *Agrostis setacea* had invaded the whole grid. The pattern was then due on the one hand to *Ulex minor* and on the other hand to the group *Pteridium aquilinum, Betula pubescens* and *Salix atrocinerea.*

The very strong pattern in the first year rapidly diminished when the unit was the 10 × 10 m quadrat. Some of the species showed no pattern as

Table 2. Pattern development of the main species in the 8 × 16 grid.

	One year after burning	Two years after burning	Three years after burning
Ulex minor	patches	patches	patches
Agrostis setacea	aggregates (strong pattern)	aggregates (strong pattern)	aggregates (weaker pattern)
Ceratodon purpureus	zones	no pattern	disappearance in wide patches
Polytrichum formosum	patches	patches	patches (less obvious)
Polytrichum juniperinum	no pattern	no pattern	no pattern
Polytrichum piliferum	zones	no pattern	disappearance in small patches

early as the second year (especially *Erica ciliaris* and *Agrostis setacea*). This can be explained by a progressive reconstitution of floristic composition over the whole 'plateau', and in accordance with prior ecological conditions. This conclusion must be modified when working on a larger scale with a 1 × 1 m quadrat.

As early as the first year, it was with this grid that the site of the 8 × 16 grid (Fig. 2) was chosen in the quadrat which seemed to be the most 'homogeneous', and the most representative of the burnt vegetation 'plateau'. This quadrat choice meant that the species responsible for pattern were either absent or rare: *Salix atrocinerea*, *Betula pubescens* and *Pteridium aquilinum*, *Erica cinerea* and *Erica ciliaris*. The moss substratum was wide-spread and only *Agrostis setacea* (and to a certain extent *Ulex minor*) was sufficiently represented.

The large scale pattern: the 8 × 16 grid

Monospecific analysis

The characteristic pattern for each species was determined on the one hand by considering the raw data and on the other hand by using the spatial autocorrelation index (Fig. 3). The higher the coefficient the stronger the pattern. Another advantage of this index is that it gives an idea of the size of the basic pattern blocks. In the chosen example the unit block size for *Ulex minor* was about 2 × 2 m, 2 × 4 m or 4 × 2 m. Consideration of raw data associated with this index provided a measure of specific pattern, and a basis for a general interpretation. The evolution of pattern for the main species observed during the 3 year study is

shown in Table 2.

The main conclusion for the 5 × 5 grid, the progressive confounding of pattern, was found again in the 8 × 16 grid, but here it was attenuated for most species. This is particularly obvious for *Agrostis setacea* (one of the 2 common grid species, with *Ulex minor*) which has a very strong pattern at 3 yr in the 8 × 16 grid whereas the pattern disappears after 2 yr in the 5 × 5 grid.

This may be explained by the change in scale of analysis which passes from description in terms of floristic pattern (5 × 5 grid) to one in terms of morphological pattern (8 × 16 grid). In other words there is a change from a measure of floristic heterogeneity to one of spatial occupation and allocation by individuals of the different species. The floristic list is progressively reestablished and converges with the condition prior to burning. The difference may be seen alternatively as one between a state of relative permanence (real in terms of floristic composition) and one of variation (real in terms of the insertion of individuals of each species).

Two types of evolution (see also Fig. 5) can be distinguished. On the one hand (Fig. 5B), *Agrostis setacea*, *Ulex minor* and *Polytrichum formosum* which are structured and *Polytrichum juniperinum* which is unstructured, have the same development, that is to say slow but regular growth. *Ceratodon purpureus* and *Polytrichum piliferum* which have invaded the whole of the grid during the second year, on the other hand (Fig. 5A), begin to structure again during the third. *Ceratodon purpureus*, dominated by the other mosses lacks light and disappears in wide patches whereas *Polytrichum piliferum* dominated in particular by higher plants

(Agrostis setacea principally, but also *Ulex minor)* disappears in small patches. Such a disappearance has also been observed using the quadrat point method (Forgeard & Touffet, 1980). Rather similar dynamics have been found in the heathlands of the Mont d'Arrée, Finistère (Clément & Touffet, 1981).

Interspecific analysis

This involves the use of Correspondence Analysis, or, exceptionally, Principal Component Analysis, when most of the data were percentage cover. The spatial patterns obtained by mapping factors are analyzed by the same spatial autocorrelation index

Factor values

```
+2 0 +2 5 +0 7 -0 1 +0 2 -0 1  0 1 -0 1 -0 1 -0 1 -0 1 -0 1 -0 1 -0 1 -0 2 -0 2

+1 1 +0 0 -0 2 -0 1 -0 1 -0 1 -0 1 -0 1 -0 2 -0 2 -0 2 -0 2 -0 1 -0 1 -0 2 -0 1

+0 3 -0 1 -0 1 -0 1 -0 0 -0 1 -0 1 -0 1 -0 1 -0 1 -0 2 -0 1 -0 2 -0 2 -0 1 -0 1

+0 7 -0 1 -0 1 -0 1 -0 1 -0 1 -0 1 -0 1 -0 2 -0 2 -0 1 -0 1 -0 1 -0 1 -0 1 -0 2

-0 1 -0 0 -0 1 -0 1  0 1 -0 1 -0 1 -0 0 -0 1 -0 1 -0 1 -0 1 -0 1 -0 1 -0 1 -0 1

-0 1 -0 1 -0 1 +0 0 -0 2 +0 2 -0 1 +0 0 -0 1 -0 1 -0 1 -0 1 -0 1 -0 1 -0 1 -0 1

-0 1 -0 2 +0 3 -0 2 +0 0 -0 2 -0 1 -0 1 +0 2 +0 1 -0 1 -0 1 -0 1 -0 1 -0 1 +0 3

+0 2 +0 2 +0 4 +0 3 +0 8 +0 2 +0 1 -0 0 -0 0 -0 0 -0 1 -0 1 +0 3 +0 6 -0 0 +1 5
```

Spatial autocorrelation matrix

```
NOMBRE DE MESURES 128
MOYENNE: -1.93626E-04
VARIANCE:  .144132
        1 *      2 *      4 *      8 *     16 *
  + 0. 00*+ 5. 71*+ 6. 14*+ 4. 63*+ 3. 73*     1  *
  + 3. 19*+ 6. 71*+ 6. 21*+ 4. 51*+ 2. 80*     2  *
  + 3. 06*+ 5. 95*+ 4. 90*+ 2. 65*- 0. 75*     4  *
  + 1. 85*+ 3. 88*+ 4. 14*+ 3. 04*+ 0. 00*     8 ' *
```

Factor map

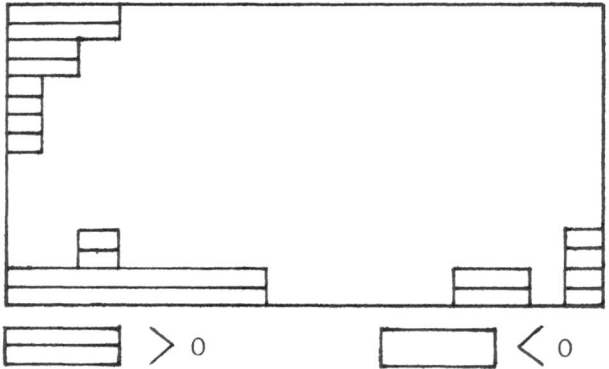

Fig. 4. Interspecific analysis: factor 1.

as for the monospecific analysis (Fig. 4). Examination of the raw data (value, and sign of the factor) associated with that of the indices, allows preparation of a map for any factor. In the chosen example,

factor 1 (which explains 54% of the total inertia) is explained for 96,6% by *Ulex minor*. The two maps, monospecific and for factor 1, are very similar. Lastly, by using the various monospecific and

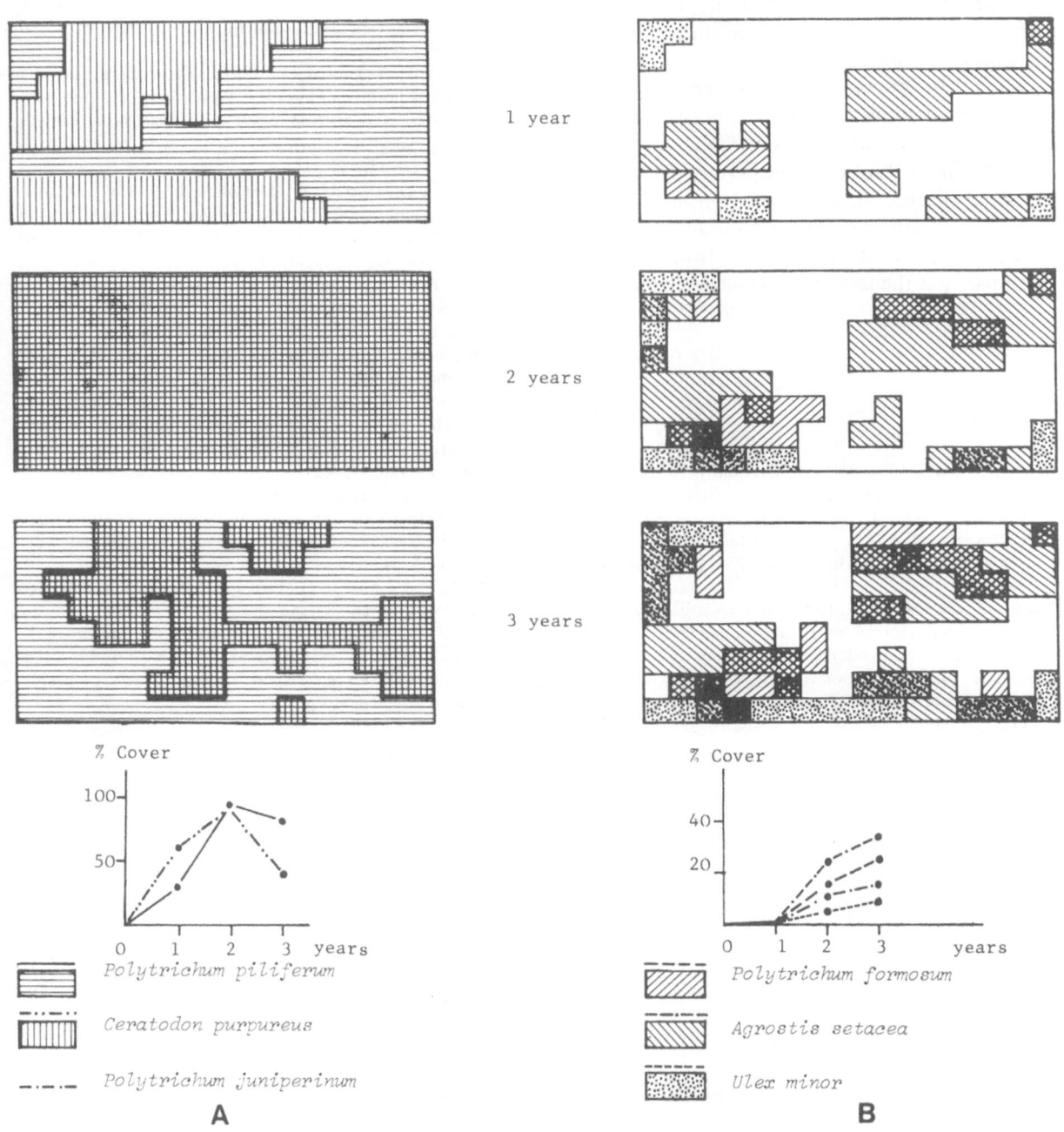

A

1 year

2 years

3 years

% Cover

Polytrichum piliferum

Ceratodon purpureus

Polytrichum juniperinum

B

% Cover

Polytrichum formosum

Agrostis setacea

Ulex minor

Fig. 5. Generalized pattern in the 8 × 16 grid.
  A. Species which grow very fast and which decline in the third year (except *Polytrichum juniperinum*): moss background;
  B. Species which grow regularly but very slowly.

factor maps (especially the 3 first factors) the schematic summaries for the permanent grids can be prepared (Fig. 5).

The first year pattern is due to *Ulex minor*, *Agrostis setacea* and *Polytrichum formosum* on a *Ceratodon purpureus*, *Polytrichum piliferum* and *Polytrichum juniperinum* background, with either *Ceratodon purpureus* or *Polytrichum piliferum* dominant. In the second year the same species are still responsible for the pattern, but this time on a uniform background of the same mosses.

Lastly, in the third year, the pattern is still due to the same species, on a *Polytrichum juniperinum*, *Polytrichum piliferum* and *Ceratodon purpureus* background, the second species disappearing in small patches, the last one in extensive patches.

References

Abbayes, H. des, Claustres, G., Corillion, R. & Dupont, P., 1971. Flore et végétation du Massif armoricain. I. Flore vasculaire. Presses Universitaires de Bretagne, Saint-Brieuc, 1227 p.

Augier, J., 1966. Flore des Bryophytes. Paul Lechevalier, Paris, 702 p.

Bachacou, J. & Chessel, D., 1979. Etude des structures spatiales en forêt alluviale rhénane. III. Dispersion interspécifique et analyse des correspondances. OEcol. Plant. 14: 371–388.

Chessel, D., 1978. Description non paramétrique de la dispersion spatiale des individus d'une espèce. In: Biométrie et Ecologie 1, Legay, J. M. & R. Tomassone éd. Soc. fr. Biométrie, Paris, 45–135.

Chessel, D. & Donadieu, P., 1977. Introduction à l'étude de la structure horizontale en milieu steppique. III. Dispersion locale, densité et niveau d'implantation. OEcol. Plant. 12: 221–240.

Chessel, D. & Gautier, N., 1979. La description des communautés végétales: exemples d'utilisation de deux techniques statistiques adaptées aux mesures sur grilles ou transects. Actes 7ème Colloque Informatique et Biosphère, Paris, mars 1979: 87–101.

Clément, B. & Touffet, J., 1981. Vegetation dynamics in Brittany heathlands after fire. Vegetatio 49: 157–166.

Clément, B., Rivière, A. & Touffet, J., 1980. Répartition des graines au sol dans les landes incendiées des Monts d'Arrée. Bull. Ecol. 11: 365–371.

Debouzie, D., Chessel, D., Donadieu, P. & Klein, D., 1975. Introduction à l'étude de la structure horizontale en milieu steppique. II. Le traitement statistique des lignes de placettes contigües. OEcol. Plant. 10: 211–231.

Estève, J., 1978. Les méthodes d'ordination: éléments pour une discussion. Biométrie et Ecologie 1: 223–250.

Forgeard, F., 1977. L'écosystème lande dans la région de Paimpont. Etude du cycle de la matière organique et des éléments minéraux. 3rd cycle thesis, Rennes, 220 p.

Forgeard, F. & Touffet, J., 1979. Les premières phases de la recolonisation végétale après incendie dans les pelouses et les landes de la région de Paimpont (Ille-et-Vilaine). Bull. Soc. Bot. Fr. 126: 473–485.

Forgeard, F. & Touffet, J., 1980. La recolonisation des landes et des pelouses dans la région de Paimpont. Evolution de la végétation au cours des trois années suivant l'incendie. Bull. Ecol. 11: 349–358.

Ozenda, P. & Clauzade, G., 1970. Les Lichens. Etude biologique et flore illustrée. Masson, Paris, 801 p.

Tutin, T. G., Heywood, V. H., Burges, N. A., Moore, D. M., Valentine, D. H., Walters, S. M. & Webb, D. A., 1964, 1968, 1972 & 1976. Flora Europaea. University Press, Cambridge, 464, 455, 370 & 505 p.

Accepted 28.7.1981.

The spatial autocorrelation matrix

D. Chessel

University of Lyon I, Laboratory of Biometry, Boulevard du 11 Novembre 1918, 69622 Villeurbanne, France

Introduction

Among the numerous approaches proposed for the analysis of spatial processes, the autocorrelation index (Geary, 1954) introduced by Cliff & Ord (1973) possesses remarkable properties. For any given variable (presence/absence, cover, quantitative measures of abundance) and random distribution of sampling points in space, this index tests the null hypothesis of absence of correlation for values recorded at two neighbouring points, using a non parametric model for equiprobability of N ! attributions of N numerical values at N sampling points.

Geary's index

Firstly Geary's index may be described to indicate the significance of its calculation. Take a series of N points $P_1, P_2, \ldots P_N$ in space, and $\mathbf{V}$ a symmetrical matrix defined by:

$$(1) \quad \begin{cases} \mathbf{V}_{ij} = \mathbf{V}_{ji} = 1 \text{ if } P_i \text{ and } P_j \text{ are neighbours} \\ \mathbf{V}_{ij} = \mathbf{V}_{ji} = 0 \text{ otherwise} \end{cases}$$

$\mathbf{V}$ is said to be the contiguity relationship matrix.

Note that $A(V)$ is the number of pairs of neighbouring points, L_i the number of points neighbour to P_i:

$$(2) \quad \begin{cases} L_i = \sum_{j=1}^{N} V_{ij} \\ A(V) = \sum_{i=1}^{N} L_i \end{cases}$$

and note that

$$D(V) = \sum_{i=1}^{N} L_i(L_i - 1)$$

Take $x_1, x_2, \ldots x_N$ N numerical values corresponding to $P_1, P_2, \ldots, P_N$ points.
The total variability of the measures is:

$$(4) \quad \begin{cases} H_T = \left[\sum_{i,j} (x_i - x_j)^2 \right] / N(N-1) = \\ = 2 \sum_{i=1}^{N} (x_i - \bar{x})^2 / (N-1) \end{cases}$$

The variability of two neighbouring points is:

$$(5) \quad H_V = \left[\sum_{i,j} V_{ij}(x_i - x_j)^2 \right] / A(V)$$

If there were no spatial pattern these two quantities would be essentially equal. Cliff et Ord give the mean and variance of the variable.

$$(6) \quad Z = H_V / H_T$$

This variable being defined in the equiprobability space of the N ! permutations of x_i (see Chessel, 1978). We have

$$(7) \quad E(Z) = 1$$

$$(8) \quad A(V)^2 N(N-2)(N-3) \text{Var}(Z) = X_1 + X_2 + X_3$$

$$X_1 = [(N^2 - 3) - (N-1)^2 B_2] A(V)^2$$

Vegetatio 46, 177–180 (1981).   0042-3106/81/0463-0177/$0.80.

with
$$
\begin{cases}
X_2 = 2\,A\,(V)\,(N-1)\,[N^2 - 3\,N + 3 \\
\quad - (N-1)\,B_2] \\
X_3 = (N-1)\,(D\,(V) + A\,(V))\,[(N^2 \\
\quad -N + 2)\,B_2 - (N^2 + 3\,N - 6)]
\end{cases}
$$

and
$$
B_2 = N \sum_{i=1}^{N} (x_i - \bar{x})^4 \Big/ \left[\sum_{i=1}^{N} (x_i - \bar{x})^2 \right]^2
$$

When a spatial autocorrelation exists, i.e. a significant resemblance between two neighbouring point values, the observed value of Z decreases and $Z = (E\,(Z) - Z) / \sqrt{\mathrm{Var}\,(Z)}$ becomes significantly positive. Cliff & Ord show that the distribution of Z approximates a normal one which makes it possible to compare the observed value with an expected one.

Subdivision into blocks

In the socio-economic field and certain other examples studied by Cliff & Ord the contiguity relationship may be imposed (measure by district) or subjected to a minor choice (relations of the castle, knight, or queen in chess). For the description of plant pattern the problem is quite different.

The fundamental problem remains that of detecting the principle scale of the phenomenon, or even of several preferred scales (cf. Mead, 1974). A series of studies have amply shown this (Greig-Smith, 1952; Kershaw, 1957; Thompson, 1958; Hill, 1973; Zalh, 1974 & Ludwig, 1979).

In the absence of a global test at every scale, a probatory test at least is needed for each of them. From an experimental point of view this test will often support the automatic treatment of large quantities of numerical results than be a real decision tool.

Most data, *in natura,* are sufficiently patterned for the recorded thresholds to allow a choice without risk of error.

Space is appreciated either by regrouping the data by blocks (Greig-Smith school, Mead, 1974; Chessel, 1978) or in describing the variability as a function of the distance between two points of measure (Matheron, 1965; Goodall, 1974; Hill, 1973). Here we have the two main methods for appreciating variation in scale. Block analysis is often criticized as being sensitive to the initial grouping, but we do not feel that this is a real ecological problem.

Neighbours by blocks

In order to apply Geary's index in the analysis of plant pattern it must necessarily be applied to particular contiguity relationships. A non parametric 'semi-variogram' could be obtained by defining the neighbour relation *'at distance h'.*

P_i and P_j are neighbours if and only if $d(P_i, P_j) = h$; However, the neighbour relationship *'to block B'* will be preferred here; P_i and P_j are neighbours if, and only if P_i and P_j are in the same block. If the data are obtained along a transect one then obtains a series of block sizes calculated in numbers of successive points and a *spatial autocorrelation vector.*

If the data are obtained on a grid, each block size chosen 'a priori' is characterised by a pair (I_1, I_2) indicating that the block regroups consecutive points on I_1 lines and J_1 columns of the scale grid. When I_1 and I_2 are varied a *spatial autocorrelation matrix* is again obtained. Note that overlapping versus separated, and equal versus incomplete blocks of any form can be chosen, and that Geary's statistic is, finally, the most general of those describing overall heterogeneity (Chessel, 1978).

Note also certain special cases. If the grid is complete one has the non parametric version of the analysis of variance proposed by Greig-Smith (1952) and when the data are binary one has the non parametric index of dispersion of Chessel & Croze (1978), and in this case a global test. In the case of inequal blocks one discovers the number of pairs statistic of Walter (1974).

Example

Gros (1978) describes the case of a wild population of raspberry canes (*Rubus idaeus*). The unit is a 1 m $\times$ 1 m square and the distance between two units is 10 m. In each unit the canes were counted and some morphological variables recorded (for example mean diameter of the canes). It is important to know the degree of spatial heterogeneity of the

A

7	2	2	1	0	9	1	3
3	2	2	2	9	8	0	8
2	0	8	7	4	15	8	12
0	5	2	6	11	10	4	11
5	9	13	12	1	5	11	11
5	4	9	22	9	20	2	11
13	16	16	11	8	15	12	7
10	10	14	12	17	8	5	4
14	0	0	18	0	0	3	8
13	3	10	16	0	23	8	8
0	0	0	1	5	0	0	2
0	10	6	2	1	0	0	22
0	11	6	13	17	0	22	15
0	26	11	19	20	9	9	10
21	4	9	6	14	8	5	7
13	19	26	18	22	7	12	4
15	19	5	7	18	14	17	27
23	1	6	21	12	10	12	18
10	10	18	16	5	12	14	4
2	9	19	20	13	10	21	13
1	16	9	17	12	9	16	25
15	14	20	11	19	17	13	19
12	8	10	14	4	12	9	3
11	4	6	10	2	10	10	0

Number of canes per quadrat

B

35.0	32.5	39.5	●	●	31.4	21.0	53.0
37.7	55.5	27.0	27.5	●	46.0	●	45.8
45.5	●	40.3	75.1	38.3	38.3	76.0	55.8
●	●	●	71.2	41.6	53.5	●	66.4
84.5	●	99.5	65.1	●	24.8	91.7	75.9
94.7	58.8	45.0	23.7	44.0	59.4	23.5	24.6
43.2	85.7	56.8	●	65.5	93.3	28.3	59.0
57.0	57.5	33.0	36.1	62.5	34.0	21.3	●
56.5	●	●	46.5	●	●	24.3	41.2
57.9	22.0	46.0	43.3	●	61.8	30.4	66.1
●	●	●	56.0	97.0	●	●	35.5
●	64.0	38.5	12.5	15.0	●	●	46.3
●	42.0	22.7	●	64.4	●	82.0	35.5
●	37.5	25.0	21.0	40.8	56.1	40.3	30.8
23.6	40.0	50.1	39.5	●	62.0	●	73.0
32.0	33.6	42.8	40.0	31.3	73.3	30.8	30.5
40.7	49.1	58.0	23.7	75.2	76.2	43.7	44.2
36.8	●	30.5	60.2	96.0	29.0	49.0	82.6
45.4	●	38.1	86.5	●	59.3	25.8	67.5
11.0	26.4	64.8	46.9	70.5	36.0	38.3	54.5
22.0	78.5	21.3	●	25.3	13.8	42.3	51.0
47.0	46.8	66.0	59.0	50.4	35.6	28.6	51.9
35.0	62.9	39.9	36.2	62.0	57.5	83.3	87.5
●	●	42.0	49.8	25.5	71.3	50.3	●

Mean cane diameter per quadrat (1mx1m)

I

1	2	4	8	$I_1 \backslash I_2$
●	2.0	4.2	6.7	1
3.9	4.4	6.8	9.8	2
2.8	4.6	7.6	11.6	4
0.7	2.4	3.7	6.9	8
4.7	6.8	10.5	19.4	12
-0.5	0.0	-0.7	●	24

1	2	4	8	$I_1 \backslash I_2$
●	1.9	3.0	2.9	1
-0.1	1.0	1.2	0.6	2
-0.4	0.1	0.2	-0.5	4
-0.3	-0.2	0.3	0.2	8
0.4	0.5	0.4	0.3	12
0.6	0.4	-0.7	●	24

II

Fig. 1. Pattern analysis in a *Rubus idaeus* population: I raw data (Gros, 1978), II Geary's index for some block sizes (I_1 x I_2).
A: number of canes per unit
B: mean diameter of the canes per unit

morphological variables in order to devise the cheapest sampling plan in a given region (estimations of production). In an empty unit the variable 'mean diameter' has no value. Geary's index allows for such lack of data.

Figure 1 shows the counting grid (A) and for 'mean-diameter' (B). The spatial autocorrelation matrices are very different, and the analysis indicates that the spatial pattern of the morphological variables is very weak.

Thus for the same precision, sampling units for counting canes must be distributed over the whole range but this is not necessary for morphological variables.

References

Chessel, D., 1978. Description non paramétrique de la dispersion spatiale des individus d'une espèce. In: Biométrie et Ecologie, Legay J. M. & R. Tomassone éd. Soc. fr. Biométrie, Paris, 45–135.

Chessel, D. & Croze, J. P., 1978. Un indice de dispersion pour les mesures de présence-absence: application à la répartition des animaux et des plantes. Bull. Ecol. 9: 19–28.

Cliff, A. D. & Ord, J. K., 1973. Spatial autocorrelation. Pion, London, 178 p.

Geary, R. C., 1954. The contiguity ratio and statistical mapping. Incorporated Statistician 5: 115–145.

Goodall, D. W., 1974. A new method for the analysis of spatial pattern by random pairing of quadrats. Vegetatio 29: 135–146.

180

Greig-Smith, P., 1952. The use of random and contiguous quadrats in the study of plant communities. Ann. Bot. N. S. 16: 293–316.

Gros, G., 1978. Structure et échantillonnage des peuplements spontanés des framboisiers (Rubus idaeus L.) dans les Vosges. 3 rd cycle thesis, INRA, Colmar, 60 p.

Hill, M. O., 1973. The intensity of spatial pattern in plant communities. J. Ecol. 61: 225–235.

Kershaw, K. A., 1957. The use of cover and frequency in the detection of pattern in plant communities. Ecology 38: 291–299.

Ludwig, J. A., 1979. A test of different quadrat variance methods for the analysis of spatial pattern. In: Spatial and Temporal analysis in Ecology. Cormack R. M. & J. K. Ord ed., international cooperative Publishing House, Fairland, 289–304.

Matheron, G., 1965. Les variables régionalisées et leur estimation. Masson, Paris, 305 p.

Mead, R., 1974. A test for spatial pattern at several scales using data from a grid of contiguous quadrats. Biometrics 30: 295–307.

Thompson, H. R., 1958. The statistical study of plant distribution patterns using a grid of quadrats. Austr. J. Bot. 6: 322–342.

Walter, S. D., 1974. On the detection of household aggregation of disease. Biometrics 30: 525–538.

Zahl, S., 1974. Application of the S-method to the analysis of spatial pattern. Biometrics 30: 513–524.

The effect of fertilization and irrigation on the vegetation dynamics of a pine-heath ecosystem*

H. Persson**

Institute of Ecological Botany, P.O. Box 559, S-751 22 Uppsala, Sweden

Abstract

The year-to-year changes in the field- and bottom layers were studied by non-destructive recording of cover percentage on permanent plots in a young Scots pine (*Pinus sylvestris*) stand, subjected to experimental treatment. The treatments were: control (0), daily irrigation (I), fertilization once a year (F) and daily irrigation plus fertilization five days a week (IF). The cover of the total field layer of the 0-, I-, F- and IF-plots increased during the period of study from 51, 40, 47 and 50% in 1973 to 70, 79, 86 and 92% in 1980. The bulk of this increase consisted of *Vaccinium vitis-idaea* with the exception of the IF-plots, where the increase especially during the later part of the period of study was due to *Chamaenerion angustifolium* and and to some extent, to *Rubus idaeus*. The cover of the total bottom layer of the 0-, I-, F- and IF-plots changed from 63, 59, 54 and 41% in 1973 to 68, 90, 37 and 36% in 1980. The most drastic change took place in the IF-plots, where the composition changed from almost exclusively lichens to bryophytes. *Pohlia nutans* and *Pleurozium schreberi* were recorded in substantial amounts filling the gap left by the vanishing lichens. There were no conclusive changes on the I-plots as compared with the 0-plots. The cover estimates, however, were generally more substantial. The overall decline in the bottom layer on the F-plots was mainly due to a decrease in the cover of *Pleurozium schreberi*.

* Nomenclature follows Lid (1974) for vascular plants, Nyholm (1954–69) and Arnell (1956) for bryophytes, Dahl & Krog (1973) for lichens, except for *Lecidia granulosa*, and the collective group *Cladonia silvatica*, where Magnusson (1929, 1952) was followed.

** I wish to thank J. G. K. Flower-Ellis for taking an active part in the planning of this work; S. Bråkenhielm for help and co-operation in the field; H. Sjörs, C. O. Tamm and many colleagues in the Swedish Coniferous Forest Project, among them F. Andersson, A. Aronsson & B. Axelsson for encouragement and support.

The work was carried out within the Swedish Coniferous Forest Project and supported by the Swedish Natural Research Council, the Swedish National Environmental Protection Board, the Swedish Council of Forestry and Agricultural Research and the Wallenberg Foundation.

Introduction

Plant ecology in its 'purest' form is descriptive and closely related to observations made *in situ* in real or manipulated ecosystems. One important feature of study in all living systems is change – both 'natural change' as a result of fluctuations in the physical environment (*inter alia* those initiated by climatic factors) and man-made disturbances. Furthermore, all living organisms act as part of the ecosystem and, consequently, they may influence its physical and biotic environment (cf. e.g. van der Maarel & Werger, 1978). However, no species has such an over-riding effect on its environment as *Homo sapiens*. Disturbances within an ecosystem vary continuously. A quantitative view of the behaviour of the system is, therefore, essential and one of the main objectives for the plant ecologist is to identify, quantify and ultimately explain the changes in definable units (cf. Sjörs, 1979).

Cover percentages of plants are readily estimated in a set of sample quadrats and the figures may be used to calculate the arithmetical mean (mean cover $\overline{C}$) (Persson, 1975) of the species population and of collective species groups (see Table 3 below). Cover estimates may also be used as a substitute for dry weight estimates since it is possible to establish weight-to-cover regression relationships for many types of ecosystems (Persson, 1975, 1979a). Non-destructive recording of cover % may, from many respects, be a useful complement to destructive sampling. Furthermore the analyses may be carried out more extensively because they are less time-consuming. An increased knowledge of changes in the plant components of an ecosystem is, from many aspects, essential to the basic understanding of its processes and to the prediction of future conditions (cf. Flower-Ellis & Persson, 1980).

Examples of previous applications of non-destructive recording of cover % on permanent plots for the purpose of detecting long and short-term vegetational changes are to be found in Persson (1975), Bråkenhielm (1977) and Bråkenhielm & Persson (1980). In practice, the technique permits the study of experimental manipulation of the study area if independent measurements are carried out simultaneously in untreated reference areas (control or 0-plots).

This paper reports on results obtained on permanent sample quadrats in a young Scots pine *(Pinus sylvestris)* stand subjected to experimental treatment (cf. Aronsson et al., 1977) with the intention of eliminating water and mineral nutrients as growth-limiting factors. The aims and strategies of these studies are outlined in Bråkenhielm & Persson (1980), Flower-Ellis & Persson (1980). The present paper is restricted to results obtained up to 1980; the investigations are still going on.

Material and methods

The investigation was conducted in a young *Pinus sylvestris* stand at Ivantjärnsheden, Jädraås in C Sweden (60°49′N, 16°30′E, 185 m above

M.S.L.). The stand was regenerated from seed trees which remained after clear-cutting in 1957 and were cut down in 1962. Parts of the stand were cleaned in August 1972, i.e. before investigation started. A less extensive cleaning was also undertaken late in the winter of 1972-1973 in an attempt to create the same density throughout the area. There is a marked gradient in age and size of the trees from the edges to the centre of the stand (Flower-Ellis et al., 1976). The area has previously been described from various respects in Flower-Ellis et al. (1976); Aronsson et al. (1977); Persson (1978, 1980a, 1980b); Axelsson & Bråkenhielm (1980); Bråkenhielm & Persson (1980). The experimental area was divided into plots treated differently with respect to water and mineral nutrient supply (Table 3). Details of the irrigation and fertilization programmes are given in Aronsson et al. (1977); Aronsson & Elowson (1980). The annual mean precipitation in the Jädraås region (1931-60) is 607 mm. However, considerable variation from 413-776 mm has been recorded (cf. Axelsson & Bråkenhielm, 1980). Due to the sandy soil the water storage capacity in the root zone is limited in relation to the annual precipitation (ca 70 mm). The intention of the experimental treatment was to eliminate water and mineral nutrients as growth-limiting factors. The treatments were: irrigation every day during the growing season (I), fertilization with solid fertilizers once a year (F), and daily irrigation plus fertilization five days a week (only irrigation was supplied on Saturdays and Sundays) during the growing season (IF).

The irrigation and fertilization started in 1974. The irrigation water was taken from a drilled well and sprinkled from a height of 0.5 m above ground level. Irrigation normally took place every day, except after heavy rains when the programme was modified. Periods of irrigation and fertilization from 1974-1980 are shown in Table 1. The mean precipitation during the period of irrigation from May – August was about 250 mm. The water supplied by irrigation was about $3 \, l \, m^{-2} \, day^{-1}$ and the total amounts of nitrogen supplied to the IF-plots were 7, 10, 15, 15, 20, 20 and 20 g · m^{-2} during the years 1974-80, resp. All other essential elements were given in proportion to nitrogen according to Ingestad (1979). The fertilizers distributed in the IF-plots dissolved in the irrigation water. The rate of nutrient supply was adjusted to match the uptake

Table 1. Periods of irrigation and fertilization of the I, IF and F-plots during 1974–1980. Data supplied by Aronsson *et al.* (1977, pers. comm.).

Year	Irrigation	Liquid fertilization	Solid fertilization
1974	July 12 – Sept. 19	July 12 – Aug. 9	July 15–16
1975	May 23 – Sept. 12	May 23 – Aug. 22	June 6
1976	May 20 – Sept. 18	May 20 – Aug. 20	June 3–4
1977	May 16 – Aug. 30	May 16 – Aug. 19	May 16–17
1978	May 23 – Sept. 12	May 23 – Sept. 12	May 30
1979	May 28 – Sept. 11	May 28 – Sept. 7	May 29
1980	May 27 – Sept. 5	May 27 – Sept. 5	May 27

rate by the vegetation. The total amount of nitrogen supplied yearly to the F-plots was 8 g · m^{-2} from 1974–80. Additional amounts of phosphorous (3 g · m^{-2}) and potassium (6 g · m^{-2}) were supplied in 1974, 1977 and 1980. Furthermore boron (0.25 g · m^{-2}) was supplied in 1978 and 1980. The permanent 0.5 × 0.5 m quadrats used for the cover analyses were distributed randomly within each experimental plot inside larger 10 × 10 m quadrats. The sampling scheme was based on a broad division, into *Calluna* and non-*Calluna* (strata I and II). The number of permanent quadrats in each stratum (I and II) was 7 in the 0-, I- and IF-plots and 8 in the F-plots. The areas covered by the two strata were estimated each year by transect inventory (cf. Persson, 1980b; Bråkenhielm & Persson, 1980). The results from the repeated transect inventories are given in Table 2.

The above-ground cover (i.e., the proportion of the ground area covered by the vertical projection of above-ground shoot parts of all species and separate layers (litter, bottom- and field layers) in the quadrats was directly estimated as percentages, using methods described in Persson (1975, 1979a) and Bråkenhielm (1977). On the basis of the non-destructive cover estimates, apparent cover ($\overline{C}_a$) and shading percentage ($\overline{Sh}$) could easily be calculated. Other related concepts, readily calculated from the data are, for instance, characteristic cover ($\overline{C}_c$) and frequency (see Persson, 1975, pp. 29–34 for an explanation of these terms). For a more complete description of the vegetation structure in the experimental plots and its variation, these concepts are very useful.

In order to schedule the sampling programme, it was necessary to find out at which point during the season the major species groups were quantitatively most developed regarding their standing crop (cf. Persson, 1979a). The most suitable time for measuring year-to-year fluctuations in the present case was found to be at a fairly late stage of the growing season (*viz.* in the end of August/beginning of September), when the current year's shoots of the dwarf shrub species had attained their definitive weight. In this study so-called 'pioneer' lichens were defined as those lichens which tend to colonize the bare, disturbed soil surface, among others species with cup- or needle-like podetia such as *Cladonia cornuta*, *C. deformis*, *C. fimbriata*, *C. pyxidata*. *Cetraria islandica* and *Lecidea granulosa* were also included in this group. 'Reindeer' lichens were defined as those species which successfully competed in a more stable vegetation (cf. Bråkenhielm & Persson, 1980), such as *Cladonia rangiferina*, *C. silvatica* and *C. alpestris*.

Table 2. The proportion of the total area (%) accounted for by stratum I (*Calluna*). The area was yearly estimated by transect inventories (cf. Persson 1980b) which were repeated at the same time as the cover analyses (end of August/beginning of September).

Year	0-plots	I-plots	F-plots	IF-plots
1973	49.8	48.0	50.4	46.3
1974	47.5	47.2	48.1	55.2
1975	47.1	46.5	47.6	53.4
1976	37.5	41.2	35.9	47.5
1977	38.1	44.6	36.4	42.6
1978	45.3	52.5	43.3	46.3
1979	49.8	47.5	49.7	19.5
1980	53.1	59.1	43.8	8.0

Results

Data from the repeated sampling of cover % (Table 3; Figs. 1–2) indicate considerable structural variations from year to year if dominant species and collective species groups such as field and bottom layers, bryophytes, lichens, 'pioneer' lichens, etc. are taken into consideration. No essential differences in the structural composition of the field- and bottom layers were observed in 1973 before experimental treatment. The differences between the experimental plots after the experiment had started in 1974 are, therefore, likely to be related to experimental impact.

Table 3. Mean cover ($\overline{C}$) ± s.e. of the field and bottom layer species at Ivantjärnsheden, Ih II, the O, I, F and IF-plots. The analyses were carried out each year during August-October on randomly distributed permanent 0.5 × 0.5 m quadrats. The sampling scheme was based on two broad divisions, *Calluna* and non-*Calluna* (strata I and II) (cf. Persson 1980b, Bråkenhielm & Persson 1980). The number of quadrats in each stratum was 7 in the O, I and IF plots and 8 in the F-plots. Cover of litter (twigs, leaf and needle remnants) is also indicated – denotes absence and (O) (0.5) denotes presence.

	O-plots								I-plots							
	1973	1974	1975	1976	1977	1978	1979	1980	1973	1974	1975	1976	1977	1978	1979	1980
LITTER	18±6	28±7	28±5	33±5	21±4	31±4	33±3	35±5	17±3	30±4	30±5	36±5	36±6	28±4	32±5	28±4
TOTAL FIELD LAYER	45±3	47±6	49±2	32±5	40±4	52±3	67±6	62±4	39±4	42±3	48±7	48±6	53±6	67±5	69±6	71±6
TOTAL BOTTOM LAYER	64±7	58±7	61±6	72±6	71±6	71±6	71±5	69±4	64±5	62±5	75±5	79±3	82±3	86±4	90±4	88±4
Trees and shrubs < 1 m																
Betula pubescens	1±1	1±1	0±0	0±0	0±0	0±0	0±0	0±0	–	–	–	–	–	–	–	–
Picea abies	–	–	0±0	–	–	–	–	0±0	–	–	–	0±0	0±0	0±0	0±0	0±0
Pinus sylvestris	3±3	1±1	1±1	1±1	2±1	2±1	4±2	5±3	–	1±1	1±1	1±1	1±1	2±1	3±2	3±2
Dwarf shrubs																
Calluna vulgaris	41±3	42±2	39±2	25±4	25±4	34±4	39±5	40±5	35±4	37±3	39±3	26±5	29±7	39±5	40±7	44±8
Vaccinium myrtillus	–	–	–	0±0	0±0	–	0±0	0±0	0±0	0±0	1±1	1±1	1±1	1±1	1±1	1±1
V. vitis-idaea	6±1	6±1	11±2	15±2	16±2	22±2	24±7	25±3	6±1	7±1	12±5	23±3	25±3	36±5	30±4	31±3
Herbaceous species																
Chamaenerion angustifolium	–	–	–	–	–	–	–	–	–	–	–	–	–	–	–	–
Dryopteris spinulosa	–	–	–	–	–	–	–	–	–	–	–	–	–	–	–	–
Rubus idaeus	–	–	–	–	–	–	–	–	–	–	–	–	–	–	–	–
Trientalis europaea	–	–	–	–	–	–	–	–	–	–	–	–	–	–	–	–
FIELD LAYER, TOTAL	51±4	50±2	51±3	41±4	42±4	58±3	67±5	70±6	40±4	44±3	52±3	54±7	56±6	73±6	76±10	79±7
Bryophytes																
Brachythecium reflexum	–	–	–	0±0	–	–	–	–	–	–	–	–	–	–	–	–
Dicranum fuscescens	–	–	0±0	0±0	1±1	1±1	1±1	–	1±1	1±1	2±1	1±1	1±1	1±1	1±1	1±1
D. polysetum	1±1	1±1	0±0	0±0	1±1	1±1	1±1	3±1	2±1	1±1	1±1	2±1	5±2	6±3	6±2	7±3
Hylocomium splendens	–	–	–	–	–	–	–	–	–	–	–	–	–	–	–	–
Pleurozium schreberi	21±6	18±5	14±4	13±8	13±3	15±4	17±5	18±5	15±6	15±5	15±4	14±4	16±4	22±5	20±6	29±6
Pohlia nutans	0±0	0±0	0±0	–	1±1	1±1	1±1	0±0	1±1	1±1	1±1	1±1	1±1	1±1	1±1	1±1
Polytrichum commune	–	–	–	–	–	0±0	0±0	0±0	–	–	–	–	–	–	1±1	1±1
P. juniperinum	0±0	0±0	–	–	0±0	0±0	0±0	–	1±1	1±1	2±1	1±1	1±1	1±1	2±1	1±1
Bryophytes, total	22±5	19±5	15±5	14±3	15±3	18±5	20±5	21±5	20±6	20±6	21±5	21±4	26±5	32±6	34±6	41±6

Table 3. (Continued)

	0-plots								I-plots							
	1973	1974	1975	1976	1977	1978	1979	1980	1973	1974	1975	1976	1977	1978	1979	1980
Lichens																
Cetraria islandica	–	–	–	–	–	–	–	–	1±1	1±1	2±1	1±1	4±2	3±2	4±2	3±2
Cladonia alpestris	–	–	0±0	–	–	0±0	0±0	–	–	–	–	–	–	–	–	–
C. botrytes	1±1	1±1	0±0	0±0	0±0	0±0	0±0	–	1±1	1±1	2±1	1±1	0±0	0±0	0±0	0±0
C. cenotea	2±1	1±1	0±0	1±1	0±0	0±0	–	–	1±1	1±1	–	1±1	1±1	0±0	0±0	0±0
C. cornuta	2±1	2±1	3±1	3±1	3±1	2±1	2±1	2±1	5±2	5±2	6±2	5±1	5±1	5±1	3±1	2±1
C. crispata	0±0	0±0	1±1	0±0	0±0	0±0	0±0	1±1	1±1	1±1	4±1	3±1	3±1	1±1	2±1	2±1
C. deformis	2±2	2±1	2±1	4±1	5±1	6±2	6±2	5±2	5±1	5±1	5±1	9±1	8±1	6±1	5±1	4±1
C. fimbriata	2±1	2±1	0±0	4±1	3±1	2±1	3±1	2±1	3±1	3±1	2±1	7±1	7±2	5±1	4±1	3±1
C. gracilis	2±1	1±1	0±0	1±1	2±1	2±1	1±1	0±0	1±1	1±1	1±1	3±1	3±2	2±1	2±1	1±1
C. macilenta	0±0	0±0	1±1	1±1	1±1	1±1	1±1	1±1	0±0	0±0	1±1	2±1	1±1	0±0	0±0	0±0
C. pyxidata	2±1	3±1	3±1	1±1	1±1	1±1	1±1	0±0	4±1	6±1	7±2	3±1	2±1	2±1	1±1	1±1
C. rangiferina	15±5	14±4	23±4	26±5	26±5	21±4	20±3	19±2	10±3	10±4	18±5	20±5	20±5	19±4	19±5	19±5
C. silvatica coll.	13±6	13±5	12±5	18±6	19±16	16±5	16±5	16±5	2±1	3±1	3±1	10±3	13±3	11±2	12±2	11±2
Lecidea granulosa	1±1	2±1	3±1	3±1	4±2	2±1	2±1	2±1	5±2	4±2	2±1	3±2	4±2	2±1	2±1	2±1
Lichens, total	40±7	39±6	49±5	62±6	63±5	53±3	52±4	47±4	39±5	42±6	54±5	69±6	70±5	55±4	52±6	49±4
'Pioneer' lichens, total	12±2	12±2	14±2	18±4	19±4	17±4	17±4	12±3	27±5	29±4	31±4	36±4	35±4	24±2	21±2	15±1
BOTTOM LAYER, TOTAL	63±8	58±8	64±6	76±6	78±5	71±4	72±4	68±5	59±7	62±6	75±5	90±6	97±6	87±4	86±6	90±5
LITTER	27±7	47±7	39±5	42±5	43±6	34±3	37±2	35±3	22±3	44±5	31±7	30±5	32±4	41±4	58±6	51±6
TOTAL FIELD LAYER	44±3	46±2	49±2	41±3	51±5	66±4	73±3	71±4	46±5	48±4	68±5	78±6	78±7	88±3	90±7	76±8
TOTAL BOTTOM LAYER	59±7	44±7	39±6	51±7	51±7	42±6	40±6	35±5	49±6	47±5	53±4	63±6	47±7	40±8	40±9	42±9
Trees and shrubs < 1 m																
Betula pubescens	0±0	0±0	–	–	–	–	–	–	–	0±0	0±0	1±1	1±1	–	–	–
Picea abies	–	0±0	–	0±0	0±0	0±0	0±0	0±0	0±0	0±0	0±0	1±1	2±1	2±1	3±1	4±3
Pinus sylvestris	0±0	1±1	2±1	2±2	3±2	1±1	1±1	1±1	2±1	4±3	5±4	4±4	5±5	1±1	0±0	0±0
Dwarf shrubs																
Calluna vulgaris	40±3	41±2	37±1	17±2	20±3	33±3	35±4	32±5	40±2	39±5	44±5	33±5	25±4	18±7	14±7	3±2
Vaccinium myrtillus	–	–	–	–	–	–	–	–	0±0	0±0	1±1	5±3	2±1	1±1	2±2	3±3
V. vitis-idaea	6±1	8±1	14±2	26±2	29±2	36±3	39±3	31±3	8±1	8±1	23±3	41±5	34±6	17±5	16±6	5±2
Herbaceous species																
Chamaenerion angustifolium	0±0	0±0	–	0±0	4±3	4±2	7±4	12±4	–	–	–	10±5	22±6	68±8	66±9	63±13
Dryopteris spinulosa	–	–	–	–	–	–	–	–	–	–	–	0±0	–	0±0	0±0	0±0
Rubus idaeus	–	–	–	–	–	–	–	–	–	–	–	4±3	1±1	1±1	9±5	15±9
Trientalis europaea	–	–	–	–	–	–	–	–	–	–	–	–	–	–	–	0±0
FIELD LAYER, TOTAL	47±3	51±2	52±3	48±3	57±5	73±4	82±4	86±7	50±3	52±3	73±5	99±9	94±10	106±5	110±7	92±12

Table 3. (Continued)

	F-plots								IF-plots							
	1973	1974	1975	1976	1977	1978	1979	1980	1973	1974	1975	1976	1977	1978	1979	1980
Bryophytes																
Brachythecium reflexum	–	0±0	–	–	–	–	0±0	–	–	–	–	0±0	0±0	0±0	0±0	1±1
Dicranum fuscescens	0±0	0±0	–	–	0±0	–	0±0	–	2±1	1±1	2±1	4±1	9±5	3±2	6±3	5±4
D. polysetum	3±1	2±1	1±1	0±0	0±0	0±0	1±1	0±0	3±2	3±2	2±1	2±1	2±1	2±1	1±1	6±4
Hylocomium splendens	–	–	–	–	–	–	–	–	–	–	–	–	–	–	1±1	1±1
Pleurozium schreberi	13±5	13±4	2±1	2±1	2±1	2±1	2±1	2±1	5±2	5±2	12±3	19±6	13±4	19±6	19±7	14±7
Pohlia nutans	2±1	1±1	1±1	0±0	0±0	2±1	5±3	6±3	1±1	2±1	8±3	7±3	10±3	12±3	12±3	9±2
Polytrichum commune	0±0	0±0	–	–	–	–	–	–	0±0	–	–	–	–	–	1±1	–
P. juniperinum	1±1	0±0	1±1	0±0	0±0	0±0	–	1±1	–	0±0	1±1	1±1	–	1±1	2±1	–
Bryophytes, total	19±5	16±5	5±2	3±1	3±1	5±2	9±3	10±3	10±3	12±3	25±4	33±7	33±6	37±7	40±9	36±9
Lichens																
Cetraria islandica	0±0	0±0	0±0	0±0	0±0	0±0	0±0	–	–	–	–	–	–	–	–	–
Cladonia alpestris	0±0	0±0	–	–	–	–	0±0	–	–	–	–	–	–	–	–	–
C. botrytes	0±0	0±0	1±1	1±1	1±1	0±0	0±0	0±0	1±1	0±0	0±0	0±0	–	–	–	–
C. cenotea	1±1	1±1	0±0	0±0	0±0	–	–	0±0	1±1	1±1	0±0	–	–	0±0	0±0	–
C. cornuta	4±1	4±1	5±1	3±1	4±2	3±1	3±1	2±1	2±1	1±1	2±1	2±1	1±1	–	–	–
C. crispata	1±1	1±1	1±1	1±1	1±1	1±1	1±1	1±1	1±1	0±0	1±1	0±0	–	–	–	–
C. deformis	5±1	4±1	5±1	6±2	7±1	5±1	5±1	4±1	2±1	1±1	3±1	5±2	3±1	1±1	1±1	–
C. fimbriata	4±1	3±1	3±1	6±2	7±2	5±1	3±1	3±1	2±1	1±1	2±1	2±1	1±1	–	–	–
C. gracilis	1±1	1±1	0±0	1±1	1±1	1±1	1±1	1±1	0±0	0±0	0±0	1±1	–	–	–	–
C. macilenta	–	0±0	0±0	1±1	0±0	0±0	0±0	0±0	0±0	0±0	0±0	1±1	–	–	–	–
C. pyxidata	3±1	3±1	3±1	2±1	1±1	1±1	1±1	1±1	3±1	3±1	3±2	1±1	1±1	1±1	–	–
C. rangiferina	7±2	8±1	15±4	13±4	12±5	11±3	10±2	9±2	10±4	8±3	17±4	11±3	8±3	2±1	–	–
C. silvatica coll.	4±1	5±2	4±2	10±3	9±3	8±3	6±2	7±2	5±3	2±1	1±1	7±3	2±2	–	–	–
Lecidea granulosa	4±1	4±2	3±1	4±1	3±1	1±1	1±1	1±1	2±1	2±1	3±1	–	–	–	–	–
Lichens, total	35±5	33±4	47±5	47±7	49±6	37±5	32±4	27±5	31±7	20±4	34±6	30±7	14±1	3±1	1±1	–
'Pioneer' lichens, total	23±4	21±4	21±4	26±5	26±4	17±1	15±1	13±1	14±3	9±1	15±2	13±3	5±2	1±1	1±1	–
BOTTOM LAYER, TOTAL	54±6	49±7	42±5	51±7	52±6	42±5	41±6	37±6	41±5	32±4	59±4	64±6	47±7	40±7	41±9	36±9

With regard to *Calluna vulgaris* (cf. Fig. 1), there were some common features in the fluctuations during the years of study in the 0-, I. and F-plots, *viz.* fairly high initial cover estimates during 1973–75, followed by a decline in cover in 1976 and 1977 and fairly high estimates during 1978–80. From the start a similar cover trend was also observed on the IF-plots. However, there was no final rise there during 1978–80. Instead, *Chamaenerion angustifolium*, which first appeared on the IF-plots in 1976 (Table 3) had a substantial increase in cover in 1978 and maintained a fairly high level during the subsequent years (around 65% in 1979 and 1980) replacing *Calluna* in the field layer almost completely. No *Chamaenerion* was observed in the 0- and I-plots. Nevertheless, it was found in limited amounts on the F-plots, even before fertilization had begun (cf. Table 3). There was no corresponding rapid increase of *Chamaenerion* on the F-plots, the highest estimate recorded being 12% in 1980.

As regards *Vaccinium vitis-idaea*, there was a gradual increase in cover during the period of study (ct. Fig. 1) from 6% in 1973 to 25% in 1980 on the 0-plots and from 6% in 1973 to 31% in 1980 on the I- and F-plots (Table 3). This increase was more pronounced in the I- and F-plots. In the IF-plots there was a substantial increase in cover from 8% in 1973 to a maximum of 41% in 1976; followed by a gradual decline down to only 5% in 1980.

The cover of the total field layer of the 0-, I-, F- and IF-plots increased during the period of study from 51, 40, 47 and 50% in 1973 to 70, 79, 86 and 92% resp. in 1980. The bulk of this increase in the 0-. I- and F-plots concerned *V. vitis-idaea*. However, in the IF-plots a considerable part of this increase during the later part of the study period 1976–1980 was due to *Chamaenerion* and, to some extent, to *Rubus idaeus*.

The cover estimates of certain bottom-layer species were closely related to those of certain field-layer species. This was especially evident in the case of *Pleurozium schreberi*, which preferred to grow within the *Calluna*-clones (stratum I). Stratum I was gradually expanding into stratum II (the non-*Calluna* stratum) in all experimental treatments. *P. schreberi* also expanded in the bottom layer (cf. Figs. 1 and 2). On the other hand, the 'pioneer' lichens occurred inversely proportional to *C. vulgaris*, and were numerous in stratum II, which was

continously being invaded by the *Calluna* clones. On the IF-plots, this invasion had come to an end in 1980 when *C. vulgaris* almost disappeared. The Calluna clones (stratum I) then covered only 8% of the total area (Table 2).

All lichens appeared to die back on the IF-plots in connection with the more regular occurrence of *Chamaenerion*. Initially there was an upsurge in lichen cover during 1974–76, characterized by monstrous forms of 'pioneer' lichens *inter alia Cladonia cornuta, C. crispata, C. deformis, C. fimbriata, C. gracilis, C. macilenta, C. pyxidata* and by *Cladonia rangiferina* (Table 3). These species, however, gradually died back from 1977 onwards and during the last inventory in 1980, no lichens were found at all.

This gradual disappearance of lichens, which was characteristic for the IF-plots, was not observed on the F-plots. The invasion of *Chamaenerion*, however, took place much more slowly on these plots, leaving plenty of space for the lichens. Furthermore, the indigenous forest mosses *Dicranum polysetum* and *Pleurozium schreberi* died back substantially on the F-plots. On the other hand, *Pohlia nutans*, which also increased substantially on the IF-plots, invaded fairly quickly. The latter species occurred in a special shade-form under the dense field-layer canopy.

Besides *Pohlia nutans, Pleurozium schreberi* was also recorded in substantial amounts on the IF-plots, especially in stratum II, where this species seemed to have intruded and filled the empty gap left by the vanishing lichens. Characteristic of the IF-plots were furthermore the following mosses: *Brachythecium reflexum, Dicranum fuscescens, D. polysetum* and *Hylocomium splendens*. These species occurred only sporadically on the other experimental plots.

There were no conclusive changes recorded in the cover of the bottom layer on the I-plots. Although some common features could be recognized in the year-to-year fluctuations, cf. *Cladonia rangiferina* and the 'pioneer' lichens in Table 3, it is evident that the cover figures for the I-plots were generally greater than those for the 0-plots. Thus, the cover of the bryophytes total increased from 20% in 1973 to 41% in 1980, and that of the lichens from 39% in 1973 to 49% in 1980. On the 0-plots, on the other hand, no substantial changes were recorded.

The cover of the bottom layer total of the 0-, I-, F- and IF-plots changed from 63, 59, 54 and 41% in

188

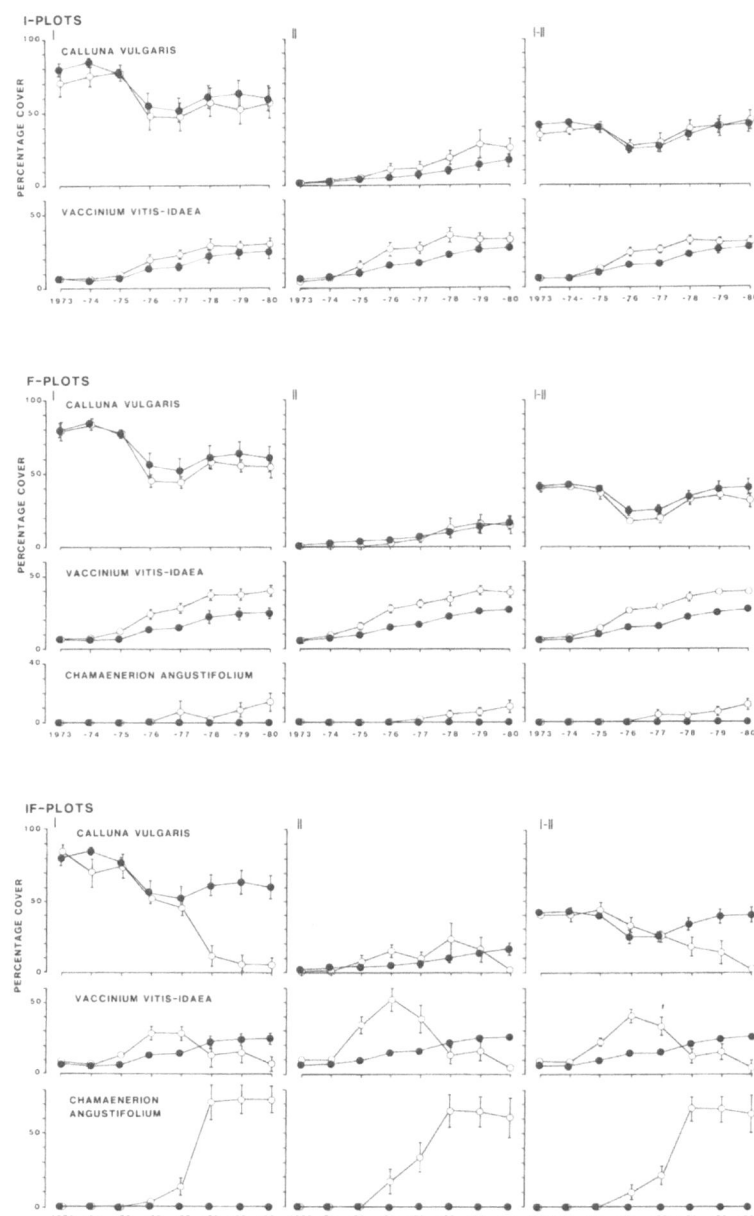

Fig. 1. Annual change in the mean cover of some field layer species during the years 1973–80 on the I-, F- and IF-plots as compared with the control (the I-, F- and IF-plots are indicated by unfilled and the control plots by filled symbols). Estimates are indicated ± one standard error. The cover analyses were carried out on permanent 0.5 × 0.5 m quadrats (14 in each of the 0, I and IF-plots and 16 in the F-plots), equally distributed in two strata (I and II), where stratum I = *Calluna,* II = non-*Calluna,* and I-II = the total area. The distribution of the strata was estimated by transect inventories (cf. Table 2).

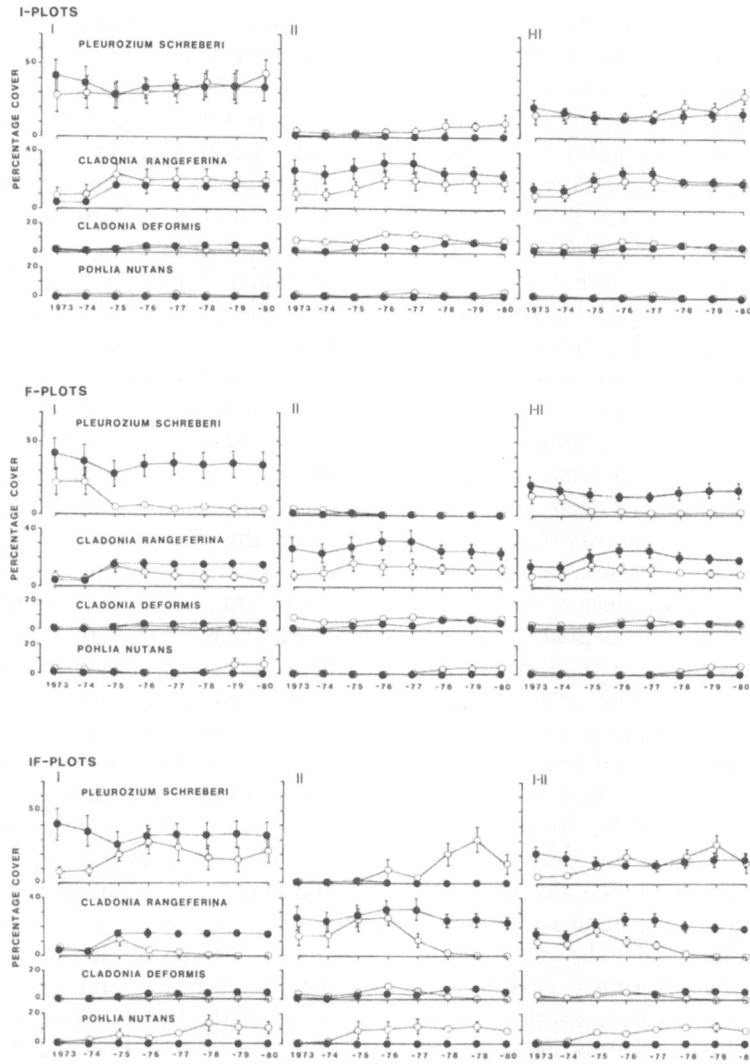

Fig. 2. Annual change in the mean cover of some bottom layer species during the years 1973–80 on the I, F and IF-plots compared with the control. For additional information cf. Figure 1.

1973 to 68, 90, 37, and 36% resp. in 1980 (Table 3). The increase in the I-plots, as indicated above, was mainly in moss cover while the decreases in the F- and IF-plots were mainly in *Pleurozium schreberi* and lichens. The composition of the bottom layer changed on the IF-plots from almost exclusively lichens to bryophytes. Thus the bryophyte total increased from 10% in 1973 to 36% in 1980, while the lichen total decreased from 31% in 1973 to 0% in 1980.

Discussion

Observations of quantitative fluctuations in plant populations, may serve as a firm basis for simplifying the ecosystem structure and identifying, factors, which most strongly control its functional organization (cf. Flower-Ellis & Persson, 1980). Although a fairly simple ecosystem was chosen in the case of the present study, there was a considerable number of environmental variables and species interactions to consider defying in many cases direct analyses of cause and effect relationships.

First of all the experiment was carried out with the intention of eliminating water and mineral nutrients as growth-limiting factors (cf. Aronsson *et al.,* 1977), by supplying water and mineral nutrients and thus modifying natural conditions. Since both the field- and bottom-layer species are located in the upper soil layer (cf. Persson, 1980a, 1980c) there is no reason to believe that shortage of water, mineral nutrients or both, should limit growth on the I-, F- and IF-plots resp. The effect of root competition between the tree root systems and those of the ground flora was also of limited importance, due to their different vertical distribution in the soil. Thus, the superficially distributed fine roots of the field-layer species had their largest annual turnover and supply of dead roots in the F/H layer, whereas the fine roots of *Pinus sylvestris* grew into the mineral soil (Persson, 1980a, 1980c).

The composition and luxuriance of the field and bottom layer vegetation is to a great extent dependent on the density of the tree canopy (cf. Hill, 1978; Bråkenhielm & Persson, 1980; Persson, 1980b). Initially, tree individuals were relatively small and their canopies were not dense enough to affect the development of the field and bottom layers (cf. Flower-Ellis *et al.,* 1976). However, as the trees develop and their needle mass increases (especially, on the F- and IF-plots), the intensity of the light which reaches the ground will gradually diminish. On the IF-plots, which showed optimized growth conditions, the vigorously growing field layer (mainly consisting of *Chamaenerion angustifolium* and *Rubus idaeus)* also prevented light from penetrating and reaching the bottom layer. The latter was reduced to a few moss species which could survive, *inter alia: Dicranum fuscescens, D. polysetum, Pleurozium schreberi* and *Pohlia nutans.*

The purpose of the fertilization and irrigation programme on the IF-plots was to ensure that enough nutrients were present in the soil medium in amounts required for an adequate uptake rate (cf. Ingestad, 1974). The mineral nutrient supply was then adapted to the expected requirements of the trees. The water requirements of the trees were, however, probably not completely satisfied during periods of heavy transpiration. Thus, the field- and bottom-layer species served as an effective barrier preventing water and mineral nutrients from reaching the tree roots. The trees on the IF-plots had also developed a luxuriant needle mass which did not correspond to an equivalent amount of fine roots (cf. Persson, 1979b, 1980d).

The fertilization of the F-plots was accomplished with a traditional experimental technique, *viz.* the addition of a constant amount of solid fertilizers once a year. Since mineral nutrient requirements varies between from species to species considerably (cf. Ingestad, 1974, 1979) and since a toxic level may be reached with a fairly moderate nutrient supply, there is a risk of nutrient toxicity if the supply is exaggerated. This may be the reason for the distinct decrease in cover of *Pleurozium schreberi* on the F-plots as compared with the 0-plots. However, on the IF-plots, where the mineral nutrients were supplied in small doses dissolved in the irrigation water this high toxic level of mineral salts was probably never attained. Thus, it was found from the inventory that a substantial increase in cover had taken place of *P. schreberi* on the IF-plots. For many other species, *inter alia Calluna vulgaris, Vaccinium vitis-idaea* and most lichens, substantial increases in cover were also indicated but only temporarily. Later there occurred a decline in the cover of these species, which was probably related to a deterioration in light conditions (see above). All lichens then died back, as the cover of the tree and field layers increased.

There was no experimental evidence of conclusive changes in the field- and bottom layers on the I-plots when compared with the 0-plots. The cover on the I-plots, however, was generally more substantial than that on the 0-plots. However, summer drought does not appear to be a factor of great importance for most species in North European heath ecosystems. Beijerinck (1940) and Bannister (1964) both express the view that *C. vulgaris* is less inhibited by summer drought than by winter cold. Havas (1965, 1971) claims that winter survival seems to be important for the dynamics of *Vaccinium myrtillus,* and Gates (1914) propounds similar views about certain North American ericoids.

Natural disturbances have traditionally been defined in terms of major catastrophic events originating from the physical environment, such as wind-storm, ice-storm, temperate fluctuations, precipitation variability etc. (cf. White, 1979). The significance of the disturbance regime on the vegetation changes, as part of the overall climatic fluctuations, may be studied, as in the case of the present study, from extensive empiric data collected over a long period of time. However, there are no straight-forward ways of separating these fluctuations *sensu stricto* from the inherent growth pattern of the vegetation. Thus, in heathlands, inherent cyclic changes may be derived from the phasic development of *C. vulgaris* (cf. Watt, 1947, 1955; Gimingham, 1972). Furthermore, in forest-heath-ecosystems, certain developmental stages may be distinguished which to a great extent, are determined by the age and development of the tree stand (Bråkenhielm & Persson, 1980).

References

Arnell, S., 1956. Illustrated moss flora of Fennoscandia. I. Hepaticae. Lund: Gleerup, 315 pp.

Aronsson, A., Elowson, S. & Ingestad, T., 1977. Eliminating of water and mineral nutrition as limiting factors in a young Scots pine stand. 1. Experimental designs and some preliminary results. Swed. Conif. For. Proj. Techn. Rep. 10, 38 pp.

Aronsson, A. & Elowson, S., 1980. Effects of irrigation and fertilization on mineral nutrients in Scots pine needles. In: Persson, T. (ed.) Structure and function of northern coniferous forests – An ecosystem study. Ecol. Bull. Stockholm 32: 219–228.

Axelsson, B. & Bråkenhielm, S., 1980. Investigation sites of the Swedish Coniferous Forest Project – biological and physiographical features. In: Persson, T. (ed.) Structure and function of northern coniferous forests – An ecosystem study. Ecol. Bull. Stockholm 32: 25–64.

Bannister, P., 1964. The water relations of certain heath plants with reference to their ecological amplitude. III. Experimental studies: General considerations. J. Ecol. 52: 499–509.

Beijerinck, W., 1940. Calluna; A monograph on the Scotch heather. Verh. Akad. Wet., Amst. (3 rd sect.) 38, 180 pp.

Bråkenhielm, S., 1977. Vegetation dynamics of afforested farmland in a district of South-eastern Sweden. Acta Phytogeogr. Suec. 63, 106 pp.

Bråkenhielm, S. & Persson, H., 1980. Vegetation dynamics in developing Scots pine stands in central Sweden. In: Persson, T. (ed.) Structure and function of northern coniferous forests – An ecosystem study. Ecol. Bull. Stockholm 32: 139–152.

Dahl, E. & Krog, H., 1973. Macrolichens. Oslo-Bergen-Tromsø: Univers. forl. 185 pp.

Flower-Ellis, J. G. K., Albrektson, A. & Olsson, L., 1976. Structure and growth of some young Scots pine stands: (1) Dimensional and numerical relationships. Swed. Conif. For. Proj. Techn. Rep. 3, 98 pp.

Flower-Ellis, J. G. K. & Persson, H., 1980. Investigation of stuctural properties and dynamics of Scots pine stands. In: Persson, T. (ed.) Structure and function of northern coniferous forests – An ecosystem study. Ecol. Bull. Stockholm 32: 125–138.

Gates, F. C., 1914. Winter as a factor in the xerophily of certain evergreen Ericads. Bot. Gaz. 57: 445–489.

Gimingham, C. H., 1972. Ecology of heathlands. London: Chapman & Hall, 266 pp.

Hill, M. O., 1978. The development of flora in even-aged plantations. In: Ford, E. D., Malcolm, D. C. & Atterson, J. (eds.) The ecology of even-aged forest plantations, pp. 175–192. Inst. of Terr. Ecology, Cambridge.

Havas, P. J., 1965. Pflanzenökologische Untersuchungen im Winter. 1. Aquilo, Ser. Bot. 4: 1–36.

Havas, P. J., 1971. The water economy of bilberry (Vaccinium myrtillus) under winter conditions. Rep. Kevo Subarct. Res. Stn. 8: 41–52.

Ingestad, T., 1974. Towards optimum fertilization. Ambio 3: 49–54.

Ingestad, T., 1979. Mineral nutrient requirement of Pinus silvestris and Picea abies seedlings. Physiol. Plant. 45: 373–380.

Lid, J., 1974. Norsk og Svensk Flora. Andre Utgåva, Oslo. Det Norske Samlaget, 808 pp.

Magnusson, A. H., 1929. Flora över Skandinaviens busk och bladlavar. Stockholm. P.A. Nordstedt & Söner, 127 pp.

Magnusson, A. H., 1952. Key to the species of Lecidea in Scandinavia and Finland. 2. Non-saxicolous species. Sv. Bot. Tidskr. 46: 313–323.

Maarel, E. van der & Werger, M. J. A., 1978. On the treatment of succession data. Phytocoenosis 7: 257–277.

Nyholm, E., 1954, 1956, 1958, 1960, 1965, 1969. Illustrated Moss Flora of Fennoscandia. 2. Musci. Lund: Gleerup, 799 pp.

Persson, H., 1975. Deciduous woodland at Andersby, Eastern Sweden: field-layer and below-ground production. Acta Phytogeogr. Suec. 62, 72 pp.

Persson, H., 1978. Root dynamics in a young Scots pine stand in central Sweden. Oikos 30: 508–519.

Persson, H., 1979a. The possible outcomes and limitations of measuring quantitative changes in plant cover on permanent plots. In: Hytteborn, H. (ed.) The use of ecological variables in environmental monitoring. Swed. Nat. Eviron. Protection Board, Rep. PM 1151: 81–87.

Persson, H., 1979b. Fine-root production, mortality and decomposition in forest ecosystems. Vegetatio 41: 101–109.

Persson, H., 1980a. Spatial distribution of fine-root growth, mortality and decomposition in a young Scots pine stand in central Sweden. Oikos 34: 77–87.

Persson, H., 1980b. Structural properties of the field and bottom layers at Ivantjärnsheden. In: Persson, T. (ed.) Structure and function of northern coniferous forests – An ecosystem study. Ecol. Bull. Stockholm 32: 153–163.

Persson, H., 1980c. Death and replacement of fine-roots of a mature Scots pine stand. In: Persson, T. (ed.) Structure and function of northern coniferous forests – An ecosystem study. Ecol. Bull. Stockholm 32: 251–260.

Persson, H., 1980d. Fine-root dynamics in a Scots pine stand with and and without near-optimum nutrient and water regimes. Acta Phytogeogr. Suec. 68: 101–110.

Sjörs, H., 1979. What are the criteria of an environmental change - natural and man-made? In: Hytteborn, H. (ed.) The use of ecological variables in environmental monitoring. Swed. Nat. Environ. Protection Board, Rep. PM 1151: 25–36.

192

Watt, A. S., 1947. Pattern and process in the plant community. J. Ecol. 43: 490–506.

Watt, A. S., 1955. Bracken versus heather, a study in plant sociology. J. Ecol. 43: 490–506.

White, P. S., 1979. Pattern, process, and natural disturbance in vegetation. Bot. Rev. 45: 229–299.

Accepted 22.5.1981.

Conservation of *Calluno-Vaccinietum* heathland in the Belgian Ardennes, an experimental approach*

A. Froment
Department of Botany, University of Liège, Sart Tilman, 4000 Liège, Belgium

Keywords: Experimental plots, Calluna, Heath, Heathland management, Nature conservation, *Vaccinium*

Abstract

It is now difficult to manage the semi-natural groupings of heathland because of the tendency for succession towards forest stages.

Several experiments were conducted in the Hautes Fagnes region (Belgium). After the vegetation was mapped, plots were treated by mowing, burn-beating, and burning.

The changing floristic composition was followed from 1972 to 1980 in permanent quadrats. Burn-beating is an excellent method of management for old heath, because it reestablishes the properties of the heath ecosystem.

Introduction

Heathlands in W. Europe are now regressing rapidly because the traditional agro-pastoral practices which were responsible for their formation and maintenance have been abandoned.

These semi-natural landscapes have nowadays a great importance for nature conservation. As Gimingham (1972) rightly pointed out: 'Heaths are of value aesthetically and in connection with various forms of recreation; they are also of great importance for scientific research and as reservoir of wild life'. One could also add a didactic and historical interest, for such landscapes date from a recent but bygone period, during which there was greater harmony between man and the available natural resources than there is today. The preservation of the surviving heath is only justified in nature reserves where an appropriate management is possible.

Yet the measures that must be taken in order to

replace the old practices must first be tried on small areas before being adopted in a management scheme for the whole protected area. The results of some experiments which were started in the Hautes Fagnes region in 1972 (Froment, 1973) are presented here.

The Hautes Fagnes is a high plateau in NE Belgium, along the Eifel. The plateau spreads to the south-west as far as the town Spa. Located in this area is the 'Fagne James' (30 ha) which includes interesting heath vegetation belonging to the sub-association *vaccinietosum uliginosi* (Schwic., 1933) of the association *Calluno-Vaccinietum vitis-idaeae* (Büker, 1942) (according to Schumacker's phytosociological revision, 1973).

The Hautes Fagnes consist of only about 5 000 ha of open areas called 'fagnes' remaining after a century of extensive plantation of spruce (*Picea abies*), which took place when several practices of the old rural economy had become out-of-date. Most of the open landscape consisting of bogs, wet heath, semi-natural grassland and heathland, now belongs to the nature Reserve of the Hautes Fagnes, itself included in the German-Belgian natural Park

---

* Nomenclature follows de Langhe *et al.*, 1978. Nouvelle flore de la Belgique, du Grand-Duché de Luxembourg, du Nord de la France et des Régions voisines, 2e éd.

Vegetatio 47, 193–200 (1981). 0042-3106/81/0471-0193/$1.60.
© Dr W. Junk Publishers, The Hague.

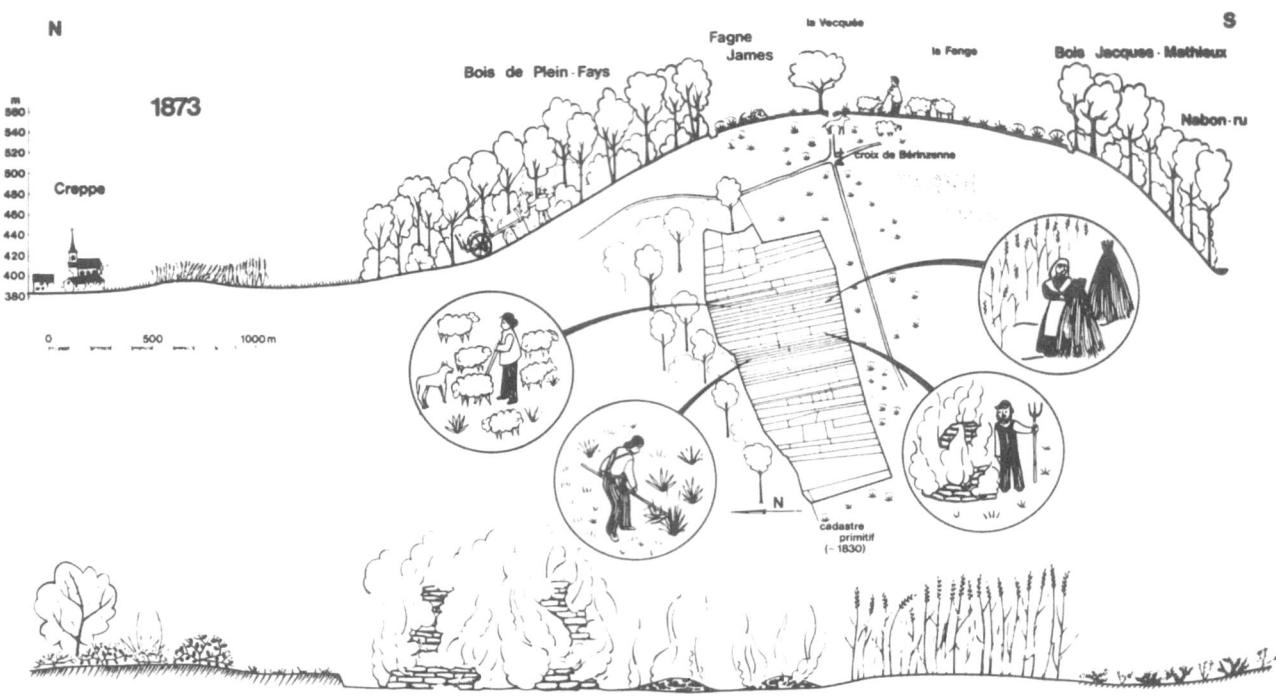

Fig. 1. Transect illustrating the characteristics of the landscape in 1873, with the old agro-pastoral practices.

'Hautes Fagnes-Eifel' (60 000 ha).

The general transformation of the landscape in a hundred years can be illustrated by transects through the heath area in the neighbourhood of the experimental plots.

Experimental plots

Figure 1 illustrates the various agro-pastoral practices in operation in 1873 in the 'Fagne James', a rather exceptional enclave consisting of about a hundred private small patches in the middle of large heath area (Froment & Robert, 1968). It is necessary to describe these practices briefly in order to understand the problems of management in the present situation.

In addition to mowing and grazing by sheep, there was occasional 'burn-beating' cultivation, so that at intervals the people gained an additional harvest. Sods were cut, dried and burnt, and the ashes were spread over the soil as fertilizer for a rye crop sown after preparing the field by creating ridges.

A century later (Fig. 2), only a few private patches remained and both the beech forest on the slopes and the heathland on the plateau have been almost entirely replaced by *Picea abies.*

The various kinds of heath show a spontaneous succession towards natural forest. According to the different kinds of soil and their use, one can distinguish between the ridged heathland, the dry heathland covered with *Calluna* and *Vaccinium* spp. and the wet-heathland with *Erica tetralix, Scirpus cespitosus, Juncus squarrosus,* etc.

The ridged heathland is very much like the dry heathland. It is rich in *Vaccinium myrtillus* and *V. uliginosum,* but in the furrows more grasses appear such as *Molinia caerulea, Deschampsia flexuosa* with various kinds of wet-heath plants.

Four experimental plots were set up in 1972 in this type of heathland:
- plot A (150 m²) : mowing and litter left on the spot
- plot B (150 m²) : mowing and litter removed
- plot C (500 m²) : burning
- plot D (250 m²) : 'burn-beating' (= écobuage)

The experimental area was mapped beforehand,

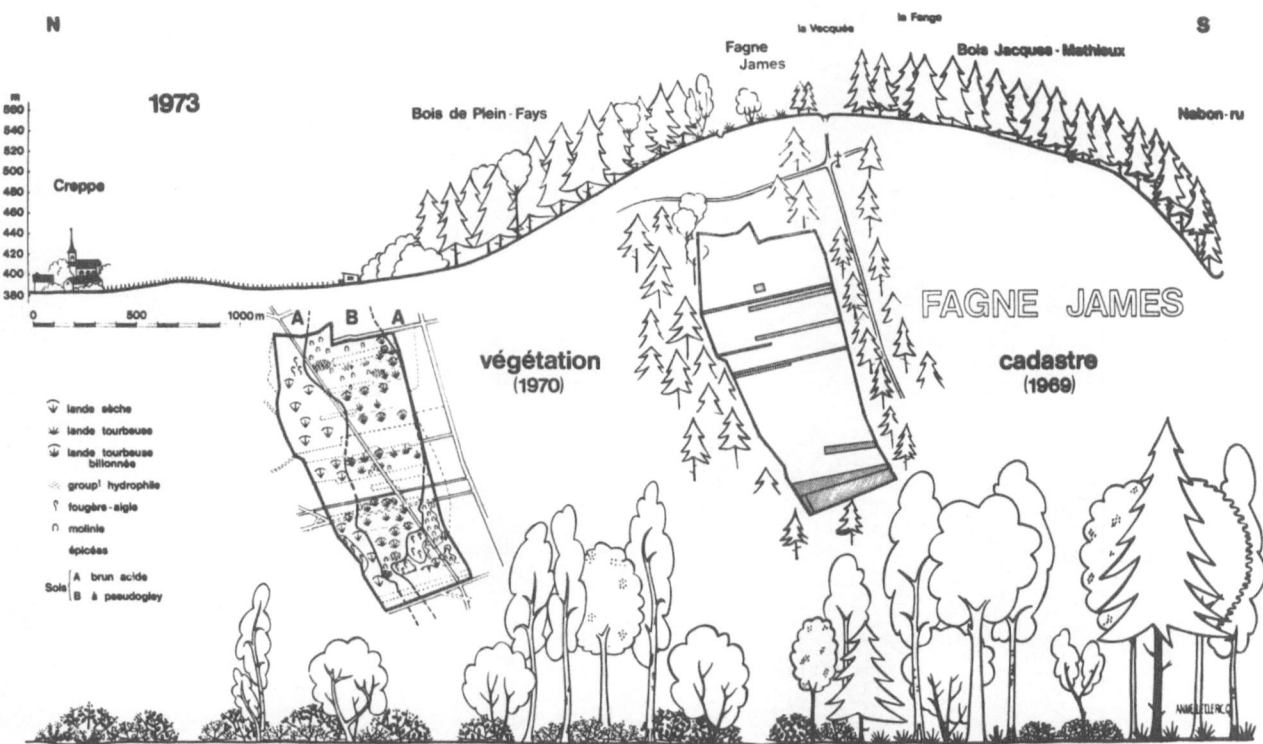

Fig. 2. Transect illustrating the landscape at present and the various stages in the successional development of the heathland of the 'Fagne James'.

and again in 1980. Permanent quadrats were set up in every plot to follow in detail the development of the vegetation.

Results

Burn-beating

Burn-beating eliminates the humus layer almost entirely and by so doing brings about a very different type of development from the other treatments.

There is a big change in the heath vegetation after burn-beating (Fig. 3). In 1972, the greatest part of the experimental area was occupied by a mixed vegetation of *Vaccinium myrtillus* and *V. uliginosum* with some *Calluna vulgaris*. *Vaccinium* areas are fringed with areas where *Molinia caerulea* predominates.

In 1980, however, the vegetation was characterized by two species each covering large areas: *Calluna vulgaris* and the moss *Polytrichum commune*.

The first year after burn-beating the total vegetation cover amounted to about 50% (ranging from 20 to 70%). Two permanent quadrats were established in contrasted 'facies'. Quadrat 1 in the left part of the experimental area being on drier ground, and quadrat 2 in the right part on wetter ground. *Juncus squarrosus,* accompanied by *J. effusus* became particularly abundant during the first years; in both quadrats numerous seedlings of *Calluna vulgaris* were observed. The differentiation between the two quadrats was noticeable up to 1980 (Table 1). At that time quadrat 1 was dominated by *Calluna vulgaris* and *Sarothamnus scoparius* with some *Luzula multiflora* subsp. *congesta*; quadrat 2 was dominated by *Polytrichum commune*. This species appeared in 1976 and quickly replaced the grasses (*Molinia caerulea* and *Deschampsia flexuosa*) as well as *Erica tetralix* and *Sphagnum,*

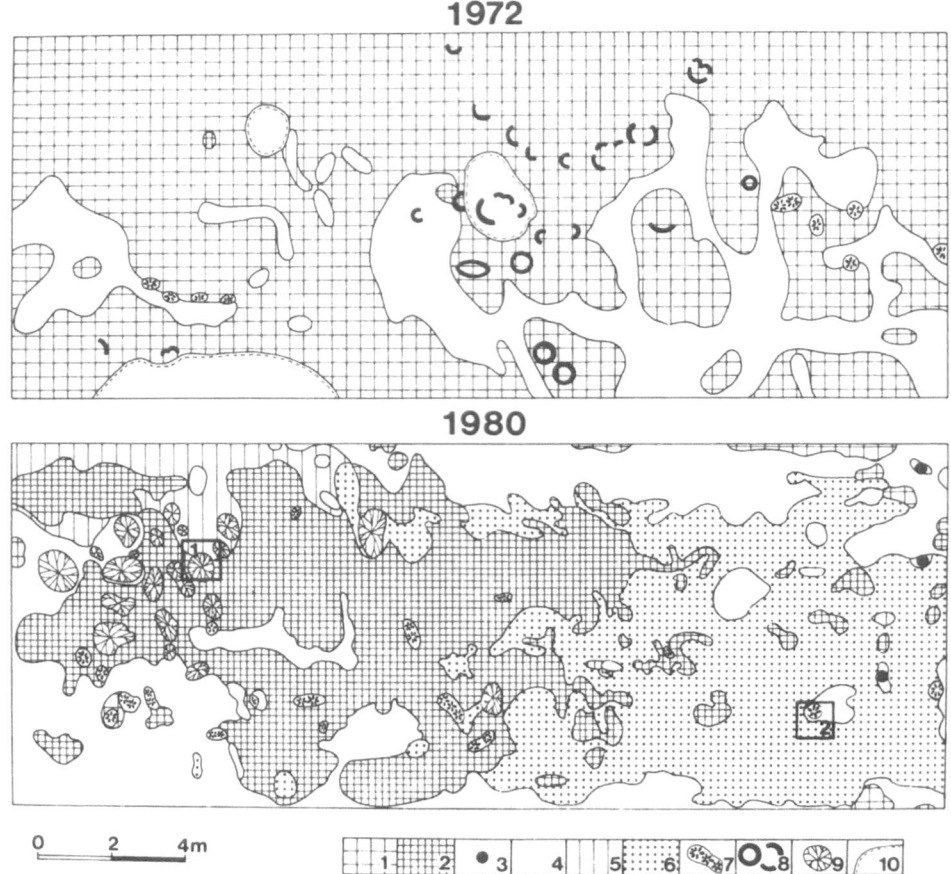

1972

1980

0   2   4m

| 1 | 2 | • 3 | 4 | 5 | 6 | 7 | O 8 | 9 | 10 |

Fig. 3. Comparison of the vegetation before (1972) and after (1980) the operation of burn-beating (plot D). 1: *Vaccinium myrtillus* and *V. uliginosum* mixed; 2: *Calluna vulgaris;* 3: *Erica tetralix;* 4: *Molinia caerulea;* 5: *Deschampsia flexuosa;* 6: *Polytrichum commune;* 7: *Sphagnum* sp.; 8: *Leucobryum glaucum;* 9: *Sarothamnus scoparius;* 10: no vegetation.

Table 1. Vegetation composition in the permanent quadrats (1 and 2) established after the burn-beating treatment.

	QUADRAT 1						QUADRAT 2					
	1973	1975	1976	1977	1978	1980	1973	1975	1976	1977	1978	1980
TOTAL Cover (in %)	20	100	100	100	100	100	70	100	100	100	100	100
Calluna vulgaris	2 (g)	2	3	3$^+$	3	2/3	2 (g)	+	1	1$^-$	1	–
Sarothamnus scoparius	1	1	1	3	2	2	–	–	–	–	–	–
Juncus squarrosus	2	4	4	2	1/2	+	4	4/5	4	3	1	1
Juncus effusus	2	4	4	2/3	1	+	1/2	1	2	2$^-$	1	+
Erica tetralix			+	+	1	1	–	–	2	2$^+$	2	1/2
Molinia caerulea	1	1/2	2	2	1/2	2	2	2/3	2/3	2/3	1	1
Deschampsia flexuosa	–	1	1	1	1/2	2/3	–	–	–	–	–	–
Luzula multiflora subsp. congesta	–				–	1	–	–	–	–	–	–
Sphagnum fimbriatum	–				–	–	1	1/2	2	3	4	2
Polytrichum commune	+	2			–	1/2	–	–	+	1	2	4

g = germination.

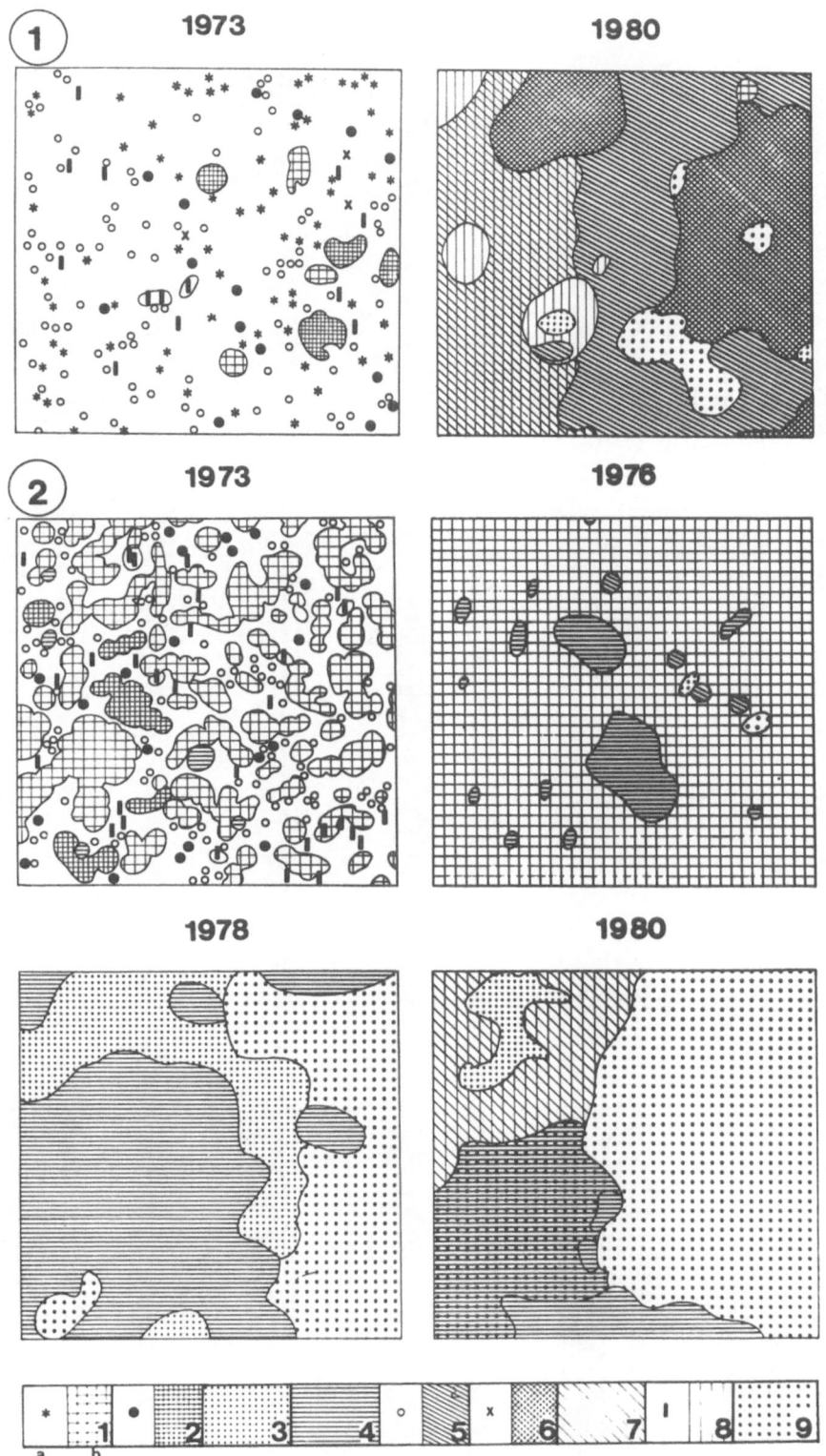

Fig. 4. Comparison of the vegetation in the permanent quadrats 1 and 2 (1 m²) established in two recolonisation 'facies' after burn-beating (plot D). 1: *Juncus squarrosus* (a: isolated; b: in spots); 2: *Juncus effusus;* 3: *Erica tetralix;* 4: *Sphagnum* sp.; 5: *Calluna vulgaris;* 6: *Sarothamnus scoparius;* 7: *Deschampsia flexuosa;* 8: *Molinia caerulea;* 9: *Polytrichum commune.*

198

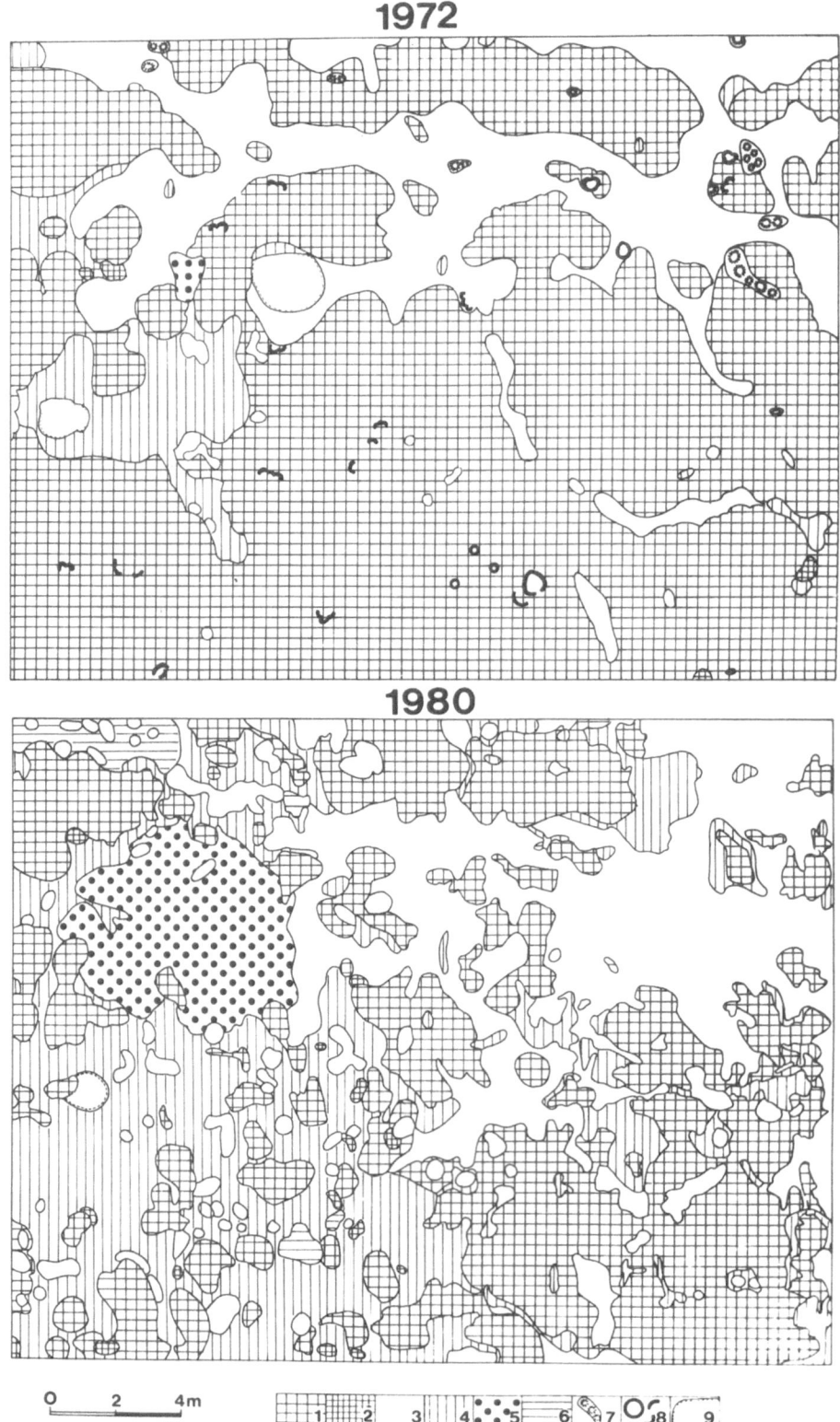

1972

1980

0   2   4m

1 2 3 4 5 6 7 8 9

Fig. 5. Comparison of the vegetation before (1972) and after (1980) burning (plot C). 1: *Vaccinium myrtillus* and *V. uliginosum* mixed; 2: *Calluna vulgaris*; 3: *Molinia caerulea*; 4: *Deschampsia flexuosa*; 5: *Calamagrostis canescens*; 6: *Agrostis canina*.

which were still the predominant species in 1978 (Fig. 4).

The other treatments

The burnt area (Fig. 5) showed a development of *Molinia caerulea* and *Deschampsia flexuosa* with the extension of small patches of *Agrostis canina* and *Calamagrostis canescens*. The regression of *Vaccinium myrtillus* and *V. uliginosum* patches was considerable, as well as their fragmentation, so the general aspect at the present day is more of a mosaic than before the burning. *Calluna vulgaris*

formed a few small isolated patches.

The mown areas with litter left on the spot or removed developed in much the same way as the burnt area (Fig. 6). However, the extension of grasses was less noticeable, compared with the burnt area (Table 2). *Leucobryum glaucum* almost completely disappeared after the burning.

Conclusions

The experiments designed to find an adequate method for the management of old *Calluna vulgaris*

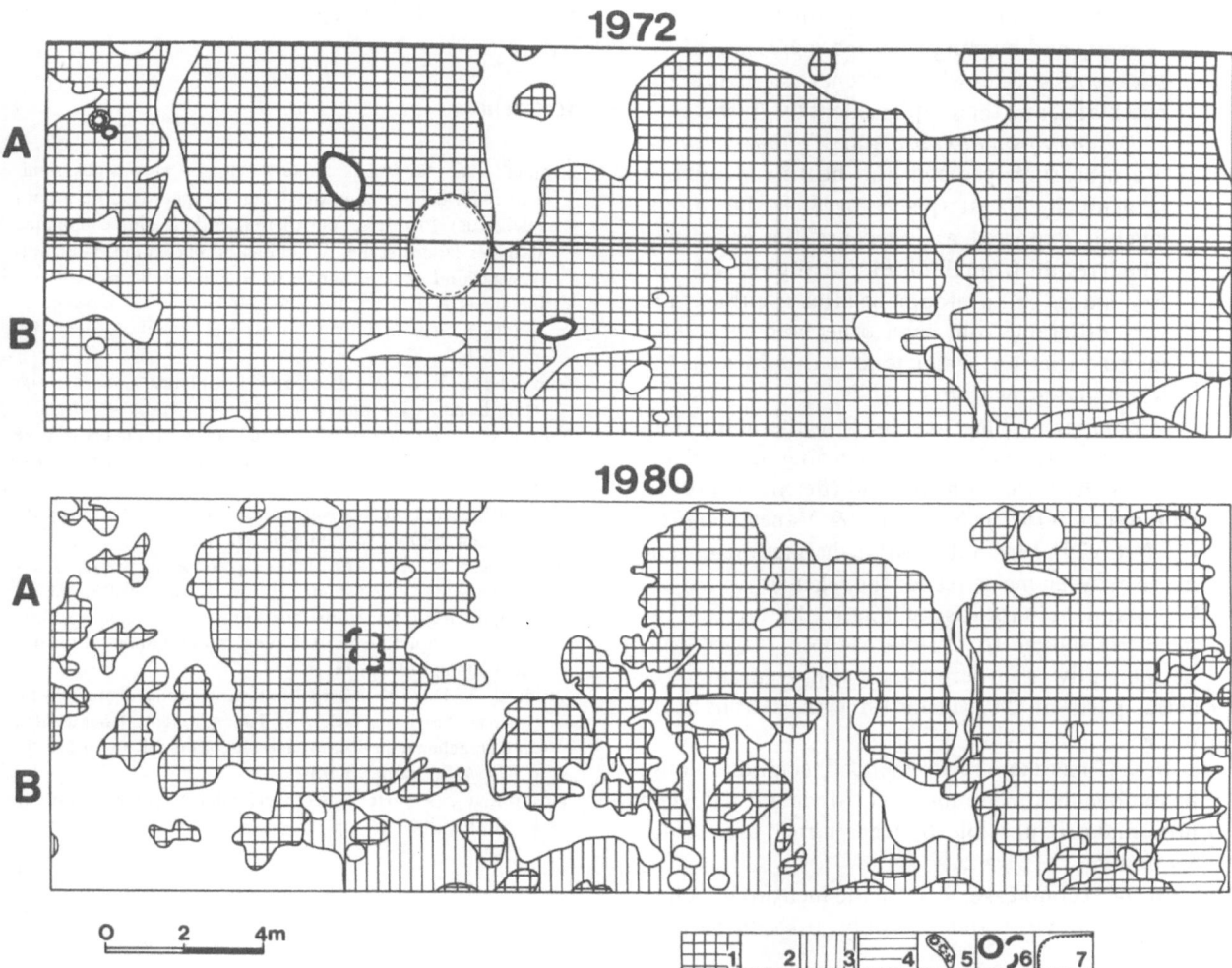

Fig. 6. Comparison between the vegetation before (1972) and after (1980) mowing, litter left on the spot (A) and litter removed (B). 1: *Vaccinium myrtillus* et *V. uliginosum* mixed; 2: *Molinia caerulea;* 3: *Deschampsia flexuosa;* 4: *Agrostis canina;* 5: *Sphagnum* sp.; 6: *Leucobryum glaucum;* 7: no vegetation.

200

Table 2. Cover in 1972 and in 1980 of the predominant species in the mown (A and B) and in the burnt (C) plots (in %).

Experimental plots	Chief species										
	Vaccinium		Molinia		Deschampsia		Calamagrostis		Agrostis		Other species
	1972/1980		1972/1980		1972/1980		1972/1980		1972/1980		1972/1980
A	80	53	18	45	–	2	–	–	–	–	2	<1
B	88	36	7	26	3	36	–	–	–	2	2	<1
C	67	29	22	26	6	38	<1	6	–	1	5	<1

and *Vaccinium* heathland in the Hautes Fagnes, lead us to draw some interesting conclusions.

1) Mowing and burning lead to a regression of the ericaceous species which are characteristic of the heathland, and their replacement by grasses such as *Deschampsia flexuosa* and *Molinia caerulea*.

2) 'Burn-beating' protects the heathland against invasion by forest species because the humus layer is exported and because the ecosystem which results from this is drastically altered. The first stages in development after treatment are very different from each other and represent micro-variations which tend to disappear as the heathland grows older.

In comparing these results with the many data appearing in the literature concerning the management of the *Calluna* heathland (de Smidt (1979), Gimingham (1972), Noirfalise & Vanesse (1976), R.I.N. (1979), Westhoff (1961)), distinction must be made between measures for the regeneration of old heathland, as in the case of 'Fagnes James', and the measures taken to maintain these associations permanently by periodically rejuvenating the aerial parts of the vegetation with various techniques.

In the first case, a recommended technique would be removal of the humus layer. Mechanical devices now make it possible to treat larger areas with economy.

In the second case, less drastic measures such as grazing, mowing, and, possibly, fire can be used with success according to the characteristics of the vegetation to be managed.

References

Froment, A., 1973. Les landes, pelouses et prairies semi-naturelles des plateaux des Hautes Fagnes et d'Elsenborn (Belgique). Problèmes et expériences de gestion écologique dans les landes sèches à Calluna et Vaccinium. Coll. Int. Végétation Landes d'Europe occidentale, Lille, p. 37–45.

Froment, A. & Robert, F., 1968. Un intéressant paysage de landes à Spa: la Fagne James. Revue Belge Géogr. 92, p. 123–148.

Gimingham, C. H., 1972. Ecology of heathlands. Chapman and Hall, London, 266 pp.

Noirfalise, A. & Vanesse, R., 1976. Les landes à bruyère de l'Europe occidentale. Conseil de l'Europe, Strasbourg, 54 pp.

R.I.N. (Rijksinstituut voor Natuurbeheer), 1979. Levensgemeenschappen. Pudoc, Wageningen, 392 pp.

Schumacker, R., 1973. Les landes, pelouses et prairies semi-naturelles des plateaux des Hautes Fagnes et d'Elsenborn (Belgique). Aspects floristiques, phytosociologiques et phytogéographiques. Coll. Int. Végétation Landes d'Europe occidentale, Lille, p. 13–36.

Smidt, J. T. De, 1979. Origin and destruction of north west european heath vegetation. In: Tüxen, R. (ed.): Über werden und vergehen von Pflanzengesellschaften, Rinteln (1978), p. 411–435. Cramer, Vaduz.

Westhoff, V., 1961. Het beheer van heidereservaten. Natuur en Landschap, 14-4, 27 pp.

Accepted 8.7.1981.

Grassland dynamics under sheep grazing in an Australian Mediterranean type climate*

M. P. Austin, O. B. Williams & L. Belbin**
Division of Land Use Research, Institute of Earth Resources, CSIRO Canberra City, ACT, 2601, Australia

Keywords: Numerical classification, Ordination, Plant demography, Population dynamics, Sheep grazing, Succession, Vegetation dynamics

Abstract

Grassland dynamics in a degraded disclimax grassland dominated by *Danthonia caespitosa* Gaudich. are examined using both demographic and multivariate approaches in an experiment designed to determine the effect of grazing intensity and exclosure on pasture dynamics. The experiment ran for 20 years from 1949 to 1968, using permanent quadrats at 3 grazing intensities and within exclosures. Demographic studies of some perennial grass species demonstrated markedly different responses to grazing; *Danthonia caespitosa* was unaffected by grazing but responsive to seasonal rainfall differences. *Enteropogon acicularis* survived only on protected sites. Numerical classification of total species set (121 species) for six observation periods demonstrated that community types were sensitive to differences in winter rainfall, and time since the start of experiment. Principal component analysis of permanent quadrat observations for individual years demonstrates quadrat trajectories which confirm this and indicate progressive divergence of the successional trends of the grazed and ungrazed quadrats. Repeated analysis on grazed quadrats only, shows that three components of pasture dynamics can be recognized; these are trend (succession?) and seasonal differences, each of which account for about 20% of the variance, and differences due to soil heterogeneity in the experimental paddock (8% of variance accounted for). No effect of grazing intensity was detected. Multivariate techniques can provide a clear partitioning of types of dynamic behaviour present in grassland communities. It is concluded that partitioning of environmental heterogeneity prior to demographic studies would increase their sensitivity.

Introduction

There are few detailed studies published of community dynamics, covering a sufficient period for the differential effects of climatic, environmental and experimental factors to be separated. This paper is an initial attempt at a descriptive analysis of such a study where some partitioning of the different effects can be achieved.

*Nomenclature follows Jacobs, S. W. L. & J. Pickard, 1981. Plants of New South Wales. New South Wales Government Printer, Sydney.
**We thank C. Helmann and A. Howard for their help with data preparation and analysis.

Currently there are two approaches to vegetation dynamics, the observation of total density, cover, or frequency for a quadrat over time, and the determination of the survival time of individual plants, together with the construction of life-tables. The latter, a demographic approach, is only feasible for a few species in any community because of the need to recognize individuals and the heavy labour requirement. Williams (Williams, 1970; Williams & Roe, 1975) has applied this approach to certain perennial grasses in the community studied here but only limited study of the behaviour of other species has been achieved (Williams, 1969). Not until multivariate techniques became available was there

Vegetatio 47, 201–211 (1981).   0042-3106/81/0471-0201/$2.20.
© Dr W. Junk Publishers, The Hague.

the possibility of examining community trends using all species simultaneously (Austin, 1977). This allows the summary and descriptive partitioning of various effects on total density or presence/absence.

The application of both approaches to a grazing experiment on a pasture in an Australian Mediterranean type are briefly considered here. The experimental site is approximately 20 km NE of Deniliquin (lat. 35°33'S long. 145°0'E) in SW New South Wales, Australia and is part of the extensive riverine plain associated with the Murray-Darling basin.

Plant community

The plant community is degraded disclimax grassland dominated by *Danthonia caespitosa* Gaudich. in which 136 annual and perennial grasses and non-grasses have been recorded (see also Moore, 1953a, b; Groves & Williams, 1981). Many of these species were introduced accidentally to Australia following the European invasion of the early 1800s and are now naturalized. The basal area of perennial species is generally less than 2% and the carrying capacity is about 1 Merino sheep per 0.8 ha. The original dominant was probably *Atriplex nummularia,* which with *A. vesicaria* and *Maireana aphylla* formed a chenopodiaceous low shrubland or shrub-steppe; in present-day remnants of this community *D. caespitosa* is a minor constituent and is represented by small plants. However when the experiment we describe was started, this grassland was defined (Beadle, 1948) as a *Chloris truncata – Danthonia semiannularis* (syn. *caespitosa*) consocies within a *Chloris – Danthonia* association.

Previous publications

Since 1946 the *Danthonia caespitosa* Gaudich. grassland at Deniliquin has been the subject of several research papers, on plants (Williams, 1961), soils (Williams, 1956), the reaction of plants to grazing (Williams, 1974), to fertilizer (Tupper, 1978), dietary selection by Merino sheep (Robards *et al.,* 1967) and the effect on animal production of reintroducing indigenous shrubby species into the grassland (Leigh *et al.,* 1970).

The previous papers on this experiment have concentrated on treatment effects on the dominant grass *Danthonia caespitosa,* survivorship of cohorts of the dominant and *Chloris* (syn. *Enteropogon*) *acicularis,* and density changes due to grazing treatments and to drought (Williams, 1968, 1969, 1970, 1974). Selected life-tables have been published for *Danthonia caespitosa, Enteropogon acicularis* and *Stipa variabilis* (Williams & Roe, 1975).

Environment

The topography of the region is a level plain with a slope of 3 cm per km to the SW. Local soils were described as Riverina and Billabong clays and the variable occurrence of the gilgai microrelief was noted (Smith, 1945). The gilgai microrelief on the experiment has been described (Williams, 1955) as having 3 levels, the intermediate level or shelf with a massive level surface showing few cracks, the depression which may be as much as 10–20 cm below the shelf and which has a massive surface structure with occasional cracks, and the puff which rises as mounds 30–50 cm above the level of the shelf and is coarse granular and self mulching. A more subtle type of gilgai pattern in this same grassland has been described by Warren Wilson & Leigh (1964). The soils of the research station were mapped by Johnston & Butler (1946). Two of their categories were mapped on the experimental site, R (Riverina clay) and H (Billabong clay). The soil moisture characteritics of significance have been

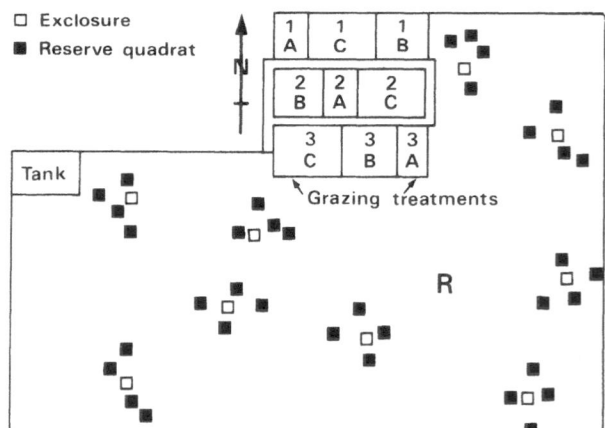

Fig. 1. Diagrammatic plan of experimental paddock. A, heavy grazing; B, medium grazing; C, light grazing; R, reserve with medium grazing.

described by Williams (1955, 1956) and Warren Wilson & Leigh (1964).

The experimental site lies between the 300 and 450 mm annual rainfall isohyets in an area with predominantly winter rainfall (approximately 60%) but with occasional effective summer rains. Only 4% of years receive less than 125 mm in winter and rainfall reliability is high from May to October, but 28% of summers receive less than this. Summer rains are less effective than winter rains because evaporation rates are much higher (50 mm July, 250 mm January). January is the hottest month with a mean maximum of 32.0 °C and a mean minimum of 16.1 °C; July is the coldest month with a mean maximum of 14.0 °C and a mean minimum of 3.9 °C. The average frost-free period is 252 days. On average there are 4 severe frosts (screen temperature below 0 °C) and 18 light frosts (between 0 °C and 2.2 °C).

Methods

Original experimental objectives and design

The experimental objective in the March 1949 preschedule was 'to determine quantitatively the change in condition, the persistence and succession of species, and some measure of a *Danthonia semiannularis* pasture when it is subjected to various levels of grazing intensity by sheep'. 'Condition' was defined as 'the relationship of the current state of the pasture to its maximum potential productivity'. A quantitative evaluation of condition 'requires the measurement of *density of the stable components of the pasture and their age distribution as an index of vigour*'.

The experimental layout consisted of a large reserve paddock of 38.3 ha (96 ac) on which 60 Merino wethers were grazed for 24 days of each 28-day cycle (Fig. 1). For the remaining 4 days of this cycle, 45 of the wethers (15 remained on reserve) were divided into 9 groups of 5 and these groups were placed in plots of 0.36, 0.53 and 0.69 ha; these plots were replicated 3 times. The calculated stocking rates on an annual basis were one sheep to 0.49 (heavy), 0.73 (medium) and 0.97 ha (light) in the small plots and one sheep to 0.73 ha (medium) on the reserve paddock. There was one exclosure of 0.1 ha with 10 × 1 m² permanent

quadrats in each plot, and 9 exclosures with 4 × 1 m² quadrats in the reserve paddock. The plots contained 20 × 1 m² permanent quadrats (heavy), 18 × 1 m² (medium) and 16 × 1 m² (light); the reserve paddock had 4 × m² permanent quadrats near each exclosure. There were 324 permanent quadrats in the experiment.

Grazing pressure treatments gave those levels accepted as average for this grassland and were maintained from 1949 to 1968 with one important exception – a drought period – which was so severe that not even a 2 sheep flock could be maintained on the 38.3 ha. Sheep were not replaced in the experiment at the breaking of the drought in autumn 1968 so that the recovery of the flora after drought could be assessed (see Williams, 1974).

The seedlings and original plants of the perennial grass species were charted with a pantograph to measure their survival. In order to provide the age distribution data the perennial grasses were charted in November 1949, July 1950, January and August 1951, January 1953, April and November 1954, November 1957, May 1961, October 1965 and October 1968. Occasional counts of all 136 species were made; it is these occasional counts that are the principal subject of the present paper.

Data and methods for this study

The multivariate data set consisted of density (plants/m²) estimates for 121 species (excluding those which occurred only once), for 324 quadrats for six years (November 1949, August 1951, November 1954, September 1961 (247 only recorded), October 1965, October 1968). The total data set was subjected to divisive information analysis (Lance & Williams, 1968) using presence/absence and principal component analysis (PCA) of the variance-covariance matrix of log transformed density data, i.e. an R-analysis of the species values was carried out. A discussion of the use of principal components analysis in succession studies is provided by Austin (1977) and of the use of divisive information analysis by Austin (1980).

By treating the data recorded for a quadrat in a particular year as observations, the PCA produces an ordination in which the records for individual years of each of the permanent quadrats appear as points (i.e. 1967 separate observations). These points can be joined up to indicate the trajectory of a permanent quadrat through time in the floristic space defined by the axes of the PCA. The trajectories can then be examined for patterns of behaviour among the permanent quadrats. Subsequently the exclosured and grazed quadrats were analysed separately because of their markedly different trajectories.

Results

Population dynamics

For purposes of comparison, the demographic results of Williams (1970) for two of the perennial grasses are briefly reported here.

Danthonia caespitosa

Differences in density between the light, medium, and heavy grazing treatments for the earlier years of the experiment were small, so the data for each of the crops of seedlings in the 198 grazed quadrats were pooled. Data for the crops in the 126 protected quadrats were also pooled.

Wide differences in mortality were observed between pooled crops. Data which represent short and long-lived crops are shown in Figure 2. Mortality progressed with a decreasing exponential rate with age for the short-lived 1950 crop, and this population exhibited a half-life of 2–4 months. In the 1952 crop, the survival was much greater, with a half-life of ca. 20 months. Ca. 12 plants/100 m² belonging to the 1952 crop were still alive in the grazed and protected quadrats after 160 months, whereas few plants of the 1950 crop in either grazed or protected quadrats survived more than 60 months.

Most of the remaining single and grouped crops display patterns of survival which are similar to the 1952 crop. In general terms, the overall half-life of these crops lies between 13 and 18 months. There is a tendency in all crops for grazing to favour survival. Relative to 1949, there has been a substantial decline in plants in grazed quadrats from 72 to 33/m², and in protected quadrats from 66 to 5/m².

Enteropogon acicularis

The density data for the various grazing treatments were combined because the number of plants in each treatment was too small for a separate

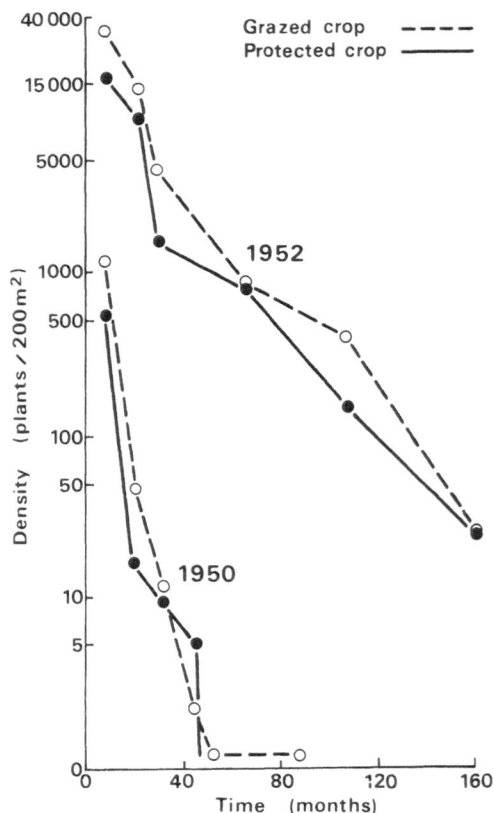

Fig. 2. Survival of *Danthonia caespitosa* in grazed and protected disclimax grassland for the 1950 and 1952 crops.

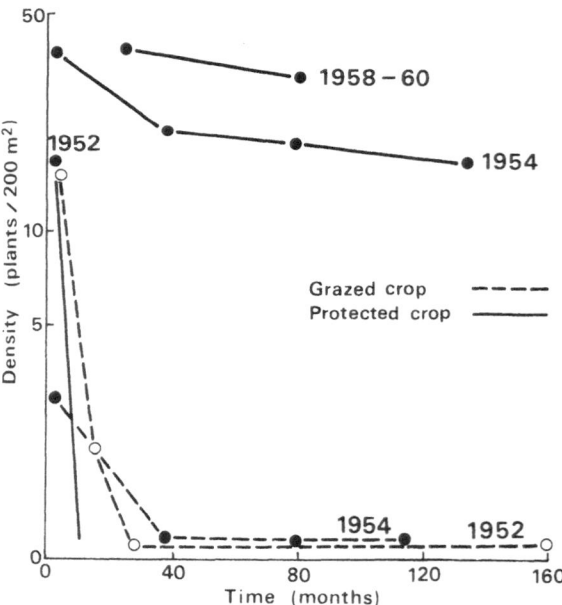

Fig. 3. Survival of *Enteropogon acicularis* in grazed and protected disclimax grassland.

analysis. Representative data for the pooled grazed and pooled protected crops for 1952 and 1954, and the protected crops of 1958–60 are set out in Figure 3.

Every grazed crop exhibited heavy mortality in the juvenile stage. The half-life was 2 months. This heavy mortality appeared in the protected crops only in the first 5 years of the experiment. There were a few long-lived survivors in both treatments. From 1954 onwards, the protected crops were larger, and the initial mortality was reduced. The half-life of these later populations was ca. 24 years.

Relative to 1949, there has been a substantial decline in the total number of plants within grazed quadrats from 20 to 8/100 m², and an increase within the protected quadrats from 12 to 48/100 m².

Classification analysis

The results of classifying the full data set with a divisive information analysis are shown in Figure 4, where the pattern of 'communities' associated with particular years is indicated. The initial division is on *Plantago varia,* a native, annual herb which characterized the wetter years and became generally distributed in the later years of the experiment. Introduced herbs, *Medicago polymorpha* and *Hypochaeris radicata* are other important division species. There is no evidence to show any marked effect of soil or grazing treatment on these groups. 'Community' D (–*P. varia* +*M. polymorpha* +*Avena fatua*) is characteristic of exclosures in the later years of the experiment though it is not confined to the exclosures. Each year is different (Table 1) though some years show similarities

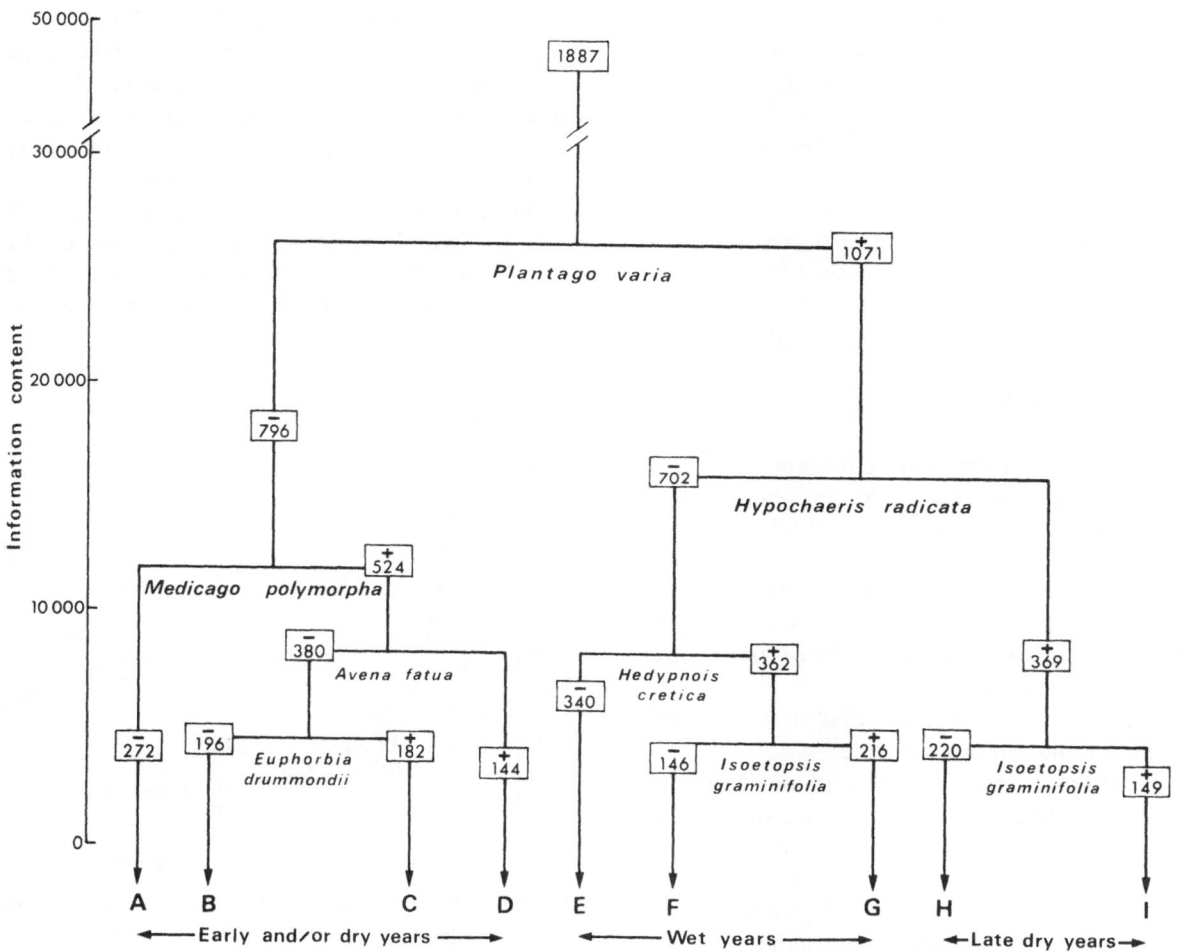

Fig. 4. Divisive information analysis classification of all permanent quadrats recorded on six separate occasions during the period 1949–1968.

Table 1. Relationship between numerical classification communities and years. Showing frequency of each community in a year as a percentage of total occurrences of the community.

Community		year 1949	1951	1954	1961	1965	1968
A	−*Plantago varia,* −*Medicago polymorpha*	<u>80.1</u>	2.2	9.6	1.5	6.6	–
E	+*P. varia,* −*Hypochaeris radicata, Hedypnois cretica*	*12.4*	<u>72.7</u>	4.7	0.6	1.8	8.3
B	+*P. varia,* +*M. polymorpha, Avena fatua, Euphorbia drummondii*	*23.7*	10.3	<u>36.1</u>	11.3	17.0	1.5
C	−*P. varia,* +*M. polymorpha, A. fatua,* +*E. drummondii*	*8.8*	–	<u>89.6</u>	1.6	–	–
I	+*P. varia,* +*H. radicata,* +*Isoetopsis graminifolia*	–	–	6.0	<u>59.1</u>	*24.2*	10.7
H	+*P. varia,* +*H. radicata,* −*I. graminifolia*	–	–	3.2	42.4	<u>49.8</u>	4.6
D	−*P. varia,* +*M. polymorpha,* +*A. fatua*	–	0.7	7.6	20.8	<u>52.1</u>	*18.8*
F	+*P. varia,* −*H. radicata,* +*H. cretica, I. graminifolia*	–	*23.6*	9.7	2.1	26.4	<u>38.2</u>
G	+*P. varia,* −*H. radicata,* +*H. cretica,* +*I. graminifolia*	–	7.9	1.9	0.5	4.7	<u>85.0</u>
	Rainfall for May–July (mm)	264	715	166	276	293	449

Year of primary occurrence underlined.
Year of secondary occurrence italics.

which can be correlated with the winter rainfall. This can be shown in Figure 5 where 4 communities are plotted against rainfall.

Each 'community' can be associated with either an early or late stage in the experiment, and/or with a particular level of early winter rainfall. In particular, the two highest rainfall years for which species records are available show markedly different 'community' composition. The classification suggests that a trend (succession?) exists which is relatively independent of the yearly differences, but that these seasonal differences are relatively important compared to soil and grazing treatments.

Principal components analysis

The behaviour of the different years is indicated diagrammatically in Figure 6. We observe a trend in the grazed plots which progressively diverge from the exclosures, a result which becomes particularly marked in 1968. Note the progressive elongation of the cloud of points for the years 1961, 1965 and 1968. A complex pattern is formed from three distinct dynamic components:

(1) a trend (succession) from 1949 to 1968 in both grazed and ungrazed quadrats
(2) a difference between grazed and ungrazed quadrats (elongation of the single year clouds)
(3) a seasonal response to weather conditions in the winter of the particular year; note the adjacent positions of 1951 and 1968 (years with wet winters) in the upper right-hand quadrant. This occurred even though the grazed plots were ungrazed in the winter of 1968.

The complexity can be exposed further by examining the behaviour of individual quadrats in relation to components two and three (Fig. 7). Here one diagonal relates to seasonal differences while the other indicates that exclosed quadrats behave very differently from the grazed plots. Note the consistency of the two grazed quadrats which were similar in soil type and gilgai position (Fig. 7).

When the above ordination is repeated but with only the grazed quadrats, one dimension of the variability (effect of exclosure) is removed and other effects can then be observed (Fig. 8). The first

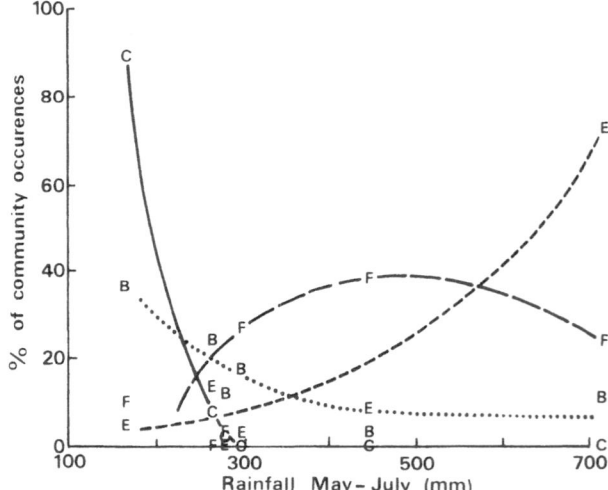

Fig. 5. Community frequency for four communities in relation to 'early winter' rainfall. Lines have been drawn subjectively. Frequency in a year is measured as the percentage of total occurrences of the community. See Table 1 for meaning of letters.

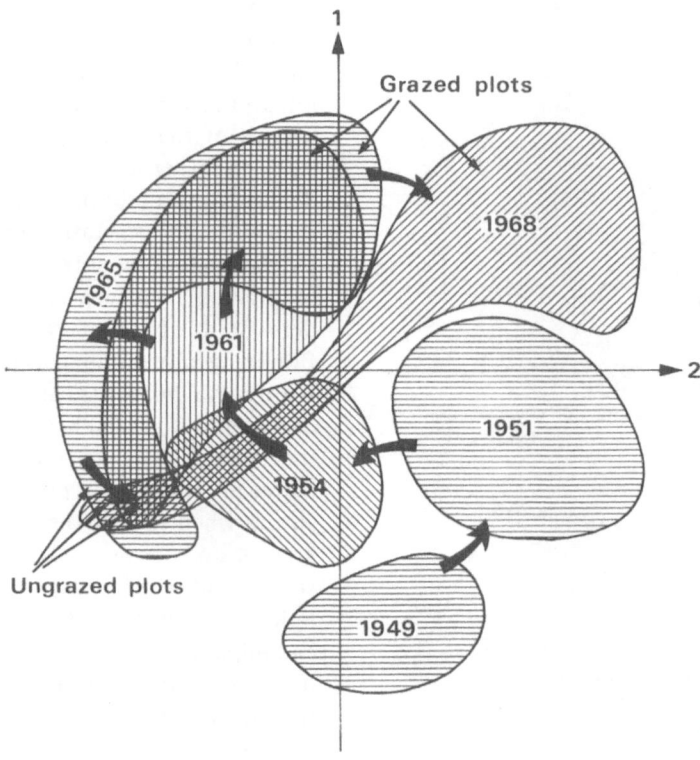

Fig. 6. Diagrammatic distribution of quadrats from different years on the first two principal component axes.

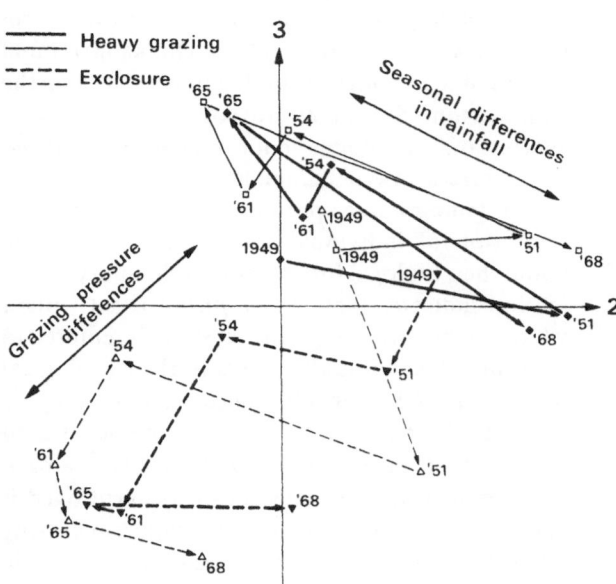

Fig. 7. Trajectories for selected individual quadrats, from the heavy grazing plot (1A) on soil type H.

and second components which are nearly equal again reflect trend and seasonality with the interpreted dimensions being at an angle to the mathematical axes as in Figures 6 and 7. The third axis (Fig. 8) now reflects soil differences. The conclusions are:

(1) for trend (succession?) and seasonal effects to be approximately equal; the first component (predominantly trend) accounts for 21.2% and the second (predominantly seasonality) for 17.2% of the total variance.

(2) These effects are three times as important as the third (soil) component which accounts for 6.8% of the variance.

(3) Grazing intensity treatment effects are less important than the soil effect; no treatment effects were detected in relation to the first three components.

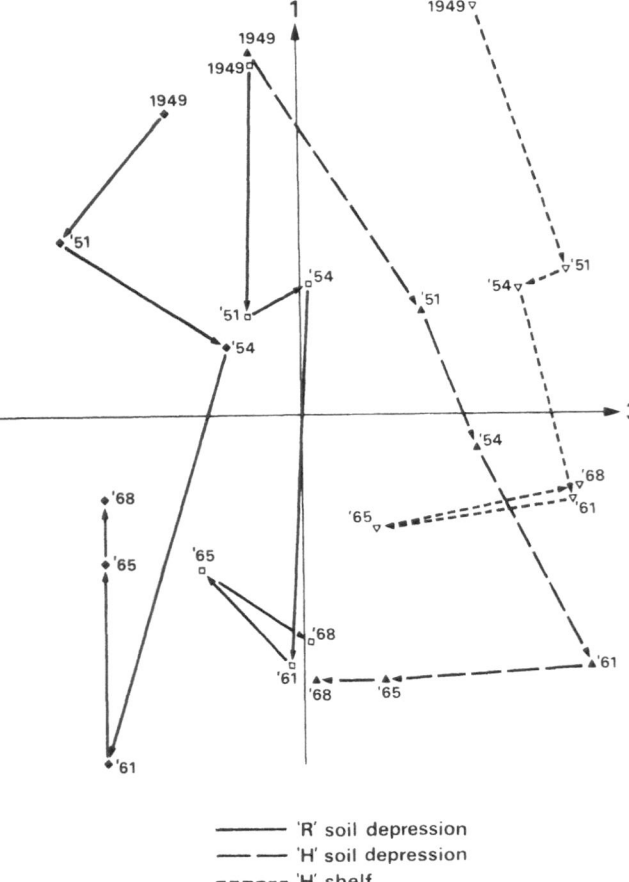

Fig. 8. Trajectories for selected individual quadrats for components one (trend) and three (soil difference) after eliminating quadrats from exclosures. Quadrats were selected from the most extreme soils, i.e. 'H' soil with gilgai shelf and 'R' soil with gilgai depression; trajectories of most other quadrats fall between these reflecting a soil continuum.

Discussion

In this Australian semi-arid environment, differences in the seasonal conditions between years are of major significance in determining vegetation dynamics. An important determinant is the time of onset of the first effective rains, whether in late summer, autumn or early winter, and the nature of the follow-up rains after this initial fall; soil differences can be observed to modify the type of response. In spite of this variability, markedly different trends (succession?) can be discerned in exclosures compared with grazed plots throughout the period of study; for example, *D. caespitosa* can no longer establish in exclosures.

There is marked contrast in the population dynamics of the two species, *Danthonia caespitosa* and *Enteropogon acicularis,* one of the few remaining representatives of the original climax species of the region; protection of *E. acicularis* leads to larger and longer-lived populations. The demography of both grasses is an expression of their different vegetative and flowering phenology in relation to seasonal rainfall (Williams, 1961), and to grazing by sheep. *E. acicularis* is repeatedly defoliated whenever rain falls in summer, and flowering is prohibited; its growth in winter is poor. Although *Danthonia caespitosa* is a preferred item of diet and is grazed heavily (Robards *et al.,* 1967) the sheep cannot remove all photosynthetic material. This intense pressure is reduced in winter-spring when 60% of the diet is made up of annual grasses and herbs – plants which grow in dense stands and which compete with *D. caespitosa* for light and water. The respite from grazing encompasses anthesis and seed set providing an opportunity for reproduction and hence persistence; in addition, *D. caespitosa* can produce leaf in all seasons.

Williams (1961) has discussed the relation between climate and phenology in this grassland. The vegetation has four major groups of species whose phenology can be described in terms of separate suites of perennial and annual species that take advantage of rain in cool seasons or warm seasons, a core of 'original' or climax perennial species that do not grow, or do not grow well, in the cool season (Specht & Rayson, 1957; Holland, 1968), and abundant cool season naturalized (Mediterranean type) annuals.

The demographic approach allows detailed consideration of individual species and speculation about their biology; for example preliminary examination indicates that *D. caespitosa* is a C_3 plant while *E. acicularis* is C_4. Further study is required to determine what role this plays in their ecological behaviour. Whether difference in demographic response in relation to the now demonstrated factors of soil and season can be detected remains to be examined. If these effects can be partitioned it may be feasible to explore whether different grazing intensities may have an influence on demographic behaviour.

To determine the demographic characteristics of all 121 species would not be practical. In order to better explore such large data sets for insight into the vegetation dynamics they recorded, multivariate techniques are necessary.

The ordinations demonstrate a partitioning between:

(a) Seasonal differences between years
(b) Trend in vegetation composition from 1949–1968
(c) Difference between no grazing and grazing in both seasonal response and trend
(d) Differences between soil types in type of trend.

Further differences due to grazing level and interactions between grazing pressure (heavy, medium, light) and season are suspected. In addition, there are known differences between quadrats based on whether they occur in the depressions or shelf of the gilgai microrelief (Williams, 1955; Warren Wilson & Leigh, 1964).

These results have several implications for the design of vegetation dynamics studies.

(1) Climatic fluctuations and their interactions with environment and treatment are an integral part of the interpretation of analysis and not optional extras. Major climatic events preceding the imposition of treatments can be influential.

(2) Stratification based on the initially observed properties of the vegetation in the permanent quadrats may improve standard statistical analysis by removing otherwise unsuspected heterogeneity and allowing differences in dynamics on the different soils to be detected (cf. Williams, 1969).

(3) The degree of environment and treatment dependency (climate, grazing factor) may be assessed from the trajectories observed for permanent quadrats recorded in different plots.

The ecological status of the plant community, the demographic and phenological performance, and production of its components can then be interpreted in terms of these trajectories. One expects marked differences between disclimax grasslands in which annual species and short-lived perennials dominate compared with climax grasslands such as those dominated by species of the perennial grasses *Astrebla, Triodia, Heteropogon* and *Themeda*. The demographic information already available, for example in Williams & Roe (1975), Williams (1977), Crisp & Lange (1976), Crisp (1978), Michalk & Herbert (1978), Davies & Walsh (1979) for Australian species, and in Young & Evans (1974), Wright & Van Dyne (1976),

Norton (1978) and West *et al.* (1979) for north American species can then be used to extrapolate to related communities. The review by Hodgkinson & Williams (1980) and in particular the field studies by Jones & Mott (1980) which elucidate the demographic picture for sown pasture species in new environments also indicate how other information can be incorporated. The descriptive use of multivariate techniques combined with such studies using permanent quadrats could provide an extremely useful combination for the analysis of long term experiments.

Other studies, in particular that by Norton (1978) of permanent quadrats in which plant species were mapped over a 40-year period 1935–1975 have shown that plant species may behave quite contrary to accepted range management theory: 'Inherent plant longevity, opportunity for plant replacement and differential response to climatic pattern may be more influential factors than grazing stress' (Norton, 1978).

(4) In the modelling of vegetation dynamics, if year to year variation is as important as trends (Figs. 6 and 8) then the transition matrix approach cannot work unless the transitions can be expressed as a function of seasonal differences (cf. Usher, 1979, 1981; Austin, 1980; Austin & Belbin, 1981). If simple physiological models are developed, then for communities with a large annual component as in the Australian climate studied here, the critical phase will be seedling establishment under different seasonal conditions rather than the growth processes of mature plants which can be assumed similar from year to year. Functions for such establishment phenomena may be less easily determined than photosynthetic responses particularly because establishment will depend on the perennial component of the vegetation, to quote Young & Evans (1974): 'an adequate perennial-grass understory is a dampening influence on extreme seedling establishment by shrubs or annual grasses'. In climatic zones like that of Deniliquin, where species with markedly different temperature responses are occurring together, models will need to include additional variability (cf. Shugart & Noble, in press).

(5) Where environmental (climatic) factors assume the importance described here, vegetation dynamics studies required to sample significant ecological events are unlikely to be short-term. New

experimental designs for permanent quadrat studies which simulate these extreme events are needed if experiments are not to take a lifetime.

We conclude that studies for pasture management could be improved by the application of vegetation dynamics concepts allied with appropriate experimental designs (cf. Austin (this symposium)).

References

Austin, M. P., 1977. Use of ordination and other multivariate descriptive methods to study succession. Vegetatio 35: 165–175.

Austin, M. P., 1980. Exploratory analysis of grassland dynamics: an example of a lawn succession. Vegetatio 43: 87–94.

Austin, M. P., 1981. Permanent quadrats: an interface for theory and practice. Vegetatio 46: 1–10.

Austin, M. P. & Belbin, L., 1981. An analysis of succession along an environmental gradient using a lawn as a test area. Vegetatio 46: 19–30.

Beadle, N. C. W., 1948. The vegetation and pastures of western New South Wales, with special reference to soil erosion. Government Printer, Sydney. 281 p.

Crisp, M. D., 1978. Demography and survival under grazing of three Australian semi-desert shrubs. Oikos 30: 520–528.

Crisp, M. D. & Lange, R. T., 1976. Age-structure, distribution and survival under grazing of the arid zone shrub Acacia burkittii. Oikos 27: 86–92.

Davies, S. J. J. F. & Walsh, T. F. M., 1979. Observations on the regeneration of shrubs and woody forbs over a two-year period in grazed quadrats on Mileura Station, Western Australia. Aust. Rangeland J. 1: 215–224.

Groves, R. H. & Williams, O. B., 1981. Natural Grasslands. In: (R. H. Groves, (ed)) Australian Vegetation, Cambridge Univ. Press, Cambridge. p. 293–316.

Hodgkinson, K. C. & Williams, O. B. in press. Adaptation in forage plants: adaptation to grazing. In: R. A. Bray and J. G. McIver (eds.) Genetic resources of forage plants. CSIRO, Melbourne.

Holland, P. G., 1968. Seasonal growth of field layer plants in two stands of mallee vegetation. Aust. J. Bot. 16: 615–622.

Johnston, E. J. & Butler, B. E., 1946. Report on soils of 'Pine Lodge' estate, parishes of Yalgadori, Woperana and Ulupna, Counties of Townsend and Denison, New South Wales. Council for Science and Industrial Research Div. of Soils Div. Rep. 8/46.

Jones, R. M., 1980. Survival of seedlings and primary tap roots of white clover (Trifolium repens) in subtropical pastures in Southeast Queensland. Tropical Grasslands 14: 19–22.

Jones, R. M. & Mott, J. J., 1980. Population dynamics in grazed pastures. Tropical Grasslands 14: 218–224.

Lance, G. N. & Williams, W. T., 1968. Note on a new information-statistic classificatory program. Computer J. 11: 195.

Leigh, J. H., Wilson, A. D. & Williams, O. B., 1970. An assessment of the values of three perennial chenopodiaceous

shrubs for food production of sheep grazing semi-arid pastures. In: Proc. of the XI Intern. Grassland Congress, Surfer's Paradise, pp. 55–59. Qld. Univ. Press, St. Lucia.

Michalk, D. L. & Herbert, P. K., 1978. The effects of grazing and season on the stability of Chloris spp. (windmill grasses) in natural pasture at Trangie, New South Wales. Aust. Rangeland J. 2: 106–111.

Moore, C. W. E., 1953a. The vegetation of the south-eastern Riverina, New South Wales. I. The climax communities. Aust. J. Bot. 1: 485–547.

Moore, C. W. E., 1953b. The vegetation of the south-eastern Riverina, New South Wales. II. The disclimax communities. Aust. J. Bot. 1: 548–567.

Norton, B. E., 1978. The impact of sheep grazing on long-term successional trends in salt desert shrub vegetation of south western Utah. Proc. of the First Intern. Rangeland Congress, Denver, Colo. pp. 610–613.

Robards, G. E., Leigh, J. H. & Mulham, W. E., 1967. Selection of diet by sheep grazing semi-arid pastures of the Riverine Plain. IV. A grassland (Danthonia caespitosa) community. Aust. J. Exp. Agric. and Anim. Husbandry 7: 426–433.

Shugart, H. H. & Noble, I. R. in press. A computer model of succession and fire response of the high altitude Eucalyptus forests of the Brindabella Range, Australian Capital Territory. Aust. J. Ecol. 6.

Smith, R., 1945. Soils of the Berriquin Irrigation District, N.S.W. Bull. Commonw. Scient. Ind. Res. Org. 189.

Specht, R. L. & Rayson, P., 1957. Dark Island Heath (Ninety-Mile Plain, South Australia). 1. Definition of the ecosystem. Austr. J. Bot. 5: 52–85.

Tupper, G. J., 1978. Effects of nitrogen and phosphorus fertilizers and gypsum on a Danthonia caespitosa - Stipa variabilis grassland. I. Response to fertilizer applications over four successive years. Aust. J. of Exp. Agric. and Anim. Husbandry 18: 253–261.

Usher, M. B., 1979. Markovian approaches to ecological succession. J. Anim. Ecol. 48: 413–426.

Usher, M.B., 1981. (This symposium).

Warren Wilson, J. & Leigh, J. H., 1969. Vegetation patterns on an unusual gilgai soil in New South Wales. J. Ecol. 52: 379–389.

West, N. E., Rea, K. H. & Harniss, R. O., 1979. Plant demographic studies in sagebrush-grass communities of south western Idaho. Ecology 60: 376–388.

Williams, O. B., 1955. Studies in the ecology of the Riverine Plain. I. The gilgai micro-relief and associated flora. Aust. J. Bot. 3: 99–112.

Williams, O. B., 1956. Studies in the ecology of the Riverine Plain. II. Plant-soil relationships in three semi-arid grasslands. Aust. J. Agric. Res. 7: 127–129.

Williams, O. B., 1961. Studies in the ecology of the Riverine Plain. III. The phenology of a Danthonia caespitosa Gaudich. grassland. Aust. J. Agric. Res. 12: 247–259.

Williams, O. B., 1968. Studies in the ecology of the Riverine Plain. IV. Basal area and density changes of Danthonia caespitosa Gaudich. in a natural pasture grazed by sheep. Aust. J. Bot. 16: 565–578.

Williams, O. B., 1969. Studies in the ecology of the Riverine Plain. V. Plant density response of species in a Danthonia caespitosa Gaudich. grassland to sixteen years of grazing by Merino sheep. Aust. J. Bot. 17: 255–268.

Williams, O. B., 1970. Population dynamics of two perennial grasses in Australian semi-arid grassland. J. Ecol. 58: 869–875.

Williams, O. B., 1974. Vegetation improvement and grazing management. In: (A. D. Wilson ed.). Studies of the Australian Arid Zone. II. Animal production. CSIRO, Melbourne pp. 127–143.

Williams, O. B., 1977. Reproductive wastage in rangeland plants, with particular reference to the role of herbivores. Proc. of the 2nd United States/Australia Rangeland Panel, Adelaide, 1972. pp. 227–248.

Williams, O. B. & Roe, R., 1975. Management of arid grasslands for sheep. In Managing terrestrial ecosystems. Proc. Ecol. Soc. Aust. 11: 142–156.

Wright, R. G. & Van Dyne, G. M., 1976. Environmental factors influencing semi desert grassland perennial grass demography. The Southw. Natur. 21: 259–74.

Young, J. A. & Evans, R. A., 1974. Population dynamics of green rabbitbrush in disturbed big sagebrush communities. J. Range Managem. 27: 127–132.

Accepted 20.6.1981.

Rates of change in vegetation during secondary succession*

R. Bornkamm

Institute of Ecology, TU Berlin, Rothenburgstr. 12, D 1000 Berlin 41, B.R.D.

Keywords: Community Coefficient, Percentage Similarity, Rates, Succession

Abstract

For a number of experiments on secondary succession on different soils (lasting 9–23 yr) calculations were made of: a) rates of floristic change as measured with the community coefficient of Sørensen (CC), b) rates of cover-based change by the percentage similarity coefficient of Dahl & Hadač (PS), c) rates of change of stages (dominant growth forms). In the calculation of CC and PS the first year of observation and the preceding year have been used as reference years for a given year. Rates of $>65\%$ have been ranked as very rapid, rates of 35–65% as rapid, rates of 5–35% as slow, whereas rates of $<5\%$ (similarity $>95\%$) indicate temporal stability. Rapid changes of CC are to a large extent confined to the first year only, rapid changes of PS may occur in the following years too. After 5 yr, in many cases CC and PS are as high as ca. 90%, but the values do not exceed 95%. The examples show that rates of change are a useful tool in the description of the succession process.

Introduction

Succession is a time depending process. Many vegetation parameters change with time. Since the beginning of succession research it is known that successions may be 'slow' or 'fast' (see Lüdi, 1930). Only few papers try to evaluate the speed of succession. This may be due to the fact that it is difficult to generalize the different rates of change of different vegetation properties.

Major (1974a) listed kinds of changes in vegetation and their durations (see also Major, 1970, 1974 b, c, d, e). Three groups can be distinguished:
1. Successional (unidirectional) changes of the succession process as a whole
2. Change of seral stages

3. Phasic or cyclic changes within a stable community or a succession.

For all types of changes we can describe chronofunctions and calculate rates of change. A great number of parameters can be used in order to detect rates: plant cover, species number (modal values, see Yodzis, 1978), biomass and productivity (Major, 1974 a, c), and chemical constituents as well (Major, 1974 d, e).

The durations of sequences from the beginning of a succession to the terminal stage (the hypothetical climax stage) can be very different (Major, 1974 b). If we use a chronofunction $v = f(t)$ for the description of vegetation, a stable situation at the end of a succession is indicated by $dv/dt \rightarrow 0$. It is supposed that the terminal stage is reached when $\dfrac{dv}{dt} < 5\%$ (Major, 1974a, see also Olsen, 1958; Tagawa, 1964). As a consequence of time limitation these problems cannot be solved through the study of permanent plots and therefore

* Nomenclature follows F. Ehrendorfer: Liste der Gefäss-pflanzen Mitteleuropas, 2. ed., Stuttgart 1973.

will not be discussed further here.

The sequence of stages of a sere (Knapp, 1974b) is defined by floristic change and by change of the dominant life forms, i.e. mainly by qualitative categories. Rates of change at the community level can be determined by measuring the time, needed for transitions of given stages into next ones. An analysis at the population level makes clear that the transition process consists of a multiplicity of small-scale changes ('Kleinsukzessionen', Born-kamm, 1961a). These changes may be uni-directional or fluctuating (cyclic or irregular, see van der Maarel, 1980); they may be quantitative only or at the same time qualitative (replacement, Horn, 1976).

In phasic or cyclic changes within a stable community or a succession (Knapp, 1974a) rates of change may refer to the processes within a cycle, or to the frequency of cycles. Cyclic changes give rise to special problems, which will not be discussed here.

Succession processes in C. European vegetation concern – with very few exceptions - secondary succession, i.e. sequences following a disturbance or the termination of a state of disturbance. Here stages are often characterized by the dominance of resp. annuals, biennials, perennials, shrubs, pioneer trees, and trees of later stages. In the present paper for a number of experiments rates of change for both stages and small scale processes within the stages will be discussed.

Examples

Means of comparison

A few examples of succession experiments are discussed. These experiments are of different design; they last 10-20 yr. Quantitative and qualitative changes of vegetation are described with the help of two indices: The presence/absence based community coefficient according to Sørensen, 1948, $CC = \dfrac{2a}{2a+b+c}$, where a is the number of species occurring in both the given and the reference year, b the number occurring only in the given year, and c the number occurring only in the reference year. The reference year may be the first year of the succession, or the preceding year (year-to-year values).

The percentage similarity according to Dahl & Hadač, 1941, $PS = \dfrac{2\Sigma min\,(x_i, y_i)}{\Sigma(x_i + y_i)} \times 100$, is based on the cover of species. Here x_i and y_i are the cover % values of given species in years x and y; and $min\,x_i$ or $min\,y_i$ is the lower cover value of species occurring in both years.

Experiments

The first experiments were located in the Institute's experimental garden at Berlin-Dahlem, where three different soil types were brought in from elsewhere: sand, loam and clayey loam. It comprised 15 plots of 1 m² on each soil type. The vegetation was left alone from 1968-1976; it was recorded yearly. The mean values of the replicates were used for calculation (for details see Born-kamm & Hennig, in press). Since the succession started with a ruderal weed vegetation it will be referred to as the ruderal succession experiment.

The second experiment comprised only ten 1 m² plots on horticultural soil in the experimental field of the Botanical Institute of the University of Cologne, recorded (and harvested) year by year without replicates. During the experiment (1967-1976) a vegetation of high forbs developed (for details see Bornkamm, 1981). It will be called the horticultural succession experiment.

The third experiment was carried out in a dry meadow near Göttingen set up in 1953 in two parts: Grassland experiment I started with destruction of the vegetation on a 4 m² plot; grassland experiment II was set up by transplanting a very small plot (0.7 × 0.8 m²) from a moist meadow into a dry meadow (for details see Bornkamm, 1961b, 1974, 1975).

Stages

The succession stages detected in the different years are plotted in Figure 1. It shows, that in the ruderal experiment on loam and clayey loam an annual vegetation (characterized by *Senecio vulgaris* and *Conyza canadensis*) merged in the second year into a perennial stage dominated by *Solidago canadensis*. On sand *Solidago* was not as successful; for this reason the annual stage lasted longer and comprised higher contributions of

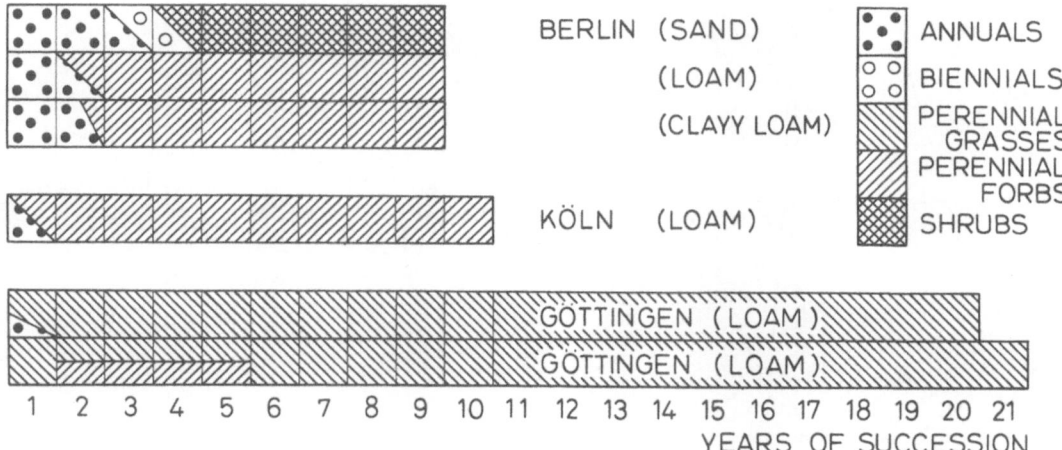

Fig. 1. Dominant growth forms characterizing the stages of different succession experiments. Berlin = ruderal experiment; Köln = horticultural experiment; Göttingen = grassland experiment.

biennial species (esp. *Oenothera biennis*). The plots then suddenly changed into a shrub stage dominated by *Sarothamnus scoparius*.

In the horticultural experiment the perennials started at the same time as the annuals, and were already dominant in the second year (esp. *Urtica dioica*).

In the grassland experiment I a small number of annuals after the removal of vegetation established only in the first year, whereas perennial grasses were dominant nearly from the beginning. Grassland experiment II showed a replacement of grasses of moist meadows (e.g. *Arrhenatherum elatius, Lolium perenne, Dactylis glomerata*) by grasses of dry meadows (esp. *Bromus erectus*). During this process herbs were for a number of years successful in competing with the grasses. Even after more than 20 yr there is no sign of a shrub layer. This is probably due to rabbit grazing.

Indices of similarity

In the ruderal experiment on sand the community coefficient is < 60% in the second year as compared with the first one, decreasing to 23% in 1975 (CC, Fig. 2a, full points). The year-to-year values are > 75% already after two years, indicating a rather slow rate of change (Fig. 2a, circles). The percentage similarity values (PS) decrease much faster than the CC values: The similarity between year 3 and 1 is already < 10%

(Fig. 2b, full points)! The similarity of one year to the preceding year's value increases slower with the PS values (Fig. 2b, circles) than with the CC values. But finally, after 9 yr, a very high value of more

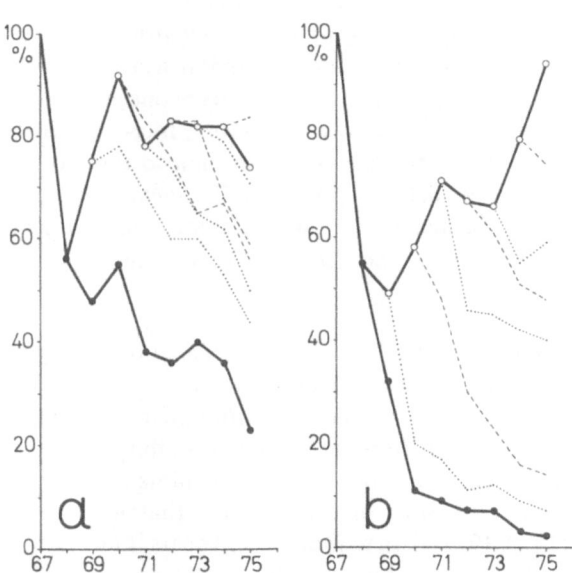

Fig. 2. Indices of similarity in the ruderal experiment on sand.
a. Community Coefficient;
b. Percentage Similarity. Full points: Reference to the first year; circles: Reference to the preceding year (year-to-year values). Secondary reference lines indicate the relation to the other years. In order to obtain a clearer distinction reference to years 2,4 and 6 is given by dotted lines, reference to years 3,5 and 7 by broken lines.

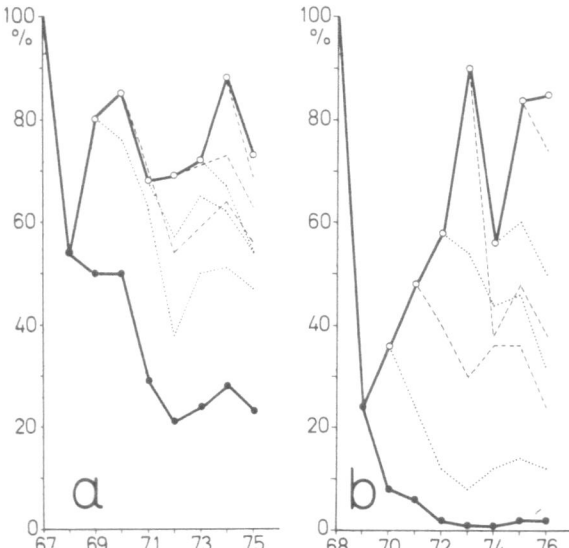

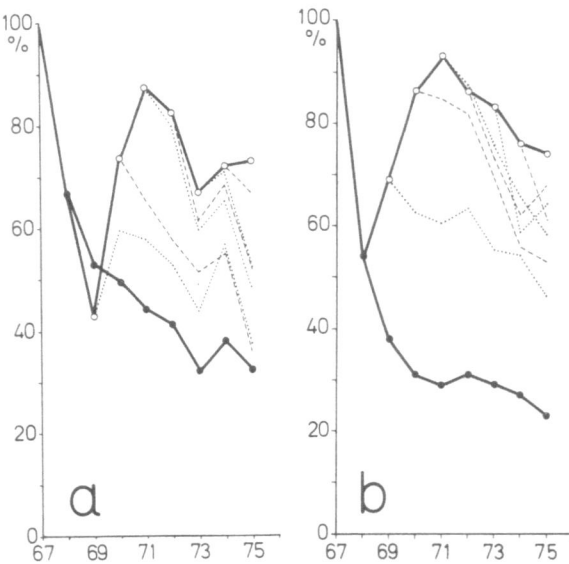

Fig. 3. Indices of similarity in the ruderal experiment on loam. a. Community Coefficient; b. Percentage Similarity. Further explanation see Figure 2.

Fig. 4. Community Coefficients in the horticultural experiment. a. Community coefficient; b. percentage similarity. Further explanation see Figure 2.

than 90% is reached, indicating an extremely slow rate of change. This is caused by the stable dominance of *Sarothamnus scoparius.* The secondary reference lines show that in a given year the plots are more similar to the plots of one year before than to plots of two years before, more similar to the plots of two years before than to the plots of three years before, and so on. This means that there is no influence of the different climatological conditions in the different years, e.g., dry and wetness.

On loam the CC values related to the first year (Fig. 3a) are comparable to those of sand. The values related to the preceding year show more variation. In some cases the secondary reference lines cross: In those years the plots were more similar to plots of earlier years that were moist (1970, 1972) than to more recent years that were dry (1971, 1973). Thus the effect of the climatological conditions is more apparent on loam than on sand. With the PS-values the similarity to the first year is already < 10% by the third year.

On clayey loam we recognize the same trends, but here exists a very high year-to-year stability caused by the early and steady dominance of *Solidago canadensis* (data not presented here).

In the horticultural experiment the decrease of CC-values related to the first year (Fig. 4) is slower than in the ruderal experiments. This is due to the fact that the dominating perennial forb, *Urtica dioica,* was present from the beginning. The CC-values related to the preceding year show a rather stable situation after 4–6 yr, caused by the dominance of *Urtica.* The decrease after 7–9 yr is caused by a methodological problem: In plots 1–7 *Urtica* was dominant, in plots 8–10 *Symphytum officinale* was more dominant than *Urtica.* Since one plot was harvested every year, starting with plot 1, only the remaining plots could be used for vegetation analysis. For this reason the cover of *Symphytum* became more and more apparent in the calculations (for details see Bornkamm, 1981) The decline of the PS-values is also not as rapid as in the ruderal experiment; still > 20% after 9 years.

In the grassland experiments the picture is different. In experiment I the vegetation was completely removed in one half of the plot, in the other half the two most dominant grasses, *Bromus erectus* and *Brachypodium pinnatum,* were left. The CC-values (Fig. 5a) show two phases. In phase I (up to the early sixties) the decrease of values related to the first year is faster than in phase II (1964–1973). This is not unusual. The CC-values related to the preceding year are 78,5 ± 1,4% in phase I (average for 1954–1963) and 85,5 ± 2,6% in phase II (average for 1964–1973), the difference being significant with P < 0,05. The first phase is characterized by the spread of the most common dry meadow plants and the occurrence of some annual weeds. The second phase is characterized by the occurrence of seedlings of woody plants (which do not succeed); a more complicated pattern combined with a tendency towards a more homogeneous distribution of the species. These tendencies result in a lower degree of floristic change. The secondary reference lines of both phases are well separated. The similarity to the dry years of 1957 and 1959 is mostly lower than to the more distant, but moister years of 1956 and 1958. The PS-values (Fig. 5b) are rather high because the most frequent species, *Bromus erectus,* was present from the beginning. The values related to the first year oscillated between 50 and 70% in phase I, and between 45 and 60% in phase II. The values related to the preceding year oscillate between 70 and 90%. Here, too, it can be seen that the similarity to dry years like 1957 and 1959 is mostly lower than to moist years like 1956 and 1958.

In the grassland experiment II greater changes take place (see Fig. 7a, b). In the first phase (1953–1958) most of the transplanted species of moist meadows are still present (little floristic change, high CC-values). But there is already a change in cover by species already present that grow in moist *and* dry meadows (decrease of PS-values related to the first year). In the second phase (for details see Bornkamm, 1974) these species are most frequent, but species of dry meadows invade. This results in a decrease of CC-values related to the first year. The third phase (since 1962) is characterized by a stable dominance of *Bromus erectus,* leading to high year-to-year CC-values (80–90%). In the PS-values the secondary reference lines show two groups with a gap at about 1962/63 and a close parallelity since 1966. Since here the vegetation was changing from a moister to a dryer type, apparently the dry years were 'pushing' the succession. CC- and PS-similarity to the earlier but dry year 1957 is greater than to the more recent but moist year 1958, and with the CC-values it is greater for 1959 than for 1960. In most cases the plots in the dry years (e.g. 1959, 1963, 1969) were less similar to the plots in the preceding years than in the moist years.

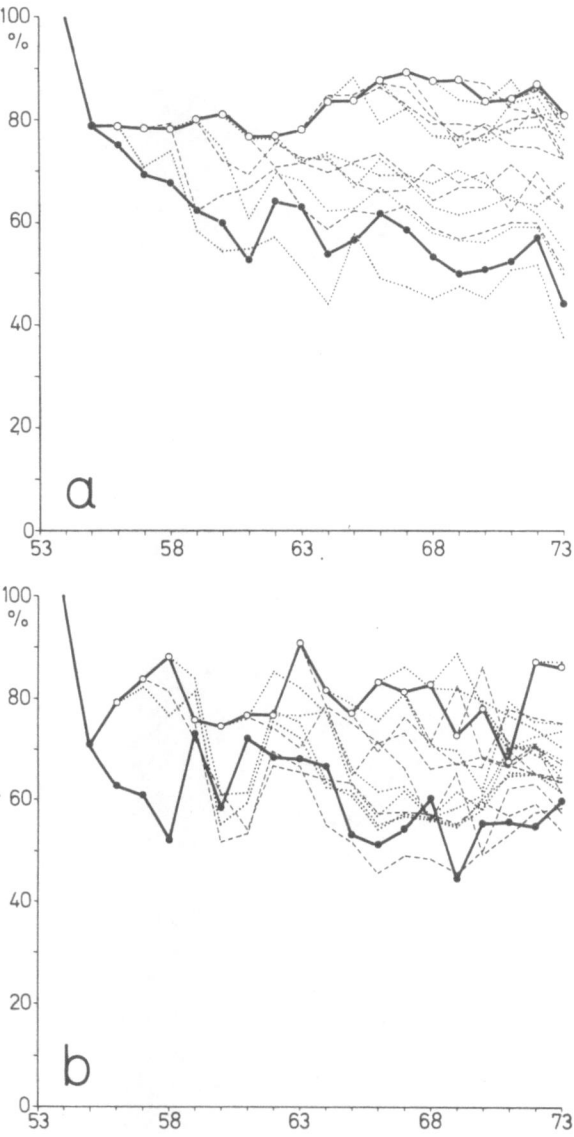

Fig. 5. Indices of similarity in the grassland experiment I. a. Community Coefficient; b. Percentage Similarity. Further explanation see Figure 2.

Conclusions

The rates of change discussed in this paper are measured as floristic change (CC) and as change of species cover (PS). In the examples used two types of dependence have been found. One type (ruderal and horticultural experiment) is outlined in Figure 7, where the mean values for the three chosen parts of the ruderal experiment are plotted. It shows a

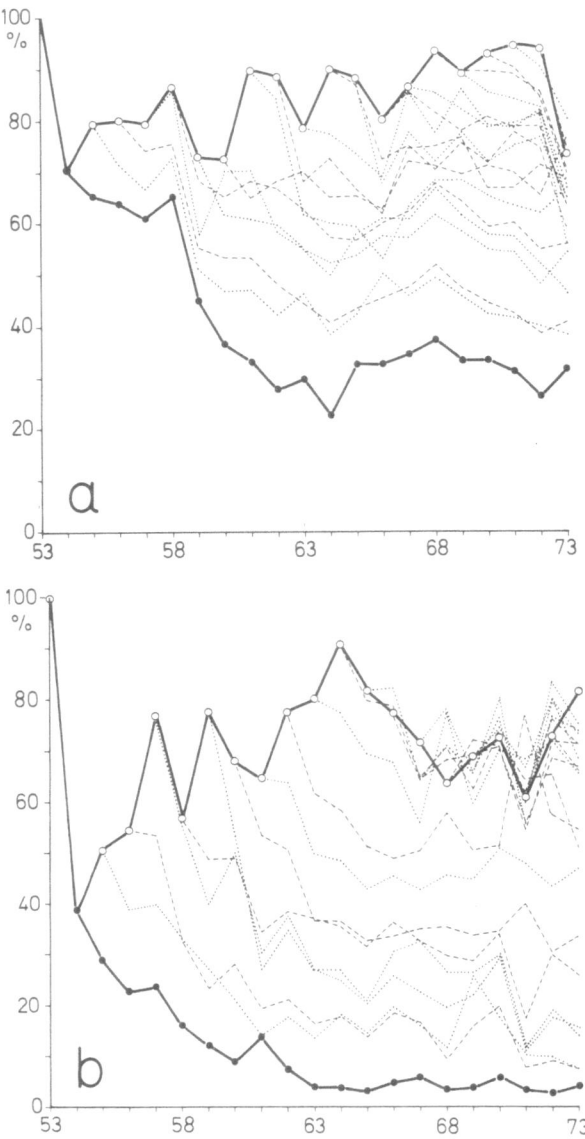

Fig. 6. Indices of similarity in the grassland experiment II. a. Community Coefficient; b. Percentage Similarity. Further explanation see Figure 2.

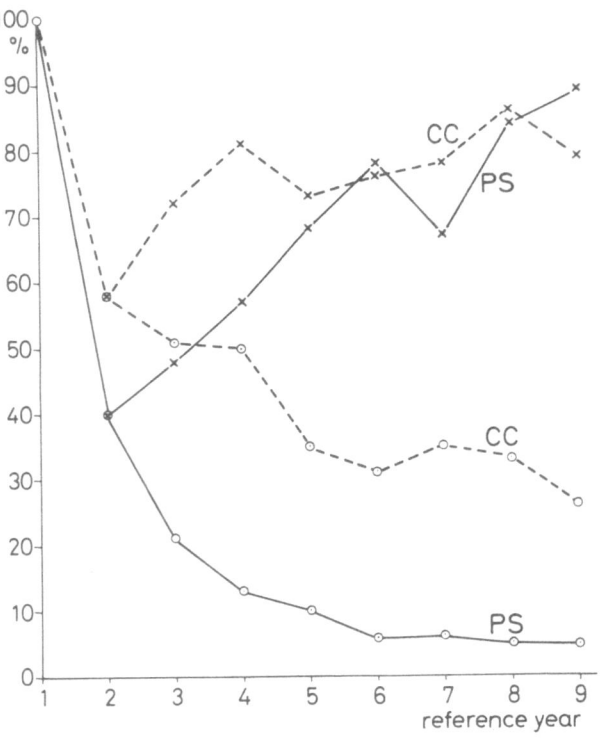

Fig. 7. Indices of similarity, mean values for the three ruderal experiments. CC (broken lines) = Community Coefficient; PS (solid lines) = Percentage Similarity; Circles: reference to the first year; X: Reference to the preceding year.

rapid floristic change in the first years, and the CC-value within 10 years decreases to ca. 30%. During this time the floristic composition becomes more and more stable and the similarity between two subsequent years is about 80%. This change is partly due to the progress of succession, (the similarity to the first year is still declining). It is also due to the variation in dry and wet years. The rate of cover based change is even more pronounced. After

5 yr the PS-value related to the first year is < 10%. Here year-to-year changes of 20% occur, while the similarity to the first year decreases very little further. This type corresponds to the dominance controlled communities, according to Yodzis, 1978.

In the grassland experiment I the dominating perennials contributed to the vegetation composition and cover from the beginning (Fig. 5). Here the rates of change are not as high. The year-to-year variation is more prominent than the succession process itself. This type corresponds with the 'founder controlled' community, according to Yodzis, 1978.

Grassland experiment II holds an intermediate position (Fig. 6). In the first phases unidirectional floristic and cover changes are detectable. In the last phase (since 1968, see Fig. 6b) the situation resembles experiment I very much.

Frequently only quantitative changes of vegetation have been called fluctuations in contrast to the qualitative and quantitative changes between the succession stages, which are called succession in a strict sense (Knapp, 1974a; Rabotnov, 1955, 1974; Barkman, 1958; Braun-Blanquet, 1964; Miles, 1979; van den Bergh, in press). From an analytical point of view it is important to distinguish yearly fluctuations from the trend of the succession process as a whole. But in fact, fluctuations are integrated parts of this process. The fluctuation of one year, even if it is governed by the weather (e.g. dry years) influences the development of the next year. (Van der Maarel, this symposium) Fluctuations may sometime change species cover to zero if annual species do not germinate in dry years, but emerge in the next moist year (Runge, 1963). In this case we have qualitative changes. The succession is the total of all fluctuations and not something additional.

It has been shown that parameters of similarity are a means of evaluating rates of change. These rates are different, if different parameters are used. If we call rates of change of 5–35% per year slow, 35–65% rapid and >65% very rapid we can rank the events of the experiments discussed (Fig. 8). Rapid changes occur in the first years of the ruderal and horticultural experiment, and likewise in the grassland experiment II, whereas grassland experiment I shows slow rates. Schmidt (in press) determined rates of change in succession experiments in old fields with different treatments. He found rapid changes of the PS-values in the first year under all treatments, of the CC-values in the first year with shading only. In all other cases rates of change were slow.

If we compare the change from stage to stage, the grassland experiment develops very rapid, because the annual stage hardly exists. In the ruderal experiment the change to the perennial stage is slow, but the conversion into shrubs is rapid. On the contrary the investigated grassland is nearly stable as a perennial grassland and does not show any signs of woody vegetation.

In conclusion we may say that it is possible to speak of succession rates if it is clearly indicated which parameter of change is chosen.

	CC	PP	stages perennials	stages shrubs
stable for CC + PS: > 95%	—	—	—	Gr I Gr II
slow for CC + PS: 65–90%	Ho S_3-S_n Ru S S_2-S_n L S_2-S_n C S_2-S_n Gr I S_1-S_n II S_1-S_n	Ho S_2-S_n Ru S S_5-S_n L S_5-S_n C S_3-S_n Gr I S_1-S_n II S_6-S_n	Ru S	Ho Ru L Ru C
rapid for CC + PS: 35–65%	Ho S_1+S_2 Ru S S_1 L S_1 C S_1	Ho S_1 Ru S S_1-S_3 L S_2-S_4 C S_1-S_2 Gr II S_1-S_5	Ho Ru L	Ru S
dramatic for CC + PS: < 35%	—	Ru L S_1	Gr I Gr II	—

Fig. 8. Classification of succession rates.
Ru = ruderal experiment; Ho = horticultural experiment; Gr. I, II = grassland experiment I, II; S = sand; l = loam; C = clayey loam; $S_1 \ldots \ldots S_n$ = years of succession.

References

Barkman, J. J., 1958. On the ecology of cryptogamic epiphytes with special reference to the Netherlands. Belmontia 2.

Bergh, J. P., van den (in press). Interactions between plants and population dynamics. Verh. Ges. f. Ökologie 10.

Bornkamm, R., 1961a. Über die Rolle der Durchdringungsgeschwindigkeit bei Kleinsukzessionen. Veröff. Geobot. Inst. Rübel Zürich 37: 16–26.

Bornkamm, R., 1961b. Zur Konkurrenzkraft von Bromus erectus. Bot. Jahrb. Syst. 80: 466–479.

Bornkamm, R., 1974. Zur Konkurrenzkraft von Bromus erectus II. Bot. Jahrb. Syst. 94: 391–412.

Bornkamm, R., 1981. Zusammensetzung, Biomasse und Inhaltsstoffe der Vegetation während zehnjähriger Sukzession auf Gartenböden in Köln. Decheniana 134: 34–48.

Bornkamm, R., 1975. Zwanzig Jahre Vegetationsentwicklung in einem mitteleuropäischen Halbtrockenrasen. In: W. Schmidt (ed.) Sukzessionsforschung, pp. 535–537. Cramer, Vaduz. 622 pp.

Bornkamm, R. & Hennig, U., in press. Experimentell-ökologische Untersuchungen zur Sukzession von ruderalen Pflanzengesellschaften auf unterschiedlichen Böden. I. Zusammensetzung der Vegetation. Flora.

Braun-Blanquet, J., 1964. Pflanzensoziologie. 3. Aufl. Springer Wien & New York, 865 pp.

Dahl, E. & Hadač, E., 1941. Strandgesellschaften der Insel Ostoy im Oslofjord. Nytt. Mag. Naturv. 82: 251–312.

Horn, H. S., 1976. Succession. In: R. M. May (ed.): Theoretical ecology. Blackwell, Oxford. 317 pp.

Knapp, R., 1974a. Cyclic successions and ecosystem approaches in vegetation dynamics. Handb. Veget. Sci. 8: 91–100. Junk, The Hague.

Knapp, R., 1974b. Some principles of classification and of terminology in successions. Handb. Veget. Sci. 8: 167–178. Junk, The Hague.

Lüdi, W., 1930. Die Methoden der Sukzessionsforschung in der Pflanzensoziologie. Handb. biol. Arbeitsmeth. XI: 5: 527–728.

Maarel, E. van der, 1980. Towards an ecological theory of nature management. Verh. Ges. f. Ökol. 8: 13–24.

Maarel, E. van der, 1981. Fluctuations in a coastal dune grassland due to fluctuations in rainfall: experimental evidence. Vegetatio 47: 259–265.

Major, J., 1970. Essay review of Rodin and Bazilevich: The illusive mineral equilibrium. Ecol. 51: 160–163.

Major, J., 1974a. Kinds and rates of changes in vegetation and chronofunctions. Handb. Veget. Sci. 8: 7–18. Junk, The Hague.

Major, J., 1974b. Differences in duration of successional seres. Handb. Veget. Sci. 8: 155–160. Junk, The Hague.

Major, J., 1974c. Biomass accumulation in successions. Handb. Veget. Sci. 8: 195–204. Junk, The Hague.

Major, J., 1974d. Nitrogen accumulation in successions. Handb. Veget. Sci. 8: 205–214. Junk, The Hague.

Major, J., 1974e. Accumulation of ash elements and pH changes. Handb. Veget. Sci. 8: 215–218. Junk, The Hague.

Miles, J., 1979. Vegetation dynamics. Chapman & Hall, London, 80 pp.

Olson, J. S., 1958. Rates of succession and soil changes on Southern Lake Michigan sand dunes. Bot. Gaz. 119: 125–130.

Rabotnov, T. A., 1955. Fluctuations of meadows. Bjull. Mosk. O. I. Prir. O. Biol. 60 (3).

Rabotnov, T. A., 1974. Differences between fluctuations and successions. Handb. Veget. Sci. 8: 19–24. Junk, The Hague.

Runge, F., 1963. Die Artmächtigkeitsschwankungen in einem nordwestdeutschen Enzian-Zwenkenrasen. Vegetatio 11: 237–240.

Schmidt, W., (in press). Ungestörte und gelenkte Sukzession auf Brachäckern. Scr. Geobot. 15.

Sørensen, T., 1948. A method of establishing groups of equal amplitude in plant sociology based on similarity of species content. Biol. Skr. K. danske Vidensk. Selsk. 5: 1–34.

Tagawa, H., 1964. A Study of the volcanic vegetation in Sakurajima, South-West Japan. I. Dynamics of vegetation. Mem. Fac. Sci. Kyushu Univ. Ser. E. Biol. 3: 165–228.

Yodzis, P., 1978. Competition for space and the structure of ecological communities. Lecture Notes in Biomathematics 25. Springer, Berlin, Heidelberg & New York. 191 pp.

Accepted 28.7.1981.

Temporal variation of species composition and species diversity in permanent grassland plots with different fertilizer treatments*

P. van Hecke[1], I. Impens[1] & T. J. Behaeghe[2]**

[1] *Departement Biologie, Universitaire Instelling Antwerpen, Universiteitsplein 1, B-2610 Wilrijk, Belgium*
[2] *Leerstoel voor Plantenteelt, Fakulteit van de Landbouwwetenschappen, Rijksuniversiteit Gent, Coupure Links 533, B-9000 Gent, Belgium*

Keywords: Analysis of Correspondences, Fertilization, Grassland, Permanent plots, Principal Components Analysis, Species reduction, Temporal diversity

Abstract

In 1963 a set of seven treatments × three replicates, situated in East-Flanders (Belgium), was sown with a seed mixture of eight grass species and one clover species. Seven combinations of mineral fertilizers are administered annually. Samples are taken in winter applying the 25 sq. cm-frequency method of De Vries. F% is calculated for each species from presence-absence data. To allow an efficient description of the annual changes in the different treatments, some multivariate ordination techniques are applied, namely R-mode Principal Components Analysis and Analysis of Correspondences. Some aspects of diversity are considered by means of four indices as well as a dominance-diversity curve. Results obtained are discussed comparing them with the F% changes of the most important species in three treatments. Some problems on species reduction are also examined.

Introduction

The longest-running experiment on the dynamics of grasslands, viz. the Park Grass Experiment at Rothamsted (U.K.), began in 1856 (e.g. Williams, 1978; Silvertown, 1980, and many earlier papers). Only a few others have been established since in Europe: notably Cambridge (1936–1970; Watt, 1971), Melle (1939–1980; Behaeghe & Slaats, 1968; Behaeghe & Cottenie, 1976; Behaeghe, Traets *et al.*: reports published every two years), Wageningen (1958–1978; Van den Bergh, 1979) and Skriduklaustur (Iceland; Björnsson, 1978). With the exception of Cambridge the main purpose was the investigation of the effect of fertilizers on the

* Nomenclature follows De Langhe e.a. (1978).
** We are indebted to the I.W.O.N.L. for financial support, to Mr. David Wouters for writing a computer program to calculate the log-linear distribution and to Dr. Zahar Samsuddin for reading an earlier draft of this paper.

botanical composition and the yield of the sward, merely based on an agricultural viewpoint, at least at the beginning. Sooner or later it has been recognized that both the farmer and the botanist can take advantage of these long term observations (e.g. Thurston, 1969, 1976). Somewhat comparable experiments have been started in Poland in the sixties (Borowiec *et al.* 1974; Traczyk *et al.*, 1976; Traczyk & Kotowska, 1976), however recent information is not available.

This paper primarily deals with part of the Belgian grassland data mentioned above. A Principal Components Analysis is applied to detect the time trajectories of the different treatments completed with a two-way Analysis of Variance (ANOVA) for eventual significant between-treatment and between-year differences. By means of the Analysis of Correspondences (also known as Reciprocal Averaging) the association or fidelity of the most important species with regard to the different

Vegetatio 47, 221–232 (1981). 0042-3106/81/0471-0221/$2.40.
© Dr W. Junk Publishers, The Hague.

treatments will be discussed as well as the yearly fluctuations of the quantities of those species. Finally some long term trends concerning grasses, legumes and herbs as distinct components and some diversity indices and dominance-diversity curves are examined.

Materials and methods

The experiment

In 1939 two sets (M.39.1 and M.39.2) of seven experimental treatments, each represented by three replicates, were established at Melle at the experimental farm of the State University of Gent (East-Flanders, Belgium). The trial is still in progress under control of the third author (T.J.B.) who provided the data on which this paper is based. The plots of both sets cover areas of 40 or 35 sq. m and are located on a sandy loam soil: sand 34% ($>50\,\mu$), loam 54% (2–$50\,\mu$) and clay 12% (0–$2\,\mu$).

The seven treatments are: unfertilized (control plot), NPKCa-, -PKCa-, N-KCa-, NP-Ca-, NPK-- and NPKCaMg (full manured plot). Details on the fertilizer applications are given in Table 1.

Samples are usually taken in the period November-January with the 25 sq. cm-frequency method of De Vries (1937). From each replicate (which is mown five or six times a year whereafter the harvest is always carried off) 50 samples are taken during 16 winters (except for the last but one). The presence of each species is recorded in each sample, whereafter frequency percentages (F%) are calculated as the mean of the three replicates.

This paper will only discuss the results of the M.39.1-set. From 1939 to 1962 these plots were provided with a crop rotation of pure cultures of grasses, maize, potatoes, *Trifolium pratense, Lolium multiflorum,* lucerne and tobacco. On May 2 1963 the same plots were resown with a seed mixture of eight grass species and one clover species, namely: *Festuca pratensis* 10, *Lolium perenne* 4.25, *Festuca rubra* and *Trifolium repens* 3, *Phleum pratense* and *Dactylis glomerata* 2.2, *Poa pratensis* 1.5, *Poa trivialis* 1 and *Agrostis stolonifera* 0.5. The quantities applied are expressed in kg ha^{-1}. Up to 1978 30 species are obtained (13 grasses, 3 legumes and 14 herbs).

Species reduction

Since in such grassland experiments it is expected that only a few species are really important in terms of frequency or other abundance parameters, an attempt is made to examine in an objective way the consequences of reducing the number of species. In order to rank species by a sum of squares criterion, the method of vector projection followed by stress analysis (Orlóci, 1978) as well as the method of identification of diagnostic species based on variance ratios (Jancey, 1979) were carried out. In the former method the best species sample is selected just before the point where the stress curve starts to oscillate; in the latter the 5% level of significance determines the size of the species subset.

Multivariate analyses

Austin (1977) in his paper on different appro-

Table 1. Fertilizer applications expressed in kg ha^{-1} year^{-1} (mean dose). An = ammonium nitrate, As = ammonium sulphate, D = dolomite, F = fertifos, K = kieserite, Ml = magnesium lime, Pc = potassium chloride, Ps = potassium sulphate, Sl = Slags, Sp = superphosphate, Ts = triplesuper.

periods	N		P_2O_5		K_2O		CaO			MgO	
	dose	fertilizer	dose	fertilizer	dose	fertilizer	lime	with NP	total	dose	fertilizer
1939–1947	50	As 20	90	Sp 18	120		150	135	285	60	Ml
1948–1957	133	An 20.5	133	Sp 18 or Sl 18	150		150	344	494	60	
1958–1961	220	An 20.5 or 22.5	180	Sp 18 or F 38	160	Pc 40	150	435	585	60	D
1962	315	An 22.5	150	TS 43	400		0	385	385		
1963–1965	307	Urea 45	140		400		167	70	237	30	D
1966–1967	300		150		400		275	80	355	100	
1968–1971	300	An 33	150	Ts 40	500	Ps 50	690	80	770	60	K
1972–1978	300		150		400		460	80	540	70	

aches studying succession pointed out the value of multivariate methods.

First we applied an R-mode Principal Components Analysis (PCA; Seal, 1968) on a variance-covariance matrix. This matrix allows a much better interpretation as compared to the correlation matrix, which limitations were discussed e.g. by Swaine & Greig-Smith (1980). The variance-covariance matrix is generated from a three-dimensional matrix of treatment $\times$ species $\times$ time as well as from a species $\times$ time matrix for each treatment. In order to represent both treatments and species in the same two-dimensional diagrams, an Analysis of Correspondences (AC; Lebart & Fénelon, 1971) is used on a species $\times$ treatment matrix for each year.

Diversity indices and dominance-diversity curves.

We attempted to describe alpha diversity with some of the most commonly used indices, despite of their shortcomings and without any particular preference. It concerns Shannon-Wiener's H' ($= - \Sigma \, p_i \log_2 p_i$), Pielou's J' ($= H'/\log_2 S$), Simpson's C ($= \sum\limits_{i=1}^{S} p_i^2$) and species richness S, where p_i is the proportion of the total frequencies consisting of the ith species and S is the number of species for each year in each treatment.

The application of dominance-diversity curves provides detailed information on fluctuations of the contribution of some species with respect to dominance. As relatively few species are involved in each treatment, the geometric series of Motomura is applied, as expressed by a log-linear distribution. This distribution is entirely determined by the value of m (Motomura's constant) which is the antilogarithm of the slope of the distribution. As the latter becomes steeper, diversity diminishes together with m, which never exceeds unity (Motomura, 1932). Curves are constructed from each species $\times$ time matrix referring to the treatments separately.

Computations are carried out on the PDP 11–45 computer of the Universitaire Instelling Antwerpen. Beforehand the frequencies are standardized by annual arithmetic total to eliminate the effect of annually different frequency totals. Applying no standardisation yields only very minor differences; as such the general pattern of ordination is not affected. Dry matter production has been discussed by Behaeghe & Cottenie (1976).

Results

Species reduction

The approaches of Orlóci (1978) and Jancey (1979) provide different results, although both ranking orders are significantly correlated with each other (Spearman rank correlation coefficient, $p < 0.01$).

Considering the 5%-level in the Jancey-method, the species subset is reduced to 18 and on the 0.1%-level to 14, the same number as selected by the Orlóci-method. This means that at the 0.1%-level species like *Hypochoeris radicata, Leontodon autumnalis* and *Ranunculus repens* are only present in the Jancey-selection, and at the 5%-level species like *Holcus lanatus, Prunella vulgaris, Rumex acetosa* and *Taraxacum* spec. remain absent from the Orlóci-selection.

Both procedures reduce the original set of 30 species to 18 (Jancey) and 14 (Orlóci) respectively. Results obtained with these three species subsets when applying PCA revealed negligible differences in the eigenvalues, the principal component loadings and the principal component scores.

Multivariate analyses

The first three axes of the R-mode PCA account for 88% of the total variance (51%, 27% and 10% respectively).

Figure 1 shows that, according to the time trajectories, three groups of treatments evolve, namely: 1) control plot + 'NP-Ca-' + 'N-KCa-', 2) '-PKCa-' and 3) 'NPKCa-' + 'NPK--' + 'NPKCaMg'. Each treatment seems to develop an own direction in its successional trend except for the plots 'full manuring' and 'NPKCa-' which cannot be separated along these and neither along the other components. High rates of change are striking in the first 6 or 7 years in the plots 'N-KCa-', 'NPKCa-', 'NPK--' and 'NPKCaMg', but only in the first three years in the 'NP-Ca-'-plot and virtually never in the '-PKCa-'-plot. All the plots seem to have reached a less dynamic situation, particularly reflected by distances becoming progressively shorter, except for the control plot. A striking change of direction in successional trends

224

seems to have occurred in the treatments 'N-KCa-',
'NP-Ca-' and 'NPK--' after 6 or 7 years. For 'N-
KCa-' and 'NPK--' this change corresponds to the
beginning of the less dynamic period mentioned
above. The reasons for those changes are obvious:
in the 'N-KCa-'-plot the frequencies of *Poa trivialis*
and *Trifolium repens* suddenly decrease in favour
of *Agrostis tenuis,* the frequencies of *Lolium perenne* sharply decrease in the 'NP-Ca-'-plot again in
favour of *Agrostis tenuis,* and once again the
frequencies of *Poa trivialis* suddenly diminish accompanied by an abrupt and gradual increase
respectively of the F% of *Dactylis glomerata* and
Agrostis tenuis ('NPK--'-plot).

The drought of the 1976 summer has caused
radical but only temporal disturbances in 3 of the 7
treatments: the F% of *Agrostis tenuis* increase and
those of *Trifolium repens* decrease in the control
plot, the F% of *Poa pratensis* increase whilst those
of *Poa trivialis* and *Trifolium repens* decrease in the

plot of N-deficiency, and the F% of *Lolium perenne*
decrease in the 'NPK--'-plot.

To test the possible differences between the
treatments and especially those between the subsequent years, a two-way ANOVA is carried out
using the projections of the centered individual
scores on the first two component axes. Therefore
the scores on PC 1 (multiplied by the ratio between
the values of the first and second eigenvalue) and
those on PC 2 are summed up. This was done
because each point or year within a treatment is
determined by two scores or co-ordinates (Fig. 1)
which in their turn are unequally proportioned by
the variances of the first two axes. As a consequence
each year is represented in the ANOVA by one
single value. This ANOVA reveals a highly significant difference (p < 0.001) between the treatments.
Moreover, both Scheffé's Multiple Comparisons
and Duncan's New Multiple Range Test reveal
that, within the group control plot + '-PKCa-' +

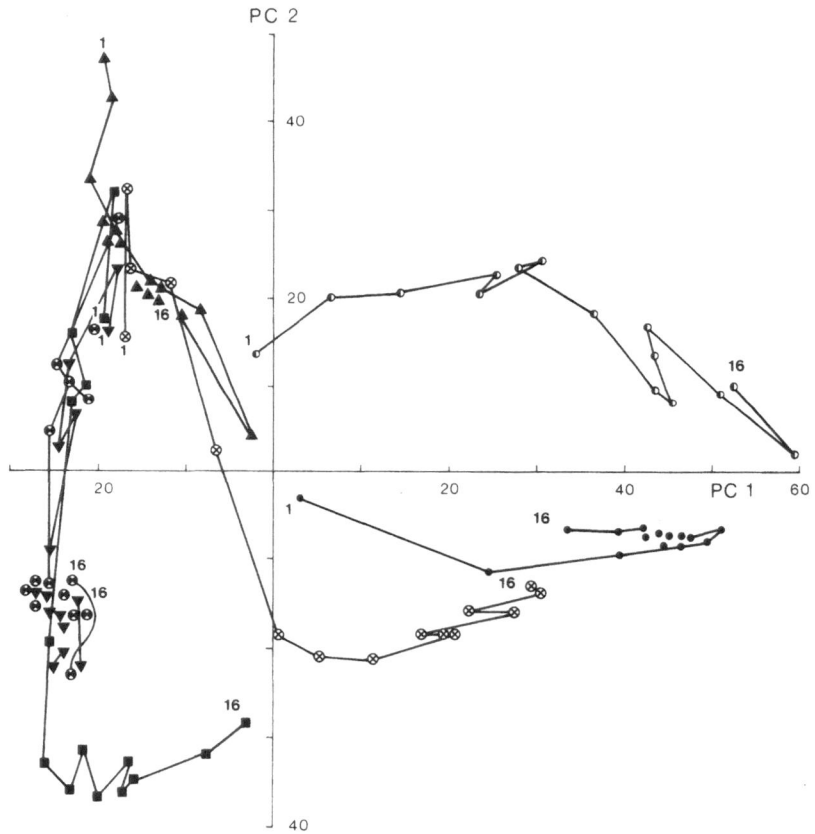

Fig. 1. Principal Components Analysis: ordination of seven treatments through time; 1 and 16 refer to the first and the last year
respectively. ○ = control plot, = NPKCa-, ▲ = -PKCa-, ⊘ = N-KCa-, ● = NP-Ca-, ■ = NPK--, ⊘ = NPKCaMg.

Table 2. Correlation coefficients between the original variables (30 species) and the first three orthogonal axes, derived from an R-mode Principal Components Analysis, analysing a variance-covariance matrix.

Original variables (species)	principal components		
	1	2	3
Agrostis tenuis (Agt)	0.962	−0.045	−0.238
Anagallis arvensis (Ana)	0.006	0.023	0.081
Bellis perennis (Bep)	0.051	0.258	0.066
Cynosurus cristatus (Cyc)	−0.081	0.033	0.032
Dactylis glomerata (Dag)	−0.545	−0.672	0.384
Agropyron repens (Agr)	−0.029	−0.025	0.141
Equisetum arvense (Eqa)	0.089	0.063	−0.034
Festuca pratensis (Fep)	−0.269	0.369	0.008
Festuca rubra (Fer)	0.769	−0.164	0.523
Geranium molle (Gem)	−0.042	0.050	0.072
Holcus lanatus (Hol)	0.319	0.008	−0.291
Hypochoeris radicata (Hyr)	0.232	0.239	−0.248
Leontodon autumnalis (Lea)	0.109	0.196	−0.152
Lolium multiflorum (Lom)	0.163	0.041	−0.129
Lolium perenne (Lop)	−0.668	−0.560	−0.007
Lotus uliginosus (Lou)	0.156	0.067	−0.174
Medicago lupulina (Mel)	0.156	0.067	−0.174
Phleum pratense (Php)	−0.501	0.059	−0.052
Plantago media (Plm)	0.013	0.052	0.080
Poa annua (Poa)	−0.530	0.200	0.072
Poa pratensis (Pop)	−0.066	−0.220	0.579
Poa trivialis (Pot)	−0.745	0.304	0.290
Polygonum aviculare (Pga)	−0.007	0.063	0.013
Prunella vulgaris (Prv)	0.215	0.112	−0.194
Ranunculus repens (Rar)	0.384	0.118	−0.405
Rumex acetosa (Rua)	0.222	0.131	−0.201
Spergula arvensis (Spa)	−0.013	0.078	0.016
Taraxacum spec. (Tar)	0.034	0.273	−0.135
Trifolium repens (Trr)	−0.243	0.934	−0.198
Cerastium spec. (Cer)	0.180	0.280	−0.158

'N-KCa-' + 'NP-Ca', all differences are significant except for the means of control plot and 'NP-Ca-' on the one side and of 'N-KCa-' and '-PKCa-' on the other side. The same is true within the group 'NPKCa-' + 'NPK--' + 'NPKCaMg' for the mutual differences between the three treatments.

The principal component loadings representing the correlations between the original variables and the orthogonal axes (Table 2) give an idea which species to consider as the most important ones in this experiment of fertilizer applications. As far as PC 1 is concerned, *Agrostis tenuis* (0.96) and *Festuca rubra*; *Phleum pratense*, *Poa annua* and *Poa trivialis* (−0.75) are acting in a reverse direction. The latter do need P and K as a combination. The highest positive loading of 0.93 on PC 2 identifies

Trifolium repens as closely related to the second axis, indicating that this species does not need the addition of mineral nitrogen. Species like *Lolium perenne* and *Dactylis glomerata* take up an intermediate position on the negative parts of both orthogonal axes.

Table 3 presents amounts of variance accounted for by each treatment separately. It is clear that the proportion referring to the control plot is smaller than that to the 'NP-Ca-'-plot on PC 1; the proportions of 'NPKCa-' and 'NPKCaMg' are very comparable on the first three component axes.

The two-dimensional scatter diagram obtained with the Analysis of Correspondences shows how the distances between the species and the different treatments fluctuate with time (Fig. 2). In the first year (Fig. 2a) the plots control, 'NP-Ca-' and '-PKCa-' are already distinctly separated from each other and from the closed group of NK-plots. Note the position near the origin of axes of *Dactylis glomerata*, *Phleum pratense* and *Poa annua*, the relatively large distances of *Festuca rubra* and *Agrostis tenuis* with regard to the plots control and 'NP-Ca', and the rather numerous less important species in the high negative scored part of the first factor. The latter ones practically disappear after four years (Fig. 2b), whereas the separation of the 'N-KCa-'-plot has begun, particularly due to a sharp increase of the frequencies of *Agrostis tenuis* and *Festuca rubra*. This latter species has now joined the 'NP-Ca-'-plot for many years, and *Dactylis glomerata* and *Phleum pratense* are now found

Table 3. Seven treatments and their variances accounted for the first three component axes. Results are taken from the same Principal Components Analysis which gives rise to Table 2 and Figure 1. Variances are calculated for each axis separately by summing up the squared individual scores in each treatment and for the seven treatments combined. The proportion of each treatment is then computed.

treatment	Principal component		
	1	2	3
control plot	26.7	10.0	29.2
NPKCa-	10.6	9.5	3.1
-PKCa-	4.7	26.4	6.0
N-KCa-	6.8	13.2	1.8
NP-Ca-	32.9	2.2	35.2
NPK--	7.3	29.9	21.7
NPKCaMg	10.9	8.9	2.9

226

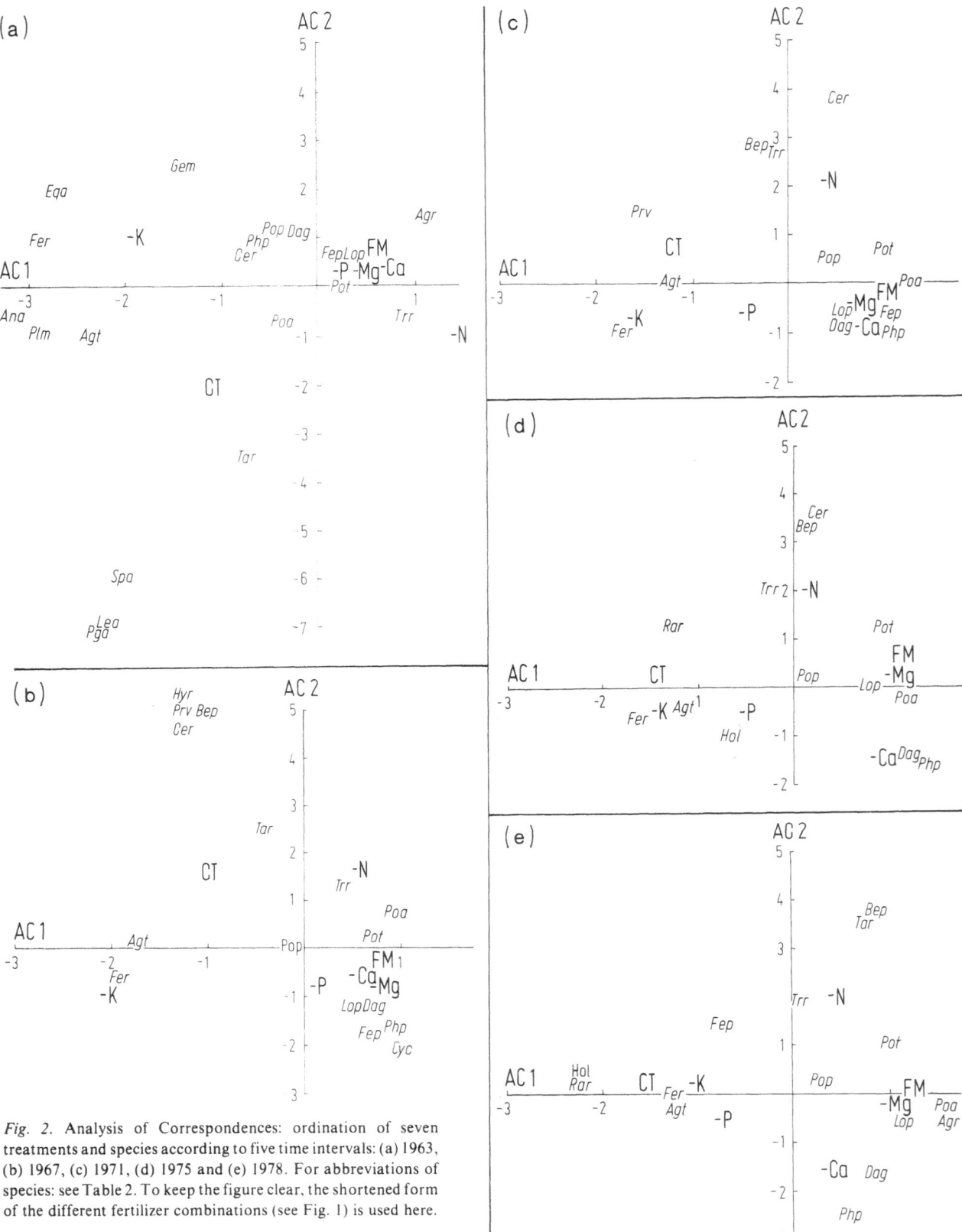

Fig. 2. Analysis of Correspondences: ordination of seven treatments and species according to five time intervals: (a) 1963, (b) 1967, (c) 1971, (d) 1975 and (e) 1978. For abbreviations of species: see Table 2. To keep the figure clear, the shortened form of the different fertilizer combinations (see Fig. 1) is used here.

in the neighbourhood of the NPK-plots. In the ninth year (Fig. 2c) the plots control, 'N-KCa-' and 'NP-Ca-' are obviously converging and, in its turn, the separation of the plot 'NPK--' has started. After 13 years (Fig. 2d) *Dactylis glomerata* and *Phleum pratense* prefer to be more closely allied to the 'NPK--'-plot and *Festuca pratensis* virtually disappears. At the end of the investigation period (Fig. 2e), the treatments 'NPKCa-' and 'NPKCaMg' remain together. Other conclusions can still be drawn: *Poa pratensis* turns out to be one of the most indifferent species, and irregularly immigrating species seem to prefer those plots permanently deprived of NPK.

Diversity indices and dominance-diversity curves

Table 4 summarizes the data on the four diversity indices, Figure 3 (a-g) the trends in the dominance-diversity curves. In general the fluctuations of the constant of Motomura (*m*) agree well with those of the Shannon-Wiener index (*H'*), which can normally be expected, but with the exception of the plots 'NPKCa-' and '-PKCa-' (t-test for simple linear regression). The slopes of the curves become always steeper between the first and the fifth year, even in an extraordinary way with the 'NP-Ca-'-plot, where the number of species falls to nearly one-third (all of them are grasses). In this plot too 1) is the lowest *m*-value obtained less than half the first-year value, and 2) dominance abruptly in-

creases with about three to four times, exclusively for account of *Festuca rubra*, *Agrostis tenuis* and *Poa pratensis*. The NK-plots 'N-KCa-', 'NPK--', 'NPKCa-' and 'NPKCaMg' are characterized by 1) a steadily decreasing diversity, 2) a small to moderate increasing dominance 3) beside a rather stable even distribution. Whereas herbs are always present in small amounts or even absent ('NPKCa-'-plot), they completely disappear at least between the fifth and ninth year. Nearly the same thing happens with *Trifolium repens* in the treatments with P- and Ca-deficiency, but this species succeeds to re-establish in the NPKCa-plots.

Typical for the control plot are the gradually decreasing diversity and increasing dominance (with a factor of about five times). The latter refers to only two species: *Agrostis tenuis* and *Trifolium repens*.

The remaining treatment, '-PKCa-', is really a sideslip in which species richness hardly changes and species diversity (*H'*) gradually increases (see also discussion).

Participation of grasses, legumes and herbs

These three components are represented in this experiment by 13, 3 and 14 species respectively. The three legumes considered here are *Trifolium repens*, *Lotus uliginosus* and *Medicago lupulina*. The two latter species occur very sporadically and only in the control plot. This low number of legumes can be

Table 4. Simpson index (*C*), Shannon-Wiener index (*H'*), Pielou's evenness (*J'*) and species number (*S*), according to seven treatments and five time intervals.

	Control plot				NPKCa-				-PKCa-				N-KCa-			
	C	H'	J'	S	C	H'	J'	S	C	H'	J'	S	C	H'	J'	S
1963	.12	3.29	.79	18	.19	2.67	.80	10	.42	1.80	.50	12	.18	2.82	.79	12
1967	.26	2.32	.63	13	.18	2.74	.79	11	.26	2.32	.65	12	.14	3.01	.87	11
1971	.41	1.78	.56	9	.22	2.45	.74	10	.21	2.55	.77	10	.23	2.33	.73	9
1975	.51	1.43	.48	8	.24	2.26	.75	8	.17	2.79	.81	11	.25	2.25	.80	7
1978	.58	1.24	.44	7	.25	2.29	.76	8	.18	2.73	.79	11	.28	2.20	.78	7

	NP-Ca-				NPK--				NPKCaMg			
	C	H'	J'	S	C	H'	J'	S	C	H'	J'	S
1963	.11	3.38	.85	16	.19	2.71	.73	13	.20	2.62	.71	13
1967	.38	1.72	.67	6	.19	2.63	.79	10	.20	2.58	.78	10
1971	.41	1.52	.59	6	.33	2.00	.67	8	.21	2.48	.78	9
1975	.34	1.69	.65	6	.39	1.80	.64	7	.28	2.13	.71	8
1978	.31	1.83	.65	7	.28	2.13	.71	8	.27	2.22	.79	7

228

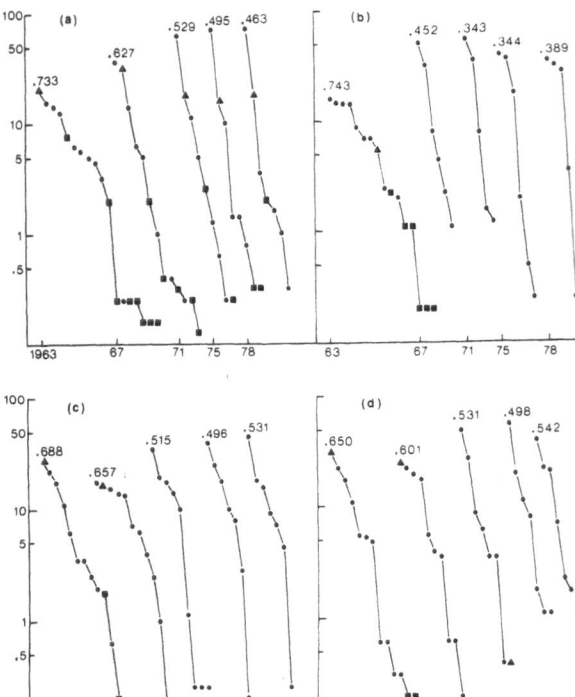

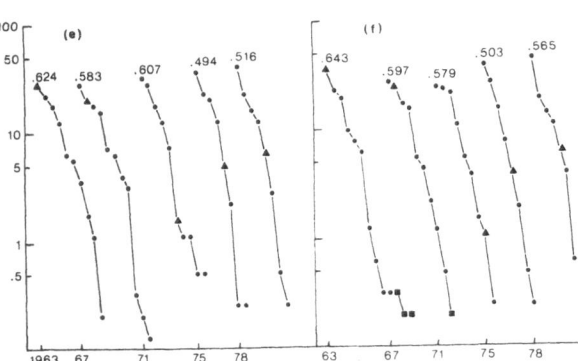

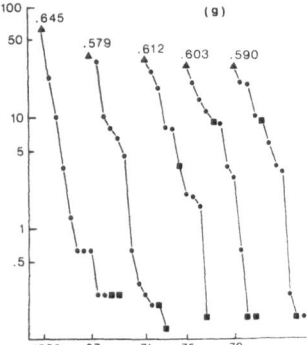

explained by a combination of different factors. The high mowing frequency, the quantitative amounts of minerals applied in the various treatments, as well as the poor and rather acid sandy loam soil disfavour legumes, not at least in a qualitative way. Once more, the plots 'NPKCa-' and 'NPKCaMg' (Table 5) show a high degree of conformity: the grasses never reach 100% and legumes never disappear. The treatments 'N-KCa-', 'NP-Ca-' and 'NPK--' also show a high conformity: as the grasses become completely predominant the legumes are the last component to be eliminated, first with the potassium deficiency, finally with the calcium deficiency treatment. The grasses never reach such an extreme predominant status in the plots control and '-PKCa-' where legumes are favoured. The latter differs from all other treatments in that the % participation of herbs increases, particularly due to *Bellis perennis*.

Discussion

In his study on the Park Grass Experiment (PGE) Silvertown (1980) has drawn two important conclusions which are mainly attained by using matrices of correlation coefficients and linear regression. Firstly the major components, viz. grasses, legumes and miscellaneous species, and their separate constituent species may achieve equilibrium at a different time and by different limiting factors. Secondly the endogenous factors which regulate the different species populations are operating within individual years.

Thurston (1969) has already pointed out that 1) species not previously present never invade the different fertilized plots and 2) all grasses are well encouraged by the application of nitrogen. Thurston *et al.* (1976) emphasized in a more elaborated study on the modern developments in the PGE 1) the slow changes in botanical composition of many plots despite the unaltered fertilizer treatments, 2) the

Fig. 3. Successional trends of dominance-diversity curves over five time intervals according to seven treatments: (a) control plot, (b) NP-Ca-, (c) N-KCa-, (d) NPK--, (e) NPKCa-. (f) NPKCaMg, (g) -PKCa-. The value at the top of each curve = constant of Motomura *(m)*. Dot = grass, triangle = legume, square = herb. For further explanations: see Whittaker (1975: 92).

Table 5. % participation of Grasses (G), Legumes (L) and Herbs (H) according to seven treatments and five time intervals.

	control plot			NPKCa-			-PKCa-			N-KCa-		
	G	L	H	G	L	H	G	L	H	G	L	H
1963	67.7	21.0	11.3	71.1	28.9	–	39.3	59.9	.8	69.8	28.1	2.1
1967	65.3	31.1	3.6	80.8	19.2	–	63.3	36.3	.4	82.1	17.8	.1
1971	78.7	18.5	2.8	98.5	1.5	–	63.1	33.2	3.7	100.0	–	–
1975	82.4	16.6	1.0	94.9	5.1	–	63.1	27.5	9.4	100.0	–	–
1978	80.3	17.7	2.0	93.9	6.1	–	63.5	27.0	9.5	100.0	–	–

	NP-Ca-			NPK--			NPKCaMg		
	G	L	H	G	L	H	G	L	H
1963	89.1	5.5	5.4	69.0	30.4	.6	68.5	30.4	1.1
1967	100.0	–	–	75.4	24.6	–	75.2	24.5	.3
1971	100.0	–	–	99.4	.6	–	98.9	1.1	–
1975	100.0	–	–	100.0	–	–	96.0	4.0	–
1978	100.0	–	–	100.0	–	–	93.4	6.6	–

plots can be botanically better distinguished since about the 7th year and 3) the importance of the cutting regime on the relative abundances of either early flowering grasses or low-growing other species.

Similar to the PGE the species number also declines in Melle, although hardly in the case of N-deficiency. Bakelaar & Odum (1978) stated that, in contrast with productivity, biomass and dominance, the number of species will diminish after addition of nutrients in either aquatic or terrestrial ecosystems.

'Wageningen' (Van den Bergh, 1979) and 'Melle' (Fig. 4a) have control plots characterized by one or two dominant species. Either experiment concerns a very different soil type, while the cutting regime is much less intensive than in Melle. The PGE however does not have dominant species on the unmanured plots (Thurston, 1969): the vegetation is considered as diverse (Thurston *et al.* 1976) presumably because this experiment was established on species-rich grassland.

Van den Bergh (1979) distinguished three groups of species in unmanured plots: 1) species with no special trend, 2) species with a declining F% and 3) species with rise-and-fall curves of long duration. *Festuca rubra* and *Festuca pratensis* belong to the first and second group respectively (Fig. 4a). Species of the third group are absent in our control plot but, contrary to 'Wageningen', *Agrostis tenuis* is

characterized by a regularly increasing F%. Apparently this species has benefited from the withdrawal firstly of *Festuca pratensis, Lolium perenne* and *Poa trivialis,* later of *Trifolium repens* which has severely (be it temporarily) suffered from the extreme drought of 1976 in the plots which did not receive nitrogen. Under normal circumstances legumes do not need nitrogenous fertilizers (Thurston, 1969). It is apparent from Table 5 that the fluctuations of one component have occurred at the expense of others in competition for the same limiting factor. Unlike 'Wageningen', *Trifolium repens* does not disappear in Melle where it still has a level of nearly 20% at the end of the period.

As stated above the '-PKCa-'-plot is unique. The very uneven species-distribution at the start is changing to a much more even one within at least five years. Species number as well as species composition are hardly affected. For example, *Poa pratensis* and *Bellis perennis* (absent from the seed mixture) increase in F% (Fig. 4b), *Poa trivalis* is showing no trend but *Trifolium repens* a gradually diminishing importance. In addition, a dramatically decreasing F% following the exceptionally dry summer of 1976 is recorded in both latter species, whilst *Poa pratensis* and *Bellis perennis* seem to take advantage of it. However all those disturbances are temporary, but are likely expressions of a strong interspecific competition. This is also clearly reflected in the ratio of the three main

230

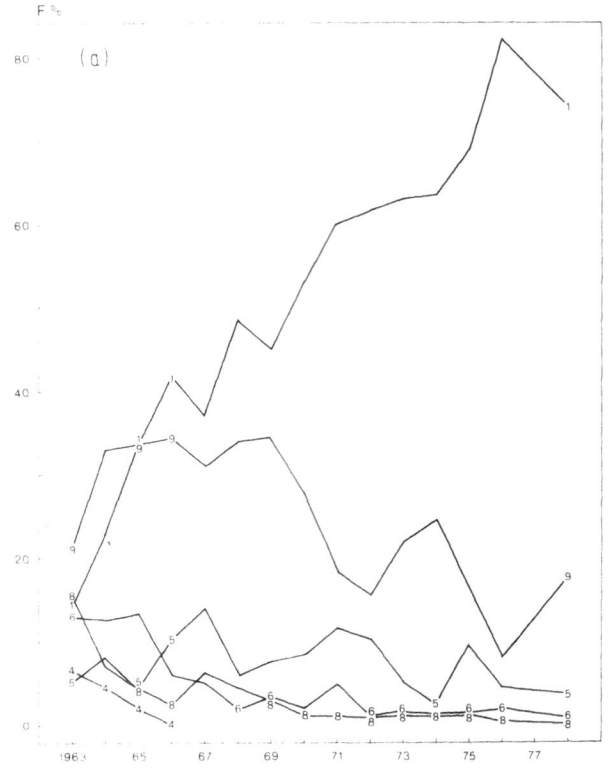

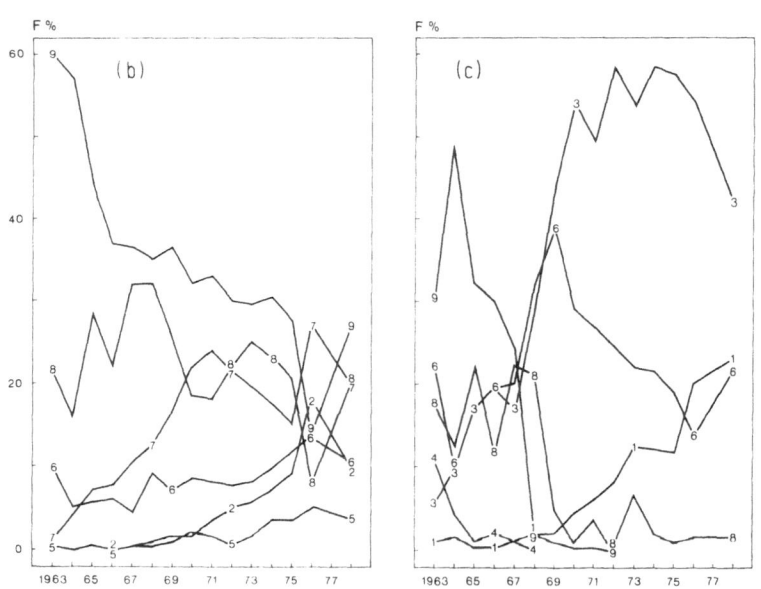

Fig. 4. Frequency percentages of some species in three treatments plotted against time: (a) control plot, (b) -PK Ca-. (c) NPK--. Numbers are referring to: 1 = *Agrostis tenuis*, 2 = *Bellis perennis*, 3 = *Dactylis glomerata*, 4 = *Festuca pratensis*, 5 = *Festuca rubra*, 6 = *Lolium perenne*, 7 = *Poa pratensis*, 8 = *Poa trivialis*, 9 = *Trifolium repens*.

components: the F% values recorded are 67.2/13.9/18.9 for grasses/legumes/herbs.

When calcium is permanently omitted both *Festuca pratensis* and *Trifolium repens* disappear after either 5 or 9 yr. As in plots with phosphorus and potassium deficiency both legumes and herbs disappear rather fast (see also Rabotnov, 1977): the grasses evidently get the upper hand at the expense of the former inasmuch as they are competing for the same limiting resource. As to the group of grasses, Figure 4c strongly gives the impression that different species populations are succeeding and/or suppressing each other. *Lolium perenne* has a lowest point at the moment when *Trifolium repens* reaches a peak; when the latter is disappearing the former is able to restore. At the same time the F% of *Dactylis glomerata* is steadily increasing and, exceeding a certain level in the seventh year, the F% of *Lolium perenne* regularly declines whilst that of *Agrostis tenuis* increases. *Lolium perenne* does not seem to recover at the expected rate (maybe reinforced by the negative influence of the very dry summer of 1976), although *Dactylis glomerata* decreases since 1975, whereas the increase of *Agrostis tenuis* persists. There is evidence that *Lolium perenne* is really competing with *Dactylis glomerata* but not with *Agrostis tenuis,* although the exact role of some endogenous factors remains obscure just like in other studies (Van den Bergh, 1979; Silvertown, 1980). Unlike the conclusion of Rabotnov (1977) *Agrostis tenuis* ought not to be dominant in the early stages to show later an increasing participation in a NPK-treatment.

In the 'NPKCa-'- and 'NPKCaMg'-plots (not presented here) *Trifolium repens* is dominated by *Lolium perenne.* As deficiency carries on as in the plots control and 'NP-Ca-'- (not presented here), *Lolium perenne* practically disappears whilst *Poa trivialis* at least diminishes. *Agrostis tenuis* benefits most from this process. At the beginning it shows a vegetative reproduction in a nearly explosive way like in the plots control, 'NPK--', 'N-KCa-' and 'NP-Ca-' (both the latter not presented here). This expansion is later delayed by obtruding species as *Festuca rubra* and *Poa pratensis,* which is most apparent in the treatments 'N-KCa-' and 'NP-Ca-'. As control, 'N-KCa-' and 'NP-Ca-' are converging it will be seen that in future 'NPK--' will apply to them if fertilizer conditions do not radically change. At this moment *Agrostis tenuis* is explicitly the only

species in that treatment with regularly increasing F%.

Obviously we are observing an invasion of *Agrostis tenuis,* being not part of the original seed mixture, in 4 out of 7 treatments. This does not happen at different times but at a different rate: very fast in 'NP-Ca-', gradually in the control plot and 'N-KCa-', slowly in 'NPK--'. The vegetative expansion, displayed by *Agrostis tenuis* with its stolons, has been made possible only by the retreat of other species, although the former easily tolerates P- and K-deficiency (Thurston, 1969).

The 'NPKCa-'- and '-PKCa-'-swards do not yield a significant correlation between diversity and Motomura's constant. Both treatments are characterized by the smallest fluctuations in species richness as well as Motomura values (Table 4 and Fig. 3, e & g). In the '-PKCa-'-treatment evenness and dominance are increasing and decreasing respectively; consequently, diversity has to increase, presumably due to the increasing importance of weeds with time. The latter fact is unique in this set of experimental plots. No direct evidence however is available to explain the same lack of correlation in the 'NPKCa-'-plots. The t-value is lying not far from significance level (t-test for simple linear regression: $0.15 < p < 0.1$).

The application of nitrogen in 5 out of 7 treatments causes in 4 of them a drastical decline of species richness and species diversity. Particularly the herbs are reduced and finally eliminated, but also some grasses as *Dactylis glomerata, Poa trivialis, Lolium perenne* and *Festuca pratensis,* naturally dependent on the fertilizer combination. Changes of some species quantities due to minor elements as e.g. aluminium and manganese cannot be discussed here but their importance has been pointed out by Thurston *et al.* (1976).

To understand such highly complex relationships between the three components and between any particular species within a single component, the different effects of external origin (e.g. the change of both concentrations and ratios of the mineral nutrients, Van den Bergh & Braakhekke, 1978) as well as internal origin, eventually resulting in a more or less equilibrated competition, can only give an incomplete idea of it. This multivariate approach describes in an efficient way only what is going on. Consequently, it is hard to believe that it would be valuable to predict quantitative changes in populations of grasslands, even experimentally established. This was also emphasized by Austin (1977) and Van den Bergh (1979).

References

Austin, M. P., 1977. Use of ordination and other multivariate descriptive methods to study succession. Vegetatio 35: 165–175.

Bakelaar, R. G. & Odum, E. P., 1978. Community and population level responses to fertilization in an old-field ecosystem. Ecology 59: 660–665.

Behaeghe, T. & Slaats, M., 1968. Kompetitie tussen plantensoorten in jong grasland. I.W.O.N.L.-verslagen over navorsingen 35.

Behaeghe, T. & Cottenie, A., 1976. Aspects botaniques et analytiques de l'évolution à longue échéance de l'état nutritif du sol. Ann. Agron. 27: 819–836.

Behaeghe, T., Traets, J. et al. Beknopte verslagen van de resultaten bekomen in het raam van het Nationaal Centrum voor Grasland- en Groenvoederonderzoek. Rijksuniversiteit Gent.

Bergh, J. P. van den & Braakhekke. W. G., 1978. Coexistence of plant species by niche differentiation. In: A. H. J. Freysen & J. W. Woldendorp (eds.), Structure and functioning of plant populations. Verh. Kon. Ned. Akad. Wet., Afd. Natuurk. 2de reeks, 70: 125–138. North-Holland Publ. Co., Amsterdam.

Bergh, J. P. van den, 1979. Changes in the composition of mixed populations of grassland species. In: M. J. A. Werger (ed.), The study of vegetation, p. 57–80. Junk, The Hague.

Björnsson, H., 1978. Analysis of a series of long-term grassland experiments with autocorrelated errors. Biometrics 34: 645–651.

Borowiec, S., Skrzyczynska, J. & Kutyna, I., 1974. The effect of fertilization and liming on the constancy of occurrence and numbers of weeds on sandy soils on loam. Ekol. Polska 22: 319–337.

Jancey, R. C., 1979. Species ordering on a variance criterion. Vegetatio 39: 59–63.

Langhe, J. E. De, et al., 1978. Nouvelle Flore de la Belgique, du Grand-Duché de Luxembourg, du Nord de la France et des Régions voisines. Edit. Patrim. Jard. Bot. Nat. Belgique, Bruxelles, 899 pp.

Lebart, L. & Fénelon, J. P., 1975. Statistique et informatique appliquées. Dunod, Paris, 439 pp.

Motomura, I., 1932. A statistical treatment of association (in Japanese). Japan. J. Zool. 44: 379–383.

Neave, H. R., 1978. Statistics Tables. Allen & Unwin, London, 88 pp.

Orlóci, L., 1978. Multivariate analysis in vegetation research. Junk, The Hague, 451 pp. 2nd ed.

Rabotnov, T. A., 1977. The influence of fertilizers on the plant communities of mesophytic grasslands. In: W. Krause (ed.), Application of vegetation science to grassland husbandry (Handbook of vegetation science 13), p. 459–497. Junk, The Hague.

Seal, H. L., 1968. Multivariate statistical analysis for biologists. Methuen, London, 209 pp.

Siegel, S., 1956. Nonparametric statistics for the behavioral sciences. McGraw-Hill Kogakusha Tokyo, XVII + 312 pp.

Silvertown, J., 1980. The dynamics of a grassland ecosystem: botanical equilibrium in the Park Grass Experiment. J. Appl. Ecol. 17: 491–504.

Swaine, M. D. & Greig-Smith, P., 1980. An application of principal components analysis to vegetation change in permanent plots. J. Ecol. 68: 33–41.

Thurston, J. M., 1969. The effect of liming and fertilizers on the botanical composition of permanent grassland and on the yield of hay. In: I. H. Rorison (ed.), Ecological aspects of the mineral nutrition of plants, Brit. Ecol. Soc. Symp., p. 3–10. Blackwell, Oxford.

Thurston, J. M., Williams, E. D. & Johnston, A. E., 1976. Modern developments in an experiment on permanent grassland started in 1856: effects of fertilisers and lime on botanical composition and crop and soil analyses. Ann. Agron. 27: 1043–1082.

Traczyk, T., Traczyk, H. & Pasternak, D., 1976. The influence of intensive mineral fertilization on the yield and floral composition of meadows. Pol. Ecol. Stud. 2: 39–47.

Traczyk, T. & Kotowska, J., 1976. The effect of mineral fertilization on plant succession of a meadow. Pol. Ecol. Stud. 2: 75–84.

Vries, D. M. de, 1937. Methods of determining the botanical composition of hayfields and pastures. Rep. 4th Intern. Grassl. Congr., Aberystwyth: 474–480.

Watt, A. S., 1971. Factors controlling the floristic composition of some plant communities in Breckland. In: E. Duffey & A. S. Watt (eds.), The scientific management of animal and plant communities for conservation, p. 137–152. Blackwell, Oxford.

Weber, E., 1972. Grundriss der biologischen Statistik. Fischer, Jena, 706 pp.

Whittaker, R. H., 1975. Communities and ecosystems. 2nd ed. Macmillan, New York, XVIII + 387 pp.

Williams, E. D., 1978. Botanical composition of the Park Grass plots at Rothamsted 1856–1976. Rothamsted Exp. Station, Harpenden, 59 pp.

Accepted 8.7.1981.

The effect of cutting and fertilizing on the floristic composition and production of an *Arrhenatherion elatioris* grassland*

M. J. M. Oomes & H. Mooi**

Centre for Agrobiological Research, Postbox 14, 6700 AA Wageningen, The Netherlands

Keywords: *Arrhenatherion elatioris*, Cutting, Fertilizing, Floristic composition, Grassland Management, Production

Abstract

Results are presented of an eight year old management experiment in a wet *Arrhenatherion elatioris* grassland on a heavy clay soil. The treatments were different cutting dates and frequencies or N (PK) fertilizing combined with June cutting. The treatments June with or without a second cut in September or only one cut in August gave a relatively stable vegetation. Cutting in early May produced a grassy dense sward and a second cut in September stimulated, moreover, some low growing dicots. The biomass that is produced after a first cut in May or June prevents lower growing species from persisting or spreading. Cutting later than the beginning of August produces a vegetation dominated by some rougher species. The number of species increases from 52 per 100 m² to 55, when the vegetation is cut in May, May and September or June; never cutting caused a decrease to 38 per 100 m². The annual dry matter production in two cuts is 5.6–6.1 ton/ha, fertilizing increased this level with 1.0–2.0 ton. Up till now, the number of species has not diminished. The results are discussed in relation to some data on growth physiology and growth strategy of plant species. The importance of a second cut or grazing in September is demonstrated and explained. Some practical implications are given for management aimed at maintaining or regenerating this grassland type.

Introduction

In 1972 we started a management experiment to gain some information about the effect of different cutting dates and -frequencies on the botanical composition and productivity of an *Arrhenatherion elatioris* grassland, and to obtain a basis for management advice. Management aims at stabilizing or improvement from a floristic point of view, and at regenerating this grassland type from former more intensively used agricultural grassland.

As was expected some of the treatments had an adverse effect on the botanical composition. They were still carried out because the results were intended to be used in explaining the effects of the treatments on individual species. While there are many data from descriptive research at different places and times and various investigations on the effects of fertilizers on the botanical composition of natural grasslands (Williams, 1978; Van den Bergh, 1979), there are only few experiments in which cutting date and -frequency are compared as to their effects on changes in the vegetation.

The experiment was carried out in a wet *Arrhenatheretum elatioris alopecuretosum* grassland (Westhoff & den Held, 1969; Klapp, 1965, Ellenberg, 1978), with elements of the *Filipendu-*

* Nomenclature of species is according to Heukels & van Ooststroom (1975).

** Acknowledgements
We are grateful to dr. ir. J. P. van den Bergh and ir. Th. A. de Boer for criticizing the manuscript and to mrs. A. H. van Rossem for translating the text. We thank all people who analysed the yearly 4000 frequency samples.

Vegetatio 47, 233–239 (1981). 0042-3106/81/0471 0233/$1.40.

Table 1. Changes in the floristic composition of an *Arrhenatherion elatioris* grassland on a heavy clay soil, caused by different cutting treatments. Only those species are shown which occurred sufficiently frequent (> 2%) to be described by means of a frequency analysis. In the first column the frequency percentage is given at the beginning of the experiment in 1972. Column (a) shows the average level of the frequency percentages in the last two years (1978 and 1979). Column (b) gives the average trend in the changes in the four plots.

Symbols: (=) unchanged, (+) increase, (++) large increase, (–) decrease, (--) large decrease, (?) uncertain trends. Where no symbol occurs the species did not occur at frequency percentage > 2.

Species occurring at percentages < 2:

Molinio-Arrhenatheretea species: Achillea ptarmica, Deschampsia cespitosa. Equisetum palustre. Lysimachia vulgaris. Ophioglossum vulgatum. Prunella vulgaris. Trifolium pratense.

Arrhenatherion elatioris species: Arrhenatherum elatius. Dactylis glomerata. Daucus carota. Heracleum sphondylium. Trifolium dubium.

Filipendulion species: Lythrum salicaria. Stachys palustris. Thalictrum flavum. Valeriana officinalis.

Calthion palustris species: Bromus racemosus. Carex disticha.

Agropyro-Rumicion crispi species: Elytrigia repens, Juncus effusus, Potentilla anserina, Ranunculus flammula. Rorippa sylvestris, Rumex crispus, Stellaria palustris.

Remaining species: Carex acuta, C. riparia, Cirsium palustre, C. vulgare, Crepis biennis, Festuca rubra, Galium palustre, Glyceria maxima. Hordeum secalinum, Mentha arvensis, Myosotis scorpioides, Polygonum aviculare, Stellaria graminea. S. media, Veronica serpyllifolia.

	1972	June Sept.		Aug. Sept.		June		May Sept.		May		Sept.		uneven year June		never cut		June NPK		June N	
	freq. %	a	b	a	b	a	b	a	b	a	b	a	b	a	b	a	b	a	b	a	b
Molinio-Arrhenatheretea																					
Cerastium fontanum	15	5	=	5	=	5	=	5	=	5	=	5	=	5	+	0	--	0	--	0	--
Anthoxanthum odoratum	50	20	=	50	-+	50	-+	10	=	10	=	10	=	0	--	0	--	10	--	10	--
Plantago lanceolata	30-50	75	++	10	==	20	+-	10	+-	30	+	0	=	0	=	0	--	0	--	0	--
Vicia cracca	20	20	=			20	+-	10	+-	30	+	10	+	10	+	5	--	10	--	10	--
Cardamine pratensis	20	20	=		?	10	=	5	=	5	=	5	=	5	=	5	--		==		==
Rumex acetosa	80	80	=		=	30	=	30	=	30	--	40	=	50	-+	20	=		==		==
Holcus lanatus	60-80	60	=	90	+		=		=		=	50	-+		=	15	=		==		==
Centaurea pratensis	60	30	--	40	--	15	=	15	=	15	=	0	=	5	=	0	--	5	--	5	--
Arrhenatherion elatioris																					
Lathyrus pratensis	20-30	15	==		=		=	10	=	10	=		=	10	=	5	--	10	--	10	--
Festuca pratensis	10-20		==	5	--		=	0	=	0	=	0	=	0	=	0	--	.5		5	--
Chrysanthemum leucanthemum	0-10	0	--		?		=		=		=	0	=	0	=	0	--	0	--	0	--
Bromus mollis	10-20	30	+	5	--		=	0	=	0	=	0	=	5	-+	0	--	10	+-+	15	+
Alopecurus pratensis	50-70	90	+	80	=	70	+	80	+	80	+	80	+	80	+	80	+	80	+	90	+
Lysimachia nummularia	0-5	10	+		--		=		=	0	=	0	-		=	0	--	10	+	10	+

Table 1. (Continued)

	freq. %	1972 June Sept. a	b	Aug. a	b	June a	b	May Sept. a	b	May a	b	Sept. a	b	uneven year June a	b	never cut a	b	June NPK a	b	June N a	b
Lolium perenne	80–90	70	–	70	–	40	–	60	–	25	–	10	–	5	–	0	–	20	–	20	–
Ranunculus acris	50–70	0	=		=	30	–	50	–	10	–	20	–	10	–	5	–	20	–	20	–
Bellis perennis	5–10		=		=		=		=	0		0	=	0	–	0		0		0	
Phleum pratense	5–10	5	–		=		=		=					0		5		5		5	
Cynosurus cristatus	0–10	0	–	60	++		=		=	0	–	5	–	0		0		0		0	
Leontodon autumnalis	0–5				=																
Symphytum officinale	0–10	10	=	10	+		=	10	+	10	+		=		=		=	5		0	
Trisetum flavescens	0–5	5	=	5	+	10	+		+					10	+		++		+		
Filipendulion																					
Phalaris arundinacea	0–5											5	+	10	=	10	+	5	+		
Filipendula ulmaria	40–50	65	+	70	+	70	+	80	++	70	+	90	++	90	++	80	++	65	+	65	+
Calthion palustris																					
Lychnis flos cuculi	0–10	20	+	40	++	15	=+		=		=		=		=	0			=		=
Lotus uliginosus	0–10	10	=	10	+				=		?				=						
Rhinanthus serotinus	0–10	0		0	+–		=														
Agropyro-Rumicion crispi																					
Poa trivialis	80–90	20	–	20	–	15	–	20	–	10	–	80	–+	10	–	20	–	30	–	20	–
Trifolium repens	10–20	0	–	20	+–	0	=		=	0		0	–	0	–	0		0	–	0	–
Ranunculus repens	30–40	50	+	40	+–	70	+	30	+–	20	+–	5	–	10	–	5		10	–		=
Agrostis stolonifera	20–30	50	=	40	+–	70	+	80	++	50	+	40	+–	30	+	10	+–	60	+–+	60	+–+
Poa pratensis	5–10	20	=		+	20	+	15	+	40	++		=	15	+	10	=	30	++	10	+
Agrostis canina	0–5	5		5			=	10	+	10	+	20	++		=						
Carex hirta	0–5						=														
Remaining species																					
Taraxacum 'officinale' s.l.	60–80	10	=–	50	+–	5	–	5	–	0		0		0	–	0		0		5	–
Polygonum amphibium	0–10	10	=+		=		=	15	=+	15	=+		=	0	–	0	=+		=		=
Glechoma hederacea	10–20	40	+	30	+	70	+	60	++	70	++	10	+–	25	+–	15	+–	60	++	60	++
Number of species	52	52		53		55		55		55		53		49		38		52		52	

lion and *Calthion palustris* alliances. Dominating species are *Alopecurus pratensis, Filipendula ulmaria, Holcus lanatus, Lolium perenne* and *Ranunculus acris* (Table 1). The average number of species was 52 on 100 m². The grassland is located in the river-clay district between the rivers Rhine and Waal.

The soil is a heavy basin-clay (clay fraction $< 2 \times 10^{-3}$ mm = 47% of dry soil and organic matter = 20%, pH-KCl = 4.9), with a low P and K status (11 mg P_2O_5 and 16 mg K_2O per 100 g of dry soil). From mid-November until the end of March the vegetation is flooded some 5–10 cm. The average dry matter production, harvested in two cuts, is 6–7 ton/ha/yr. Up to 1970 the grassland was fertilized at a level of 50 to 75 kg N/ha/yr, and after a hay-cut in June grazed by cattle in September. Both the application of fertilizer and the fluctuation of the water level are supposed to be responsible for the relatively high dominance of some *Agropyro-Rumicion crispi* species.

Materials and methods

Each management treatment was carried out on four plots of 100 m² each, distributed at random in the field. To describe changes in the floristic composition frequency analysis was used with a complementary biomass estimation of species not present in the samples. Monocotyledons were sampled by taking 50 samples of 25 cm² each per plot (De Vries, 1937), dicotyledons by 50 samples of 400 cm² each per plot. In a preliminary trial this size was found to be large enough for a 70% score of the dicotyledons present. The plots were analysed each year at the beginning of May. We can only describe changes in more or less frequently occurring species; for most species not occurring in the frequency analyses, the period of 8 yr was too short to describe changes. A tendency to increase, constancy or decrease in time was estimated from four replicates per species per treatment.

Dry matter production was determined on 20 m² on each replicate.

The fertilizer treatments were NPK (N=50 kg/ha, P = 20 kg/ha. K = 20 kg/ha) and N (N = 50 kg/ha), N supplied as ammonium nitrate.

The treatment: cutting in June with a second cut in September may be considered as a continuation of the past management.

Results

The results are summarized in Table 1. The first column shows the average frequency % of all the plots at the beginning of the experiment in 1972. For each treatment the average % has been given for the last two years (a) and the tendency to change up till June 1979 (b). Changes in biomass are not given in the table, only some distinct changes of species not occurring in the frequency samples are mentioned in the text.

a) Frequency percentages

Especially cutting in June with or without a second cut in September, and only one cut in August gave a relatively stable vegetation. However, we expect that the very low spread of *Filipendula ulmaria, Symphytum officinale* and *Holcus lanatus* in the August cutting treatment will cause a rougher and denser vegetation and in the long run a loss of some species. Very specific was the increase in *Plantago lanceolata* and *Bromus mollis* when the vegetation was cut in June and September. Striking in the August treatment was that some *Calthion palustris* species and also *Cynosurus cristatus* and *Crepis biennis* increased.

In comparing cutting in June and September with only one cut in September, we conclude that leaving the biomass until September prevents some lower growing species from persisting or spreading (*Plantago lanceolata, Lysimachia nummularia, Lolium perenne, Bellis perennis, Ranunculus repens, Taraxacum officinale, Glechoma hederacea*). Moreover, one cut in September produced an increasingly rough vegetation by the fast increase of *Filipendula ulmaria* during the first three years.

The decrease of lower growing species and increase of *Filipendula ulmaria* also occurred when the vegetation was cut only once in June in two years, and more extreme when the vegetation was never cut. By this management some species not occurring in the frequency samples at the beginning of the experiment increased: *Galium palustre, Carex acuta* and *C. riparia*.

Cutting in early May produced a grassy dense sward with *Agrostis stolonifera, A. canina, Poa pratensis*, and a second cut in September stimulated, moreover, some low growing species (*Cynosurus cristatus, Trifolium repens, Ranunculus repens, Bellis perennis, Cardamine pratensis, Lysi-*

machia nummularia). An interesting phenomenon was the delay in flowering and production of viable seed of some species until the end of August (*Holcus lanatus, Centaurea pratensis, Symphytum officinale, Plantago lanceolata*). Up til now this delay led to an increase in *Symphytum officinale*.

Both low levels of fertilization resulted already after two years in a dense vegetation with only few species occurring at a high frequency % (*Holcus lanatus, Alopecurus pratensis, Agrostis stolonifera, Filipendula ulmaria, Rumex acetosa*). Most other species decreased and some nearly disappeared.

b) Number of species

The number of higher plants at the beginning of the experiment was 52 on 100 m². Only the treatment: never cutting caused a decrease to 38. In the plots cut in May, May and September and June the number increased to 55. The fertilizer application did not cause a decrease up till now in the number of species.

c) Dry matter production

Figure 1 shows the dry matter production of some treatments from 1972 to 1978. In the very dry summer of 1976 the level in all the treatments was 1,5–2,5 ha. Comparing the level in the June and August cutting treatment it may be noticed that despite the longer growing season, dry matter production was not higher and the difference with the September treatment was neither as high as could be expected. The average of one cut in June was 4.6 ton ha. From determinations after 1978 the level of the second cut can be estimated at 1.0–1.5 ton/ha, so the total annual production in two cuts on this clay soil with a very low P and K status is 5.6–6.1 ton/ha. Fertilization gave a higher production level of 1.0–2.0 tons. It appeared from the N-treatment that in the first year of the experiment P and K were not limiting. N-fertilization, therefore, caused accelerated P and K removal from the soil. However, the advantage of soil exhaustion was obviated by the adverse effect on the vegetation as described under a.

Discussion

The changes observed in our experiments relate

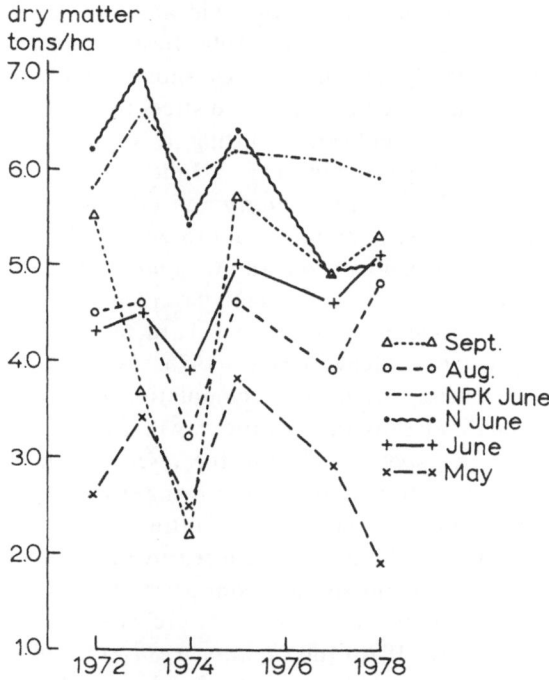

Fig. 1. Dry matter production of an *Arrhenatherion elatioris* grassland at different cutting dates and fertilizing treatments. Production in 1976 was not taken into account because of drought.

to a period of 8 yr. For many species this may be too short to show eventual cyclic changes (Van den Bergh, 1979), which may indeed be possible with *Bromus mollis, Anthoxanthum odoratum, Agrostis stolonifera* and *Vicia cracca*. It may be that this cyclic behaviour depends on the stage in which the vegetation is cut.

The results show that the growth stage at which the vegetation is cut also determines the botanical composition. Especially in long term research on succession, cutting the vegetation for a few years at a somewhat later stage than usual may considerably affect the results.

Cutting especially early in the growing season, is a drastic interference in the growth cycle. The response of the different species in a vegetation to this treatment, therefore partly explains the influence of the cutting time and frequency on development of a vegetation. A generalizing classification of growth strategy types is given by Grime (1979). Besides a ruderal growth strategy, he distinguishes a competitive and a stress tolerant strategy. Characteristic for the latter strategies is the ability to store carbohydrates in rhizomes, thickened roots or stubbles combined with rapid and slow overground production, respectively. In a permanent pasture nearly all the species show a strategy intermediate to competitive and stress tolerant.

In the treatment-series: cutting in August, September, and every odd year in June, *Alopecurus pratensis, Filipendula ulmaria* and other rougher species increase in dominance. Moreover, *Phalaris arundinacea, Carex acuta* and *C. riparia* increased appreciably in the treatment: never cutting. All these species show some distinct characteristics of a competitive strategy (Grime, 1975, 1979; Wolf, 1979). Cutting when the accumulation of carbohydrates is already well advanced, will favour these competitive species, because the reserves can be used in spreading in the next growing season.

Unfortunately, the effect of cutting on the distribution of carbohydrates and regrowth is known for only few plant species from permanent grasslands. Some examples of weeds are *Sonchus arvensis* (Pegtel, 1976) and *Cirsium arvense* (Hoogerkamp, pers. comm.). These studies showed that at least three cuts per growing season are required to reduce the overground biomass of these species. Moreover, the effect is influenced by the growth stage at the time of cutting. *Elytrigia repens* is not reduced, because the tillers rapidly grow selfsupporting in the carbohydrate supply (Cussans, 1973). Growth rate and some morphological characteristics of 132 species were studied by Grime (1975). By comparing plants in eco-physiological research with respect to growth characteristics, a better understanding will be obtained of the observations in the research on vegetation succession (Bazzaz, 1979; Grime, 1979). Moreover, the effect of management can be better predicted.

Much more is known about the growth of grasses and the effect of cutting (May, 1960; Youngner, 1973; Harris, 1978), from which some tendencies observed in this experiment can be explained.

The dominance of *Alopecurus pratensis* in all the treatments can partly be explained from its early and relatively rapid growth (Bommer, 1959). Therefore, it is one of the first species to benefit from the space and minerals present.

The grassy dense sward occurring after cutting in early May is due to the elimination of apical dominance (Laidlaw & Berrie, 1974) or to induction of tillering by reduced shading (Mitchell, 1954; Luxmoore & Millington, 1971). In this other

management experiments it has been demonstrated that a second cut, even if the yield is low, will decrease dominance of grasses and density of the sward, while the number of species will increase. This aftermath contains a high percentage of carbohydrates (Sonneveld, 1962; Smith, 1973). When it has not been cut, it stimulates regrowth in spring (Behaeghe, 1979), thus giving the grasses a lead over most of the dicots.

This second cut will also produce a low vegetation structure and an open sward in autumn and spring, two main elements of the regeneration niche (Grubb, 1977) which stimulates establishment of seedlings (Bakker *et al.,* 1980) and the vegetative spread of plant species. Together with the time of cutting, dependent on the time of seed production of some expected or desired plant species, it is important for the regeneration of this vegetation. This may explain the increase of *Holcus lanatus, Rhinanthus serotinus* and *Cynosurus cristatus* in the August cutting treatment and of *Plantago lanceolata* and *Bromus mollis* in the June + September cutting treatment. Cutting in May increased the number of species from 52 to 55 per 100 m^2. Despite most species not being able to produce viable seed and the grassiness of the sward, the establishment of seedling is successful, probably because of the low vegetation structure in early summer. At present, we are studying the influence of vegetation structure on germination and establishment.

On the one hand, early cutting is necessary for establishment, but on the other hand, a later cut is of importance. The later cut will enable the seed production of species which are required to spread. Moreover, many young tillers will die from light deficiency (Mitchell, 1954) thus producing an open sward with more regeneration niches. For regeneration of grasslands poor in species to those rich in species it is important that fields with a low number of species which are cut early, are adjacent to fields with more species which are cut later.

The number of species decreases (in 3–4 yr), only if the vegetation is not cut at all. This is due to competition of some dominant taller growing species and to the formation of a dense layer of litter. Al-Mufti *et al.* (1977) suggested that the number of species would decrease by fertilizing, because the production is much higher than 7.0 ton/ha. However, we think this is doubtful since time and frequency of cutting also determine the number of species.

When an *Arrhenatherion elatioris* grassland is to be regenerated from a more intensively used grassland, management should be directed in the first place at the greatest possible removal of minerals (soil exhaustion). It is therefore important to know that one cut in September hardly produces more dry matter than a cut in June. In this case a cut in June or July is preferable, because of the longer regrowth period which enables a higher yield in the second cut and prevents species from dying out in a crop that is too dense. Cutting later to enable the seed production of species desired is only justified, when the yield of that cut is lower than 4.0 ton/ha of dry matter. Cutting or grazing in early May and a second cut in September seems a possibility to combine both effects to some extent.

References

Al-Mufti, M. M., Cydes, C. L., Furness, S. B., Grime, J. P. & Band, S. R., 1977. A quantitative analysis of shoot phenology and dominance in herbaceous vegetation. J. Ecol. 65: 759-791.

Bakker, J. P., Dekker, M. & de Vries, Y., 1980. The effect of different management practices on a grassland community and the resulting fate of seedlings. Acta Bot. Neerl. 29: 469-482.

Bazzaz, F. A., 1979. The physiological ecology of plant succession. Ann. Rev. Ecol. Syst. 10: 351-371.

Behaeghe, T. J., 1979. De seizoenvariatie in de grasgroei. Thesis, Gent, 272 pp.

Bergh, J. P. van den, 1979. Changes in the composition of mixed populations of grassland species. In: M. J. A. Werger (ed.), The study of vegetation. p. 59-80. Junk, The Hague.

Bommer, D., 1959. Ueber Zeitpunkt und Verlauf der Blütendifferenzierung bei perennierenden Gräsern. Z. Acker-Pfl. bau. 109: 95-118.

Cussans, G. W., 1973. A study of the growth of Agropyron repens (L) Bauv. in a ryegrass ley. Weed Res. 13: 283-291.

Ellenberg, H., 1978. Vegetation Mitteleuropas mit den Alpen in ökologischer Sicht. Eugen Ulmer, Stuttgart, 733 pp.

Grime, J. P. & Hunt, R., 1975. Relative growth-rate: its range and adaptive significance in a local flora. J. Ecol. 63: 393-422.

Grime, J. P., 1979. Plant strategies an vegetation processes. John Wiley and sons, Chichester.

Grubb, P. J., 1977. The maintenance of species-richness in plant communities: the importance of the regeneration niche. Biol. Rev. 52: 107-145.

Harris, W., 1978. Defoliation as a determinant of the growth, persistence and composition of pasture. In: J. R. Wilson (ed.), Plant relations in pastures. p. 65-85. C.S.I.R.O., East Melbourne.

Heukels, H. & van Ooststroom, S. J., 1975. Flora van Nederland. P. Noordhoff, Groningen, 925 pp.

Klapp, E., 1965. Grünlandvegetation und Standort. Paul Parey, Berlin, 143 pp.

Laidlaw, A. S. & Berrie, A. M. M., 1974. The influence of expanding leaves and the reproductive stem apex on apical dominance in Lolium multiflorum. Ann. Appl. Biol. 78: 75-82.

Luxmoore, R. J. & Millington, R. J., 1971. Growth of perennial ryegrass (Lolium perenne L.) in relation to water, nitrogen, and light intensity. I. Effects on leaf growth and dry weight. Plant Soil. 34: 269-281.

May, L. H., 1960. The utilization of carbohydrate reserves in pasture plants after defoliation. Herb. Abstr. 30: 239-245.

Mitchell, K. J., 1954. Growth of pasture species. I. Short rotation and perennial ryegrass. N.Z.J. Sci. Technol. 36A: 193-206.

Pegtel, D. M., 1976. On the ecology of two varieties of Sonchus arvensis L. Thesis, Groningen, 148 pp.

Smith, D., 1973. The nonstructural carbohydrates. In: G. W. Butler & R. W. Bailey (eds.), Chemistry and biochemistry of herbage. p. 106-155. Academic press. London.

Sonneveld, A., 1962. Distribution and re-distribution of dry matter in perennial fodder crops. Neth. J. Agric. Sci. 10: 427-444.

Vries, D. M. de, 1937. Methods of determining the botanical composition of hayfields and pastures. Resp. 4th Intern. Grassl. Congr. Aberystwyth, 474-480.

Westhoff, V. & den Held, A. J., 1969. Plantengemeenschappen in Nederland. W. J. Thieme, Zutphen, 324 pp.

Williams, E. D., 1978. Botanical composition of the Park Grass Plots at Rothamsted, 1856-1976. Rothamsted Exp. Sta. Harpenden, pp. 1-61.

Wolf, G., 1979. Veränderungen der Vegetation und Abbau der organischen Substanz in aufgegebenen Wiesen des Westerwaldes. Schriftenreihe für Vegetationskunde, 13. Bonn-Bad Godesberg, 50 pp.

Youngner, V. B., 1973. In: V. B. Youngner & C. M. McKell (eds). The biology and utilisation of grasses. p. 292-303. Acad. Press. New York.

Accepted 10.7.1981.

From intensively agricultural practices to hay-making without fertilization
*Effects on moist grassland communities**

L. van Duuren**, J. P. Bakker & L. F. M. Fresco***

Department of Plant Ecology, Biological Centre, University of Groningen, P.O. Box 14, 9750 AA Haren (Gn), The Netherlands

Keywords: Cluster analysis, Grassland, Hay-making, Hydrology, Nutrients, Vegetation dynamics

Abstract

In the valley of the Dutch brook 'Drentsche A' formerly fertilized lots came under hay-making without fertilization practices. The vegetation at the beginning of the experiments varied depending on the intensity of former agricultural practices.

Clustering revealed unidirectional successional lines in all lots with relatively large changes in the first four years. Changes still occur after 30 years of unfertilized hay-making. Changes concerning vegetation types as well as increasing or appearing and decreasing or disappearing species suggest a diminishing availability of nutrients as the main environmental process. Therefore it is concluded that after a period of fertilizing some restoration of plant communities associated with former agricultural practices is possible.

Variation already present within lots nearly always remained and in one lot a divarication of successional lines could be recorded, probably correlated with hydrological differences.

Introduction

In primeval forests on moist peaty or sandy/peaty soils (*Alnion glutinosae* and *Salicion cinereae*) large herbivores probably maintained grassy vegetation on open places. Tree cutting and superficial drainage by man enhanced the further development of nutrient poor grasslands, for hay- or straw-cropping. These unfertilized grasslands (*Juncion acutiflori* Br-Bl 1947) reached their max-imal extension already in Medieval times (Ellenberg, 1978) and were maintained until the beginning of this century. From that time onwards they gradually disappeared. They became fertilized on a small scale and changed into annually hayed and sometimes grazed grasslands (*Calthion palustris*). Moreover *Calthion palustris* communities originated by drainage of *Magnocaricion* communities and by drainage plus fertilizing of *Caricion curto-nigrae* communities. Subsequent heavy fertilization often changed these seminatural communities into the most intensively used cultivated grasslands: the pure pastures (*Agropyro-Rumicion crispi* with *Poo-Lolietum*) and the haypastures (*Cynosurion cristati* with *Lolio-Cynosuretum*) (Klapp, 1965; Ellenberg, 1978).

Today many of such grasslands have been abandoned. The subsequent succession in former hayfields resulted in felted swards with tall grasses, sedges or herbs and in former pastures resulted in woodland (Borstel, 1974; Böcker, 1978; Ellenberg,

* The nomenclature of taxa is according to Heukels & van Ooststroom (1977), that of syntaxa according to Westhoff & den Held (1969), Ellenberg (1978) and Everts *et al.* (1980).

** Present address: Centraal Bureau voor de Statistiek, afd. Landbouwstatistieken, P.O. Box 959, 2270 AZ Voorburg, The Netherlands.

*** The authors would like to thank Drs. N. Schotsman, Drs. A. Zeevalking-van Yperen, Drs. L. Rohof, Drs. P. Struyk, Drs. G. Boedeltje, Mrs. E. Hermans and Mr. W. van der Lans for recording the permanent plots, to Mr. E. Leeuwinga for drawing the figures and to Dr. S. Daan for correcting the English text.

Vegetatio 47, 241–258 (1981). 0042-3106/81/0471–0241/$3.60.
© Dr W. Junk Publishers, The Hague.

1978). In order to prevent felting or tree shooting the effects of burning, mulching, hay-making and grazing have been studied experimentally, usually on a small scale and in short-term studies (Holger, 1978; Riess in Ellenberg, 1978; Oomes, 1977, 1981; Bakker, 1978; Bakker *et al.*, 1980; Willems, 1979, 1980; Campino, 1980; Dierschke, 1981; Schreiber, 1981). Especially in the Netherlands attempts to restore plant communities connected with former agricultural practices are studied on a larger scale by nature management authorities (P. Bakker, 1979, Rijksinstituut voor Natuurbeheer 1979).

Hay-making without fertilization together with a stable groundwater regime possibly induces a succession from cultivated grassland communities via *Calthion palustris* communities into *Juncion acutiflori* communities. This study records the changes during nine years of unfertilized hay-making on formerly cultivated grasslands.

Material and methods

Lots acquired by the government in the nature reserve 'Stroomdallandschap Drentsche Aa' (The

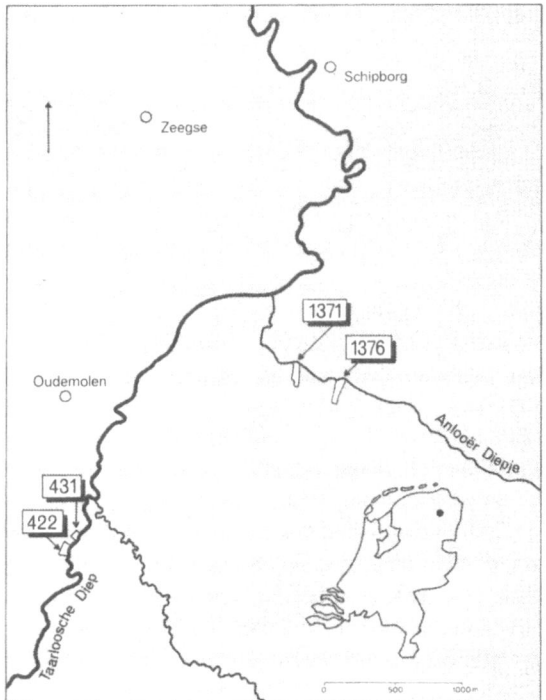

Fig. 1. Situation of lots along the 'Drentsche A'.

Netherlands) have been hayed without fertilizing for many years, and the aftermath sometimes been grazed. The lots studied (Fig. 1) were situated in the middle reaches of the brook-valley (53°05′ NL, 6°40′ EL). The groundwater regime of the main stream with peaty soils (lots 422 and 431) mainly depended on deep groundwater flow, that of the secondary stream with sandy soils (lots 1371 and 1376) on subsurface flow (Grootjans, 1980).

In each lot permanent plots (4 m²) were laid down in a transect on sandy/peaty soil perpendicular to the brook.

The lots 422, 431 and 1376 were hayed and their vegetation recorded from 1972 onwards (in lot 1376 the aftermath has been grazed in 1976). Lot 1371 has been hayed from 1946 onwards and its vegetation recorded from 1974 onwards. Only the relatively dry part of lot 1371 (permanent plots 5–9) has been reclaimed i.e. the soil had been ploughed superficially prior to 1946. In 1975 the lots were not surveyed with exception of lot 1371.

The vegetation in these permanent plots was recorded in May or June until 1976, using a slightly modified scale of Braun-Blanquet ($2 m$ = cover <5%, many individuals; $2a + b$ = cover 5–25%), from 1976 onwards using the decimal scale (Londo, 1976). The relevés recorded with the scale of Braun-Blanquet were transformed to the decimal scale as follows: $r = .1, + = .2, 1 = .4, 2 m = .4, 2a + b = 1, 3 = 4, 4 = 6, 5 = 9$. Those parts of the decimale scale were chosen that agreed with the middle of the parts of the Braun-Blanquet scale (Londo, 1976, Table 1). The cover values (both scales), expressed in cover %, were transformed by calculating the integer part of the square root of these values.

Dissimilarities between relevés were calculated by means of a dissimilarity index based on Sørensen (1948, Goodall, 1973).

$$D(a,b) = (1 - \frac{2\Sigma \min (a_i, b_i)}{\Sigma a_i + \Sigma b_i}) \times 100\%$$

a_i = transformed value of species i in relevé a. When used for presence-absence data $a_i = 1$ if species i is present in relevé a, else $a_i = 0$.

Dissimilarities between consecutive years were calculated for each permanent plot, as well as dissimilarities between each of the years and the initial year for each permanent plot.

Table 1. (A-C). Species composition of the clusters. Average cover of species when frequency is ≧50%. When frequency is <50% the symbol + is used. Accompanying species also includes less important character species.

A

Average number of species	17	22	24	20	19	22	19
Number of relevés	10	14	6	7	5	9	4
Clusternumber	1.1	1.5	1.3	1.2	2.1	1.4	1.6
Cynosurion cristati							
Cynosurus cristatus		1.3	1.7	2.5	+	1.4	
Bellis perennis	+	+		2.6	1.0	+	
Leontodon autumnalis						1.1	
Poo-Lolietum							
Poa pratensis	3.4	1.0	+	3.0	5.0	1.0	1.0
Poa trivialis	3.6	11.4	6.2	4.6		8.0	5.5
Ranunculus repens	7.0	21.2	15.7	7.4	13.2	8.8	30.0
Trifolium repens	+	2.2	1.5	+	7.0	1.1	1.0
Lolium perenne				+	3.3		
Agrostis stolonifera		5.1	1.2			5.0	
Molinio-Arrhenatheretea							
Deschampsia cespitosa	+		+	+	+		11.7
Trifolium pratense	2.8			+			
Festuca pratensis	11.2	2.1	5.8	11.7	5.2	5.6	1.2
Holcus lanatus	18.4	20.1	1.3	10.6	44.0	10.7	8.3
Festuca rubra	+	1.2	1.2	4.6	5.2	3.2	5.5
Rumex acetosa	2.9	1.5	2.2	5.7	11.2	4.4	2.2
Ranunculus acris	2.8	1.1	2.3	3.7	19.2	4.1	1.2
Cardamine pratensis	3.6	1.3	1.7	2.4	3.6	1.3	2.0
Cerastium holosteoides	1.4	1.0	+	2.1	3.2	1.2	0.8
Plantago lanceolata			+			5.8	
Taraxacum spec.	+	+	+	2.4	2.0	2.3	0.8
Lotus uliginosus							1.0
Juncus effusus	+	1.0	0.8	1.7	3.3	1.0	
Cirsium palustre		+				+	
Rhinanthus serotinus		+	17.0	+		7.1	1.7
Equisetum palustre	2.2	1.0	1.0	+	+	+	+
Accompanying species							
Alopecurus geniculatus	4.4	2.1	1.0	1.7	3.5	1.0	+
Glyceria fluitans	5.8	5.2	6.0	+	+	1.2	3.2
Juncus articulatus	+	1.1	1.0			+	
Myosotis palustris	+	1.0	+		+		
Ranunculus flammula	+	1.2	1.5		+		
Equisetum fluviatile	+	+	1.0				
Carex nigra		+		+		1.7	
Carex ovalis		+	+	+	+	+	+
Cirsium arvense				+	2.0		
Rumex obtusifolius	+			1.2	1.2		
Anthoxanthum odoratum	2.6	2.9	9.5	5.7	4.0	9.1	1.7
Stellaria alsine	+	+	+	+	2.0		+
Eleocharis palustris	+	+	+	+	+	+	

Addenda: other accompanying species

Potentilla anserina, Agrostis tenuis, Sagina procumbens, Filipendula ulmaria, Montia fontana, Carex panicea, Mentha aquatica, Veronica serpyllifolia, Elytrigia repens, Phalaris arundinacea, Rumex crispus, Polygonum spec.

Table 1. (Continued)

B

Average number of species	21	25	35	30	32	33	24	36	34
Number of relevés	3	9	6	29	11	23	2	5	8
Clusternumber	5.5	5.4	5.7	5.1	5.8	5.2	5.9	5.6	5.3
Calthion palustris									
Caltha palustris	6.0	+	17.3	5.6					
Glyceria maxima	3.3	13.8	4.0	+	8.4	1.1	10.0	4.8	1.5
Lychnis flos-cuculi	2.7	2.7	4.0	1.9	2.3	1.1	+	2.2	1.0
Carex acutiformis	16.0	6.7	19.0	15.4	+	8.1	5.5	19.6	14.6
Crepis paludosa	+	2.1	3.7	9.1	2.0	5.0		4.7	7.8
Myosotis palustris	2.7	1.8	3.0	0.8	3.6	0.9	+	3.2	0.6
Bromus racemosus	+	+	+	1.2		+			1.1
Molinio-Arrhenatheretea									
Festuca pratensis	+	2.6	2.0	1.4	7.9	1.5	1.0	2.8	2.5
Holcus lanatus	2.7	3.8	5.0	1.2	8.9	8.5	3.0	5.2	5.9
Rumex acetosa	2.7	3.6	4.0	1.9	4.0	1.7	3.0	5.2	1.6
Cerastium holosteoides	2.0	2.9	4.0	1.0	3.3	1.1	3.0	3.2	1.0
Cardamine pratensis	+	2.2	3.7	1.1	2.0	1.0	1.5	3.2	1.1
Ranunculus acris	2.0	2.2	3.7	8.1	2.4	3.3	3.0	3.0	1.9
Festuca rubra	+	+	4.0	1.4	14.8	6.6		15.5	2.0
Plantago lanceolata	+	2.6	3.3	2.0	3.1	3.9	6.0	6.4	4.8
Trifolium pratense	+	+	1.7	3.8	6.4	7.9		3.3	1.3
Taraxacum spec.	2.0	2.0	1.5	1.3	2.0	1.1	2.0	2.0	1.2
Vicia cracca		2.0	1.2	+	+	1.2		+	1.2
Cirsium palustre		+	1.3	0.6	1.7	1.5		1.3	1.1
Lotus uliginosus		+	2.3	+	+	5.2		18.0	2.4
Rhinanthus serotinus	+	+	4.5	4.1	+	3.7		+	1.8
Juncus effusus		+	+	+				+	
Achillea ptarmica		+	+	+					
Lysimachia vulgaris					+	+		+	
Galium uliginosum					+	+		+	
Equisetum palustre				+	4.5	1.8	7.0	3.2	1.0
Trifolium dubium				+	6.5	2.3		+	+
Stellaria graminea			+	+	+	+		+	
Bromus mollis		+	+	+		+			
Anthriscus sylvestris	+	+	+	+	+	+	6.0	4.4	1.7
Veronica chamaedrys	+	+	5.3	+	+	+	+	1.3	1.0
Phyteuma nigrum			+	+		+			+
Cynosurion cristati									
Cynosurus cristatus				+	3.3	2.3	+	2.8	1.4
Bellis perennis					1.3	1.0	1.0	1.3	+
Leontodon autumnalis					+	+			
Phleum pratense				+				+	
Poo-Lolietum									
Lolium perenne					2.1	+	1.0		+
Poa trivialis	8.0	7.3	2.6	3.8	7.4	3.0	2.0	3.0	7.9
Poa pratensis	2.0		6.4	+	2.7	1.0		4.4	
Ranunculus repens	+	2.1	2.2	1.6	3.9	1.2	3.0	2.8	2.5
Trifolium repens		+	+	1.2	3.1	2.3		+	+
Plantago major					+	+			
Agrostis stolonifera			+			+			

Table 1. (Continued)

B

Average number of species	21	25	35	30	32	33	24	36	34
Number of relevés	3	9	6	29	11	23	2	5	8
Clusternumber	5.5	5.4	5.7	5.1	5.8	5.2	5.9	5.6	5.3

Filipendulion

Filipendula ulmaria	10.0	22.7	25.0	16.7	2.4	7.1	7.0	14.8	20.5
Valeriana officinalis	8.0	+	6.5	+					
Lythrum salicaria	+	+	1.7	+	+	+		+	+
Hypericum tetrapterum					+	+		+	

Accompanying species

Festuca arundinacea	2.0	+	1.3	+		+		1.5	+
Juncus articulatus				+	+	+			
Rumex crispus			+	+	1.7	+	1.5	1.3	+
Equisetum fluviatile	2.0	+	2.4	1.0	3.3	1.6	50.0	5.2	5.1
Anthoxanthum odoratum	+	2.2	3.7	5.1	2.7	1.0		4.4	
Veronica arvensis	+	2.0	4.0	1.0	2.7	1.0	+	3.6	1.0
Ajuga reptans		+	1.7	1.3		+		+	1.5
Lysimachia nummularia					+	+	+	+	+
Urtica dioica		+	+		+	+	2.0	1.7	1.2
Glechoma hederacea				+		+	+	1.7	1.1
Stellaria media		+	1.7	,+	+	+			
Stellaria palustris	1.7	+	+	+		+			
Carex nigra			+		+	+		+	+

Addenda: other accompanying species

Quercus robur, Stellaria alsine, Alnus glutinosa, Chrysoplenium alternifolium, Galium aparine, Melandrium rubrum, Galium palustre, Eleocharis palustris, Phalaris arundinacea, Phragmites australis, Mentha aquatica, Sagina procumbens, Rumex obtusifolius, Veronica serpyllifolia, Polygonum spec., Rhinanthus minor, Betula spec., Carex disticha, Valeriana dioica, Ranunculus ficaria, Epilobium hirsutum, Epilobium spec., Lysimachia thyrsiflora, Holcus mollis, Calamagrostis canescens, Potentilla anserina, Prunella vulgaris, Viola palustris, Alopecurus geniculatus.

C

Average number of species	31	33	31	34	37
Number of relevés	10	5	20	21	7
Clusternumber	4.1	4.3	4.2	3.1	3.2

Juncion acutiflori

Juncus acutiflorus	40.0	36.0	30.0		1.2
Juncus subuliflorus	3.2		4.9	1.0	
Luzula multiflora		+	1.4		
Carex panicea	+		+	+	
Succisa pratensis				+	

Calthion palustris

Caltha palustris				1.2	1.7
Lychnis flos-cuculi	2.0	0.8	0.9	1.4	1.1
Crepis paludosa	+	2.8		!.1	1.6
Myosotis palustris	2.0	+		+	1.0
Orchis majalis	1.6	1.2	2.4	2.8	1.3

Table 1. (Continued)

C					
Average number of species	31	33	31	34	37
Number of relevés	10	5	20	21	7
Clusternumber	4.1	4.3	4.2	3.1	3.2
Caricion curto-nigrae					
Carex nigra	2.3	1.4	1.5	6.9	2.7
Carex echinata	+		+	+	6.1
Agrostis canina	+		+	+	
Viola palustris	+	+	+	2.4	1.3
Menyanthes trifoliata				40.8	1.8
Ranunculus flammula				+	+
Potentilla palustris				+	
Molinio-Arrhenatheretea					
Trifolium pratense	+	1.5	+	7.1	3.0
Holcus lanatus	8.0	3.6	4.8	4.0	3.3
Rumex acetosa	3.8	1.6	3.1	2.0	2.3
Cerastium holosteoides	2.0	1.0	0.9	1.2	1.3
Cardamine pratensis	2.0	1.0	1.1	1.3	1.4
Festuca pratensis	2.2	1.0	+	+	1.3
Ranunculus acris	2.4	2.2	1.1	3.0	1.4
Festuca rubra	4.4	1.2	1.6	3.9	6.4
Plantago lanceolata	13.0	8.0	15.6	5.0	2.1
Deschampsia cespitosa	3.0		1.0		
Prunella vulgaris	2.7	0.7	1.2	+	+
Taraxacum spec.					
Cirsium palustre	2.4	1.4	4.3	1.4	1.1
Juncus effusus	1.9	+	+	1.8	10.9
Lotus uliginosus	3.6	1.0	1.1	2.4	1.0
Rhinanthus serotinus	3.0	12.4	3.4	4.0	10.4
Equisetum palustre	2.0	1.0	0.9	3.0	2.7
Lysimachia vulgaris		+	1.2	+	1.0
Achillea ptarmica	+		+		+
Galium uliginosum	+	+		1.4	1.2

C					
Average number of species	31	33	31	34	37
Number of relevés	10	5	20	21	7
Clusternumber	4.1	4.3	4.2	3.1	3.2
Stellaria graminea	1.4		0.9	1.4	1.2
Dactylis glomerata	+	0.7	1.3		
Trifolium dubium	+		+		
Veronica chamaedrys			+	+	
Poo-Lolietum					
Lolium perenne			+		
Poa pratensis	2.5	1.0	+	1.5	2.5
Poa trivialis		2.8	3.2	2.1	2.3
Ranunculus repens	3.0	1.0	1.7	1.7	12.0
Trifolium repens	2.4	2.4	1.2	5.6	20.4
Agrostis stolonifera		+	3.1	+	
Filipendulion					
Filipendula ulmaria	2.0	1.4	+	+	2.3
Lythrum salicaria		+	+	+	+
Valeriana officinalis		+	+	1.4	1.3
Accompanying species					
Cynosurus cristatus	4.0	5.0	1.2	6.9	11.1
Luzula campestris	3.3	1.0	+	1.3	
Agrostis tenuis	8.1	+	15.3	+	+
Equisetum fluviatile	2.3	1.0	+	2.0	3.9
Anthoxanthum odoratum	9.4	4.6	1.2	2.8	2.9
Ajuga reptans	1.2	1.0	1.1	1.2	1.2
Carex ovalis	+	+	+		
Myosotis discolor		1.0	+		
Orchis maculata			+	0.8	+
Carex paniculata				2.0	+

Addenda: Other accompanying species

Epilobium spec., Mentha aquatica, Lysimachia thyrsiflora, Alnusglutinosa, Quercus robur, Betula spec., Stellaria alsine, Rumex crispus, Galium aparine, Potentilla erecta, Achillea millefolium. Carex pulicaris, Galium palustre.

To obtain local vegetation types and to describe their succession an agglomerative group average clustering (Everitt, 1974) was carried out based on the quantitative dissimilarities. The average dissimilarity to a newly formed cluster was defined with the formula $D (a + b, c) = \frac{1}{2} (D(a,c) + D(b,c))$. The results are presented in a dendrogram. The dendrogram method was applied to group the 214 relevés in clusters on two levels of importance. These were chosen arbitrarily because the average within-cluster sum of dissimilarities increased gradually. The first level with 5 clusters was chosen at a dissimilarity level of 50%, the second level with 21 subclusters at 38%. The number of relevés, the average number of species and the average cover per species are presented for each cluster. Each annual relevé of each permanent plot was ascribed to one of the subclusters. According to this analysis, the succession of vegetation was described schematically in terms of clusters (e.g. Noordwijk-Puijk *et al.*, 1979). Clusters and subclusters were interpreted as vegetation types. The number of vegetation types in each of the years and the number of transitions of relevés from one type to another were deduced from this scheme.

The disappearance and appearance of each species in each permanent plot is defined by the following criteria. A species was considered to be newly appearing when it had been observed for the first time in the third year or later, with a possible interruption of one year. An observation in the last year only was left out of consideration. The criterion for disappearance was defined analogously.

The successional response of each species was analysed by comparing the observed series of cover % in time to 9 (= as much as number of years) standardized time series. In each of these standardized series, maximal cover occurred in a different year (in the first series the maximum in the first year etc.). The comparison was carried out by adding rank numbers to cover values of both actual observations and standardized series. The correlation of these series was expressed by Spearman's coefficient, and a significance level of 2% was used. If correlation with more than one series was significant the species in the permanent plot was ascribed to the standard series to which it showed the highest correlation. The succesional response of the cover values was roughly divided into 5 groups.

I. decreasing cover;
II. increase followed by decrease;
III. increasing cover;
IV. successional response not equal for all permanent plots;
V. cover value more or less constant.

Disappearance and appearance were considered as a special case of resp. decrease and increase.

Results

Vegetation

The dendrogram of 214 relevés is shown in Figure 2. The 5 main clusters chiefly represent the separate lots. Because cluster 2 represents a deviating year (1974) within cluster 1, these two are treated together.

The species composition and average cover for all clusters is shown in Table 1 A-C. The clusters 1 and 2 consist of relevés from lot 1376 and include both species of the *Poo-Lolietum* (*Agropyro-Rumicion crispi*) and of the *Lolio-Cynosuretum* of the alliance *Cynosurion cristati*. The vegetation is relatively poor in species. *Glyceria fluitans* and *Alopecurus geniculatus* are restricted to these clusters. *Holcus lanatus*, *Festuca pratensis* and *Ranunculus repens* often have a high cover.

Cluster 5 includes the relevés of lots 422 and 431 and can be assigned to the community of *Carex acutiformis* (cf. *Angelico-Cirsietum oleracei*) (van Tooren, 1979; Everts *et al.*, 1980) of the alliance *Calthion palustris*. The species diversity is variable. This cluster is characterized by the presence and abundance of *Glyceria maxima*, *Phalaris arundinacea*, *Urtica dioica*, *Stellaria media*, *Rumex crispus* (affinity to *Agropyron-Rumicion crispi*), *Carex acutiformis*, *Crepis paludosa*, *Caltha palustris* (*Calthion palustris*), *Filipendula ulmaria*, *Festuca arundinacea*, *Valeriana officinalis* (tall *Molinietalia* species), *Anthriscus sylvestris*, *Veronica arvensis*, *Veronica chamaedrys*, *Bromus racemosus*, *B. mollis*, *Trifolium dubium*, *Glechoma hederacea*

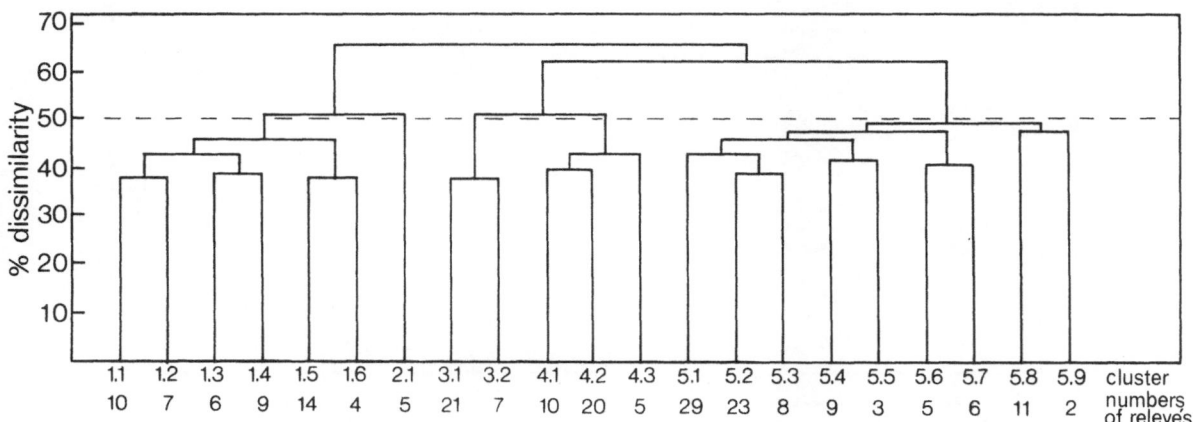

Fig. 2. Simplified (all dissimilarities below 38% are not distinguished) dendrogram of all permanent plots recorded from 1972 up to and including 1980.

(*Arrhenatherion elatioris*), *Chrysoplenium alternifolium, Phyteuma nigrum, Melandrium rubrum* (woodland species with affinity to the *Arrhenatherion elatioris*).

Cluster 3 comprises the relevés of the not reclaimed part of lot 1371 an can be assigned to the *Senecioni-Brometum racemosi* of the alliance *Calthion palustris*. The vegetation is rich in species and characterized by *Menyanthes trifoliata, Carex nigra, Viola palustris* (*Caricion curto-nigrae* communities) and by *Galium uliginosum, Cynosurus cristatus* and *Carex paniculata*.

Cluster 4 comprises the relevés of the formerly reclaimed part of lot 1371 and can be assigned to the alliance *Juncion acutiflori* (syn. *Junco-Molinion*). The vegetation is rich in species and characterized by the abundance and presence of *Juncus acutiflorus, J. subuliflorus, Luzula multiflora* (*Juncion acutiflori*), *Luzula campestris, Agrostis tenuis* (*Cynosurion*), *Myosotis discolor, Prunella vulgaris, Plantago lanceolata, Cirsium palustre* and *Dactylis glomerata*.

A more detailed view of the vegetation can be obtained by describing the 21 subclusters. Their presence in space and their succession will be considered (Table 1 A-C, Table 2, Fig. 3 A-C).

A. On relatively wet places of lot 1376 the transitions 1.1 → 1.5 → 1.3 (Fig. 4, plot 1), on relatively dry places the transitions 1.2 → 2.1 → 1.4 (Fig. 4, plot 3) and near the brook, probably the best drained area, the transitions 1.2 → 2.1 → 1.6 were observed.

B. In the rugged plots of lot 422 the transitions 5.4 → 5.7 → 5.1 (Fig. 5), in the most rugged plot the transitions 5.5 → 5.7 → 5.1 were recorded. On the lowest and relatively wettest parts of lot 431 the transitions 5.9 → 5.6 → 5.3, on the highest and relatively dry parts the transitions 5.8 → 5.2 (Fig. 6) were recorded.

C. In the relatively dry and formerly reclaimed part of lot 1371 the transitions 4.1 → 4.2 (Fig. 7, plot 8), near a ditch the transitions 4.1 → 4.3 were observed. In the not reclaimed wet part of lot 1371 near a ditch cluster 3.1 (Fig. 7, plot 2), in the rest of this part cluster 3.2 has been recorded from 1974 onwards.

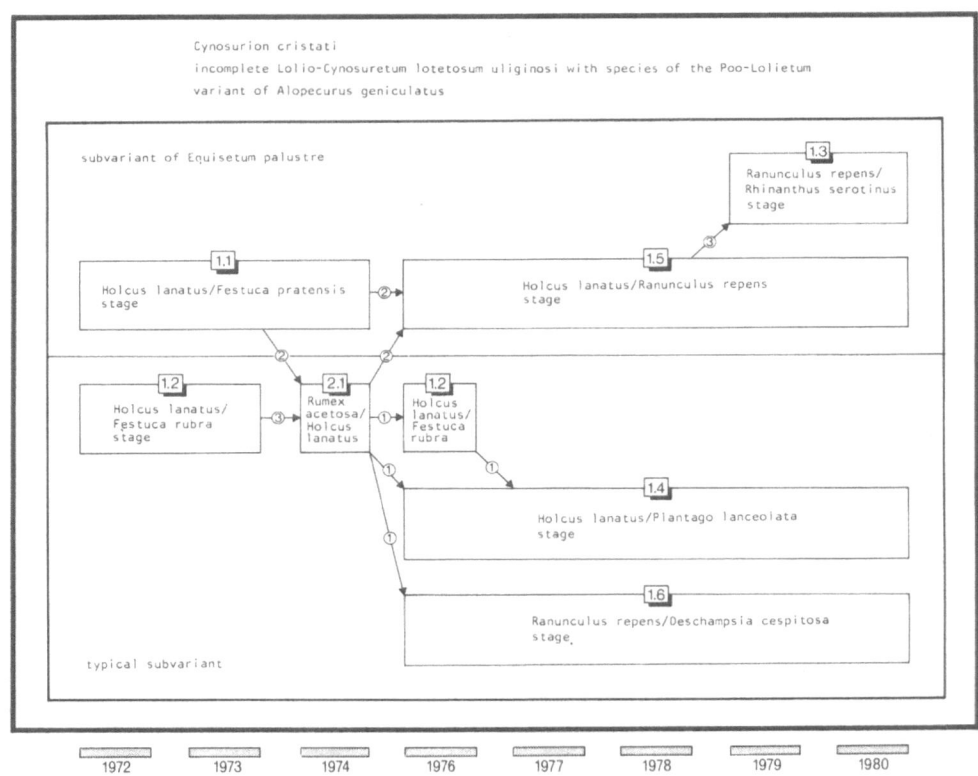

Fig. 3A.

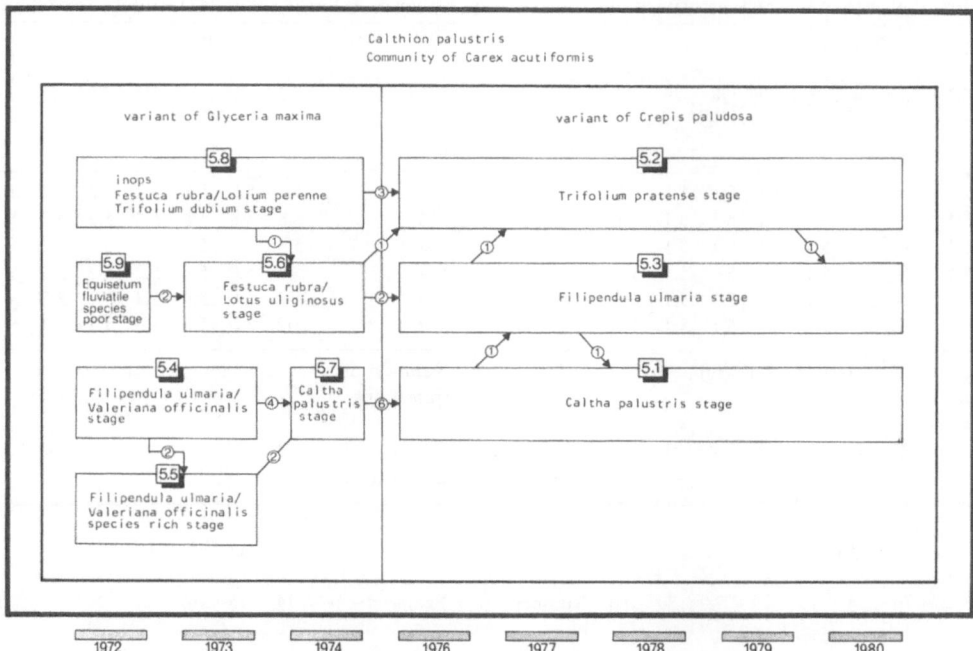

Fig. 3B.

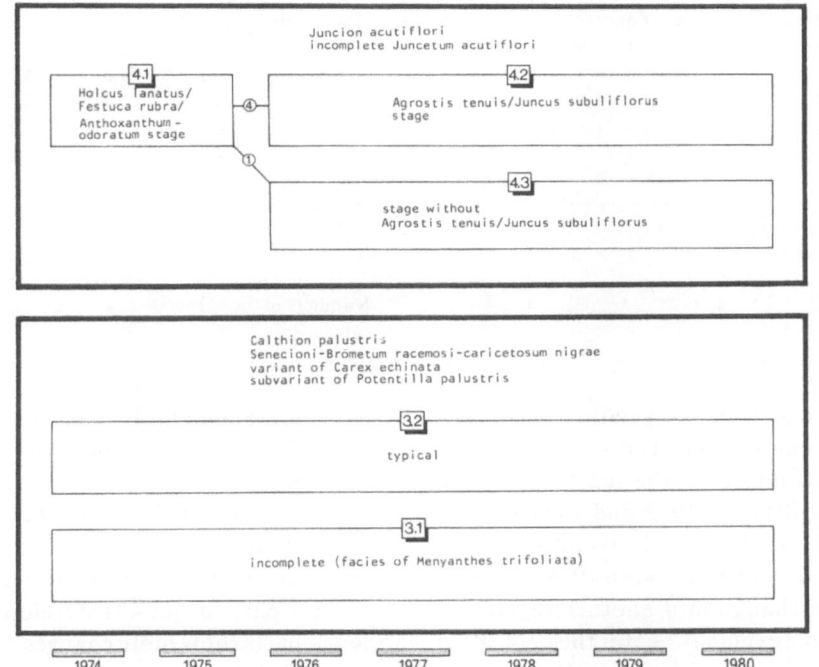

Fig. 3C.

Fig. 3. (A-C). Diagram of cluster transitions in the permanent plots. Numbers in arrows indicate numbers of transitions. Subclusters are described in Table 1.

Table 2. Cluster number to which each permanent plot each year was assigned. A: lot 422; B: lot 431; C: lot 1376; D: lot 1371.

A: Lot 422

Years Plot	72	73	74	76	77	78	79	80	Numbers of clusters
1	5.4	5.4	5.7	5.1	5.3	5.1	5.1	5.1	4
2	5.4	5.4	5.7	5.1	5.1	5.1	5.1	5.1	3
3	5.4	5.4	5.7	5.1	5.1	5.1	5.1	5.1	3
4	5.4	5.5	5.7	5.1	5.1	5.1	5.1	5.1	4
5	5.5	5.5	5.7	5.1	5.1	5.1	5.1	5.1	3
6	5.4	5.4	5.7	5.1	5.1	5.1	5.1	5.1	3
Numbers of transitions	1	6	6	1	1	0	0		
Numbers of sub clusters	2	2	1	1	2	1	1	1	

B: Lot 431

Years Plot	72	73	74	76	77	78	79	80	Numbers of clusters
1	5.8	5.8	5.8	5.2	5.2	5.2	5.2	5.2	2
2	5.8	5.8	5.8	5.2	5.2	5.2	5.2	5.2	2
3	5.8	5.8	5.8	5.2	5.2	5.2	5.2	5.2	2
4	5.8	5.8	5.6	5.2	5.2	5.2	5.2	5.2	3
5	5.9	5.6	5.6	5.3	5.2	5.2	5.2	5.3	4
6	5.9	5.6	5.6	5.3	5.3	5.3	5.3	5.3	3
Numbers of transitions	2	1	6	1	0	0	1		
Numbers of sub clusters	2	2	2	2	2	2	2	2	

C: Lot 1376

Years Plot	72	73	74	76	77	78	79	80	Numbers of clusters
1	1.1	1.1	1.1	1.5	1.5	1.5	1.3	1.3	3
2	1.1	1.1	1.1	1.5	1.5	1.5	1.3	1.3	3
3	1.2	1.2	2.1	1.4	1.4	1.4	1.4	1.4	3
4	1.1	1.1	2.1	1.5	1.5	1.5	1.5	1.5	3
5	1.2	1.2	2.1	1.2	1.4	1.4	1.4	1.4	2
6	1.1	1.1	2.1	1.5	1.5	1.5	1.3	1.3	4
7	1.2	1.2	2.1	1.6		1.6	1.6	1.6	3
Numbers of transitions	0	5	7	1	0	3	0		
Numbers of sub clusters	2	2	2	4	2	3	4	4	

D: Lot 1371

Years Plot	74	75	76	77	78	79	80	Numbers of clusters
1	3.1	3.1	3.1	3.1	3.1	3.1	3.1	1
2	3.1	3.1	3.1	3.1	3.1	3.1	3.1	1
3	3.1	3.1	3.1	3.1	3.1	3.1	3.1	1
4	3.2	3.2	3.2	3.2	3.2	3.2	3.2	1
5	4.1	4.1	4.3	4.3	4.3	4.3	4.3	2
6	4.1	4.1	4.2	4.2	4.2	4.2	4.2	2
7	4.1	4.1	4.2	4.2	4.2	4.2	4.2	2
8	4.1	4.1	4.2	4.2	4.2	4.2	4.2	2
9	4.1	4.1	4.2	4.2	4.2	4.2	4.2	2
Numbers of transitions	0	5	0	0	0	0		
Numbers of sub clusters	3	3	4	4	4	4	4	

Table 2 shows to which vegetation type each permanent plot belongs from year to year and thus describes the development of the vegetation in the permanent plots. Between 1973 and 1974 all permanent plots in lot 1376 and 422 pass into an other vegetation type. Moreover, between 1974 and 1976 most plots again change into another vegetation type. Three possible explanations for the 1974–1976 changes are (i) the very dry summer of 1976, (ii) the plots were not recorded in 1975 and, moreover, (iii) the decimal scale was introduced in 1976 replacing the Braun-Blanquet scale. In general the changes are uni-directional (Table 2, Fig. 3). Only in three cases there was a reversal: in lot 431 permanent plot 5, 5.3 → 5.2 → 5.3, in lot 422 permanent plot 1, 5.1 → 5.3 → 5.1, and in lot 1376 permanent plot 5, 1.2 2.1 → 1.2. The cause of these few exceptions is not known.

The successional lines in the different lots show marked differences (Fig. 3). In lot 422 a small decrease in number of subclusters takes place. The highest parts of lot 431 develop slowly, while the lower parts show more changes. The reclaimed part of lot 1371 develops slowly and the other part of this lot remains constant.

In lot 1376 a divarication of the successional lines occurs which give an increase of the number of vegetation types, probably representing the differences in substrate.

The relatively large changes in the lots in the first

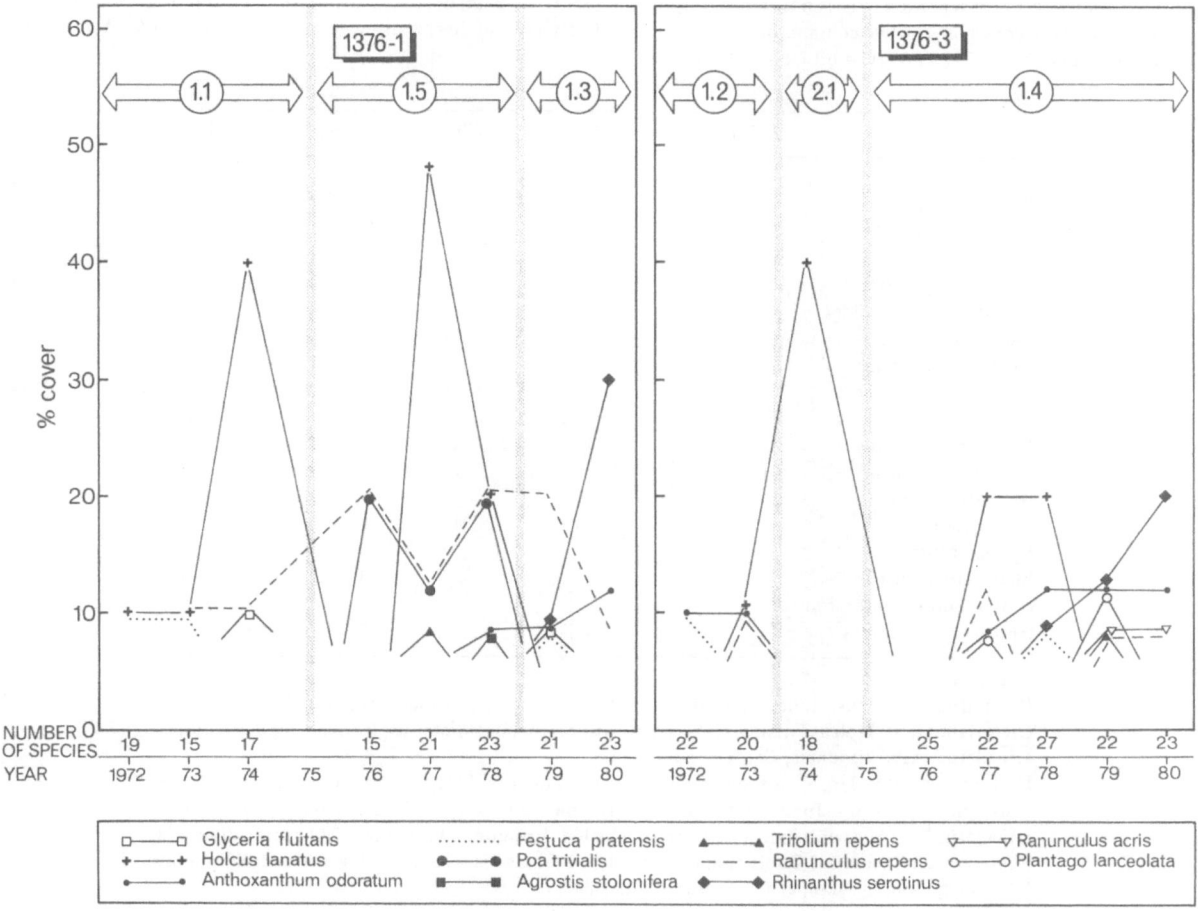

Fig. 4. Cluster transitions, cover of the main species and number of species in two permanent plots in lot 1376. Species are shown when cover is $\geq 8\%$.

years of the sample period compared with the last years can be seen in a comparison between the successive years (Fig. 8). It is clear that shifts in the abundance of species already present cause larger distances than shifts in the species composition.

Species

The year with optimal cover was determined for each species per permanent plot. In lot 422 *Glyceria maxima, Cerastium holosteoides, Ranunculus repens, Holcus lanatus, Poa pratensis* indicating a relatively high P-status (Kruijne *et al.*, 1967) decrease or disappear (Table 3A). On the other hand *Rhinanthus serotinus, Ranunculus acris, Bromus racemosus, Trifolium pratense, Festuca rubra* and *Ajuga reptans* indicating a relatively lower P-status appear or increase.

In lot 431 *Ranunculus repens, Glyceria maxima, Rumex acetosa, R. crispus, Cerastium holosteoides, Poa pratensis* and *Cardamine pratensis* indicating relatively a high P-status decrease or disappear (Table 3B). On the other hand *Trifolium pratense, Lotus uliginosus, Filipendula ulmaria, Rhinanthus serotinus, Anthoxanthum odoratum* and *Ajuga reptans* indicating a relatively lower P-status appear or increase.

In lot 1376 *Bellis perennis, Poa pratensis, Alopecurus geniculatus, Cerastium holosteoides* and *Rumex obtusifolius* indicating a relatively high P-status decrease or disappear (Table 3C). On the other hand *Anthoxanthum odoratum, Rhinanthus serotinus, Festuca rubra* and *Cynosurus cristatus* indicating a relatively lower P-status appear or increase.

Table 3. (A–D). Frequency of years in which each species reaches its maximum ($\alpha \leqq 0.02$). Only species occurring in 50% of the permanent plots are shown (* = not recorded, n.s. = not significant). I decrease, II increase/decrease, III increase, IV response not equal in all permanent plots, V constant. The two columns on the right present frequencies of disappearance (D) and appearance (A). A. lot 1376 (n = 7.) B. lot 422 (n = 6.) C. lot 431 (n = 6.) D. lot 1371 (n = 9).

A	72	73	74	75*	76	77	78	79	80	n.s.	total	D	A
I													
Bellis perennis	2									2	4	2	
Poa pratensis	1	5								1	7		
Alopecurus geniculatus		4								3	7	2	
Equisetum palustre		3								3	6		
Cerastium holosteoides	1	2								4	7	1	1
Taraxacum spec.		2								5	7	1	1
Rumex obtusifolius		2								3	5	2	
III													
Cynosurus cristatus						1	1	1		4	7		2
Ranunculus repens							2	1		4	7		
Agrostis stolonifera							2	2		2	6		6
Poa trivialis							1	1		5	7		
Festuca rubra	1							3		3	7		2
Eleocharis palustris								1	1	2	4		3
Anthoxanthum odoratum									2	5	7		1
Rhinanthus serotinus									7		7		7

Addenda:

IV. Trifolium repens, Juncus effusus V. Trifolium pratense, Rumex acetosa, Stellaria alsine, Glyceria fluitans, Holcus lanatus, Galium palustre, Cardamine pratensis, Ranunculus acris, Juncus articulatus, Agrostis tenuis, Carex ovalis.

D. Juncus effusus 1/6, Carex ovalis 1/4, Mentha aquatica 1/1, Ranunculus flammula 1/2. A. Trifolium repens 3/7 Juncus effusus 2/6, Glyceria fluitans 1/7, Galium palustre 3/5, Juncus articulatus 2/5, Agrostis tenuis 1/4, Carex ovalis 1/4, Leontodon autumnalis 2/2, Carex nigra 1/3, Phalaris arundinacea 1/1 Equisetum fluviatile 1/2, Plantago lanceolata 1/2, Ranunculus flammula 1/2, Polygonum hydropiper 1/2, Lotus uliginosus 1/1.

B	72	73	74	75*	76	77	78	79	80	n.s.	total	D	A
I													
Glyceria maxima	5	1									6	6	
Cerastium holosteoides	4	1								1	6		
Ranunculus repens	2									4	6		
Stellaria palustris	2	1								1	4	2	
Holcus lanatus	2	4									6		
Myosotis palustris		3	1							2	6		1
Festuca arundinacea		2								4	6	2	
Achillea ptarmica		2								2	4		
Equisetum fluviatile		2								4	6		1
Lotus uliginosus		2							1	1	4	1	1
Vicia cracca	1	1								2	4	2	
Veronica chamaedrys		1	1							3	5	1	
Poa pratensis			3							3	5	2	
Lythrum salicaria			2							4	6	2	
II													
Stellaria graminea		1				2				2	5		1
Carex acutiformis					1	1				4	6		

Table 3. (Continued)

B	72	73	74	75*	76	77	78	79	80	n.s.	total	D	A
III													
Rhinanthus serotinus							4			2	6		5
Ranunculus acris							3	2	1		6		
Bromus racemosus							2		1	3	6		4
Trifolium repens							1		1	2	4		2
Festuca rubra							1		1	4	6		3
Trifolium pratense							2	3		1	6	3	
Crepis paludosa								1	4	1	6		1
Ajuga reptans									2	4	6		4

Addenda:
IV. Filipendula ulmaria, Anthriscus sylvestris. V. Rumex acetosa, Festuca pratensis, Stellaria media, Taraxacum spec., Plantago lanceolata, Veronica arvensis, Cardamine pratensis, Lychnis flos-cuculi, Anthoxanthum odoratum, Veronica serpyllifolia, Cirsium palustre, Caltha palustris, Stellaria alsine, Bromus mollis, Poa trivialis, Phalaris arundinacea.
D. Anthriscus sylvestris 4/5, Stellaria media 1/4, Phalaris arundinacea 2/4, Urtica dioica 1/2, Juncus effusus 1/2. A. Plantago lanceolata 1/6, Veronica arvensis 1/6, Anthoxanthum odoratum 1/6, Cirsium palustre 3/6, Caltha palustris 1/6, Bromus mollis 1/5, Phragmites australis 1/2, Juncus effusus 1/2, Valeriana officinalis 1/3, Polygonum amphibium 1/1, Sagina procumbens 1/1.

C	72	73	74	75*	76	77	78	79	80	n.s.	total	D	A
I													
Ranunculus repens	4									2	6	1	
Glyceria maxima	4	1	1								6	3	
Equisetum palustre	3	1	1							1	6		
Rumex acetosa	1	3	1							1	6		
Cerastium holosteoides	2	4									6		
Equisetum fluviatile	1	1								4	6		
Melandrium rubrum	1	1								3	5	1	
Cynosurus cristatus	1	1								4	6		
Poa pratensis		4	2								6	3	
Veronica arvensis		3								3	6		
Myosotis palustris		2	1							3	6		
Cardamine pratensis		2	2							2	6		
Rumex crispus	1		2							3	6	3	
Anthriscus sylvestris	1		1							3	5	1	
II													
Plantago lanceolata			1		1	1				3	6		
III													
Trifolium pratense							3			3	6		
Poa trivialis		1					2			3	6		
Trifolium repens							2			4	6		1
Lotus uliginosus			1				2			3	6		2
Bromus racemosus							2	1		1	4		2
Filipendula ulmaria							1	1		4	6		
Carex acutiformis					1		1		1	3	6	2	
Rhinanthus serotinus			1					5			6		5
Anthoxanthum odoratum						1		2		3	6		1
Ajuga reptans						1		2	1	4		2	
Cirsium palustre	1							1	1	3	6		1
Crepis paludosa								1	1	4	6		

Addenda:
IV. Holcus lanatus, Festuca rubra, Bellis perennis, Mentha aquatica. V. Festuca pratensis, Trifolium dubium, Lolium perenne, Taraxacum spec., Galium uliginosum, Festuca arundinacea, Ranunculus acris, Lychnis flos-cuculi, Veronica chamaedrys.
D. Holcus lanatus 1/6, Lolium perenne 3/6, Festuca arundinacea 1/4, Veronica chamaedrys 1/5, Plantago major Rumex obtusifolius 1/3, Stellaria alsine 1/1, Lythrum salicaria 1/3, Urtica dioica 1/3, Phleum pratense 1/1. A. Festuca rubra 1/6, Trifolium dubium 1/1, Vicia cracca 2/3, Bromus mollis 1/1, Stellaria graminea 1/3, Agrostis stolonifera 1/1, Carex nigra 1/2.

Table 3. (Continued)

D	74	75	76	77	78	79	80	n.s.	total	D	A
I											
Cerastium holosteoides	4							5	9		
Juncus effusus	4	1					1	3	9	6	
Equisetum palustre	4				1			4	9		
Trifolium repens	3	1		1				3	8		
Holcus lanatus	3							6	9		
Equisetum fluviatile	3							5	8		
Prunella vulgaris	3							6	9		2
Anthoxanthum odoratum	2							7	9		
Myosotis palustris	2							4	6		2
Ranunculus repens	2							7	9		
Lotus uliginosus	2							7	9		
Lychnis flos-cuculi	2							7	9	1	
Stellaria graminea	2							6	8	1	
Filipendula ulmaria	2							4	6	1	
Festuca pratensis	2							6	8		
Rhinanthus serotinus		3					1	5	9		
II											
Plantago lanceolata			1	3				5	9		
Agrostis stolonifera			3	1				2	6		
Agrostis tenuis			2					7	9		
Orchis majalis					5		1	3	9		2
Lysimachia vulgaris					3		1	2	6		1
III											
Poa trivialis						5		4	9		4
Dactylis glomerata						3		2	5		3
Luzula multiflora				1		2	1	1	5		

Addenda:

IV. Carex echinata, Valeriana officinalis, Luzula campestris. V. Galium palustre, Galium uliginosum, Trifolium pratense, Viola palustris, Cynosurus cristatus, Carex nigra, Ajuga reptans. Taraxacum spec., Rumex acetosa, Cardamine pratensis, Cirsium palustre, Juncus articulatus, Festuca rubra, Poa pratensis, Juncus subuliflorus.
D. Carex echinata 1/6, Galium uliginosum 1/5, Trifolium pratense 1/5, Poa pratensis 2/9, Juncus subuliflorus 1/7, Achillea millefolium 1/2, Rumex obtusifolius 1/1, Lolium perenne 1/3, Rumex crispus 1/1, Deschampsia cespitosa 1/2, Cirsium arvense 1/1. A. Valeriana officinalis 1/6, Carex nigra 1/8, Ajuga reptans 1/9, Taraxacum spec. 1/7, Myosotis discolor 2/4, Crepis paludosa 1/4, Stellaria palustris 1/2, Ranunculus flammula 1/2, Potentilla erecta 1/3, Lythrum salicaria 1/4, Carex panicea 1/3.

In lot 1371 *Cerastium holosteoides, Trifolium repens, Holcus lanatus, Juncus effusus, Ranunculus repens* and *Festuca pratensis* indicating a relatively high P-status decrease or disappear (Table 3D). No species indicating a low P-status appear or increase.

Discussion

Each lot had different vegetation types during the study period. Due to the hay-making without fertilization from 1946 onwards in lot 1371 com-munities indicating conditions relatively poor in nutrients could be expected (Klapp, 1965; Ellenberg, 1978; Westhoff & Den Held, 1969). The lots 422, 431 and 1376, however, all had been under agricultural practices until 1972. According to the agricultural grassland mapping in 1965 all lots had little agricultural value (Proefstation voor de Akker- en Weidebouw 1965, 1966). The species composition and abundance in 1972, however, suggest a relatively high intensity of former agricultural practices in lot 1376, as judged from e.g. *Festuca pratensis, Alopecurus geniculatus* and *Glyceria fluitans,* all indicating a high P-status of the soil

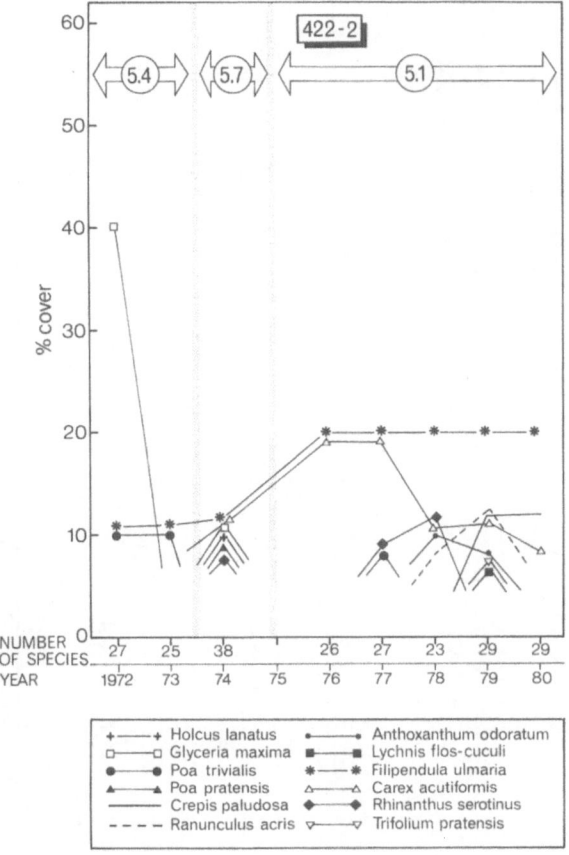

Fig. 5. As Fig. 4. Cluster transitions in lot 422.

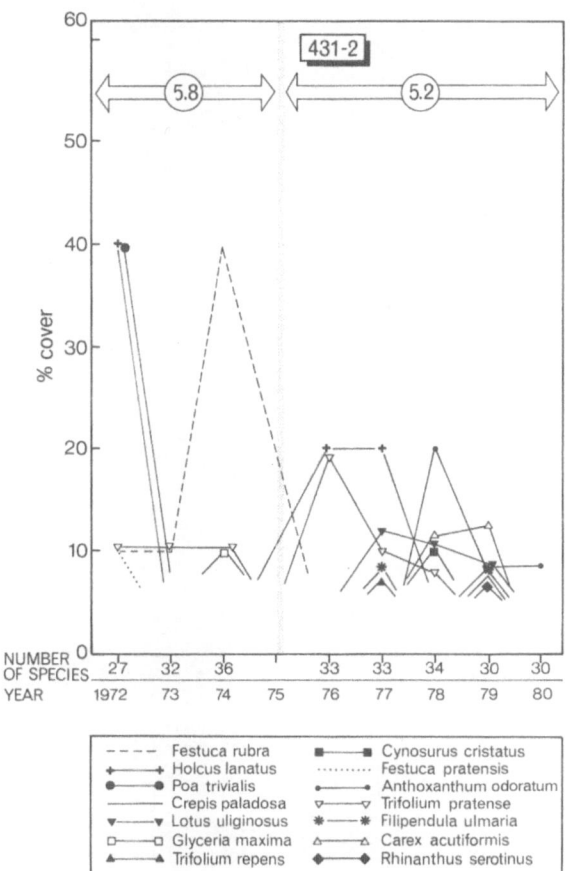

Fig. 6. As Fig. 4. Cluster transitions in lot 431.

(Kruijne *et al.*, 1976). This could be due to grazing which has been practiced everywhere on sandy soils in the brook valley, in contrast to the peat soil where especially hay-making was practised. The lots 422 and 431 suggest less intensive former agricultural practice, as judged from e.g. *Festuca rubra, Equisetum palustre, Filipendula ulmaria, Plantago lanceolata, Myosotis palustris* and *Achillea ptarmica* indicating a low P-status of the soil.

The *Carex acutiformis* communities are accompanied by species of drier and lightly fertilized *Arrhenatheretalia* communities e.g. *Bromus racemosus, Anthriscus sylvestris, Bellis perennis, Veronica arvensis* and *Bromus mollis* (Table 1) (Everts *et al.*, 1980). Due to the continuous seepage in the middle reaches the peatbody grew convex and dried out superficially during the summer period resulting in some N-mineralization (Grootjans, 1979, 1980).

The sequence of vegetation types nearly always

was the same: no reversals of vegetation types occurred (Table 2, Fig. 3) probably pointing into unidirectional environmental processes induced by management practices. Figure 8 illustrates this view: the differences between successive years become gradually smaller, but always contribute to an increasing deviation from the beginning situation. The unidirectional environmental processes possibly include a diminishing availability of nutrients from the soil. By hay-making without fertilization N-total will diminish quickly. However, no reliable plant indicators for N exist. Meyer (1957), Kovacs (1969) and Ellenberg (1974) give N-mineralization rates for separate species. In fact the N-mineralization rate implies some kind of growth rate that cannot always be effectuated under field conditions (Grootjans, 1979). The P-status is an important criterion to determine the intensity of agricultural practices (Kruijne *et al.*, 1967; de Boer & de Gooijer, 1979). Judging from the decrease or dis-

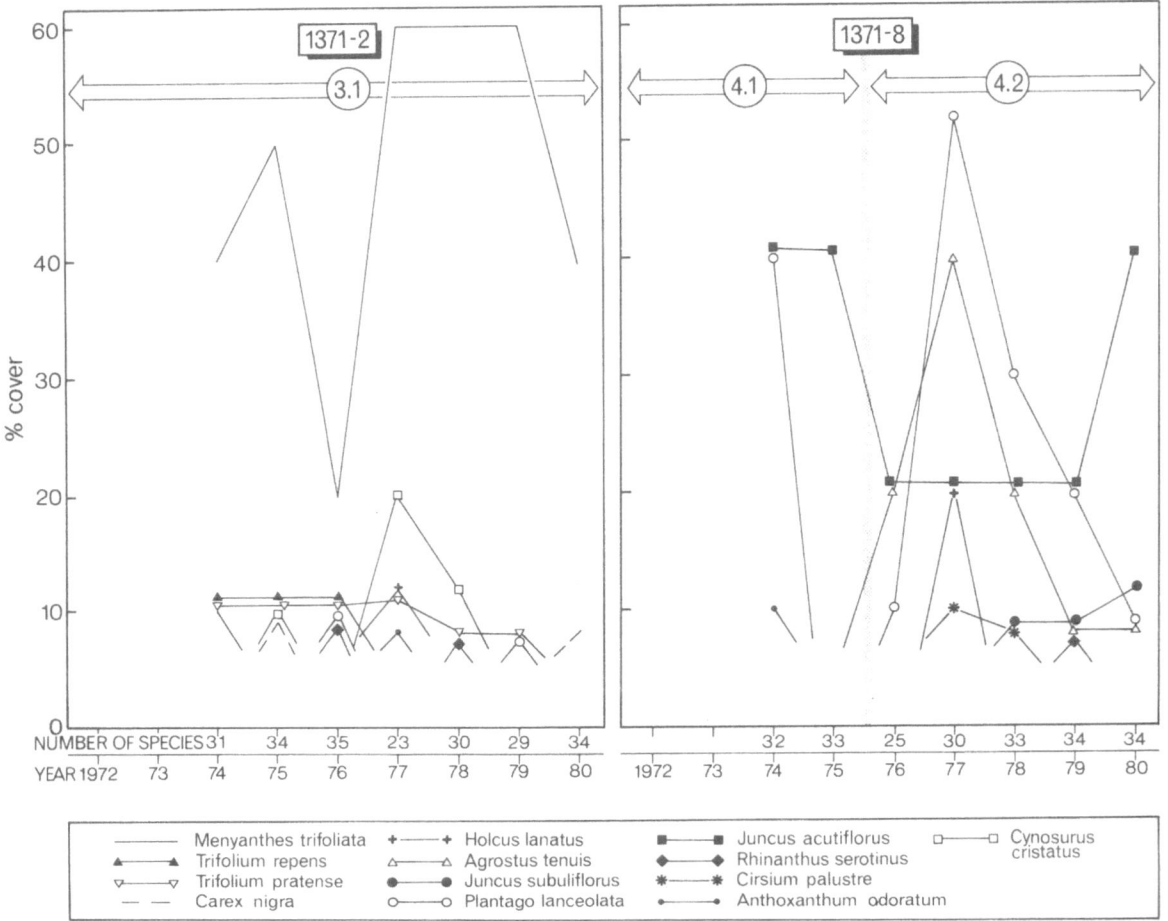

Fig. 7. As Fig. 4. Cluster transitions in lot 1371 (plot 2 no reclaimed, plot 8 reclaimed).

appearance of species indicating a relatively high P-status and the increase or appearance of species indicating a lower P-status, we assume a diminishing nutrient availability. The dominance of *Glyceria maxima* in 1965 according to the grassland vegetation mapping (Proefstation voor de Akker- en Weidebouw 1965) also fits in this view. Despite large dissimilarities (Fig. 8) the communities in lot 1376 do not pass into other vegetation types according to Everts *et al* (1980). The appearance of *Rhinanthus serotinus,* however, indicates a stage in a secondary succession of hay-making without fertilization (Fresco, 1980).

After three years a divarication of successional lines was found in lot 1376 (Table 2, Fig. 3), resulting in a larger coverage of *Holcus lanatus* on the dry places as compared with moist places. Koene & Veerman (1978) found in the middle

reaches of the brook-valley little N-mineralization during the summer period if the groundwater table remained less than 30 cm below surface. So the larger *Holcus lanatus* cover on the drier places of lot 1376 is in agreement with those finds and Grootjans (1979, see also Williams, 1968). Moreover Elliot *et al.* (in Watt, 1978) found in experiments with mineral N applied in a *Holcus/Agrostis* community a direct response of *Holcus lanatus*.

Although some populations show a definite trend (increasers or decreasers) during the study period, we also found very rapid annual changes (Figs 4, 5, 6, 7). This holds both for the lots with changed management practices at the beginning of the experiments and for lot 1371 with unfertilized hay-making from 1946 onwards. Van den Bergh (1979) found the same phenomenon with unfertilized hay-making, but not with unfertilized gra-

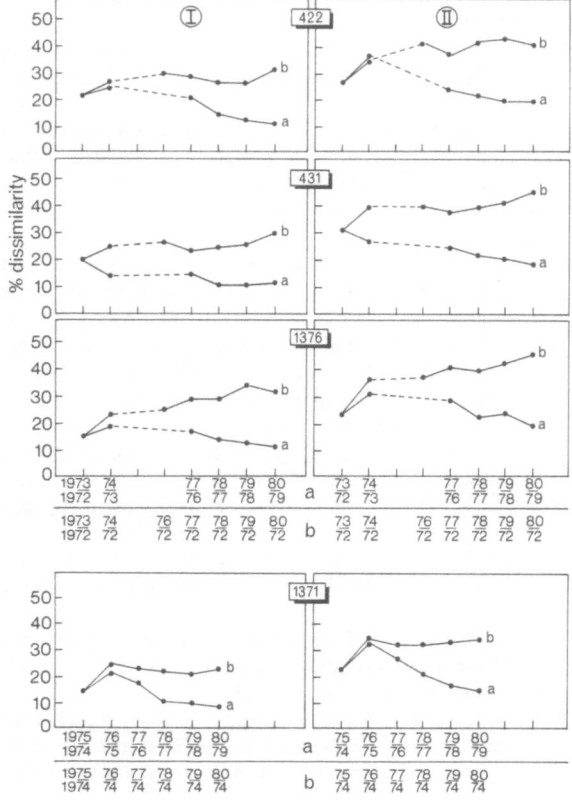

Fig. 8. Mean distances for all permanent plots in a lot between a. subsequent years, and b. the first year and subsequent years. I qualitative, II quantitative.

zing. Grazing suppresses seed production and therefore periodic changes in seed production and in the number of established seedlings. In fact the annual timing of hay-making influences the composition of the vegetation and the number of seedlings reaching the juvenile stage as well (Bakker *et al.,* 1980; Oomes, 1977, 1981). Moreover, climatological influences will play a role. In the extremely dry summer of 1976 and after the severe winter of 1978/79 (cf. 't Hart & de Vries in Watt, 1978) the population of *Holcus lanatus* declined and consequently tillering species e.g. *Ranunculus repens, Agrostis stolonifera* and *Poa trivialis* invaded the open places and reached large coverages (Bakker *et al.,* 1980; see also Rabotnov, 1975). The density of the annual hemiparasite *Rhinanthus serotinus* is also correlated to the structure of vegetation (Fresco, 1980).

References

Bakker, J. P., 1978. Some experiments on heathland conservation and regeneration. Phytocoenosis 7: 351–370.

Bakker, J. P., Dekker, M. & Vries, Y, de, 1980. The effect of different management practices on a grassland community and the resulting fate of seedlings. Acta Bot. Neerl. 29: 509–522.

Bakker, P. A., 1979. Vegetation science and nature conservation. In: M. J. A. Werger (ed.). The study of vegetation, pp. 247–288. Junk, The Hague, 316 pp.

Bergh, J. P. van den, 1979. Changes in the composition of mixed populations of grassland species. In: M. J. A. Werger (ed.) The study of vegetation pp. 57–80. Junk, The Hague, 316 pp.

Böcker, R., 1978. Vegetations- und Grundwasserverhältnisse im Landschaftsschutzgebiet Tegeler Fliesstal (Berlin West). Verh. Bot. Ver. Prov. Brandenburg 114: 1–164 (Thesis, Berlin).

Boer, Th. A. de & de Gooijer, H. H., 1979. Kartering van korte vegetaties van het cultuurlandschap. CABO karteringsverslag 185. Mimeographed report Wageningen, 27 pp.

Borstel, U. O. von, 1974. Untersuchung zur Vegetationsentwicklung auf ökologisch verschiedenen Grünland- und Ackerbrachen hessischer Mittelgebirgen (Westerwald, Rhön, Vogelsberg). Thesis, Giessen, 159 pp.

Campino, I., 1980. Beziehungen zwischen der oberirdischen lebenden Pflanzenmasse und einigen abiotischen und biotischen Faktoren in drei verschiedenen Grünlandbeständen bei unterschiedlicher Nutzungsintensität. In: O. Wilmanns & R. Tüxen Eds. Epharmonie. Berichte über die internationalen Symposium der I.V.V. 1979, pp. 191–207. Cramer, Vaduz 462 pp.

Dierschke, H., 1980. Untersuchungen auf Dauerflächen in Kalk-Magerrasen (Mesobromion) mit unterschiedlicher Nutzung. Proc. Symp. Vegetation Dynamics in grasslands, heathlands and mediterranean ligneous formations: IV-6.

Ellenberg, H., 1974. Zeigerwerte der Gefässpflanzen Mitteleuropas. Scripta Geobotanica Band 9. Goltze, Göttingen, 97 pp.

Ellenberg, H., 1978. Vegetation Mitteleuropas mit den Alpen. Ulmer, Stuttgart, 981 pp.

Everitt, B. S., 1974. Cluster analysis. Heinemann, London, 122 pp.

Everts, F. H., Grootjans, A. P. & Vries, N. P. J. de, 1980. De vegetatie van de madelanden in het stroomdal van de Drentse Aa. Part 2. Internal report Dept. of Plant Ecology, University Groningen (mimeo), 117 pp.

Fresco, L. F. M., 1980. Ecological response curves of Rhinanthus serotinus: a synecological study. Acta Bot. Neerl. 29: 533–539.

Goodall, D. W., 1973. Sample similarity and species correlation. In: R. H. Whittaker (ed.) Handbook of vegetation science. Part V. Ordination and classification of communities p. 105–156. Junk, The Hague, 737 pp.

Grootjans, A. P., 1979. Some remarks on the relation between nitrogen mineralisation, groundwater table and standing crop in wet meadows. Acta Bot. Neerl. 28: 234–235.

Grootjans, A. P., 1980. Distribution of plantcommunities along rivulets in relation to hydrology and management. In: O. Wilmanns & R. Tüxen (eds.). Epharmonie. Berichte über die

258

internationalen Symposium der I.V.V. 1979, pp. 143–170. Cramer, Vaduz, 462 pp.

Holger, G., 1978. Wirkungen einiger Landschaftspflegeverfahren auf die Pflanzenbestände und Mögligkeiten der Bestandeslenkung durch Schafweide im Bereich von Grünlandbrachen. Phytocoenologia 7: 218–236.

Klapp, E., 1965. Grünlandvegetation und ihre Standort. Parey, Berlin, 284 pp.

Koene, H. & Veerman, M., 1978. Vegetatie en bodem van nat-droog gradienten. Internal Report, Dept. of Plantecology, University Groningen (mimeo), 104 pp.

Kovacs, M., 1969. Pflanzenarten und Pflanzengesellschaften als Anzeiger des Bodenstickstoffs. Acta Bot. Scient. Hung. 15: 101–118.

Kruijne, A. A., Vries, D. M. de & Mooi, H., 1967. Bijdrage tot de oecologie van Nederlandse graslandplanten. Versl. van Landb. Onderz. 696. Pudoc, Wageningen 65 pp.

Londo, G., 1976. The decimal scale for relevés of permanent quadrats. Vegetatio 33: 61–64.

Meyer, F. M., 1957. Über Wasser und Stickstoffhaushalt der Röhrichte und Wiesen im Elbealluvium bei Hamburg. Mitt. Staat Inst. Allgem. Bot. (Hamburg) 11: 139–203.

Noordwijk-Puijk, K. van, Beeftink, W. G. & Hogeweg, P., 1979. Vegetation development on salt-marsh flats after disappearance of the tidal factor. Vegetatio 39: 1–13.

Oomes, M. J., 1977. Cutting regime experiments on extensively used grasslands (summary). Acta Bot. Neerl. 26: 265–266.

Oomes, M. J., 1981. The effect of cutting and fertilizing on the botanical composition and production of an Arrhenatherion elatioris community. Proc. Symp. Vegetation Dynamics in grasslands, heathlands and mediterranean ligneous formations IV–15 (also: Vegetatio 47: 233–239).

Proefstation voor de Akker- en Weidebouw, 1965. Een graslandvegetatiekartering van het ruilverkavelingsgebied 'Vries'. Report 186. Mimeographed, Wageningen 32 pp.

Proefstation voor de Akker- en Weidebouw, 1966. Een graslandvegetatiekartering van het ruilverkavelingsgebied 'Anlo'. Report 211. Mimeographed, Wageningen 21 pp.

Rabotnov, T. A., 1975. On phytocoenotypes. Phytocoenologia 2: 66–72.

Rijksinstituut voor Natuurbeheer, 1979. Natuurbeheer in Nederland; Levensgemeenschappen. Pudoc, Wageningen 392 pp.

Schreiber, K. F. & Schiefer, J., 1980. Vegetations- und Stoffdynamik in Grünlandbrachen. Ergebnisse fünfjähriger Bracheversuche in Baden-Württtenberg. Proc. Symp. Vegetation Dynamics in grasslands, heathlands and mediterranean ligneous formations: VI–3.

Sørensen, Th., 1948. A method for establishing groups of equal magnitude in plant sociology based on similarity of species content. Acta K. Danske Vidensk Selsk Biol. Skr. J. 5: 1–34.

Tooren, B. van, 1979. Enkele aspecten van de oecologie van Carex aquatilis en Carex acutiformis in het stroomdal van de Drentse Aa. Internal report Dept. of Plantecology, University Groningen (mimeo), 50 pp.

Watt, T. A., 1978. The biology of Holcus lanatus and its significance in grassland. Herbage Abstracts 48: 195–204.

Westhoff, V. & Held, A. J. den, 1969. Plantengemeenschappen in Nederland. Thieme, Zutphen, 324 pp.

Williams, J. T., 1968. The nitrogen and water relations of wet meadows. Veröff. Geobot. Inst. E. T. H. Stiftung Rübel 41: 69–193.

Willems, J. H., 1979. Experiments on the relation between species diversity and above-ground plant biomass in chalk grassland. Acta Bot. Neerl. 28: 235.

Willems, J. H., 1980. An experimental approach to the study of species diversity and above-ground biomass in chalk grassland. Proc. Kon. Ned. Akad. Wet. Series C. 83: 279–306.

Accepted 1.7.1981.

Fluctuations in a coastal dune grassland due to fluctuations in rainfall: Experimental evidence*

E. van der Maarel**

Division of Geobotany, University of Nijmegen the Netherlands, now at the Institute of Ecological Botany, University of Uppsala, Box 559, S 751 22 Uppsala, Sweden

Keywords: Dune grassland, Dynamics, Experiment, Fluctuation, Grazing, Rabbit, Rainfall, Vegetation

Abstract

Experiments with rainfall on a dune grassland near Oostvoorne, the Netherlands with *Festuco-Galietum* as the main syntaxon are described. Both increase in rain through additional watering and decrease in rain through catchment are presented to plots belonging to the xerosere and the mesosere. Clear changes in the floristic composition are the result, even after only two years. Typical *Festuco-Galietum* species are promoted by high rainfall, species of open habitat such as *Corynephorus canescens* are promoted by drought. The results are discussed against the background of long term permanent plot observations in the area and a relation with rabbit grazing intensity is supposed.

Introduction

The coastal dune grassland under study is situated near the Biological Station 'Weevers' Duin" of the Institute for Ecological Research, Oostvoorne, the Netherlands. It measures a few ha and covers a slightly undulating inner dune area belonging to the mediaeval zone of the Voorne dunes (Van der Maarel & Westhoff, 1964; van der Maarel, 1978). It consists mainly of plant communities of dry dunes assigned to the association *Festuco-Galietum maritimi* suballiance *Luzulo-Koelerion*. They form mosaics with damp grasslands in which taxa of the class *Molinio-Arrhenatheretea* and the alliance *Agropyro-Rumicion* predominate, with open communities of the *Erodio-Koelerion* on dune tops and with dwarf herb and moss communities of the *Nanocyperion* and *Lolio-Plantaginion* on and along pathes across the hollows (van der Maarel, 1966, 1978). The grassland is surrounded by low to medium high scrub with *Salix repens* and *Crataegus monogyna*; the latter species tends to extend its population and is invading the grassland.

Since 1963 a part of the dune grassland of c. 2000 sq.m. is analyzed regularly through description of 40 2×2 sq.m. permanent plots and a description of a grid of such quadrats covering half of the total area every 7–10 yr. From 1963–1970 the vegetation changes could be interpreted as a progressive succession from more open to more closed grassland communities in which the tall grasses *Arrhenatherum elatius* and *Helictotrichon pubescens* are the most conspicuous increasers. Still some fluctuations were recognized which were interpreted as related to fluctuations in the soil moisture conditions in spring and early summer

* Nomenclature of vascular plants follows Heukels-van Oost-stroom (1975) Flora van Nederland 18 ed., Wolters-Noordhoff, Groningen; Nomenclature of syntaxa follows Westhoff & den Held (1969). Plantengemeenschappen in Nederland, Thieme, Zutphen.
** Field work in 1978 was carried out with Frans Bongers and Marc de Lyon, in 1979 with Marc de Lyon, Pieter Meeuwissen and Guiljo van Nuland, all then MSc. students at the Division of Geobotany. Their help and the advice of Dr. Peter van der Aart, Institute of Ecological Research, Oostvoorne, are acknowledged.

Vegetatio 47, 259–265 (1981). 0042-3106/81/0472-3-0259/$1.40.

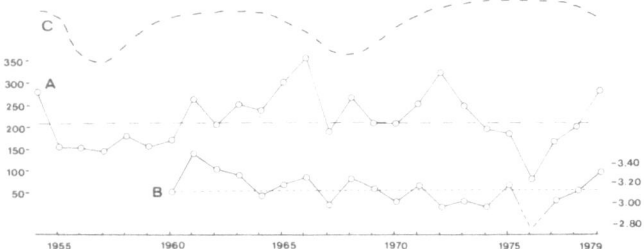

Fig. 1. A. Precipitation sum April – August for the years 1954–1979 and average (---) in mm. B. Average ground water table April – August and average (---) in m-altitude for the period of registration 1960–1979. Value for 1976 undetermined because the table sank beyond the bottom of the tube. C. Estimated density of the local rabbit population from 1954–1979. In 1955 the first wave of myxomatosis occurred.

(van der Maarel, 1978).

Since 1970 some very dry summers occurred with the 1976 summer as an extreme and by 1977 the aspect of the area had completely changed. The tall grasses largely disappeared again and even the vegetation in the hollows died off. Thus the idea of a directional change had to be abandoned. Figure 1 presents data on the summer rainfall and the average ground water from April-August and clearly shows the pattern of fluctuation, which is largely the same for the two parameters involved. Deviations such as in 1972 are explained by the uneven distribution of the rainfall throughout the year.

After a rabbit exclosure experiment had been started in 1972 to check whether the slow colonization of the grassland by shrubs and trees was due to rabbit grazing, a second experiment was began in 1978 in order to find evidence for a direct impact of dry respectively wet summers on the structure and species composition of the dune grassland. Since the rabbit population heavily increased in the seventies the precipitation experiment had to be combined with the exclosure experiment. Figure 1 shows how, according to an estimation of the density of the local rabbit population, the rabbit density follows a pattern of fluctuation.

Material and methods

Five series of 2×2 m quadrats were laid down in the study area, three sets in the dune top in the SE

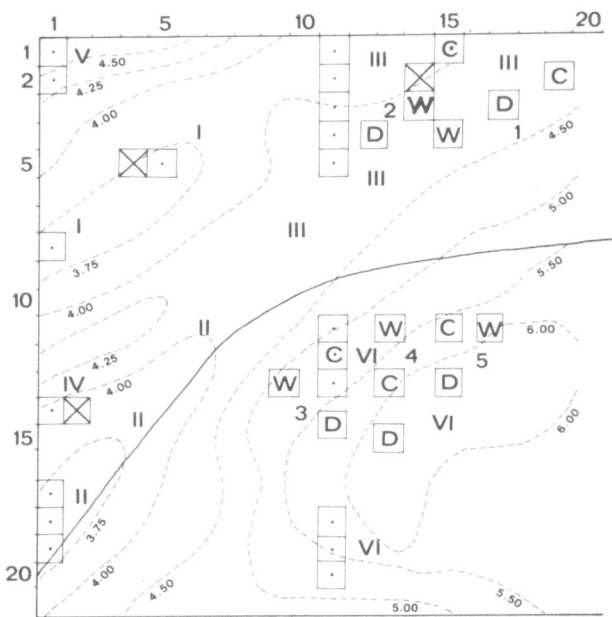

Fig. 2. Contour map of the study area in the dune grassland near 'Weevers' Duin', Oostvoorne. The situation of the five series of rainfall experiment quadrats is indicated: W = quadrat receiving more rain; D = quadrat receiving less rain; C = comparison quadrat.

Series 1 and 2 are situated in the mesosere (alt.c.4.00 m), series 3–5 in the xerosere (alt.c.5.00–6.00 m). The situation of 20 2×2 sq.m. quadrats used for yearly analysis is indicated as well (.) with respect to the vegetation zone (I–VI) they occur in. Quadrats are numbered from 1.1, 1.2, 1.8 etc. left to 17.3, 17.11 and 19.2 right. Also indicated are rabbit exclosure quadrats (X) and the course of a narrow path (-).

part at ca. 5.50–6.00 m above "N.A.P." (Amsterdam water-mark) and two sets on the flat depression in the W. part at ca. 4.00 m. Figure 2 presents a contour map of the major part of the study area, the situation of the 20 quadrats which were chosen from the original 40 quadrats for yearly research since 1970, the situation of the rabbit exclosures, and the situation of the new precipitation experimental quadrats.

Each series consists of three quadrats:

W: receives much more precipitation than normal through regularly watering with water from a nearby well. (The chemical composition of this water did not differ essentially from that of the rainfall itself; the watering was done in small amounts per time at the end of a day once every week or less; the intensity of watering was comparable to that in a heavy shower, which

Fig. 3. View on the dune grassland with plastic rainfall catchment devices in function. Nrs. 1 and 2 to the right, nrs. 3–5 to the left on the dune top. Rabbit exclosure quadrat 15.2 is visible in front of device 2. Shrubs on the foreground are *Crataegus monogyna*. (photograph by M. de Lyon).

was checked by following the moisture content in the top soil layers during 24 h after watering resp. a heavy shower.

D: receives much less rain than normal through catching the rainfall by means of a plastic cover which is spread over a frame in advance of periods of heavy rain (Fig. 3). According to measurements the light interception during the (cloudy) days was negligeable. Details of the treatments are found in Bongers & de Lyon (1979).

C: control quadrat, not treated.

The resulting amounts of water in mm of precipitation for the seasons of 1978 and 1979 and the average and extreme precipitation figures for the period 1 April – 31 August over the years 1930–1977 are as follows:

	1978		1979
	18.4 – 13.9		21.3 – 10.9
'wet'	482		647
'dry'	100		65
'control'	207		380
average 1930–'60		274	
average 1961–'77		306	
maximum 1966		454	
minimum 1976		89	

Because the difference between 'dry' and 'control' 1978 in was not as big as anticipated it was decided to lower the 1979 rainfall in the 'dry' quadrats. As the rabbit density was very high at the beginning of the rainfall experiment it was decided to fence most of the area (as shown in Fig. 3) in order to strongly reduce the influence of rabbits. Early 1979 this fence was completed around the entire area. According to excrement countings (de Lyon, Meeuwisse & van Nuland, 1980) this fencing was effective: the average number of pellets per sq.m. dropped from 500 to 100. (The statistical variance of these countings during the period May–August was very high, so the figures are an indication only).

The vegetation in all experimental, comparison and permanent plots was recorded in May and August 1978 and 1979. The May recordings were restricted to winter annuals and early flowering perennials (cf. van der Maarel, 1966). Integrated year values for both years were obtained (by taking the highest of either estimate) for total cover % of herb and moss layer, and for all vascular plants, lichens and mosses. Species values are given as ordinal transforms of the Braun-Blanquet estimates (van der Maarel, 1979) as follows r → 1; + → 2; 1 → 3; 2m → 4; 2a → 5; 2b → 6; 3 → 7; 4 → 8; 5 → 9.

Results

The results to be presented here will concentrate on the tendencies of change in species composition of wet versus dry quadrats.

Table 1. Tendency of change in the cover-abundance value of 18 species and 23 syntaxonomical groups in experimental dune grassland plots kept summer-wet and summer-dry 1978–1979 resp., in the mesosere and xerosere (see text). Only species with a clear response in either treatment are listed.
++ strong increase, + slight increase, o ± constant, – slight decrease, – – strong decrease, ( ) not occurring.

Selected species	More rain		Less rain	
	Meso	Xero	Meso	Xero
Luzula campestris	++	++		
Poa pratensis-pratensis	++	++		
Anthoxanthum odoratum	+	++	–	
Agrostis tenuis	+	++		
Festuca ovina-tenuifolia	+	++	o	
Thymus pulegioides	+	+	o	
Polytrichum juniperinum	+	+	o	
Carex caryophyllea	+	( )	–	( )
Centaurium erythraea	++	( )	o	( )
Sieglingia decumbens	o	++	o	( )
Vicia lathyroides		o	o	
Aira praecox		+	++	o
Carex arenaria		o	++	o
Rhytidiadelphus squarrosus			o	+
Corynephorus canescens	( )	–	( )	+
Erophila verna	o		++	+
Hypnum cupressiforme			++	o
Cerastium semidecandrum			++	++

Selected syntaxa

Violion caninae	+	++		
Festuca – Sedetalia			+	o

Table 1 presents such tendencies for 18 of the 72 species occurring in any of the 15 experimental plots. The other species did not (yet) show a clear tendency.

The species in Table 1 are arranged in order of their reaction from positive response to more rain, to positive response to less rain. Species with a positive response to wetter summer conditions are mainly grasses and also some herbs characterizing the *Festuco-Galietum*. This association was considered very characteristic for the area and was found in abundance in 1963 (van der Maarel, 1966), a year with both a relatively moist summer and a relatively high rabbit density. After 1970 this community gradually disappeared or became floristically impoverished.

The tendency towards restoration of this community is most pronounced in the series 3–5 in vegetation zone VI (Fig. 2) which is characterized by a mixture of *Festuco-Galietum* and *Mesobromion* species such as *Galium verum, Agrostis tenuis* and *Ononis repens* and *Helictotrichon pubescens*, and species of more open habitats, optimally developed in zone V such as *Corynephorus canescens* and *Cladonia foliacea* (de Lyon *et al* 1980). In view of the depth of the ground-water table here we can consider this zone as part of the xerosere (Westhoff, 1947; cf. van der Maarel, 1966).

In the series 1 and 2 a positive response is also found with species characterizing (locally) the *Nanocyperion*, e.g. *Centaurium erythraea*. Here the ground water table is within the reach of plant roots in normal to wet summers (Fig. 1) but inundation in winter does not occur. The corresponding zone III can be considered as mesosere. It is characterized by a mixture of *Festuco-Galietum* with *Nanocyperion* and *Violion caninae* species (examples *Festuca ovina* and *Thymus pulegioides; Centaurea erythraea* and *Sagina nodosa; Rhytidiadelphus squarrosus* and *Euphrasia officinalis*, cf. de Lyon *et al.*, 1980). The positive response of those species to more rain is generally accompanied by a negative response to dry conditions, particularly in the xerosere.

The overall response to the extra rainfall, an increase in the cover of the (low) herb layer and a darker green colour of the grasses, could be observed already in the course of the first summer.

The positive response to the dry conditions as shown by *Erodio-Koelerion* and *Festuco-Sedetalia* species such as a *Cerastium semidecandrum* and

Table 2. Cover-abundance values (ordinal transform values of Braun-Blanquet cover-abundance estimates) of the 18 species of Table 1 in ten years from 1963–1979 in dune grassland plots 15.1 (mesosere) and plot 11.12 (xerosere), two permanent plots which serve as control plots. Species number on the 4 sq.m. of the quadrat and diversity value as $(1-D_s).100^{-1}$ (D_s is Simpson's index) are given as well.

Wet (W) and dry (D) seasons		63	70	72 W	73	74	75	76 D	77 D	78	79 W
Luzula campestris	meso	( )	3	3	5	6	5	5	3	4	5
	xero	3	5	3	5	5	5	6	2	4	4
Poa pratensis-pratensis	meso	2	3	( )	3	( )	( )	( )	2	2	3
	xero	( )	5	2	( )	2	( )	( )	( )	( )	3
Anthoxanthum odoratum	meso	( )	2	3	3	4	( )	( )	( )	2	3
	xero	( )	( )	5	5	( )	3	( )	( )	2	3
Agrostis tenuis	meso	5	3	2	( )	3	4	3	3	4	5
	xero	2	3	3	3	3	4	2	( )	3	4
Festuca ovina-tenuifolia	meso	( )	5	5	5	6	6	6	4	6	6
	xero	2	5	5	5	5	5	5	2	3	5
Thymus pulegioides	meso	( )	5	5	5	5	3	2	1	2	3
	xero	3	5	5	5	3	1	2	2	2	2
Polytrichum juniperinum	meso	( )	( )	( )	( )	2	2	3	4	4	4
	xero	2	5	2	( )	2	2	4	4	4	4
Carex caryophyllea	meso	( )	( )	( )	( )	( )	( )	( )	( )	2	( )
	xero	( )	( )	( )	( )	( )	( )	( )	( )	( )	( )
Centaurium erythraea	meso	( )	2	( )	2	( )	( )	( )	( )	2	4
	xero	( )	( )	( )	( )	( )	( )	( )	( )	( )	( )
Sieglingia decumbens	meso	2	2	( )	( )	( )	( )	( )	( )	( )	4
	xero	( )	( )	( )	( )	( )	( )	( )	( )	( )	( )
Vicia lathyroides	meso	( )	( )	( )	( )	( )	( )	( )	3	( )	( )
	xero	( )	( )	3	( )	1	( )	( )	3	3	1
Aira praecox	meso	( )	( )	3	5	4	4	6	4	6	4
	xero	( )	( )	3	5	3	( )	2	( )	4	3
Carex arenaria	meso	2	3	2	( )	( )	( )	( )	( )	2	( )
	xero	2	5	5	3	2	2	3	3	3	4
Rhytidiadelphus squarrosus	meso	( )	3	( )	3	2	3	2	3	4	4
	xero	( )	5	5	3	2	( )	( )	( )	4	3
Corynephorus canescens	meso	( )	( )	( )	( )	( )	( )	( )	( )	( )	( )
	xero	2	5	5	6	5	2	2	3	2	4
Erophila verna	meso	( )	2	3	3	2	1	( )	3	3	3
	xero	( )	( )	3	3	2	2	( )	1	2	2
Hypnum cupressiforme	meso	( )	( )	3	2	5	3	7	4	4	4
	xero	8	7	7	7	7	9	9	3	9	9
Cerastium semidecandrum	meso	2	2	5	3	4	5	( )	3	4	4
	xero	( )	( )	3	3	2	2	2	2	4	4
Number of species on 4 sq.m.	meso	20	40	41	35	40	39	21	28	36	37
	xero	26	35	44	36	35	27	26	28	37	37
Diversity	meso	76	93	92	92	88	81	84	89	85	90
	xero	43	94	91	92	81	54	51	36	55	63

Corynephorus must be related to the opening up of the herb and/or moss layer and the subsequent activities of ants leading to the exposure of small bare spots. This response is accompanied by a negative response to the wet conditions which lead to a closer herb layer.

A third and peculiar reaction is shown by the moss *Hypnum cupressiforme,** and to a lesser extent also by the moss *Rhytidiadelphus squarrosus.*** *Hypnum* is generally characteristic as a dominant moss in the xerosere dune grasslands. It appears to react in

* Hedw. var. Lacunosum Brid.
** (Hedw.) Warnst.

a strongly negatively way to a high rainfall. While its positive response to dryer conditions can be explained as an occupation of open spots becoming available, the dying off under a high precipitation seems to be a direct physiological reaction.

Discussion

The general tendencies in floristic change and change in vegetation cover brought about by the experiments are in accordance with the general impression regarding the influence of fluctuations in the summer moisture conditions. (Cf. Dodd & Lauenroth, 1979 for similar results in a *Bouteloua* prairy in the USA and the discussion by French on grasslands in general) Clearly the experimental dry season is of the same order of dryness as the extreme dry season of 1976, the experimental wet season is still wetter than the extreme season of 1966. The main difference with the natural situation is of course the prolongation of the extreme conditions. Serie of very dry or very wet seasons are hardly found. However, the years 1963-1968 have an average season precipitation of 378 mm, with a minimum value of 290 mm, which is still above average. The years 1975-1977 had an April-August precipitation of only 198 mm and if we do not include August 1977, the only wetter month in the three summers, with 109 mm, the average would be only 164 mm. To check the above-mentioned impression the cover-abundance values of the same 18 species showing any response to either more or less rainfall were studied over two permanent plots representing the mesosere of zone III (plot 15.1) and the xerosere of zone VI (plot 11.12).

The results are summarized in Table 2. They are rather heterogeneous. Many *Festuco-Galietum* species, show a tendency of decrease in dry years (1976-1977) together with an increase in wet years (1972-1973, 1979). Examples are *Anthoxanthum odoratum, Luzula campestris, Thymus pulegioides*. The mesosere species, notably *Centaurium erythraea* and *Rhytidiadelphus squarrorus* follow the same pattern.

Aira praecox is one of the species showing the opposite response as in the experiments. Many more species show an unclear or even a contradictory behaviour. Species of open habitats such as *Erophila verna* and *Cerastium semidecandrum* decrease in

the dry years 1976-1977 rather than increase and *Corynephorus canescens* shows the same reaction. The moss *Hypnum cupressiforme* shows strong fluctuations, but differently in the two plots: in the mesosere the pattern is in accordance with the general impression and outcome of the experiments. In the xerosere however, where the experiments did present a less clear response, it is true, the pattern is deviant in that there is only a strong decrease in a dry period, and only in the less extreme year 1977.

The diversity figures presented in Table 2 show that the species richness is related to the general moisture conditions. The diversity figures, based on Simpson's (1949) dominance index, show that the xerosere has a higher proportion of the total cover-abundance sum concentrated in a few species, whereas the mesosere has a highly regular distribution of cover % over the species. We also see that in the xerosere the fluctuations are more pronounced with clear tops of diversity values in the wet year and clear minima during the dry years. The diversity figures of the experimental plots did not yet show a clear pattern. The wet plots tend to develop a higher richness but do not always differ from the comparison plots. A longer series of observations seems necessary here. It is planned to continue the experiment until 1981. In a subsequent study with O. van Tongeren concrete figures over the period 1978-1981 will be presented. It will be possible to statistically test the then obtained differences.

It is assumed that there are at least two factors which could explain that the clear patterns as found in the experiments are not so clear in the observation series 1963-1979. One is the influence of the conditions in the years preceding an extreme condition. The drought in 1976 was preceded by a series of relatively but not extremely moist years, whereas the experimental drought in 1978-1979 followed relatively dry years in which the general vegetation cover had become more open already. The relatively moist years 1971-1973 followed a period of normal years, whereas the moist year 1979 followed after the natural drought of 1976-1977.

The second interfering factor is the impact of the rabbit population. As Figure 1 shows both the rabbit density and the moisture conditions show a fluctuation pattern, but these patterns do not coincide. The year 1976 was extreme in both aspects: dry and with heavy grazing. The experimental dry sites were very dry but suffered hardly from rabbit grazing.

In order to unravel these two factors it would be necessary to compare the rabbit exclosures with their comparison quadrats over a period including a relatively moist period. 1979 has been a wet year, and 1980 a normal one, so we could cover all possible situations with the enclosure experiments which started in 1972. A special study of the effects of grazing and decrease in grazing is in preparation.

Moreover we would have to analyse the changes in the entire spatial distribution of species and plant communities on the basis of the full description of the area. Such descriptions are available for the relatively normal year 1963 with heavy grazing, for the comparable year 1970 (cf. van der Maarel, 1978) and for the relatively wet year 1979 with relatively heavy grazing (de Lyon *et al.,* 1980). This comparison is now in progress. It is our wish to keep the rabbit population low and repeat the full description after some years. Such a spatio-temporal analysis will form a reference for a numerical classification and ordination of the 15 experimental plots over 4 years, which is planned as well.

Conclusions

Through simple manipulation of rainfall amounts it has been possible to bring about floristic and structural changes in dune grassland plots which could, on the whole be understood against the background of the long term observations in the area since 1963 (cf. Persson, this volume for a similar approach).

There is interference between effects of drought and wetness respectively and the intensity of rabbit grazing. This interference is still not clear because observations under moist and low grazing conditions are still missing.

References

Bongers, F. & Lyon, M. de., De invloed van neerslag verandering op de vegetatie van een duingrasland bij Oostvoorne. Doct. report Division of Geobotany, Nijmegen.

Dodd, J. L. & Lauenroth, W. K., 1979. Analysis of the response of a grassland ecosystem to stress. In N. R. French (ed.) Perspectives in grassland ecology, p. 43–58. Springer, New York.

French, N. R., 1979. Principal subsystem interactions in grasslands. In: N. R. French (ed.) Perspectives in grassland ecology, p. 173–190. Springer, New York.

De Lyon, M., Meeuwissen, P. & Nuland, G. van, 1980. Dynamiek in een duingrasland-vegetatie te Oostvoorne. Doct. report Division of Geobotany, Nijmegen.

Maarel, E. van der, 1966. Over vegetatiestructuren, -relaties en -systemen (with a summary). Thesis Utrecht, 170 pp.

Maarel, E. van der, 1978. Experimental succession research in a coastal dune grassland, a preliminary report. Vegetatio 38,1: 21–28.

Maarel, E. van der, 1979. Transformation of cover-abundance values in phytosociology and its effects on community similarity. Vegetatio 30: 97–114.

Maarel, E. van der & Westhoff, V., 1964. The vegetation of the dunes near Oostvoorne, The Netherlands. Wentia 12: 1–61.

Simpson, E. H., 1949. Measurement of diversity. Nature 163: 688.

Westhoff, V., 1947. The vegetation of dunes and salt marshes on the Dutch islands of Terschelling, Vlieland and Texel. Thesis, Utrecht, 131 pp.

Accepted 20.7.1981.

Succession, diversité et amplitude de niche dans les pâturages du centre de la péninsule ibérique

F. D. Pineda, J. P. Nicolas, M. Ruiz, B. Peco & F. G. Bernaldez*
Departamento de Ecología, Universidad Autónoma, Madrid-34, Espagne

Keywords: Diversity, Grassland, Niche, Succession, Time and Space

Abstract

Some diversity and niche amplitude parameters were applied to rangeland pastures of the Central Iberian Peninsula and to their succession stages after the periodical ploughing typical of the traditional management of these areas. Four different slopes within a large area of undulating terrain were selected for the monitoring of succession as they contained the characteristical geomorphological pattern of the area (denudation, transport and accumulation sectors).

If we consider the total entropy theorem, $H(E.P) = H(E) + H(P/E)$, the total entropy of the slope $H(E.P)$ and the entropy of species $H(E)$ increase as succession progresses. As the value of the entropy of the sampling plots conditioned by the species $H(P/E)$ is affected by the number of plots utilized, we employed the expression $A = H(P/E)/\log_2$ number of plots, similar to Pielou's index for niche amplitude, $W = H(P/E)/H(P)$.

This values decreases with succession, indicating that plant species tend to occupy more definite sectors along the slope. The number of low entropy species $H(P/E)_i$ or specialist species, confined to narrow sectors also increases. When computed separately within the different sectors niche amplitude results in small values for the low slope regions (accumulation sector). This effect becomes more pronounced when succession advances.

Introduction

Ce travail fait suite à quelques études préliminaires (Bernaldez *et al.*, 1978, 1979; Pineda *et al.*, 1981; Ruiz *et al.*, 1979) sur la distribution et la structure de la végétation dans les pâturages semiarides en bordure de la Sierra de Guadarrama (Système Central, Teran 1952) et d'autres zones analogues.

Il existe plusieurs études sur l'homogénéité spatiale de la végétation, faites à différentes échelles de détail et à différents points de vue (voir e.g. Gounot,

1956). Quelques concepts de la théorie de l'information ouvrent des perspectives applicables à ce sujet. Godron (1966) a développé des idées semblables à celles de Margalef (1957) qui a appliqué cette théorie à l'écologie. La diversité constitue un moyen de décrire la structure d'un écosystème à un moment donné et peut être prise comme une mesure des possibilités d'interaction entre les espèces qui la composent. Etudiée sous forme de spectre, la diversité informe de la distribution et de l'importance des formes d'interaction entre les espèces (Margalef, 1980). On peut considérer que l'abondance de chacune d'entre elles est proportionelle à l'espace qu'elles peuvent occuper, c'est à dire à l'amplitude de leur niche écologique spatiale (MacArthur, 1968; Margalef, 1980). Les dimensions

* Nous remercions le Conseil d'Administration de La Paranza, propriétaire du Castillo de Viñuelas, et particulièrement Mrs. C. Hernandez-Ros et J. A. Léon-Vrquijò, pour les facilités qu'ils ont données à cette équipe durant la réalisation de ce travail.

Vegetatio 47, 267–277 (1981). 0042-3106/81/0472/3-0267/$2.20.

et caractéristiques de niche font partie de l'organisation des écosystèmes et peuvent varier au cours de la succession. On peut utiliser les paramètres de ces deux concepts, diversité et niche, pour étudier l'organisation dans l'espace et dans le temps des écosystèmes de pâturage.

La différenciation dans l'espace des communautés au cours de la succession et l'étude de leur hétérogénéité constitue un thème de travail fort intéressant en écologie (Margalef, 1958, 1962, 1974; Shelford, 1963). Pielou (1969) considère que l'hétérogénéité spatiale à échelle hiérarchisée se trouve présente dans la plus grande partie des communautés naturelles et constitue par elle-même une forme de diversité. On peut parler de la diversité d'un petit échantillon (diversité alpha) et d'une diversité de 'motif' (diversité bêta, 'pattern diversity') pour se rapporter à la diversité dérivant de la variation de petites structures, plus ou moins différentes les unes des autres, qui se combinent en structures plus grandes hiérarchiquement (Margalef, 1974). Les termes alpha et bêta peuvent se rapporter l'un à la répartition globale des espèces au sein d'une communauté, et l'autre à la répartition existant entre des échantillonnages faits dans cette communauté (voir Whittaker, 1970). La variabilité de l'environnement (physique, chimique) peut jouer un rôle important dans ce pattern (Galiano, 1980; Ruiz, 1980), et on a développé à ce sujet des théories et des modèles sur le recouvrement de niches utilisables pour interpréter des structures écologiques (May & MacArthur, 1972).

Pour notre part, nous avons utilisé quelques paramètres de diversité et caractéristiques de niche pour étudier les pâturages du Centre de l'Espagne, ainsi que la succession écologique qui s'établit quand on cesse les cultures itinérantes pratiquées dans quelques localités (Caxa de Leruela, 1631; Nicolas et al., 1980).

Composantes de la diversité

Dans le cas d'un écosystème donné dans lequel nous avons établi des parcelles d'échantillonnage, nous appelons entropie espèces-parcelles, $H(E.P)$, l'incertitude que l'on a sur la présence d'une espèce dans une parcelle déterminée. Cette incertitude n'existerait pas - c'est à dire, $H(E.P)$ prendrait la valeur zéro - s'il apparaissait une seule espèce dans une seule parcelle.

Supposons que nous trouvons toujours au moins un individu d'une certaine espèce par parcelle. Pour un nombre donné d'individus, espèces et parcelles d'échantillonnage, la valeur de cette entropie est minimum quand il existe dans chaque parcelle une espèce différente et seulement une espèce par parcelle, et maximum quand toutes les espèces se trouvent présentes dans toutes les parcelles avec la même proportion d'individus.

Si l'on considère le théorème de l'entropie totale, on peut employer l'expression

$$H(E.P) = H(E) + H(P/E) \qquad (1)$$

où $H(E)$ représente l'entropie des espèces présentes, $H(P/E)$ celle des parcelles conditionnée aux espèces et $H(E.P)$ l'entropie totale.

$H(E)$ mesure l'incertitude que l'on a sur l'espèce à laquelle peut appartenir un exemplaire pris au hasard en considérant l'ensemble de toutes les parcelles (c'est à dire toute l'information dont on dispose, sans analyser la variabilité interne de l'ensemble étudié, mesurable avec les parcelles d'échantillonnage). Ce paramètre dépend du nombre total d'espèces apparues dans l'ensemble des parcelles, et de l'équiprobabilité de la répartition de celles-ci dans cet ensemble, mais est indépendant de la forme de la répartition des individus de chaque espèce entre les différentes parcelles.

$H(P/E)$ représente une mesure pondérée pour toutes les espèces, de l'incertitude que l'on a sur la parcelle d'où vient un exemplaire d'une espèce déterminée; on peut considérer ce paramètre comme la mesure de la valeur moyenne de la diversité d'habitats pour l'ensemble des espèces. Au contraire de $H(E)$, elle n'est pas indépendante de la forme de la répartition des individus de chaque espèce entre les différentes parcelles. Indépendamment du nombre de parcelles considérées et des espèces présentes, $H(P/E)$ a une valeur zéro quand chaque espèce se trouve dans une seule parcelle, et maximale quand toutes les espèces sont également distribuées dans toutes les parcelles. Néanmoins, au contraire de $H(E.P)$, ce paramètre atteint aussi cette valeur maximale bien que les abondances des espèces diffèrent entre elles dans l'ensemble des parcelles. Le fait que chacune des espèces diffère beaucoup dans ses abondances respectives sur toute la zone considérée est sans effet sur cette valeur. Cette situation rend en plus la valeur de $H(P/E)$

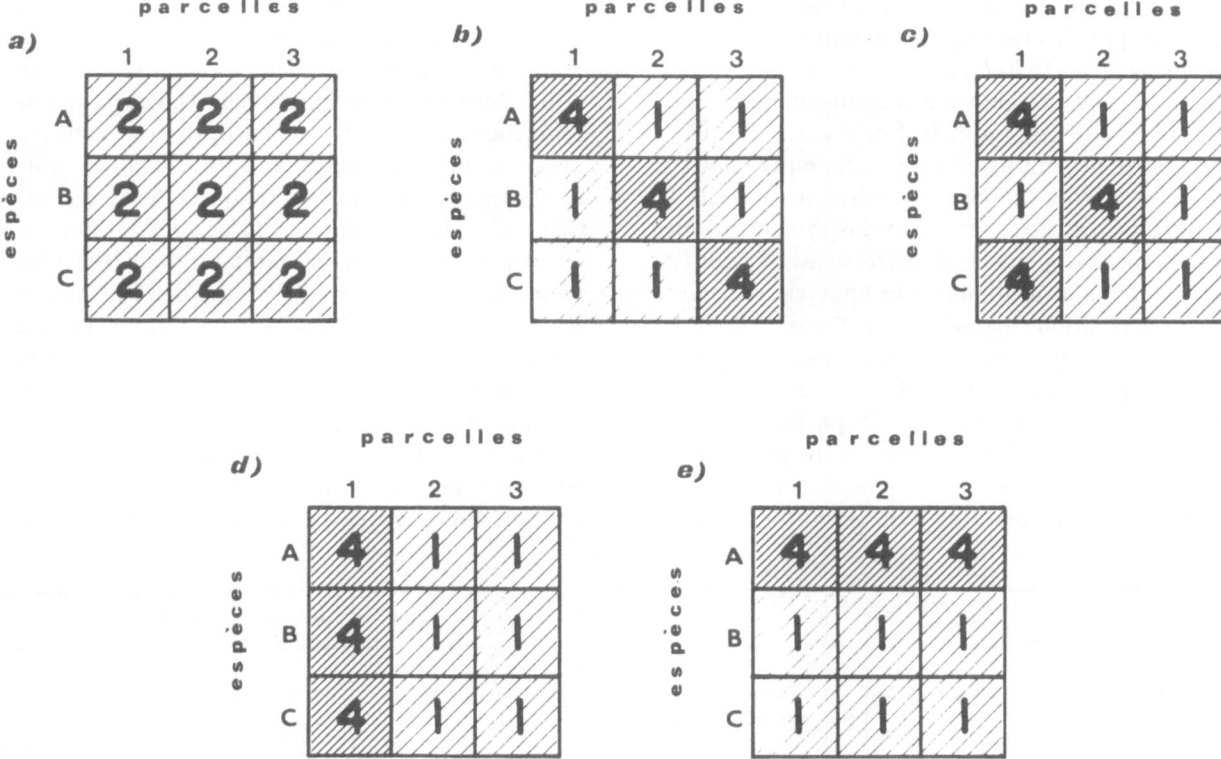

Fig. 1. Données simulées pour différentes formes de distribution des abondances de 3 espèces dans 3 parcelles d'échantillonnage.

indépendante du nombre total d'espèces rencontrées.

Bien que la valeur de *H(P/E)* ne soit pas influencée par le nombre d'espèces, elle se trouve très affectée par le nombre de parcelles considérées.

Dans l'appendice, nous avons indiqué les calculs des différentes expressions de l'entropie.

Exemple avec des donnés artificielles

La Figure 1 représente plusieurs modèles élaborés avec des données simulées. Les 5 matrices **a, b, c, d** et **e** correspondraient à autant de zones différentes où on aurait pris régulièrement 3 parcelles d'échantillonnage (colonnes) et anoté dans celles-ci l'abondance des espèces A, B et C (lignes). Pour simplifier l'exemple nous avons supposé que, dans tous les cas, chaque espèce est représentée au moins par un individu dans chaque parcelle. Bien que toutes les zones contiennent le même nombre d'individus, on observe que chaque espèce occupe l'espace d'une manière particulière.

Dans ces circonstances, on peut observer les valeurs extrêmes que peuvent prendre les différentes expressions de l'entropie commentées antérieurement. La zone **a** est totalement homogène. Les individus de chaque espèce sont également répartis dans chaque parcelle de la zone en question. Ici la valeur de n'importe laquelle des expressions de l'entropie que nous utilisons sera maximale. Il n'apparaît aucune ségrégation spatiale. L'amplitude de la niche spatiale est totale.

La zone **b** a des caractéristiques très différentes. Dans le cas présent, l'hétérogénéité spatiale est maximale d'après le type de distribution des espèces dans les différentes parcelles. Il y a un habitat spécifique pour chacune d'entre elles et la ségrégation est grande. La valeur que prend ici *H(E)*, mesurable par l'expression (2) (voir appendice) continue à être maximum, comme dans la zone **a**, et équivaut à $\log_2 3$. C'est la même chose pour *H(P)*. L'organisation spatiale différente des zones **a** et **b** peut être mise en évidence, néanmoins, avec la valeur de *H(P/E)*, qui est maintenant très faible (elle serait zéro si les 18 individus présents se répartissaient de n'importe quelle manière le long de la diagonale principale de la matrice **b**). Il en est de même avec *H(E/P)*.

La zone **c** présente une hétérogénéité spatiale intermédiaire entre **a** et **b**. Les 3 parcelles qu'elle comporte ne sont pas si différentes entre elles que celles de la zone **b**: une est très diverse (parcelle 3), l'autre peu diverse, presqu'exclusivement peuplée par l'espèce B (parcelle 2), la troisième possède des caractères intermédiaires. On apprécie seulement une ségrégation spatiale entre l'espèce B et les autres. Ce type d'occupation de l'espace par les espèces peut aussi se différencier de ceux de **a** et **b** par les valeurs que prennent les paramètres d'entropie. *H(E)* continue à être maximale, *H(P/E)* prend la même valeur que pour la zone **b**, néanmoins la valeur de *H(P)* descend notablement et

celle de *H(E/P)* augmente, étant intermédiaire entre celles de la zone **a** et **b**.

La zone **d** comprend deux parties de diversité égale; l'une très peuplée (parcelle 1), l'autre moins (parcelles 2 et 3). Dans une partie de cette zone il y aurait des conditions de milieu qui rendent possible la coincidence des niches spatiales des 3 espèces présentes. Dans ces circonstances nous continuons à trouver la même valeur maximale pour *H(E)* et la même valeur faible de *H(P/E)* des deux cas précédents. En revanche, *H(E/P)* prend maintenant une valeur maximale, comme dans la zone **a**, et *H(P)* une valeur très faible (zéro si tous les individus se trouvaient dans une seule parcelle).

Quant à la zone **e** elle est homogène, comme la première, bien que plus peuplée pour l'une des espèces. La diversité des habitats est aussi maxi-

ZONE	TYPE DE MATRICES DE DONNÉES (1)	H(E.P)	H(E)	H(P/E)	H(P)	H(E/P)	VALEURS COMPAREES		
TYPE *a*. HOMOGÈNE SANS DOMINANCE D'AUCUNE ESPÈCE.		VALEUR MAXIMUM (2) $\alpha = Log_2(r \times c)$	VALEUR MAXIMUM $\beta = Log_2 r$	VALEUR MAXIMUM $\gamma = Log_2 c$	VALEUR MAXIMUM $\gamma = Log_2 c$	VALEUR MAXIMUM $\beta = Log_2 r$	$\beta = \gamma$	(3)	$\dfrac{H(P/E)}{H(P)} = 1$
TYPE *b*. TRÈS HÉTÉROGÈNE, AVEC UNE GRANDE SÉGRÉGATION SPATIALE.		VALEUR MINIMUM σ	VALEUR MAXIMUM β	VALEUR MINIMUM ϵ (4)	VALEUR MAXIMUM γ	VALEUR MINIMUM ϵ	$\beta = \gamma$		$\dfrac{H(P/E)}{H(P)} << 1$
TYPE *c*. HÉTÉROGÈNE, AVEC SÉGRÉGATION SPATIALE ENTRE UNE ESPÈCE ET LES AUTRES.		VALEUR MINIMUM σ	VALEUR MAXIMUM β	VALEUR MINIMUM ϵ	VALEUR INTERMEDIAIRE ν	VALEUR INTERMEDIAIRE μ	$\beta > \nu$		$\dfrac{H(P/E)}{H(P)} < 1$
TYPE *d*. HÉTÉROGÈNE, SANS SÉGRÉGATION ET UNE PARCELLE TRÈS PEUPLÉE.		VALEUR MINIMUM σ	VALEUR MAXIMUM β	VALEUR MINIMUM ϵ	VALEUR MINIMUM ϵ	VALEUR MAXIMUM β	$\beta > \epsilon$		$\dfrac{H(P/E)}{H(P)} = 1$
TYPE *e*. HOMOGÈNE, AVEC UNE ESPÈCE DOMINANTE.		VALEUR MINIMUM σ	VALEUR MINIMUM λ	VALEUR MAXIMUM γ	VALEUR MAXIMUM γ	VALEUR MINIMUM ϵ	$\lambda < \gamma$		$\dfrac{H(P/E)}{H(P)} = 1$

Tableau 1. Valeurs d'expressions de l'entropie pour des zones d'hétérogénéité spatiale différente (voir Fig. 1).

(1) Le diamètre du cercle représente l'abondance de l'espèce dans la parcelle de référence.
(2) *r* indique le nombre d'espèces présentes et *c* le nombre de parcelles considérées.
(3) Dans le cas des matrices où *r = c*.
(4) Les valeurs exprimées au moyen de lettres grecques dont le calcul n'est pas indiqué, correspondraient à celles trouvées pour un nombre donné d'individus, espèces et parcelles. Par exemple, dans les matrices de la figure 1, la valeur de ϵ est de 1, 2 bits.

male, puisque les espèces sont également abondantes dans toutes les parcelles. On peut en outre considérer que toute la zone est un habitat préférentiel pour une des espèces (la plus abondante). Seulement dans ce cas, $H(E)$ prend une valeur très basse, comme $H(P)$ dans le cas précédent. $H(P/E)$ reprend la valeur maximale de la première zone, et $H(E/P)$ la valeur faible trouvée dans la zone **b**.

Dans le Tableau 1, nous regroupons quelques comparaisons interessantes qui pourraient se faire entre les valeurs de l'entropie calculées pour des zones de différente organisation spatiale des communautés (Fig. 1).

Zone d'étude

Nous avons étudié les pâturages de deux zones bordant la Sierra de Guadarrama:
- l'une située sur le pédiment granitique
- l'autre située sur les sédiments miocènes arkosiques contigus au pédiment.

Les localités sont situées à l'intérieur de la zone pilote où se déroule le travail de recherche sur les pâturages du Département d'Ecologie de l'Université Autonome (1980).

La localité à terrain arkosique (Castillo de Viñuelas, zone du Pardo) (Pineda *et al.*, 1981) permet d'observer l'effet de la succession qui a lieu à la suite de labours qui détruisent les pâturages (Caxa de Leruela, 1631; Nicolas *et al.*, 1980). La localité en terrain granitique (zone de Galapagar) constitue un pâturage qui n'a jamais été labouré depuis au moins 100 ans (Ruiz, 1980). Ces deux localités sont utilisées en pâturages extensifs parsemés de chênes-verts, *Quercus rotundifolia* Lam., ce qui correspond plus ou moins à un écosystème de 'dehesa' (Bernaldez *et al.*, 1979).

Echantillonnage

Dans les deux localités, nous avons procédé à un échantillonnage stratifié. Les secteurs (strates) correspondent à différentes modalités du dynamisme géomorphologique des pentes (Bernaldez *et al.*, 1978, 1979). Dans la zone arkosique, nous avons considéré un versant d'environ 800 ha sur lequel nous avons choisi 4 pentes voisines de même exposition, labourées il y a 1, 2, 4 et plus de 20 ans.

Sur chaque pente nous avons distingué 3 secteurs ou strates (Bernaldez *et al.*, 1978, 1979; Pineda *et al.*, 1981):
- secteur supérieur avec prédominance d'exportation de matériaux
- secteur moyen, avec prédominance de transport
- secteur inférieur, avec prédominance d'accumulation.

Dans chacun de ces secteurs, nous avons échantillonné au moyen de parcelles de 2×3 m. placées à des intervalles égaux, en suivant la direction de la pente. Le nombre de parcelles dans chaque secteur oscille de 3 à 5, dépendant de la longueur du versant. Dans chaque parcelle, nous avons noté les fréquences des espèces de pâturage au moyen de 10 carrés de 20×20 cm jetés au hasard. Nous avons considéré que la surface des carrés est adéquate pour estimer la fréquence à cause de la grande étendue de la parcelle. L'échantillonnage a été effectué deux fois dans un intervalle de 4 ans.

Dans la zone granitique, nous avons considéré un versant d'environ 40 ha où il nous fut seulement possible de distinguer deux secteurs: le secteur supérieur de dispersion et exportation de matériaux, et le secteur inférieur d'accumulation (Ruiz *et al.*, 1979).

Au moyen de 13 et 15 parcelles d'échantillonnage respectivement par secteur et d'unités élémentaires de 20×20 cm, nous avons enregistré, comme auparavant, des données de fréquence des espèces du pâturage.

Analyse des données relevées sur le terrain

Zone arkosique

Le Tableau 2 rassemble les résultats du calcul de différentes expressions de l'entropie à partir des données recueillies sur le terrain. Les colonnes correspondent respectivement aux années écoulées depuis le dernier labourage du territoire étudié. Chaque pente est divisée en secteurs d'exportation, transport et accumulation déjà mentionnés. Nous avons considéré ces secteurs comme des unités d'analyse pour étudier l'organisation de l'espace des communautés de pâturage sur la pente. Nous constatons que la diversité totale, exprimée par l'entropie espèce-secteur, $H(E.S)$, augmente avec la succession, ainsi que la diversité des espèces $H(E)$.

Tableau 2. Variation des valeurs de l'entropie de pentes labourées à différentes dates et abandonnées. Ces valeurs sont estimées en divisant ces pentes en secteurs supérieur, moyen et inférieur.

	Années écoulées après le labourage			
	1	2	4	plus de 20
Nombre d'espèces	25	33	34	49
H(E.S)	4.12	4.72	5.62	6.04
H(E)	2.76	3.52	4.42	4.94
H(S/E)	1.36	1.20	1.20	1.10
$A_s = \dfrac{H(S/E)}{\log_2 \text{ nb. secteurs}}$ {8}	0.86	0.76	0.76	0.69

L'augmentation de la valeur de cette dernière expression est due surtout à l'augmentation progressive du nombre total des espèces, mais non à une meilleure répartition des individus en espèces. Avec la succession secondaire qui a lieu dans le pâturage après le labour, nous assistons à un enrichissement continuel en espèces, mais leur distribution se fait de manière différentielle, la ségrégation spatiale augmente et la pente devient hétérogène. En effet, la diminution progressive de la valeur de la diversité des secteurs conditionnée aux espèces, $H(S/E)$, indique une colonisation différentielle des espèces de pâturage sur les secteurs de la pente au cours de la succession. Les résultats sont en accord avec ceux obtenus dans un travail préliminaire sur la succession dans ces pâturages (Pineda *et al.*, 1981).

Nous avons utilisé l'expression (8) (Tableau 2) analogue à la (7) (voir appendice), semblable à celle décrite par Pielou (1975), pour indiquer la valeur moyenne de l'amplitude de niche au cours de la succession. Cette expression diminue nettement

Tableau 3. Variation des valeurs de l'entropie de pentes labourées à différentes dates et abandonnées. Ces valeurs sont calculées en divisant ces pentes de haut en bas en parcelles contiguées de 2×3 m.

	Années écoulées après le labourage			
	1	2	4	plus de 20
Nombre d'espèces	25	33	34	49
H(E.P)	5.87	6.61	6.73	8.03
H(E)	2.76	3.52	4.42	4.94
H(P/E)	3.10	3.09	2.32	3.10
$A_p = \dfrac{H(P/E)}{\log_2 \text{ nb. parcelles}}$	0.81	0.77	0.77	0.74

avec la succession, indiquant que les espèces tendent à coloniser différemment les secteurs géomorphologiques de la pente.

Afin d'étudier le phénomène précédent à une échelle plus détaillée, nous avons réalisé le même type d'analyse en prenant comme unités élémentaires les parcelles d'échantillonnage des pentes sans les grouper en secteurs. Le Tableau 3 en montre les résultats. Nous avons pu observer une situation semblable à la précédente; la diversité totale des pentes, $H(E.P)$, et celle des espèces, $H(E)$ (qui prend logiquement les mêmes valeurs qu'avant), augmentent nettement. L'entropie des parcelles conditionnée aux espèces, $H(P/E)$, ne varie pas maintenant régulièrement au cours de la succession. Nous avons pris un différent nombre de parcelles sur les 4 pentes étudiées et cette mesure de l'entropie manque alors de valeur comparative. Pour cette raison, nous avons utilisé l'expression (7) dans le but de contrôler cet effet. On observe que cette expression diminue au cours de la succession; on peut interpréter, comme auparavant, qu'à cette nouvelle échelle de détail les espèces du pâturage tendent à occuper progressivement des positions précises le long de la pente.

On peut observer dans le Tableau 4 qu'au cours de la succession dans la pente, augmente le nombre d'espèces d'entropie faible, mesurée par: $H(P/E)i/\log_2$ *nb. parcelles*.

Les espèces de faible entropie peuvent être considérées comme des 'spécialistes' qui tendent à occuper un habitat restreint sur la pente (ce sont des espèces dominantes dans une ou quelques parcelles). Ces espèces augmentent avec la succession et plus rapidement que le nombre total des espèces. Les premières années, on tend à avoir une colonisation indifférenciée le long de la pente: par exemple, *Anchusa undulata* L., *Ajuga chamaepitys* (L.) Schreber, *Linaria spartea* (L.) Wild., *Heliotropium europaeum* L. entre autres, sont des espèces pionnières caractéristiques de la première et seconde année et présentent le comportement déjà mentionné ('généralistes') (Pineda *et al.,* 1981). Avec celles-ci on trouve quelques 'spécialistes' (voir colonne 1 du Tableau 4) comme *Paronychia cymosa* (L.) DC., qui tend à être localisée en haut de la pente, *Andryala integrifolia* L. au milieu, et *Poa annua* L. en bas. Le nombre de 'spécialistes' tend à augmenter au cours du temps. Dans des étapes plus avancées de la succession, elles deviennent abon-

Tableau 4. Apparition d'espèces de faible entropie ('spécialistes') au cours de la succession, $A_i = H(P/E)_i / \log_2 nb.\ parcelles.$

	Années écoulées après le labourage			
	1	2	4	plus de 20
Nombre d'espèces	25	33	34	49
Nombre d'espèces à faible entropie*	6	8	10	20
Rapport	0,24	0,24	0,29	0,41
	0,26 Andryala integrifolia L.	0,30 Logfia minima (S.M.) Dumort	0,24 Pterocephalus diandrus (Lag.) Lag.	0,30 Evax carpetana Lange
	0,26 Paronychia cymosa (L.) DC	0,22 Viola kitaibeliana Schultes	0,00 Biserrula pelecinus L.	0,29 Tuberaria gutata (L.) Fourr.
	0,24 Erodium cicutarium (L.) L'Hér.	0,25 Scandix australis L. subsp. microcarpa (Lange) Thell	0,30 Linaria diffusa Hoffman & Link	0,30 Dipcadi serotinum Medik L.
	0,24 Papaver somniferum L.	0,24 Agrostis castellana Boiss & Reuter	0,30 Scandix australis L. subsp. microcarpa (Lange) Thell	0,00 Nardurus tenellus Reichenb. ex Godron
	0,27 Holcus setiglumis Boiss & Reuter	0,30 Logfia arvensis (L.) J. Holub.	0,29 Campanula patula L.	0,30 Linaria diffusa Hoffmann & Link
	0,21 Poa annua L.	0,15 Sisymbrium runcinatum Lag. ex DC	0,22 Crepis capillaris L. Wallr.	0,24 Centaurea melitensis L.
		0,25 Anchusa undulata L.	0,20 Sisymbrium runcinatum Lag. ex DC	0,00 Carlina racemosa L.
		0,25 Taeniatherum caput medusae (L.) Nevski	0,00 Daphne gnidium L.	0,25 Bromus madritensis L.
			0,29 Campanula erinus L.	0,29 Logfia arvensis (L.) J. Holub.
			0,30 Bromus hordeaceus L.	0,00 Euphorbia exigua L.
				0,30 Melica ciliata L.
				0,28 Crepis capillaris (L.) Wallr.
				0,29 Ranunculus parviflorus L.
				0,30 Sherardia arvensis L.
				0,23 Valerianella eriocarpa Desv.
				0,24 Rumex angiocarpus Murb.
				0,25 Trifolium campestre Schereber
				0,22 Senecio erucifolius L.
				0,00 Lotus parviflorus Desf.
				0,23 Festuca ampla Hackel

* Nous avons indiqué les espèces à valeur de A_i inférieures ou égales à 0,30. La valeur moyenne de cette expresión oscille entre 0,50 (1 an) et 0,60 (4 ans).

Tableau 5. Amplitude de niche dans différents secteurs des pentes au cours de la succession, dans une zone de sédiments arkosiques.

Secteurs de la pente	Amplitude de niche dans les secteurs, $A_{p,s} = \dfrac{H(P/E)}{\log_2 \text{ nb. parcelles}}$							
	1 an		2 ans		4 ans		plus de 20 ans	
	Nombre d'espèces	$A_{p,s}$	Nombre d'espèces	$A_{p,s}$	Nombre d'espèces	$A_{p,s}$	Nombre d'espèces	$A_{p,s}$
Zone supérieure	9	0.82	11	0.81	25	0.82	28	0.81
Zone intermédiaire	19	0.80	22	0.79	22	0.73	26	0.81
Zone inférieure	16	0.78	24	0.79	26	0.73	39	0.74

dantes, et peuvent s'ordonner corrélativement suivant leur position le long d'un gradient exportation-accumulation: *Evax carpetana* Lange, *Tuberaria guttata* (L.) Four., etc. occupent les zones supérieures, *Carlina racemosa* L.; *Bromus madritensis* L. etc., les intermédiaires, *Melica ciliata* L. *Crepis capillaris* (L.) Wall. etc., les zones inférieures. (voir dernière colonne du Tableau 4).

Finalement nous avons essayé d'étudier l'occupation de l'espace par les espèces à l'intérieur de chaque secteur en rapport avec la succession. Sur le Tableau 5, nous indiquons les valeurs d'amplitude de niche dans les secteurs de la pente, en prenant comme unités d'analyse les parcelles d'échantillonnage. En général on observe que la zone inférieure tend à être occupée par un plus grand nombre d'espèces et, bien que le résultat soit moins net que les précédents, le plus grand nombre d'espèces des zones inférieures est accompagné d'une diminution de l'amplitude de niche dans ces zones. Cet effet, généralisé, se trouve un peu moins accusé au début de la succession qu'à des étapes plus évoluées.

Entre les versants labourés à différentes dates, il peut exister des différences spatiales qui, bien qu'elles semblent minimes, n'ont pu être entièrement contrôlées. Pour cela, nous pensons que, dans tous les cas, il est nécessaire d'observer des parcelles permanentes pour mieux interpréter les résultats obtenus avec ces indices.

Parmi les différents indices de diversité proposés, il est interessant d'en appliquer un qui vérifie le théorème de l'entropie totale. L'indice de Shannon-Weaver (1963) semble recommandable (Pielou, 1969, 1975). D'autres indices ne possédent pas comme celui-ci les propriétés que doit réunir un indice de diversité (voir Pielou, 1975; Legendre, 1979). L'indice de Simpson (1949), par exemple, a été proposé comme mesure d'hétérogénéité (Good, 1953). Il présente un biais plus petit que l'indice de Shannon (Routledge, 1980) et peut, d'autre part, indiquer la dominance, ce qui le rendrait utile pour ce travail: l'expression $S = \Sigma_{i=1} pi^2$ de Simpson mesure la probabilité de trouver des individus de la même espèce dans des échantillonnages successifs, ce qui représente une propriété opposée à la diversité (Margalef, 1980). Néanmoins, quand il est appliqué à l'expression (1), comme cela a été fait avec l'indice de Shannon-Weaver, il ne prouve pas l'égalité proposée. L'expression modifiée de Simpson, $H = \log \Sigma_i pi^2$ (Pielou, 1975) représente une entropie de second ordre qui ne la vérifie pas non plus. Ce fait empêche d'interpréter correctement les causes des changements de valeur de l'entropie totale, $H(E.P)$, c'est à dire, la mesure pour laquelle

Tableau 6. Valeurs de différentes expressions de l'entropie dans les secteurs d'érosion (supérieur) et d'accumulation (inférieur) d'une pente granitique.

Secteurs de la pente	Nb. d'espèces	$H(E.P)$	$H(E)$	$H(P/E)$	$A_p = \dfrac{H(P/E)}{\log_2 \text{ nb. parcelles}}$
Supérieur	59	6.81	5.63	1.18	0.32
Inférieur	74	7.00	6.05	0.95	0.24

l'augmentation de diversité avec la succession est due à l'augmentation du nombre des espèces, *H(E)* et à la répartition des individus dans les échantillonnages, *H(P/E)*, objectifs de ce travail.

Zone granitique

Le Tableau 6 montre le résultat des calculs réalisés pour différentes expressions de l'entropie à partir de deux secteurs, inférieur et supérieur, du versant granitique.

Les valeurs de l'entropie totale et celle des espèces sont plus élevées dans la partie inférieure du versant que dans la partie supérieure. Dans les deux secteurs il apparaît une organisation spatiale des communautés végétales différente. Margalef (1968) indique que dans n'importe quel endroit de la nature, on peut délimiter des surfaces à l'aide de frontières de séparation entre deux ou plusieurs sous-systèmes: 'ces frontières séparent souvent des sous-systèmes – même si la frontière a été délimitée arbitrairement – qui présentent différents degrés d'organisation'. La plus grande organisation est liée, en général, à l'augmentation de la diversité. Dans notre cas, nous pouvons dire qu'il existe une plus grande organisation dans la partie inférieure de la pente, où, par exemple, la diversité totale, *H(E.P)*, se trouve plus élevée. En faisant abstraction de l'organisation interne des parcelles, nous considérons que cette diversité se décompose en une variation à petite échelle, *H(E)* (qui correspond à un macro-pattern de variation), et une autre, à grande échelle, *H(P/E)*, qui recueille la diversité due à la variation entre parcelle (micro-pattern).

Comme cela est arrivé dans la zone arkosique, l'augmentation de *H(E)*, qui a lieu ici dans la partie inférieure de la pente, peut être attribuée à l'existence d'un plus grand nombre d'espèces. La valeur trouvée pour *H(P/E)* est plus faible dans le secteur inférieur, indiquant l'existence d'habitats spécifiques bien déterminés pour les différentes espèces. Les valeurs d'amplitude de niche confirment le résultat. En suivant l'idée de Margalef, nous pouvons supposer une plus grande organisation (hétérogénéité dans l'espace, absence du hasard dans la colonisation des espèces) dans la partie inférieure de la pente, car elle peut contenir un plus grand nombre d'espèces, mais en revanche leurs niches sont plus restreintes.

Conclusions

1. L'application du théorème de l'entropie totale à l'analyse de la diversité d'un pâturage sur plusieurs versants, a permis d'obtenir une idée générale de l'organisation dans l'espace de leurs communautés végétales.

Le calcul de l'entropie totale, des espèces et des parcelles d'échantillonnage conditionnées aux espèces, permet de connaître dans quelle mesure les variations de la diversité sont dues à des changements dans le nombre des espèces et à la diversité de motif ('pattern diversity'). L'emploi de l'indice de Shannon-Weaver (1963) est recommandé pour appliquer le théorème de l'entropie totale (voir aussi Pielou, 1975).

2. En étudiant la succession dans les pâturages d'un versant, nous trouvons que le nombre d'espèces, la diversité totale et la diversité des espèces augmentent avec le temps. Au contraire, la 'diversité des espèces conditionnée aux parcelles d'échantillonnage' et l'amplitude moyenne de niches, diminuent durant la succession. Le résultat indique que la proportion des espèces 'spécialistes' augmente (augmentation de la ségrégation spatiale). Les circonstances précédentes se présentent aussi quand, au lieu de parcelles, nous considérons les secteurs du versant caractérisés par leur dynamisme géomorphologique.

Pour le versant étudié nous n'avons pas pu disposer de données de la succession entre 4 années et plus de 20 écoulées depuis le dernier labourage. En observant la tendance de l'augmentation de l'entropie des années 1 à 4, on pourrait penser que les valeurs trouvées pour plus de 20 ans peuvent correspondre à une situation postérieure à un optimum compris entre 4 et plus de 20 ans (voir par exemple l'augmentation de *H(E)* les premières années, Tableau 2). Néanmoins, d'autres versants étudiés – géomorphologiquement distincts, et par conséquent non considérés dans ce travail – ne présentent jamais de valeurs de *H(E)* supérieures à 4, 5 bits pour des étapes inférieures à 20 ans. Il n'est pas non plus fréquent de trouver des valeurs si élevées de l'entropie dans des écosystèmes semblables à ceux que nous avons étudiés. Des pâturages très stabilisés, jamais labourés, dans une zone granitique du Centre de l'Espagne arrivent à atteindre une valeur de 6 bits (Ruiz, 1980). Des valeurs supérieures à 4 bits pour des pâturages de terrains abandonnés depuis 40 ans sont rares (Bazzaz, 1975). En général des valeurs d'environ 5 bits sont déjà considérées comme assez élevées (Margalef, 1980).

3. Si nous comparons entre eux ces secteurs avec différent dynamisme géomorphologique, l'amplitude moyenne de niches est toujours plus faible dans la zone inférieure du versant (accumulation de matériaux) que dans la zone supérieure (dénudation) ou l'intermédiaire (transport), ce qui indique une plus grande proportion de 'spécialistes' dans les zones d'accumulation. Ce phénomène est plus accusé dans les zones granitiques étudiées que dans les arkosiques.

Références

Bazzaz, F. A., 1975. Plant species diversity in old field successional ecosystems in Southern Illinois. Ecology 56: 485–488.

Bernáldez, F. G. et al., 1978. Reconocimiento de pastizales mediante teledetección y estudios integrados. Pastos 8: 85–93.

Bernáldez, F. G. et al., 1979. Prospection intégrée de pâturages extensifs ('dehesa') en Sierra Morena (Espagne). L'Espace Géographique 3: 241–252.

Caxa de Leruela, M., 1631. Restauración de la Grandeza de España. J. P. Le Flem, Ed. Ministerio de Hacienda, Madrid.

Equipo del Departamento de Ecología de la Universidad Autónoma de Madrid, 1980. Modelos de paisaje en áreas de montaña de Madrid. Coloquio Hispano-Francés de Areas de Montaña, Madrid.

Galiano, E. F., 1980. Detección de pautas de distribución espacial en pastizales. Tésis Doctoral. Univ. Autónoma, Madrid.

Godron, M., 1966. Application de la théorie de l'information a l'étude de l'homogénéité et de la structure de la végétation. Oecol. Plant. 1: 187–197.

Good, I. J., 1953. The population frequences of species and the estimation of population parameters. Biometrika 40: 237–264.

Gounot, M., 1956. Carte des Groupements végétaux 1: 20.000. Bull. Serv. Cartephytogéogr., Ser. B. (C.N.R.S.) 1(1): 7–17.

Legendre, L. & Legendre, P., 1979. Ecologie numérique. T. 1. Le traitement multiple des données écologiques, Masson, Paris.

McArthur, R., 1968. The theory of the niche. In: Population, biology and evolution. T. C. Lewontin, Ed. Syracuse Univ. Press, Syracuse.

Margalef, R., 1957. La teoriá de la informacion en ecologia. Mem. R. Acad. Ciencias Barcelona 32(13): 373–449.

Margalef, R., 1958. Temporal succession and spatial heterogeneity in natural phytoplankton. In: Perspectives in marine biology. Berkeley: Univ. California Press.

Margalef, R., 1962. Succession in marine populations. Advancing frontiers of plant sciences II. Inst. Adv. Sc. Cul. New Delhi.

Margalef, R., 1968. Perspectives in ecological theory. Univ. Chicago Press, London.

Margalef, R., 1974. Ecología. Omega, Barcelona.

Margalef, R., 1980. La biosfera. Entre la termodinámica y el juego. Omega, Barcelona.

May, R. M. & MacArthur, R. H., 1972. Niche overlap as a function of environmental variability. Proc. Nat. Acad. Sci. USA 69: 1109–1113.

Nicolás, J. P. et al., 1980. Description des pâturages du centre de la Péninsule-Ibérique au moyen de processus stochastiques. Symposium 'Dynamique, Végétation', Montpellier, Actes 1–4.

Pielou, E. C., 1969. An introduction to mathematical ecology. Wiley, New-York.

Pielou, E. C., 1975. Ecological diversity. Wiley Interscience Publ. London

Pineda, F. D. et al., 1981. Ecological succession in oligotrophic pastures of Central Spain. Vegetatio 44: 165–176.

Routledge, R. D., 1980. Bias in estimating the diversity of large, uncensused communities. Ecology 61: 276–281.

Ruiz, M. et al., 1979. Estructura y variabilidad de pastizales semiáridos en áreas graníticas. Pastos 9: 41–57.

Ruiz, M., 1980. Características de la variación de pastizales en áreas graniticas. Tésis Doctoral. Univ. Autónoma Madrid

Shannon, C. E. & Weaver, W., 1963. The mathematical theory of communication. Univ. Illinois Press, Urbana.

Shelford, V. E., 1963. The ecology of North America. Univ. Illinois Press, Urbana.

Simpson, E. H., 1949. Measurement of diversity. Nature 163: 688.

Teran, M., 1952. Geografía de España y Portugal. I. Montaner y Simón, Barcelona.

Whittaker, R. H., 1970. Communities and ecosystems. Macmillan, London.

Accepted 15.7.1981.

Appendice

Calcul des différentes expressions de l'entropie

On peut appeler u_{ij} ($i = 1, 2, 3, \ldots r; j = 1, 2, 3, \ldots c$) le nombre d'individus de l'espèce i trouvés sur la parcelle j. Le nombre total d'individus de cette espèce s'exprime par $\Sigma_{j=1}^{c} u_{ij} = u_{i.}$. Egalement, $\Sigma_{i=1}^{r} u_{ij} = u_{.j}$ exprimerait le nombre d'individus de toutes les espèces trouvées dans la parcelle j.

La matrice initiale de données de la biocénose étudiée aurait la forme suivante:

$$
\begin{array}{llll l}
u_{11} & u_{12} & u_{13} & \cdots\cdots u_{1c} & u_{1.} \\
u_{21} & u_{22} & u_{23} & \cdots\cdots u_{2c} & u_{2.} \\
u_{31} & u_{32} & u_{33} & \cdots\cdots u_{3c} & u_{3.} \\
\cdot & \cdot & \cdots\cdots & \cdot & \\
u_{r1} & u_{r2} & u_{r3} & \cdots\cdots u_{rc} & u_{r.} \\
\hline
u_{.1} & u_{.2} & u_{.3} & \cdots\cdots u_{.c} &
\end{array}
$$

La probabilité de trouver l'espèce i dans la parcelle j est donnée par p_{ij}

$$p_{ij} = \frac{u_{ij}}{\Sigma_i \Sigma_j u_{ij}} \; ; \; p_{i.} = \frac{u_{i.}}{\Sigma_i \Sigma_j u_{ij}}$$

où $\Sigma_i \Sigma_j u_{ij}$ représente le nombre total d'individus de toutes les espèces dans toutes les parcelles.

Le terme $H(E)$ correspond à la diversité des espèces dans l'ensemble des parcelles et on l'obtient au moyen de l'expression de Shannon-Weaver (1963):

$$H(E) = -\Sigma_{i=1}^r p_{i.} \log_2 p_{i.} = -\Sigma_{i=1}^r \frac{u_{i.}}{\Sigma_i \Sigma_j u_{ij}} \log_2 \frac{u_{i.}}{\Sigma_i \Sigma_j u_{ij}} \quad (2)$$

qui pour son calcul programmé est exprimé par:

$$H(E) = -\Sigma_{i=1}^r (\Sigma_{j=1}^c p_{ij}) \log_2 (\Sigma_{j=1}^c p_{ij}) \quad (3)$$

et également, la diversité des parcelles, $H(P)$, serait donné par:

$$H(P) = -\Sigma_{j=1}^c (\Sigma_{i=1}^r p_{ij}) \log_2 (\Sigma_{i=1}^r p_{ij}) \quad (4)$$

où c et r représentent respectivement le nombre de parcelles et d'espèces de la matrice initiale.

Selon le théorème de l'entropie totale,

$$H(E.P) = H(E) + H(P/E) = H(P) + H(E/P) \quad (5)$$

où le terme $H(E.P)$ représente l'entropie totale pour la double classification en espèces et parcelles; il s'obtient au moyen de l'expression suivante:

$$H(E.P) = -\Sigma_i \Sigma_j p_{ij} \log_2 p_{ij} \quad (6)$$

Les termes $H(P/E)$ et $H(E/P)$ représentent la moyenne pondérée de la diversité des parcelles pour les espèces et la diversité des espèces pour les parcelles.

Il s'obtient au moyen des expressions suivantes:

$$H(P/E) = \Sigma_{i=1}^r (-\Sigma_{j=1}^c p_{ij} \log_2 p_{ij}) p_{i.}$$

$$H(E/P) = \Sigma_{j=1}^c (-\Sigma_{i=1}^r p_{ij} \log_2 p_{ij}) p_{.j}$$

Si les classifications E et P étaient indépendantes, c'est-à-dire si le nombre d'espèces était égal à celui des parcelles et la distribution de celles-ci était identique, nous aurions

$$H(P/E) = H(P); \; H(P/E) = H(E/P); \; H(E/P) = H(E)$$

A l'opposé, si les deux classifications étaient totalement dépendantes, c'est-à-dire si chaque espèce se trouve seulement dans un habitat (parcelle) particulier et chaque habitat avait été occupé par une seule espèce,

$$H(E/P) = H(P/E) = 0 \text{ et } H(E.P) = H(E) = H(P)$$

La marge de variation de la valeur de la moyenne pondérée de la diversité des parcelles par espèce et de celle de la diversité des espèces par parcelle est donnée respectivement par:

$$0 \leqslant H(P/E) \leqslant H(P) \text{ et } 0 \leqslant H(E/P) \leqslant H(E)$$

De ces paramètres de l'entropie, nous avons employé dans ce travail spécialement $H(P/E)$. Cette expression semble très affectée par le nombre de parcelles considérées: il en résulte que deux zones où nous aurions utilisé un nombre différent de parcelles ne pourraient pas être comparées. Pour cette raison, nous avons aussi utilisé l'expression

$$A = \frac{H(P/E)}{\log_2 nb. \; de \; parcelles} \quad (7)$$

qui varie entre 0 et 1 et représente une mesure standardisée. Si les valeurs de A sont petites, chaque espèce tend à se confiner dans un sous-ensemble restreint d'habitats (parcelles), tandis que si ces valeurs sont élevées, chaque espèce tend à se distribuer sur la plupart des habitats.

The effect of cement factory air pollution on thermophilous rocky grassland*

J. Kubíková**
Prague Centre for Care of Monuments and Nature Conservancy, 110 00 Praha 1, Malé nám. 13, Czechoslovakia

Keywords: Air pollution, Cement factory, Frequency changes, Permanent plots, Soil pollution, Thermophilous grassland

Abstract

The thermophilous communities of calcareous rocky slopes in Radotín Valley are rather stable and present a species composition very similar to that described by Domin (1928). Some changes in their structure may nevertheless be observed due to the changing complex of human impact. The worst influence in the last 20 years is the air pollution of a cement factory. The following changes were observed on permanent plots at various distances from the source of pollution:
1. Complete disappearance of a number of species.
2. Decrease in frequency of other species and increase of frequency of a few of the most resistant species, occupying the free niche.
3. Decrease in cover of the entire community.
4. Acceleration of soil erosion, disintegration of the soil profile, enormous accumulation of Ca cations accompanied by decrease of many other mineral elements.
These results show that the effect of cement factories on the environment should not be underestimated in comparison with more serious damage caused by other air pollutants.

Introduction

Changes in vegetation caused by anthropogenic factors are receiving more and more interest recently, with the expansion of modern industry and abandonment of old traditional land cultivation. A good example for such a study is Radotín Valley, 20 km Sw of Prague, Czechoslovakia. It was formed during Quaternary in Palaeozoic calcareous rocks and extends along a brook for more than 7 km. The Tertiary plateau lies 320 m above sea level and was gradually eroded by Radotín Brook, which empties into the Berounka River at the altitude of 220 m.

Thus a very diversified landscape was formed with steep slopes facing to the four cardinal points and occupied by a rich and heterogenous vegetation. The climate is characterized by annual average precipitation of 500 mm and annual average temperature of 9 °C.

Vegetation, microclimate, and soils of this locality were studied by Domin and his collaborators (Domin, 1928; Hilitzer & Zlatník, 1928). However, the localities of Domin's relevés were indicated inaccurately and many associations were characterized only by constancy tables, so that it is virtually impossible, to locate the individual communities and compare previous and present composition and structure.

Such comparison would be most useful for the study of developmental processes going on in thermophilous grasslands, because many changes

* I am grateful to Dr. B. Holubičková for help with statistical analyses; to Ing. J. Haleš, PUDIS, Praha, for chemical analyses of soil samples; to D. Osborn for improving the English of my translation.
** Nomenclature follows Oberdorfer (1970)

have taken place under the impact of human activities. In the past centuries the bottom of the valley was gradually transformed because of several water mills operating on the brook running through the valley. Much of the original woodland was cleared for cultivation. In this way most of the grasslands, xerophilous as well as hygrophilous, were formed and maintained by mowing and grazing (sheep, goats and cows). Around 1950 the traditional management ceased; parts of the xerophilous grasslands were afforested by alien woody species (*Pinus nigra, Robinia pseudoacacia*) and in 1961 a cement factory and an extensive quarry opened. Still, a substantial and valuable part of the valley is untouched by development, having been declared a Nature Reserve over 110 ha in 1951. In that area different thermophilous grassland communities may be found which still have nearly the same composition and structure as described by Domin in 1928.

The following communities of thermophilous grassland can be differentiated in Radotín Valley, according to modern terminology: *Erysimo crepidifolii-Festucetum valesiacae* Klika 1933, *Festuco-Stipetum capillatae* Sillinger 1939, *Helianthemo cani-Caricetum humilis* Kubíková 1977, *Helianthemo cani-Seslerietum calcariae* Klika 1933, *Primulo veris-Seslerietum calcariae* Zlatník 1928, *Scabioso ochroleucae-Brachypodietum pinnati* Klika 1932.

Methods

The aim of this study was to ascertain the effect of air pollution from the cement factory on adjacent thermophilous grasslands. Since it was not possible to compare the present state with old data, three permanent plots were established at various distances from the source of the pollution. All plots are at the same elevation as the chimneys of the factory, which is situated down in the valley:

Plot I – at 300 m distance on the windward side,

Plot II – at 800 m distance on the lee side,

Plot III – at 2 000 m distance on the lee side in the Nature Reserve area.

Care was taken to find sites with a similar environment (parent rock, slope orientation and inclination, microclimate, edaphic conditions) and similar associations. Since the association with

dominant *Carex humilis,* the *Helianthemo-Caricetum humilis* occupies the greater part of the remaining grasslands, the permanent plots were chosen in this community. The size of each plot is 20 by 20 m.

On every plot the presence of all vascular plants was recorded and their frequency was sampled on 100 random quadrats (area 100 cm²) in June 1977 and June 1979. The frequency differences for individual species on different plots and in different years were tested by chi-square test.

Results and discussion

The frequency results for species growing on permanent plots are given in Table 1. The cement plant started production in 1961. Sixteen years later, in 1977, the following species belonging to the characteristic composition of *Helianthemo-Caricetum humilis* (see Kubíková 1977) were missing on Plot I: *Stipa capillata, Scabiosa canescens, Eryngium campestre, Calamintha acinos, Artemisia campestris, Stachys recta, Thlaspi perfoliatum, Lactuca perennis, Pulsatilla pratensis subsp. nigricans, Sedum album, Centaurea triumfetti.* The number of species on Plot I was substantially lower than on sample plots II and III (Table 2).

The next step was to compare the frequencies of species occurring on all three plots in the year 1977. The results are summarized in Table 3.

The frequencies were sampled again on the permanent plots in June 1979 and the values obtained were compared with these from 1977. The results are summarized in Table 4.

As may be seen in Table 4, seven species growing on Plot I, which was heavily damaged by the pollution of the cement factory, showed a significant decrease of frequency. Examples of extinction were recorded. This happened with species that seemed indifferent in 1977, and were lower or even higher in comparison with the other sample plots. *Carex humilis, Koeleria gracilis, Hieracium pilosella* were diminishing, and *Achillea collina* almost to complete extinction; *Brachypodium pinnatum, Sanguisorba minor, Seseli hippomarathrum* diminished to one half or one third of their previous frequency.

In Plot II, 800 m from the source of pollution, the frequency of only one species, *Aster amellus,*

Table 1. The frequencies of species found on permanent plots. Species with frequencies <3 in any plot have been omitted.

Plot No. Date	I 1977	I 1979	II 1977	II 1979	III 1977	III 1979
Differential species of the Helianthemo cani-Caricetum humilis						
Helianthemum canum/L./Baumg.	25	18	27	22	16	12
Carex humilis Leys.	16	1	73	68	38	31
Bothriochloa ischaemum/L./Keng	4	3	2	–	24	12
Stipa capillata L.	–	–	2	–	1	3
Anthericum liliago L.	69	73	2	9	48	42
Scabiosa ochroleuca L.	3	1	–	1	–	–
Hieracium pilosella L.	7	1	8	14	–	–
Species of Festucion valesiacae						
Festuca valesiaca Schleich. ex Gaud.	4	4	3	9	1	1
Koeleria gracilis Pers.	9	2	7	9	–	3
Potentilla arenaria Borkh.	12	13	28	34	24	5
Seseli hippomarathrum Jacq.	40	19	28	17	18	15
Euphorbia cyparissias L.	15	16	7	7	3	5
Thymus praecox Opiz	6	11	8	10	12	13
Salvia pratensis L.	1	2	4	3	6	3
Sanguisorba minor Scop.	24	9	48	56	8	15
Eryngium campestre L.	–	–	3	5	3	2
Teucrium chamaedrys L.	12	32	15	19	33	43
Achillea collina J. Beck	4	–	12	6	1	–
Calamintha acinos/L./Clairv.	–	–	–	–	11	–
Artemisia campestris L.	–	–	–	–	6	1
Arenaria serpyllifolia L.	–	–	–	–	3	–
Helianthemum obscurum/Cel./Holub	–	–	–	3	–	–
Stachys recta L.	–	–	–	–	11	5
Taraxacum laevigatum/Willd./DC.	–	–	–	3	–	–
Species of Alysso-Festucion pallentis						
Asperula cynanchica L.	1	2	5	6	7	3
Pulsatilla pratensis/L./Mill.	–	–	–	–	7	5
Melica transsilvanica Schur	1	–	–	–	6	3
Alyssum alyssoides/L./Nath.	–	–	–	–	15	–
Species of Seslerio-Festucion glaucae						
Sesleria varia/Jacq./Wettst.	7	7	12	19	–	–
Anthericum ramosum L.	–	–	40	27	–	–
Medicago falcata L.	–	–	29	21	2	–
Centaurea triumfetti All.	–	–	12	11	–	1
Chrysanthemum corymbosum L.	–	–	10	8	–	–
Inula hirta L.	–	–	3	2	–	–
Aster amellus L.	–	–	15	6	–	–
Species of Cirsio-Brachypodion						
Brachypodium pinnatum/L./P.B.	37	20	20	30	–	1
Festuca sulcata/Hackel/Nym.	2	–	5	3	–	–
Cirsium acaulon/L./Scop.	1	–	9	7	–	–
Fragaria viridis Duch.	9	4	4	11	9	18
Bupleurum falcatum L.	3	–	6	11	4	8
Lotus corniculatus L.	3	–	1	23	–	–
Leontodon hispidus L.	–	–	3	2	–	–
Anthyllis vulneraria L.	–	–	1	3	–	–
Linum catharticum L.	–	–	–	3	–	–
Woody species						
Rosa spp.	3	–	1	1	–	2

Table 2. Species numbers found on sample plots.

Plot Number	I	II	III
Species Number – 1977	27	36	35
Species Number – 1979	21	41	35

Table 3. Comparison of frequencies of species common to Plots I, II, III in the year 1977.

Indifferent species:	Helianthemum canum
	Teucrium chamaedrys
	Thymus praecox
	Koeleria gracilis
	Fragaria viridis
	Salvia pratensis
	Bupleurum falcatum
Species significantly lower on Plot I:	Carex humilis
	Bothriochloa ischaemum
	Potentilla arenaria
	Asperula cynanchica
	Sanguisorba minor
	Achillea collina
Species significantly higher on Plot I:	Anthericum liliago
	Euphorbia cyparissias
	Seseli hippomarathrum

Table 4. Comparison of frequencies of different species, sampled in 1977 and 1979.

	Number of species with frequency		
Plots	higher in 1977	higher in 1979	No change
I	7	1	18
II	1	2	29
III	2	0	24

decreased significantly.

In Plot III, 2 000 m from the source of pollution, the decrease of the frequency of two species, *Bothriochloa ischaemum* and *Potentilla arenaria* was recorded.

There were some species, however, with increasing frequency. On Plot I the frequency of *Teucrium chamaedrys,* was more than doubled, and on Plot II that of *Anthericum liliago* and *Lotus corniculatus.*

The effect of the pollution was very markedly

manifested in the cover of the entire community. Cover estimates in 1979 were only 50% on sample Plot I in comparison with 85% on Plot II and 65% on Plot III (1977 values were 60%, 80% and 65% resp.)

Since it was not possible to measure the air pollution on the sample plots directly, soil samples (5–10 cm deep) from the different plots were taken for analyses of mineral elements (Table 5). The results show an enormous accumulation of Ca cations in the soil of Plot I; it reaches nearly four times that of Plot III. There is also a substantial enrichment of soil calcium on Plot II, where the vegetation is apparently not yet damaged. Surprisingly, the other mineral elements studied were mostly higher on Plot III, the farthest site in the valley. Thus the dying off of plants cannot be attributed to toxic effects of mineral elements in the dust covering the nearby sites in the vicinity of the cement factory. However, the equilibrium of mineral nutrients in the soils on Plot I, and also Plot II has changed in favour of Ca ions, which apparently occupy the adsorption complex and cause the leaching of other mineral elements.

The study demonstrated different sensitivity of plant species composing thermophilous communities to air and soil pollution. A very marked phenomenon is the serious damage to tussock grasses and sedges such as *Carex humilis, Stipa capillata, Koeleria gracilis, Bothriochloa ischaemum,* and *Melica transsilvanica,* which are always present or even dominant in a well developed community of the type studied and play an important role in soil formation and maintenance on steep slopes. A remarkable exception is *Sesleria varia* which seems to show a high resistance to this type of pollution. The same high resistance of this species to accumulation of calcareous dust in the vicinity of a quarry in Britain was observed by Usher (pers. comm.).

After disintegration of most perennial grasses the site is dominated by perennial herbs such as *Anthericum liliago, Teucrium chamaedrys, Helianthemum canum, Seseli hippomarathrum, Euphorbia cyparissias* and *Potentilla arenaria.* Especially remarkable is the very high frequency of *Anthericum liliago,* which is twice as high as on the relatively undamaged Plot III.

The reasons for the greater persistance of just these plants cannot be given on the basis of this

Table 5. Results of analyses of mineral elements in the soils of the studied plots. The organic matter was burnt for 12 h at 400 °C, the mineral components of the soils were extracted with 20% HCl and measured with a spectrophotometer. Figures are averages of two measurements.

Mineral elements in % dry matter	Plot I	Plot II	Plot III	Averages in soils according to Vernadskij (1967)
K	0,54	0,93	1,08	1,36
Na	0,04	0,05	0,12	0,63
Mg	1,29	1,05	1,48	0,60
Ca	21,93	10,47	5,95	1,37
Fe	1,71	2,40	4,11	3,80
ppm dry matter				
Mn	770	710	710	800
Zn	68	77	159	50
Pb	36	53	75	10
Cd	3,4	2,8	2,3	0,5
As	0	0	0	0,2

study. It is only possible to speculate that the solution may lay in their individual morphology and life cycle. *Anthericum liliago* for instance is a geophyte with smooth bifaciale leaves, *Teucrium chamaedrys* is a chamaephyte able to shade damaged leaves and form new ones. *Helianthemum canum* and *Potentilla arenaria* have the same ability. *Euphorbia cyparissias* and *Seseli hippomarathrum* have thin, smooth, dissected leaves. However, these mechanisms may function only below a certain degree of air and soil pollution. The dust from the cement factory accumulates in the soil (see Table 5) and the damage to plants is increasing as may be seen from the decreasing number of species and their frequencies in the short period of two years.

Little attention has been given to the damage caused by dust of cement factories in the literature. Dässler (1976) mentions the decrease of assimilation and dry matter production in coniferous trees and also the enrichment of mineral content of soils, namely by K, Ca, and Mg ions.

References

Dässler, H. G., 1976. Einfluss von Luftverunreinigungen auf die Vegetation. G. Fischer. Jena.

Domin, K., 1928. The plant associations of the valley of Radotín. Preslia. Praha. 7: 3–68.

Hilitzer, A. & Zlatník, A., 1928. Résultats des observations microclimatiques dans les associations du terrain calcaire de la vallèe Radotínské údolí pres de Prague. Preslia. Praha. 7: 69–93.

Kubíková, J., 1977. The Vegetation of Prokop Valley Nature Reserve in Prague. Folia Geobot. Phytotax. Praha. 12: 167–199.

Oberdorfer, E., 1970. Pflanzensoziologische Exkursionsflora für Süddeutschland. Verl. Eugen Ulmer, Stuttgart. 3. Aufl.

Vernadskij, V. J., 1967. Biosfera. Mysl. Moskva.

Accepted 12.5.81.

List of lectures and poster contributions presented at the Symposium

I. Succession: theory and methods

M. P. Austin: Permanent quadrats: an interface for theory and practice.

E. Feoli, A. Alto-Belli & C. Cenci: Measures on changements of global species association structure in vegetation dynamics. Examples with data of permanent plots of grasslands under controlled grazing.

R. Gittins: Canonical analysis in studies of the dynamic status of vegetation.

J. P. Nicolas, M. A. Quintas, J. Haeger, F. D. Pineda & F. G. Bernaldez. Description des pâturages du centre de la péninsule ibérique au moyen de processus stochastiques.

L. Orlóci: Testing the overall sharpness and dimensionality of the compositional gradient in succession.

S. Persson: Vegetation as a possible sensor in a national program for environmental monitoring. A methodological study.

S. T. A. Pickett & R. T. T. Forman: Invasion and turnover patterns of populations over 22 years following old-field abandonment in New Jersey (U.S.A.).

J. Ritter: Remarques sur la structure horizontale des groupements végétaux.

M. B. Usher: Modelling ecological successions.

II. Succession in mediterranean ligneous formation

M. Abdel Gawad Ayyad: Effect of protection and controlled grazing on the vegetation of a mediterranean desert ecosystem.

C. Allier & A. Lacoste: Groupements de reconstitution et de dégradation de l'étage méditerranéen en Corse: relations floristiques et dynamiques.

G. Aubert, M. Barbero, G. Bonin, J. Gamisans, M. Gruber, R. Loisel, P. Quezel, H. Sandoz, M. Thinon & G. Vedrenne: Les taxons indicateurs de la dynamique des écosystèmes forestiers et préforestiers provençaux des étages méditerranéens et supraméditerranéens: I. Mise en évidence, II. Position dans les stades évolutifs.

M. C. Aubin, J. P. Barry & J. Gallangau: De la parcelle au lieu-dit.

P. B. Bridgewater & D. J. Backshall: Dynamics of western australian ligneous formations with special reference to the invasion of adventive species.

M. Debussche, J. Escarre & J. Lepart: Effect of plant succession of early trees attracting birds.

C. Floret: The effect of protection on steppic vegetation of the mediterranean arid zone. Dynamical study on 5 south tunisian types of vegetation.

Lj. Ilijanic & S. Hecimovic: Die Sukzessionsforschung der mediterranen Vegetation auf der Insel Lokrum bei Dubrovnik/Kroatien, Jugoslawien.

Z. Kosir: Die Veraenderung des Unterwuchses in Tannen-Buchenwaeldern des Hochkarstgebietes unter biotischen Einfluessen.

D. Lausi, R. Gerdol & F. Piccoli: On the dynamics of praealpine Ostrya-woods.

A. Z. Lovric: Coenodynamism in an adriatic half-sempervirent pseudomaquis, coastal pinewoods, and xerophytic shrublands.

D. Mueller-Dombois: Vegetation dynamics in the coastal lowland ecosystem of Hawaii.

Z. Naveh: The evolution of the mediterranean landscape as a pedogenic and anthropogenic biofunction. Theoretical and practical implications for conservation management.

J. Poissonet, P. Poissonet & M. Thiault: Evolution de la végétation et de sa valeur pastorale dans les parcelles expérimentales d'une garrigue de Chêne kermès (Quercus coccifera L.).

M. M. H. Thiault: Interface animal végétal en formation ligneuse basse (garrigue de Chêne kermès).

L. Trabaud & J. Lepart: Evolution de la composition floristique d'une garrigue de Chêne kermès soumise à des feux contrôlés.

III. Succession in heathlands

B. Clement & J. Touffet: Dynamique de la recolonisation végétale dans les landes bretonnes incendiées.

A. Froment: La conservation des landes du Calluno-Vaccinietum en Ardenne belge.

C. H. Gimingham, R. J. Hobbs & Azim U. Mallik: Community dynamics in relation to management of heathland vegetation in Scotland.

J. C. Gloaguen: Evolution de la structure horizontale de la végétation au cours de la recolonisation d'une lande armoricaine incendiée.

Vegetatio 47, 285–286 (1981).   0042-3106/81/0472/3-0285/$0.40.
© Dr W. Junk Publishers, The Hague.

M. Hossaert-Palauqui & N. Gautier: Dynamique de régénération d'une lande au cours des trois années suivant l'incendie.

J. Miles: Problems in grassland and heathland dynamics.

H. Persson: The effects of fertilization and irrigation on the vegetation dynamics of a pine-heath community.

E. Pignatti & S. Pignatti: Duaerquadratuntersuchungen über Brandwirkung in einem Calluna-Bestand bei Triest.

J. T. de Smidt & E. De Hullu: Cyclical succession or fluctuation in permanent plots in heath vegetation.

IV. Succession in herb formations

M. P. Austin: An analysis of succession along an environmental gradient.

J. P. Bakker: Impoverishing the soil and vegetation dynamics in wet grasslands.

J. Benz, G. Spatz, B. Weis & G. Ohmayer: Grundsätzliche Probleme beim Einsatz von Simulationsmodellen als Systemanalytisches Instrument in der Sukzessionsforschung.

R. Bornkamm: Succession rates: change of several vegetation properties during secondary succession.

G. Bouxin: Analyse de la structure horizontale dans des pelouses calcaires de Belgique.

H. Dierschke: Untersuchungen auf Dauerflächen in Kalk-Magerrasen (Mesobromion) mit unterschiedlicher Nutzung.

J. M. Gomes Gutierrez, A. Puerto Martin & M. Rico Rodriguez: Primary succession: colonization of granitic rocks.

G. Grabherr: The impact of trampling by tourists on a high altitudinal grassland in the Tyrolean Alps, Austria.

P. van Hecke, I. Impens & T. J. Behaeghe: Long-term temporal variation of niche and diversity from permanent grassland plots with different fertilizer treatments.

W. Heinrich: Dauerbeobachtungen in Halbtrockenrasen der jenaer Umgebung (Thüringen, DDR).

D. Hubert: Evolution d'une pelouse des Causses sous l'action de fertilisants et du pâturage.

B. Krusi: Phenological methods in permanent plot research.

J. Kubikova: The effect of cement factory air pollution on thermophilous rocky grassland.

M. N. Numata: Studies in early stages of secondary succession in Japan.

M. J. Oomes: The effect of cutting and fertilizing on the botanical composition and production of an Arrhenatherion elatioris community.

F. D. Pineda, J. P. Nicolas, M. Ruiz & F. G. Bernaldez: Succession, diversité et niche écologique dans les pâturages du centre de la péninsule ibérique.

A. Puerto Martin, J. M. Gomez Gutierrez & M. Rico Rodriguez: Secondary succession in marginal land in western Spain (siliceous soils).

W. Schmidt: Artenreichtum und Artenwechsel in Grasland-gesellschaften (Species diversity and species changes in grassland communities).

C. Tosca & A. Fabre: Les pelouses supra-forestières dans les Pyrenées centrales (France); structure interne et relations entre les groupements.

A. Valdes Amado & J. M. Gomez Gutierrez: The use of dominant species in the caracterization of herbaceous associations of highly intervened semiarid grassland.

J. H. Willems: Species composition in permanent plots in chalk grassland with different management.

O. B. Williams & M. P. Austin: Grassland dynamics in an australian bioclimate of a modified mediterranean-type.

V. Succession: dunes, marches, water's edges, seaside

W. G. Beeftink: Types of natural and man-made disturbances in salt-marsh communities.

J. J. Corre: Dynamique de la végétation le long du littoral méditerranéen: ses conséquences sur le paysage.

A. Garcia & E. Luis: Rapports entre les landes tourbeuses et les landes à Calluna, dans les cols de Panderrueda et Pandetrave (Picos de Europa) Espagne.

A. Grünig: The development of vegetation and soil on loafing and breeding islets in an artificially created shallow lake.

W. Joenje: Succession on former tidal flats.

D. van der Laan: Comparison of the ecological amplitude of dune slack species with regard to the ground water table, based on spatial distribution and temporal variation.

E. van der Maarel: Fluctuations in a coastal dune grassland due to fluctuations in rainfall: experimental evidence.

F. van der Meulen: Vegetation changes and water catchment in coastal dunes of the Netherlands.

VI. Miscellaneous

J. Lejoly: Quelques aspects écologiques de la dynamique végétale en région équatoriale perhumide.

R. M. Masalles: Notes sur la dynamique de la végétation après abandon des cultures.

K. F. Schreiber & J. Schiefer: Vegetations- und Stoffdynamik in Grünlandbrachen. Ergebnisse fünfjähriger Bracheversuche in Baden-Württemberg.

M. Tanghe & P. Duvigneaud: Huit ans d'observations phytodynamiques sur des parcelles permanentes de culture abandonnée dans la région bruxelloise.

M. Thinon: Apports de la pédoanthracologie à la connaissance de la dynamique végétale en région méditerranéenne.